Mode of Action of Herbicides

FLOYD M. ASHTON

Professor of Botany
Department of Botany
University of California at Davis

ALDEN S. CRAFTS

Emeritus Professor of Botany
Department of Botany
University of California at Davis

A Wiley-Interscience Publication

John Wiley & Sons
New York   London   Sydney   Toronto

Library of Congress Cataloging in Publishing Data

Ashton, Floyd M
Mode of action of herbicides.

Includes bibliographies.
1. Herbicides.   I. Crafts, Alden Springer, 1897–
joint author.   II. Title.
SB951.4.A84   632'.954   72–6850
ISBN 0–471–03510–6

Printed in the United States of America

10 9 8 7 6 5 4 3 2 1

Preface

Chemical weed control is a miracle of our technological age. Long known as one of the most arduous of agricultural operations, weed killing has taken on an entirely new aspect as chemical after chemical is added to our arsenal of herbicides. And with new and better compounds being synthesized and developed almost each day, it seems that the improvements will continue for the foreseeable future. Between 1959 and 1965 the acreage of weeds treated with herbicides in the United States rose from 53 to 120 million, an increase of 126%; use has increased at an even faster pace since 1965.

This country has led the world both in production and use of herbicides and as a result yields of cereals, soybeans, cotton, sugar beets, and many other crops have increased since 1945, in some cases 100% or more. Thus while use of fertilizers and new high-yielding crop varieties have contributed greatly to the "green revolution," chemical weed control has been at the forefront in technological achievement.

Although the future of chemical weed control seems bright, and continued testing and adoption of new compounds is proceeding at an accelerating rate, a new element has been thrust into the field. With the discovery that certain chlorinated hydrocarbon insecticides persist for years in organisms, and in fact, in the total environment, all pesticides are now being viewed with suspicion by people interested in protecting our world from broad-scale pollution.

Centered initially on insecticides, work on toxicity and persistence of pesticides has now turned to other materials including herbicides. When certain commercial samples of 2,4,5-T were found, upon injection into test animals, to produce teratogenic effects, registration of liquid formulations of 2,4,5-T for use around the home and on lakes, ponds and ditchbanks was suspended. Use of solid formulations around the home and on all food crops intended for human consumption was also cancelled. Fortunately the registered use of 2,4,5-T for control of weeds and brush on ranges, pastures and forests, rights-of-way and other non-agricultural areas has not been prohibited at this time. Although it has been found that the original sample of 2,4,5-T used in the above tests contained the impurity 2,3,7,8-tetrachlorodibenzo-*p*-dioxin, a known teratogenic agent, pure 2,4,5-T affects experimental animals only when injected at very high dosages. Tests are still under-way to clarify the situation.

Meanwhile the recognized persistence of several substituted ureas, uracils, *s*-triazines, the benzoic acid derivatives, picloram and other herbicides in soils has received detailed study. It is being recognized that some persistence is a necessary property of all herbicides; with no persistence, soil-applied herbicides would not control weeds. The problem then is to find and know the relative retention and persistence of all the various herbicides in soils and to use them within the permissible range of activity with consideration given to crop tolerance, soil degradation, retention against leaching, etc. Soil-active herbicides for use on non-agricultural areas such as road verges, ditch banks, fire breaks and commercial sites need to be persistent to be economical. The chances for such herbicides to enter human food sources are extremely remote.

Chemical weed control is a relatively new science that involves knowledge in the fields of chemistry and biology, some familiarity with reactions of plants to phytotoxic agents, and at least observational experience in the responses of common weeds and crops to herbicides. Weed and crop ecology and appreciation of the factors determining selectivity, tolerance and suscepti-bility are important. And finally, to be useful in sales and service one needs a vast backlog of detailed information concerning the role of weed control in practical agriculture.

This book attempts to provide a basic introduction to the physiology and biochemistry of chemical weed killers, and to summarize the body of infor-mation that has been acquired concerning the properties, commercial forms, and field use of some 150 products now available. It should serve as a text-book for advanced courses, a reference volume for research workers and a source of much detailed information that is needed from day to day by extension specialists, contract applicators, salesmen and farmers involved in the practical use of herbicides in the field.

The material in this book has been arranged in a manner to facilitate the readers finding specific information about a given herbicide. In general each chapter about a given class of herbicides is presented in the following sequential form: (1) Introduction, (2) Growth and Plant Structure, (3) Absorption and Translocation, (4) Molecular Fate, (5) Biochemical Responses and (6) Mode of Action. The Mode of Action section is essentially a concise summary of the above mentioned topics and points out what the authors believe to be the most relevant aspects of the herbicidal action of these compounds. The reader may find it advantageous to read the Mode of Action section to obtain overall orientation before reading the detailed information which precedes it. The chapters at the beginning of the book, Chapters 2 to 7, introduce the listed topics in a general manner by briefly citing selected examples of the topic which are then covered in detail in the chapters concerned with the specific class of herbicides.

In general, we have chosen not to include the research on soil-herbicide interactions unless it was particularly relevant to the discusion. This was necessary in order to adequately cover the plant aspects within the space limitations. In addition, Kearney and Kaufman's 1969 book, *Degradation of Herbicides*, covers soil-herbicide interactions.

The chemical nomenclature follows that of the Weed Science Society of America as given on the back cover page of their journal, *Weed Science*, and the 1970 edition of their Herbicide Handbook. The botanical nomenclature for weeds follows that of the Weed Science Society of America's Report of the Subcommittee of Standarization of Common and Botanical Names of Weeds, *Weed Science* **19**:435–476 (1971). The common name followed by the scientific name of the weed is given the first time it is used in each chapter, subsequent reference in the same chapter uses only the common name. The scientific name of the crops is not given since this is common knowledge.

Both of us are grateful to our wives, Phyllis Ashton and Alice Crafts, for their patience and assistance.

Davis, California

FLOYD M. ASHTON
ALDEN S. CRAFTS

Contents

CHAPTER 1

Introduction

Weeds and Weed Control

Weeds are a product of human society. Primitive man, the gatherer and hunter was not conscious of weeds in the modern sense. Contemporary man has created the concept of the weed as a plant in a place where it is not wanted. The Indians in America often started wild fires and they did not worry if thousands of acres burned; the resprouting plants provided food for deer and hunting was good on the following year. We, in contrast, deplore the ravages of fire in our forests. We term it a disaster because we need the trees for lumber and the forest for recreation.

A few million primitive people could live off the lands of the world. Our present 3 billion people demand food and clothing, recreation and living space. If we were to provide all living people today with an adequate diet almost every arable acre would be required. And as population overtakes the food supply every productive acre will be at a premium and waste by weeds will not be tolerated. When millions of people face starvation possibly past non-essential crops such as tobacco, hemp, coffee, tea and plants that provide perfumes, spices, and stimulants will become weeds. Even low yielding varieties of our staple crop species will have to yield to the pressure from hungry mouths. Thus weeds take on a new meaning for our present and future generations.

Agricultural technology is undergoing a new revolution. The mechanical revolution has completely altered agricultural methods and now the chemical revolution is carrying on to new heights of efficiency. Table 1-1 lists the manpower requirements per unit of total population required to operate the farming industry. Nowhere except possibly in air transport has there been as great an increase in efficiency.

Weeds have been with us from the beginnings of agriculture. The primitive farmer who first pulled by hand the plants that competed with his cereal crops initiated the process which has, through the years, been one of the most

1

TABLE 1-1.   Some Statistics on the Industrialization of Agriculture in the United States (Miller, J. F., 1970)

Year	Total population	Farm population	Farm as % of total	Significant events of mechanization and chemicalization	
				Year	Event
1790	3,929,214	—	> 90	1793	Thomas Jefferson invented a moldboard for a plow
1820	9,638,543	—	72	1818	Jethro Wood patented an iron plow with interchangeable parts
1840	17,069,453	9,012,000	69	1837	John Deere began manufacturing plows with steel share and smooth wrought iron moldboard
1850	23,191,876	11,680,000	64	1854	Patent granted for two-wheeled bar mower
1860	31,443,321	15,141,000	58	1856	Two-horse straddle row cultivator patented
1870	38,558,371	18,373,000	53	1878	A twine knotter for binding grain perfected by John F. Appleby
1890	62,947,714	26,379,000	43	1892	Successful gasoline tractor built
1900	75,994,575	29,414,000	38	1903	C. W. Hart and C. H. Parr established first firm devoted to manufacture of gasoline tractors
1920	105,710,620	31,614,269	27	1926	Successful light gasoline tractor developed
1930	122,775,046	30,840,350	21	1927	Mechanical cotton picker invented by John D. Rust
1940	131,820,000	30,840,000	18	1941–1945	The Second American Agricultural Revolution began during World War II
1950	151,132,000	25,058,000	11	1940's	Synthesis and development of 2,4-D
1960	180,000,000	20,827,000	9	1960	Development of 17 herbicidal chemicals (1959–1961)
1970	204,000,000	10,300,000	< 6	1959–1965	Total acres treated with herbicides increased from 53 million to 120 million, an increase of 126%

tedious of agricultural operations. In many countries this simple means of handling weeds is still in vogue being carried out often by women and children. Only within the past quarter century has it been possible to eliminate this arduous drudge work. The application of this modern technology in the developing countries is essential in order to allow adequate time for the education of children and to free women to provide a higher standard of living in the home (Holm, 1971). The man with the hoe, the classical symbol of field crop agriculture is rapidly being supplanted by chemical methods. Weeds will never again be the limiting factor in crop production that they have been in the past.

Consolidation of small farms into larger, more economical units is going on in many places and it must continue in order to increase food production. Use of machinery and chemicals is bringing about almost unbelievable changes in agriculture.

It would be wrong here to imply that all of the changes in agricultural technology are free of problems. As new chemicals are introduced and groups of weeds are put under control, other weeds very soon, being relieved of competition and being tolerant of the chemical, take over and become serious. Everyone is cognizant of the shift from broadleaf weeds to grassy ones that occurred with introduction of the chlorophenoxy compounds. Similar shifts have occurred whenever one chemical or a related group of chemicals is used continuously. This problem was met in the sugar cane plantations of Hawaii a decade ago by using a rotation of herbicides. In many situations mixtures of herbicides are used to broaden the spectrum of weeds that may be controlled.

Another problem that is threatening mechanized and chemicalized agriculture is the escape of crop plants and their gradual adoption of weedy habits. An early example is johnsongrass (*Sorghum halepense*) that soon escaped and became a noxious weed in agronomic crops. A more recent one is dallisgrass (*Paspalum dilatatum*), a forage plant that has invaded thousands of miles of irrigation ditch banks, fence lines, and roadsides. Milo, sudangrass (*Sorghum sudanense*) and field bindweed (*Convolvulus arvensis*) in corn and cotton fields are further examples. Very recently escaped sugar beets have reverted to a weedy habit and their control in cultivated sugar beets challenges the modern weed specialist. Yellow nutsedge (*Cyperus esculentus*) and purple nutsedge (*C. rotundus*), weeds throughout the tropic and subtropic countries continue to spread and continue to survive chemical after chemical. They now have the dubious honor of being named the world's most serious weeds (Holm, 1969).

Costs of weeds, always considered high, assume new proportions as labor becomes scarce and expensive, crops become more critical to our needs, and new lands non-existent. Table 1-2 lists the losses caused by weeds and the costs of control in four of the most important areas where weed control is

TABLE 1-2.   Losses Caused by Weeds and Costs of Control in the United States (Anon., 1965)

Crop or situation	Losses in yield and quality	Cost of control	Total
Agronomic crops	$1,573,024,000	$1,876,000,000	$3,449,024,000
Horticultural crops	254,281,000	307,000,000	561,281,000
Grazing lands	632,325,000	365,000,000	997,325,000
Aquatic sites and non-cropland	53,140,000	55,638,000	108,778,000
Total	$2,512,770,000	$2,603,638,000	$5,116,408,000

practiced. These figures for the decade 1950–1960 would be much higher if today's prices and wage scales should be used.

The weed control problem presents a major challenge to the most efficient farm operator because of the increasing labor and other production costs that reduce his net income. Weeds hinder complete mechanized production of many crops. In addition to lowering crop quality and yield, weeds cause many other losses, such as poisoning of livestock, inducing off-flavors in milk, and reducing flow of irrigation and drainage waters (Anon., 1965).

Table 1-3 shows the relative losses from weeds, insects, and diseases and provides figures on pesticide sales and research efforts in the USA.

TABLE 1-3.   Relative Losses from Weeds, Insects, and Diseases Compared with Pesticide Sales and Research Efforts in the United States (Furtick, 1967)

	Annual losses and costs of control $ Millions	1965 pesticide sales $ Thousands	Research support USDA and state $ Thousands
Weeds	5,064	201,753	8,707
Insects	4,298	237,317	34,368
Diseases	3,779	48,603	44,164

With the discovery of the great herbicidal potential of the chlorophenoxy compounds in the mid-forties, chemical weed control progressed at an accelerating rate. Now the manufacture and sale of herbicides is a multimillion dollar business.

The United States has led in herbicide use and production. Table 1-4 shows the production of organic herbicides for the years 1958–1968.

TABLE 1-4.   **U.S. Production of Organic Herbicides in the United States (House, W. B. et al., 1967, (1958–1966). Anon., 1967–1969)**

Year	1000 pounds		
	2,4-D and 2,4,5-T acids	Other organic herbicides	Total
1958	34,622	25,295	59,917
1959	34,829	29,756	64,585
1960	42,522	33,201	75,723
1961	50,301	46,367	96,668
1962	51,366	51,913	103,279
1963	55,402	64,626	120,028
1964	65,148	93,909	159,057
1965	74,921	111,127	186,048
1966	83,671	149,352	233,023
1967	91,691	206,759	298,450
1968	96,793	235,541	332,334
1969	52,076	272,606	324,682

Table 1-5 gives figures on world consumption of herbicides in 1968.

Paralleling the figures in Table 1-4 are values for increases in grain yields for the period 1934–1938 to 1960 compiled at the International Plant Protection Center (Table 1-6) (Furtick, 1970).

TABLE 1-5.   **Estimated 1968 World Consumption of Herbicides at the Consumer Level (Furtick, 1970)[a]**

Area	Consumption
North America	$550,000,000
Japan	70,000,000
Latin America	80,000,000
Near East, Southeast Asia, and Oceania	80,000,000
Western Europe	60,000,000
Africa	40,000,000
Total	$880,000,000

[a] Based on figures compiled by the International Plant Protection Center, Oregon State University. From industry, agricultural agency, and commerce agency sources.

TABLE 1-6. Increase in Grain Yields per Acre 1934–1938 to 1960 (Furtick, 1970)

Area	Increase, %
North America	107
Oceania	68
West Europe	38
East Europe and USSR	20
Africa	20
Latin America	10
Asia	8

While use of fertilizers and new improved varieties are involved as well as the use of herbicides, comparison of these tables shows an obvious correlation between chemical weed control and crop yields.

Table 1-7 gives some figures on the results of yield trials with herbicides on rice from experiments on rice in the Philippines. These data show clearly that the herbicides produced significant increases in yield over the untreated plots and that they were approximately equivalent to two hand weedings. At present wage scales, even in the Philippines, this represents a real saving in cost of production, as well as relief from the back-breaking toil of hand weeding.

TABLE 1-7. Effect of Granular Herbicides on the Grain Yield of Rice When Only One Application Was Made 3 Days after Transplanting (Chandler, R. F., Jr., 1969)

Treatment	Rate of application kg/ha of active ingredients	Grain yield kg/ha
Trifluralin plus MCPA	0.7 + 0.4	6831
Nitrofen plus 2,4-D	2.0 + 0.5	6778
EPTC plus MCPA	1.75 + 0.7	6725
TCE-Styrene + 2,4-D	1.00 + 0.5	6575
Two hand weedings	25 and 40 days after transplanting	6924
Untreated	—	4328

None of the treatments shown in Table 1-7 gave a yield significantly different from that of the others, but each produced significantly more grain than did the unweeded control plot.

Table 1-8 from Matsunaka (1970) provides an indication of the rising costs of weed control in transplanted rice production in Japan and it shows the tremendous savings effected when chemicals are substituted for hand labor.

TABLE 1-8. Changes in Weeding Costs in Japan (Transplanted Rice Culture) (Matsunaka, 1970)

Year	Hand-weeding cost cent/hr (1)	Weeding labor hr/ha (2)	Cost of weeding cost $/ha (1) × (2) (3)	Cost of herbicides $/ha (4)	Weeding cost in total $/ha (3) + (4) (5)	Weeding cost without herbicides $/ha (1) × 505.6 (6)	Saved money by herbicides $/ha (6) − (5) (7)	Total area transplanted 1,000 ha (8)	Total saved money million $ (7) × (8) (9)
1949	7.4	505.6	37.2	0.00	37.2	37.2	—	2,875	—
1952	8.9	357.0	31.6	0.38	32.0	44.8	12.8	2,872	36.7
1954	10.9	310.7	33.8	0.49	34.3	55.1	20.7	2,888	59.8
1956	11.9	313.5	37.2	0.72	37.9	59.9	22.1	3,059	67.5
1958	12.7	309.8	39.2	0.83	40.1	64.0	24.0	3,080	73.8
1960	14.2	267.6	38.1	1.72	39.9	72.0	32.2	3,124	100.5
1962	20.2	208.7	42.1	4.61	46.8	102.0	55.3	3,134	173.4
1964	27.4	175.7	48.2	6.61	54.8	138.6	83.8	3,126	262.0
1965	30.7	174.4	53.6	6.92	60.5	155.4	94.9	3,123	296.4
1966	34.0	164.2	55.8	8.22	64.0	171.9	107.8	3,129	337.5

These calculations would show even higher savings by 1970 because new and better herbicides for weed control have been introduced since 1966.

Inasmuch as rising expenditures for herbicides indicate increased profits for agriculture the data presented in Tables 1-6, 1-7, and 1-8 show that chemical weed control, in addition to alleviating the tremendous burden of hand weeding, has increased the real income of farmers around the world. And great as these advances have been, improvements can be expected to continue for a long time as herbicides become available and used in the less developed countries of the world.

Behind and supporting this new development in agriculture lies a large research effort, involved in synthesis, testing, development, and production of new herbicidal compounds. Techniques from almost every aspect of biology have been adopted in this activity. Laboratories of biochemistry and plant physiology in universities and Federal experiment stations as well as those of industry have carried out research on the absorption, translocation and mode of action of herbicides. Studies on the morphological effects of herbicides have been made. And laboratories of soil science, microbiology, and pesticide toxicology have been involved in studies of the fate of herbicides; adsorption, conjugation, chemical alteration, and biological degradation have been researched. Much of this effort represents the normal study required to understand the functions of herbicides in their role of weed killers. And much has been done to aid the ecologists in their work of protecting the environment from pollution. Practically all modern herbicides are organic compounds that eventually break down to CO_2, H_2O, $SO_4^=$, $PO_4^\equiv$, NO_3^-, Cl^-, Br^-, etc. Those that resist this extensive degradation and remain in soils and plant products as intermediate breakdown compounds must be studied for their toxicological properties. Many of these intermediate compounds are no more harmful than salt, baking powder or common pharmaceuticals. Those that present a hazard to human health or to the safety of the environment must be recognized and handled in such a way as to render them harmless. Through all of this work in the attempt to improve the lot of the farmer, to increase food and fiber production, and to preserve meanwhile a healthy stable environment, we must remain calm and objective in our thinking. Weed control is rapidly becoming a major field of agricultural technology, and synthesis testing and use of weed killers have assumed major roles in the modern technological drama.

REFERENCES

Anon. 1958–1969. *The Pesticide Review.* USDA Agric. Stabilization and Conservation Service. Washington, D.C.

Anon. 1965. A survey of extent and cost of weed control and specific weed problems. USDA, ARS 34–23–1, 1965.

Chandler, R. F., Jr. 1969. New horizons for an ancient crop. XI. *Internatl. Bot. Cong. 1969.* All-congress symposium. World Food Supply.

Furtick, W. R. 1967. National and international need for weed science, a challenge for WSA. *Weeds* **15**:291–295.

Furtick, W. R. 1970. Present and potential contributions of weed control to solution of problems of meeting the world's food needs. *Technical Papers of the FAO International Conf. on Weed Control.* pp. 1–6. Davis, Calif., June 22–July 1, 1970.

Holm, Leroy. 1969. Weed problems in developing countries. *Weed Sci.* **17**:113–118.

Holm, Leroy. 1971. The role of weeds in human affairs. *Weed Sci.* **19**:485–490.

House, W. B., et al. 1967. Assessment of ecological effects of extensive or repeated use of herbicides. U.S. Dept. Defense, DDC AD 824314, 369 pp. Midwest Research Inst.

Matsunaka, S. 1970. Weed control in rice. *Technical Papers of the FAO International Conf. on Weed Control.* pp. 7–23. Davis, Calif., June 22–July 1, 1970.

Miller, J. F. 1970. Sociological aspects of mechanized farming and chemical weed control. *Technical Papers of the FAO International Conf. on Weed Control.* pp. 591–604. Davis, Calif., June 22–July 1, 1970.

CHAPTER 2

Classification and Selectivity of Herbicides

Classification of Herbicides

Many chemicals of varied properties have found their way into the field of chemical weed control. Active search for phytotoxic compounds during the past 25 years has resulted in a great increase in the number of available weed killers.

Herbicides may be classified in several ways. One popular one represented by Table A-1 (appendix) lists the chemicals alphabetically by common name. The trade names, chemical names and manufacturer are also given; code designations of the manufacturers are included in a few cases. Table A-2 (appendix) lists these chemicals alphabetically by trade name and corresponding common name.

A second method exemplified by Table 2-1 groups the compounds chemically. This is useful in bringing into groups all compounds having chemical affinities.

A third classification illustrated by Table 2-2 provides a physiological grouping wherein the crop conditions for optimum results are given; the herbicides are divided into general contact and translocated foliar sprays and soil applied materials.

As illustrated in Table A-1 there are now over 150 basic compounds in use as herbicides. Inherent in this number and variety of chemicals are differences in phytotoxicity, in selectivity, in translocatability in method and mode of action, and in availability and cost. Most modern organic herbicides are

10

relatively high in phytotoxicity; this is a prerequisite determined in the screening process. Selectivity is becoming more and more exact as many of the niches in the spectrum of possible herbicides become filled. But selectivity is not an essential feature of all herbicides; chemical fallow, non-tillage and new and important industrial and other non-agricultural uses are creating an ever-increasing market for non-selective materials.

Translocatability is important and knowledge of the potential for both phloem- and xylem-mobility is essential to proper application of herbicides. For perennial weed control by foliar application, phloem mobility is necessary as is knowledge as to the response of each species to each chemical. For perennial weed control by soil application, an understanding of the behavior of a chemical in the soil and the mode of its uptake by roots is essential. Some chemicals are strongly fixed in soils whereas some are readily leached. Some are volatile and rapidly lost; with these soil incorporation is essential; others stay unchanged in the soil for weeks or months. Some herbicides break down in the soil within weeks; some persist for months or even years.

While understanding of mode of action is not required in the ordinary use of herbicides, this knowledge is needed to interpret results in the field in cases of unusual or aberrant action, and in cases of failure. Knowledge of the mode of action is basic to understanding the principles of herbicide synthesis and screening. And this information is of prime importance in work on residues and contamination of the environment.

Table 2-1 presents a chemical classification of herbicides wherein the array of known compounds is categorized on the basis of chemical affinities. While this is a useful way to list and characterize herbicides it tells us nothing concerning the methods by which they are employed.

When one considers the great variety of weed problems ranging from the broad field of agriculture through brush control and forest management, highway, park, and recreation area maintenance to industrial soil sterilization it is obvious that many different kinds of herbicides are needed.

Table 2-2 presents a scheme for classifying herbicides according to their mode of action and methods of application. It is based upon application methods that are in use in the field and each category is related to the mode of action.

Phytotoxicity of these many chemicals varies from the destruction of semipermeability of living membranes as brought about by oils (Currier, 1951; Van Overbeek and Blondeau, 1954) to complex interactions with enzyme systems. These may be exemplified by the competition of dalapon with pantoate and its inhibition of pantothenic acid synthesis in microorganisms (Hilton et al., 1959; Prasad and Blackman, 1965), and by the blocking of oxygen release during photosynthesis by urea, triazine, and uracil herbicides. Dalapon, as shown in Table 2-3 is a foliarly absorbed translocated herbicide which moves

TABLE 2-1. Herbicides: A Chemical Classification

INORGANIC HERBICIDES

Acids

Arsenic acid	Arsenic trioxide
Arsenious acid	Sulfuric acid

Salts

Ammonium sulfamate	Potassium cyanate
Ammonium sulfate	Sodium arsenate
Ammonium thiocyanate	Sodium arsenite
Borax	Sodium chlorate
Copper nitrate	Sodium chloride
Copper sulfate	Sodium dichromate
Hexafluorate	Sodium pentaborate
Iron sulfate	Tricalcium arsenate
Potassium chloride	

ORGANIC HERBICIDES

Oils

Diesel oil	Stoddard solvent
Polycyclic aromatic oils	Stove oil
Paraffinic additives	Xylene-type aromatic oils

ORGANIC HERBICIDAL COMPOUNDS

1. Aliphatics

Acrolein	Glytac
Allyl alcohol	TCA
Dalapon	

2. Amides

Alachlor	Methachlor
Butachlor	Monalide
Carbetamide	Naptalam
CDAA	Pronamide
Chlorthiamid	Propachlor
Delachlor	Prynachlor
Dicryl	R-7465
Diphenamid	Solan
Karsil	

3. Arsenicals

AMA	MAA
Cacodylic acid	MAMA
CMA	MSMA
DSMA	

TABLE 2–1 (*continued*)

4. Benzoics
   Chloramben   Tricamba
   Dicamba   2,3,6-TBA
   PBA

5. Dipyridiliums
   Diquat   Paraquat
   Morphamquat

6. Carbamates
   Asulam   Metham
   Barban   Phenmedipham
   Chlorbufam   Propham
   Chlorpropham   Proximpham
   Cycloate   Swep
   Dichlormate   Terbutol
   Karbutilate

7. Dinitroanilines
   Benefin   Oryzalin
   Isopropalin   Trifluralin
   Nitralin

8. Nitriles
   Bromoxynil   Diphenatrile
   Dichlobenil   Ioxynil

9. Phenols
   Dinosam   Fluorodifen
   Dinoseb   Medinoterb-acetate
   Dinoterb-acetate   Nitrofen
   DMPA   PCP
   DNOC

10. Phenoxys
    Dichlorprop   2,4-D
    Erbon   2,4-DB
    MCPA   2,4-DEB
    MCPB   2,4-DEP
    MCPES   2,4,5-T
    Mecoprop   2,4,5-TES
    Sesone
    Silvex

TABLE 2–1   (*continued*)

11. Thiocarbamates	
Butylate	Molinate
CDEC	Pebulate
Diallate	Triallate
EPTC	Vernolate
12. Triazines	
Ametryne	Methoprotyn
Atratone	Prometone
Atrazine	Prometryne
Aziprotryn	Propazine
Bladex	Simazine
Chlorazine	Simetone
Cyprazine	Simetryne
Desmetryne	Terbutryn
Ipazine	Trietazine
13. Triazoles	
Amitrole	Amitrole-T
14. Uracils	
Bromacil	Lenacil
Isocil	Terbacil
15. Ureas	
Benzomarc	Chlortoluron
Benzthiazuron	Cycluron
Buturon	DCU
Chlorbromuron	Difenoxuron
Chloroxuron	Diuron

with assimilates from leaves to various sinks throughout the plant. The urea, uracil, and triazine compounds on the other hand are absorbed from the soil by roots, transported to the foliage via the transpiration stream and distributed throughout the green tissues wherever transpiration is going on.

Although many of the new herbicides may act by blocking enzymes concerned in the synthesis of essential metabolites only a relatively few complete biochemical mechanisms of action have been worked out. Considering the great number of recognized herbicides that are phytotoxic there can be no doubt that many different enzyme systems are affected; few however have been studied and described. For example only now after about 25 years is it found that the phenoxy herbicides affect RNA synthesis and hence interfere in some way with protein synthesis.

In contrast to the biochemical mechanism of action, it is of more practical importance to understand the absorption and distribution of herbicides by

TABLE 2–1   (*continued*)

Fenuron	Monolinuron
Fluometuron	Monuron
Isonoruron	Monuron-TCA
Karbutilate	Neburon
Linuron	Norea
Methabenzthiazuron	Siduron
Metobromuron	Trimeturon
Metoxuron	

16. Unclassified

ACNQ	Dichlone
Bandane	Endothall
Benazolin	Euparen
Bensulide	EXD
Bentazon	Fenac
Bentranil	Flurenol
Benzadox	Glenbar
Benzazin	MH
Brompyrazon	NOA
Calcium cyanamide	Oxapyrazon
Chlorazon	Picloram
Chlorflurazol	PMA
Chlorfluorenol	Pyrazon
Chloropicrin	Pyrichlor
Cypromid	TCBA
Dazomet	Tunic
DCPA	Vorlex
Decazolin	

the plant. When the physiology of herbicidal action is known, such matters as proper time and method of application are recognized and formulation and method of application may be adjusted so that maximum effectiveness is obtained. The biochemistry of herbicides is most applicable to problems of synthesis and testing, and to handling of residues. Knowledge of the chemical mechanism of one herbicidal response may be used as a guide to discovery of other herbicides of similar chemical affinities. For example van Oorschot (1965) has used blocking of the Hill reaction as a method for indicating the herbicidal properties of many compounds having similar chemical groupings.

Classification of herbicides on a physiological basis provides a variety of contrasts (Table 2-2). An obvious contrast exists between systemic or translocated materials and non-systemic or contact compounds. Examples of the first are picloram, MH, amitrole and 2,4-D; of the second are sodium arsenite, diquat or aromatic oil.

In considering translocation one must include both phloem-mobility and xylem-mobility. Phloem-mobile translocated compounds applied to foliage move basipetally in plants to kill the roots and acropetally to kill the shoots; they are effective against perennial plants; applied to roots they move to the root tips. Xylem-mobile translocated compounds applied to foliage, move acropetally to the leaf tips; applied to roots they penetrate the cortical tissues, move into the xylem strands and proceed to the foliar organs via the transpiration stream. Metham is a contact material that moves with water in the soil and kills roots by contact; many soil fumigants do likewise when injected into the soil. In contrast, the urea, uracil, and triazine compounds are xylem mobile; when applied through the soil they rapidly penetrate the roots, ascend into the foliar region, and kill the plants through interference with the biochemistry of photosynthesis.

Of herbicides normally applied to the soil, arsenicals, boron compounds, carbon bisulfide, etc. are normally not translocated; they kill by contact ac-

TABLE 2-2.   Categories of Herbicides Physiological Basis

| Condition of Crop | Selective Herbicides | |
| | Foliar sprays | |
	General Contact	Translocated
Preplant	Dinoseb-acetate, Dinoterb-acetate, diquat, dinosam, dinoseb, paraquat, PCP, pyrichlor, weed oil	Amitrole, amitrol-T, AMA, AMS, atrazine + oil, barban, dalapon, dichlorprop, MAA, MAMA, mecoprop, MH, PBA, picloram, silvex, tricamba, 2,3,6-TBA, 2,4-D 2,4-DB, 2,4-DEB, 2,4-DEP, 2,4,5-T
Postplant, deep-rooted crops, trees, vines	Chlorflurazol, dinosam, dinoseb, dinoseb-acetate, dinoterb-acetate, diquat, paraquat, PCP, weed oil	Amitrole, amitrol-T, dalapon DBA, DSMA, erbon, flurenol MAMA, MSMA, 2,4-D, 2,4-DB
Preemergence	Dinoseb-acetate, dinoterb-acetate, dinosam, dinoseb, DNOC, PCP, pyrichlor, Stoddard solvent	Amitrole, amitrol-T, DBA, dichloroprop, flurenol, mecoprop MH, phenmedipham, sesone, 2,4-D 2,4-DB, 2,4-DEB, 2,4-DEP, 2,4,5-T

tion. For this reason these materials, in order to be effective, must be applied so that they thoroughly occupy the zone of soil in which the roots, rhizomes or seedlings are to be killed.

In contrast, when pelleted monuron, simazine, sodium chlorate or picloram is mixed with soils, all plants growing in the cultures are affected; these materials are absorbed and translocated to the tops of the plants where they bring about their toxic action. An example of the latter effect is the use of fenuron pellets in grass land or forest to control trees and brush.

Where a soil sterilant is used to control perennial weeds preparatory to laying concrete or macadam, the contact type of material should be used; if a systemic material is employed the roots will produce shoots and heave the pavement before the toxic action has killed the roots. Woestemeyer and Zick (1960) found monuron and simazine much less effective than chlorate, TCA, borates or mixtures of these in controlling bermudagrass preparatory to laying macadam pavement.

Selective Herbicides

Soil-applied

Absorbed from Soil

Benetin, butachlor butylate, calcium cyanamide, CDAA, CDEA, CDEC, cycloate, diallate, isopropalin, nitralin, pebulate, pyrazon, triallate, trifluralin, vernolate, vorlex

Bromacil, chlorthiamid, dichlobenil, diphenamid, diuron, monuron, simazine, terbacil, trifluralin

Alachlor, ametryne, arsenate-tricalcium, asulam, atratone, atrazine, aziprotryn, bandane, benefin, bensulide, bentranil, benzomarc, benzthiazuron, bladex, bromacil, brompyrazon, butachlor, buturon, chloramben, chlorazine, chlorazon, chlorbromuron, chlorbufam, chloroxuron, chloropropham, chlortoluron, cycloate, cycluron, DCPA, DCU, delachlor, desmetryne, diallate, difenoxuron, dinoseb, diphenamid, diphenatrile, diuron, EPTC, fenuron, fluometuron, fluorodifen, glenbar, ipazine, isonoruron, isopropanil, karbutylate, lenacil, linuron, maloran, medinoterb-acetate, methabenzthiazuron, methachlor, methoprotryn, metobromuron, metoxuron, molinate, naptalam, neburon, nitrofen, norea, oxapyrazon, preforan, prometone, propachlor, propham, proxipham, prynachlor, pyrizon, RH-315 (Kerb), sesone, siduron, simazine, simetone, simetryne, swep, TCA, terbacil, terbutol, triallate, trietazine, trifluralin, 2,4-D, 2,4-DEB, 2,4-DEP, 2,4,5-TES, VC438 (Tunic), vernolate

---

TABLE 2-2 *(continued)*

Condition of Crop	Selective Herbicides		
	Foliar sprays		Soil-applied absorbed from soil
	General Contact	Translocated	
Postemergence	AMA, chlorflurazol, CMA, cypromid, dicryl, dinosam, dinoseb, dinoseb-acetate, dinoterb-acetate, DMPA, DNOC, EXD, flurenol, glytac, HCA, hexaflurate, karsil, KOCN, lenacil, NOA, oryzalin, PMA, potablan, solan, tritac	AMA, amitrole, amitrol-T, atrazine + oil, barban, CMA, dalapon, DBA, dicamba, dichlorprop, DSMA, fluorenol, MCPA, MCPB, mecoprop, MH, PBA, phenmedipham, picloram, silvex, tricamba, tritac, 2,4-D, 2,4-DB, 2,4-5-T, 2,3,6-TBA	Asulam, aziprotryn, benazolin, bentazon, benzadox, benzazin, benzomarc, bromofenoxim, chloroxuron, chlorthiamid, cyprazine, DBA, decazolin, diphenamid, diuron, EXD, fluometuron, fluorodifen, hexaflurate, isonoruron, karbutylate, lenacil, linuron, MCPES, methabenzthiazuron, methoprotryn, metobromuron, metoxuron, monalide, neburon, nitrofen, norea, oryzalin, oxapyrazon, phenmedipham, prometryne, propanamide, propazine, prynachlor, pyrazon, swep, terbacil, terbutryn, tunic, tritac
Turfweed control, spray or pellets	Cacodylic acid, MAA, PMA	MCPA, MCPP, silvex, 2,4-D, 2,4,5-T	Benefin, bensulide, DCPA, siduron, terbutol

Non-selective Herbicides

	Foliar sprays		Soil applied, absorbed by roots
Mixed weeds no crop	Contact	Translocated	
Roadsides, canal banks, rights-of-way, air-fields, tank farms	Benzadox, cacodylic acid, chlorate, dinosam, dinoseb, diquat, DNOC, endothall, glytac, HCA + oil, morphamquat, paraquat, PCP	Amitrole, amitrol-T, AMA, AMS, chlorate, CMA, dalapon, dicamba, dichlorprop, DSMA, MSMA, PBA, phenmedipham, picloram, silvex, tricamba, 2,4-D, 2,4,5-T, 2,3,6-TBA	Arsenate-tricalcium, atrazine, boron salts, bromacil, chlorates. diuron, erbon, fenac, fenuron, karbutylate, isocil, metham, monuron, monuron-TCA, neburon, picloram, simazine, TCA, terbacil
Chemical fallow	Dinosam, dinoseb, diquat, fortified oil, morphamquat, paraquat, PCP	Amitrole, amitrol-T, dalapon, dicamba, erbon, tricamba, 2,4-D, 2,4,5-T	Atrazine, bromacil, diuron, erbon, fenac, fenuron, isocil, karbutylate, monuron, monuron-TCA, neburon, simazine, tandex
Woody plant control	Chlorate, cacodylic acid, diquat, weed oil	Amitrole, amitrole-T, AMS, MH, silvex, 2,4-D, 2,4,5-T, AMS, cacodylic acid, DSMA, MSMA, picloram, 2,4-D or 2,4,5-T amine	Fenuron pellets, picloram (spray or pellets), fenuron-TCA
Seed bed	Allyl alcohol, chloropicrin, methylpicrin, methylbromide		
Aquatic weed control, irrigation & drainage ditches, ponds, lakes, cooling systems	Acrolein, aromatic solvents, copper sulfate, diquat, endothall, fenac	Amitrole, fenac, silvex, 2,4-D	Atrazine, borates, bromacil, chlorate, diuron, erbon, fenac, isocil, monuron, monuron-TCA, simazine

Ahrens et al., (1970) have found a mixture of metham and dichlobenil very effective in controlling tree roots in sewer lines. When these are mixed with water and the lines filled, roots are killed within and for short distances outside the lines but little or no translocation to tops is observed.

Because the degree of systemic distribution of herbicides may vary with application method, maturity of plants, moisture supply, atmospheric humidity, and even plant species, and because the pattern of translocation depends upon the physiological condition of the plants as determined by the various factors listed above, strict classification on a physiological basis must be used with an appreciation of the variables involved. The arrangement presented in Table 2-2 has proved useful in teaching. Based on the common methods of application each category is related to the mode of action.

Selectivity of Herbicides

Selectivity of herbicides is relative, depending on concentration or dosage. All herbicides are phytotoxic; if applied in high enough dosage they kill all plants; if dosage is sufficiently low no plants die. At dosages between these extremes some plants are killed and some are uninjured.

The first selective herbicides, iron and copper salts in aqueous solution, and dilute sulfuric acid, were effective by differential wetting; containing no surfactant, the droplets of spray bounced off of the leaves of cereal crops but wet the broad leaves of cruciferous weeds. These materials were cheap and effective but demanding of labor and of ideal conditions for application.

Sodium dinitrocresylate (Sinox) was the next selective herbicide that appeared and it was superior because of its high phytotoxicity; not over 4 pounds of active ingredient per acre were required to eliminate most broadleaf weeds, and by adding a pound or so of ammonium sulfate the dosage required was reduced by 50%. But the basic mechanism was still differential wetting; if the dosage was increased the crop suffered; if a surfactant was added the normal dosage injured the crop.

With the appearance of 2,4-D a new mechanism of selectivity was introduced. Applied as a spray in aqueous solution, or in oil solution or applied through the soil, 2,4-D killed many broadleaf weeds but was relatively nontoxic to grass species. The mechanism of selectivity here resides in the protoplasm and has not yet been fully explained.

With the introduction of the many new herbicides in the past 25 years new types of selectivity have been found. Compounds of low solubility or those that are strongly fixed in soils may kill shallow rooted weeds while sparing deeper rooted crops (Figure 2-1). Examples are diuron and simazine as used in orchards and vineyards or DCPA, bensulide or neburon as used in turf, (Figures 2-2, 2-3).

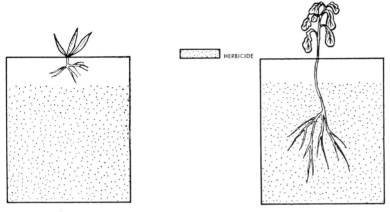

Position of herbicide in soil. Shallow-rooted crop (left) remains alive if herbicide moves beyond its rooting zone. Deep-rooted weed is killed when herbicide is leached into the deeper zones of the soil. (Ashton and Harvey, 1971.)

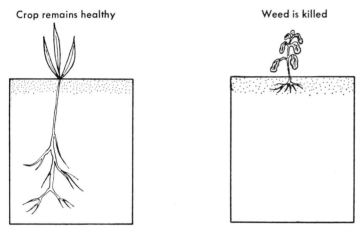

FIGURE 2-1.   Position of herbicide in soil. Deep-rooted crop (left) is not affected by herbicide which remains near soil surface. Shallow-rooted weed is killed by herbicide which stays near surface.

More subtle mechanisms are acting in many cases of herbicide selectivity. For example there occur compounds in maize, sorghum, and in other grass species that are able to replace the chlorine on the substituted triazine molecule with an OH group. This renders the molecule innocuous and the crop is spared. In the case of maize the compound has been identified as a cyclic

FIGURE 2-2.   Weed control in an almond orchard. Chemical weed control in the tree row and cover crop between the rows.

hydroxamate; in other crops it has not yet been identified. Studies on the mode of action of trifluralin have shown that when this compound is absorbed by roots there is soon an inhibition of secondary root formation. In the case of large seeded crops such as cotton, the strong tap root rapidly penetrates the shallow layer of soil holding the herbicide and is able to grow normally below this layer. Weed seedlings and seeds germinating in this layer are not able to overcome the root inhibition and so they succumb.

Hogue and Warren (1968) have shown that the selectivity of linuron between tomato, a susceptible crop and parsnip a tolerant one depends upon differential accumulation and metabolism of the chemical; tomato translocates the compound throughout the foliar portion of the plant; parsnip metabolizes the small amount that does reach the tops.

Wheeler and Hamilton (1968) found the selectivity of atrazine between susceptible wheat and oats and tolerant maize and sorghum to be closely paralleled by loss of chlorophyll by the susceptible plants. They propose that the primary site of toxic action leading to acute toxicity may be the chloroplasts.

Many other mechanisms of selectivity will undoubtedly be found as de-

FIGURE 2-3. Weed control in a vineyard. Chemical weed control in the vine row and native weeds between the rows. The latter will be disced in later.

tailed research on the many new preplant and preemergence herbicides is carried out. Knowledge of the mechanism of selectivity is extremely useful in formulating, applying and interpreting the effects of herbicides on weeds and crops.

REFERENCES

Ahrens, J. F., O. A. Leonard, and N. R. Townley. 1970. Chemical control of tree roots in sewer lines. *Jour. Water Pollution Control Federation.* **42**:1643–1655.

Ashton, F. M. and W. A. Harvey. 1971. Selective chemical weed control. Cir. 558 *Calif. Agr. Expt. Sta. and Ext. Ser.* 17 pp.

Currier, H. B. 1951. Herbicidal properties of benzene and certain methyl derivatives. *Hilgardia* **20**:383–406.

Hilton, J. L., J. S. Ard, L. L. Jansen, and W. A. Gentner. 1959. The pantothenate-synthesizing enzyme, a metabolic site in the herbicidal action of chlorinated aliphatic acids. *Weeds* **7**:381–396.

Hogue, E. J. and C. F. Warren. 1968. Selectivity of linuron on tomato and parsnip. *Weed Sci.* **16**:51–54.

Oorschot, J. L. P. van. 1965. Selectivity and physiological inactivation of some herbicides inhibiting photosynthesis. *Weed Res.* **5**:84–97.

Overbeek, J., van and R. Blondeau. 1954. Mode of action of phytotoxic oils. *Weeds* **3**:55–65.

Prasad, R. and G. E. Blackman. 1965. Studies in the physiological action of 2,2-dichloropropionic acid. III. Factors affecting the level of accumulation and mode of action. *J. Exptl. Bot.* **16**:545–568.

Wheeler, H. L. and R. H. Hamilton. 1968. The leaf concentrations of atrazine in cereal crops as related to tolerance. *Weed Sci.* **16**:7–10.

Woestemeyer, V. W. and W. H. Zick. 1960. Response of Bermuda grass to herbicides applied as pre-paving treatments. *Western Weed Control Conf. Res. Prog. Rept.* Denver, Colo. p. 66.

CHAPTER 3

Morphological Responses to Herbicides

Cellular and Structural Changes

While the formative effects of growth regulators have been recognized since the early experiments of Darwin and Went, little attention was paid to the morphological responses to herbicides until the introduction of 2,4-D. Then in quick succession were described the epinastic bending of leaves, epicotyl curvature (the bean test) excessive callus formation in roots and in stems of some plants, and secondary root stimulation on roots and stems. By 1949 and 1950 the formation of callose on the sieve plates of bean and nutsedge (*Cyperus sp*) was described (Eames 1949, 1950). In his 1950 study of the effects of 2,4-D on bean plants Eames found cell proliferation starting in the endodermis of the stem and spreading to phloem and cortex parenchyma until the phloem was completely crushed. Maleic hydrazide (MH) has been reported to cause phloem necrosis (McIlrath, 1950). Akobundu et al. (1970) found that dichlobenil treatment on purple nutsedge (*Cyperus rotundus*) plants brings about destruction of phloem in the basal regions of leaf sheaths. This results in accumulation of assimilates in the treated leaves and in tubers rather than in young underground tissues. Accumulation in tubers may have resulted from starch storage which would maintain a concentration gradient in the conducting phloem conducive to enhanced assimilate flow. Treated plants also showed increased respiration in tubers, itself conducive to a maintained assimilate gradient.

Gifford (1953) worked on the results of 2,4-D application to the cotyledons of cotton seedlings. He found that sublethal dosage affects the structure and morphology, not of the apical meristem itself but of the organs and tissues derived from it. The 2,4-D affects not only the foliage-leaf primordia present at the time of application but also some that develop subsequently.

Kaufman (1953) studied the morphological responses of the rice plant to 2,4-D at herbicidal dosages and found this plant to be very sensitive during early seedling growth, tillering, and root and panicle emergence. Symptoms included tubular leaves, small irregular panicles and injury to the primary roots. Kaufman's work was promptly put to use in establishing the most favorable time of application for the control of broadleaf weeds in rice.

Maun and Cavers (1969) treated curly dock (*Rumex crispus*) plants 12 days before anthesis, at anthesis, and 7 and 34 days after anthesis with the lithium salt of 2,4-D at 1000 ppm. Treatment 12 days before anthesis inhibited viable seed formation. Treatment at anthesis produced 2% of the seeds with minute embryos; these seeds did not germinate. In plants treated 7 days after anthesis embryos were present in 91% of the seeds but they ranged in size from 0.5 to 2.5 mm. Seed weight was reduced and only 5 to 15% germinated. Spraying inflorescenses 34 days after anthesis had no effect on number of seeds produced or their viability.

Meanwhile the effects of 2,4-D, IAA, NAA, TIBA, 2,4,5-T and other growth regulators on various plants were studied (Kiermayer, 1964). Apical meristems were found to be little affected by these materials which follows from the fact that protophloem sieve tubes do not differentiate until elongation has started; food materials, endogenous auxins and viruses must move beyond these by the slow process of diffusion. In the regions of cell elongation and differentiation profound responses to herbicides occur. Disturbances in cell shape, cell size, cell divisions and tissue morphology were found following herbicide application. Fragmentation and fusions of nuclei, blocked metaphases, giant nuclei, multinucleate cells, contracted chromosomes, aberrant spindle behavior and even prevention of cell divisions have been observed.

Although all of the compounds mentioned above are growth regulators, picloram is by far the most powerful growth regulator so far discovered. Kreps and Alley (1967) describe abnormalities throughout the root system of Canada thistle (*Cirsium arvense*) plants following treatment with picloram. There was swelling, splitting, and deterioration of parenchyma tissues with severe injury to adventitious and endogenous bud and root primordia. Cambium and phloem were destroyed. Scifres and McCarty (1968) observed accelerated procambium activity in leaves of Western ironweed (*Vernonia baldwini*) plants treated with picloram. This increased activity resulted in destruction of phloem parenchyma, sieve elements and companion cells. Comparable 2,4-D-treated plants responded much more slowly. Martin et al. (1970) report swellings of fourwing saltbush (*Atriplex canescens*) seedlings in regions immediately below cotyledonary nodes and at the soil level. Tissues were distorted, the endodermis was lacking and small parenchyma-like cells make up the tumerous tissue between the vascular tissue and the outermost cell layer.

Propham and chlorpropham have been observed to produce profound disturbances in the cytology of roots (Ennis, 1948; Ivens and Blackman, 1949; Scott and Struckmeyer, 1955; Canvin and Friesman, 1959). More recently trifluralin and nitralin have been found to inhibit lateral root development. Stanifer and Thomas (1965) found that trifluralin inhibits secondary root development on cotton seedlings and it killed johnsongrass (*Sorghum hale-pense*) seedlings if the first internode passed through the treated soil; it inhibits lateral root development from stems exposed to chemical in the soil.

Hacskayla and Amato (1968) observed inhibition of lateral root growth in both cotton and maize seedlings; cotton was more tolerant than maize, probably because it has a tap root. From cytologic studies they found that cell division was markedly inhibited but nuclear division and fragmentation was abundant. Affected cells continued to enlarge but they failed to differentiate; cell plate and cell wall formation were rarely found. Gentner and Burk (1968) observed similar inhibition of lateral root growth of maize seedlings growing in soil treated with nitralin at rates equivalent to applications of $\frac{1}{2}$, 1, 2, and 4 lb/A. When the herbicide was applied immediately after planting, the roots ranged from digitate at $\frac{1}{2}$ lb/A to globose and severely stunted at 2 and 4 lb/A. Spiral contraction and swelling of chromosomes occurred; swelling was not restricted to chromosomes; whereas normal parenchyma cells measured 50 μ treated roots produced parenchyma cells exceeding 160 μ; these giant cells were often multinucleate. Numerous chromosomes in some of these cells aggregated and fragmented into clumps that passed into the restitution phase. Spindle formation was not seen and many multinucleate cells were observed. Schweizer (1970) noted severe inhibition of root growth of sugarbeet seedlings treated with trifluralin. Mature beets were deformed from exposure to trifluralin; there was severe gall formation of the hypocotyledonary neck region. Norton et al. (1970) found that the trifluralin inhibited lateral root development on field grown pecan seedlings. On greenhouse grown plants lateral roots were inhibited, radicles were club-shaped and the tap roots failed to penetrate soils containing 1.0 ppm of trifluralin. Nuts produced on trees growing in trifluralin treated soil were undersized.

Bingham (1968) found that DCPA inhibited rooting of bermudagrass (*Cynodon dactylon*) stolon nodes. Histological studies of these root tips indicated that cell division had ceased although cell enlargement continued. Cells grew to excessive size and were irregularly shaped. In maize, radicle cell size was not affected but treated roots contained six times as many dinucleate cells as untreated; many cells were in an arrested metaphase state in treated roots. Onion root cells showed inhibition of cell division in treated roots.

The triazine, urea, and uracil compounds, though not growth regulators, have growth effects on vegetative apices. They also affect chloroplast structure. Ashton et al. (1963a) found atrazine to cause precocious vacuolation of bean chloroplasts, destruction of chloroplasts with time, reduced stomatal chamber and palisade airspace, cessation of cambial activity and decreased cell wall thickness of sieve tubes and tracheal elements. This latter effect took place only in the light and led the workers to postulate the possible production of toxins, of the nature of free radicles to account for their observations.

Pursuing this subject further Ashton et al. (1963b) used the electron microscope to examine the fine structure of chloroplasts. Using bean plants they found that atrazine treatment in the light caused chloroplasts to assume a spherical form. Frets were destroyed leading to disorganization of grana; compartments of the grana swelled and the envelope and swollen compartments disintegrated. Starch disappeared from the lamellar system. Since most of these effects did not take place in the dark they were not simply the result of diminished nutrition following cessation of photosynthesis.

Hill et al. (1968) made electron microscope studies on the chloroplasts of barnyardgrass (*Echinochloa crusgalli*), treated with atrazine. During the 1- to 3-leaf stage degradation of the chloroplasts starts as a swelling of the fret system followed by swelling and disruption of the granal discs. As breakdown proceeds the membranes of the grana and chloroplast envelope rupture. There were fewer starch grains formed as the treatment exceeded 4 hours. Mitochondria remained normal throughout these tests and the ultrastructural changes preceded any macroscopically observable symptoms. The above changes in barnyardgrass took place in from 2 to 8 hours and became advanced by 12 hours. The results described above for Ashton et al. (1963b) required 30 hours or more in bean plants.

Geronimo and Herr (1970) treated tobacco plants with pyriclor and found changes in chloroplasts similar to those described for atrazine in bean (Ashton et al. 1963b) and barnyardgrass (Hill et al. 1968). Initial changes included assumption of spherical form, swelling of the fret system and initiation of loss of starch. Subsequently all starch disappeared, membranes became more swelled and disorganized both in the frets and in the grana; chloroplast envelopes were ruptured. Mitochondria were not changed but they appeared to increase in numbers. Appearance of visual symptoms of phytotoxicity (chlorosis) correlated with the disruptive changes in chloroplast structure.

Liang et al. (1969) treated grain sorghum plants 6 inches tall in 1966 and 2 to 18 inches tall in 1967, growing in the field, with 2,4-D, atrazine and atrazine plus 2,4-D. They examined them for chromosome aberrations in pollen cells and found aneuploidy, polyploidy, stickyness, multinucleoli, and lack of orientation. The ester of 2,4-D produced more chromosome aberrations than its amine counterpart; oil in the formulation enhanced the ability of the herbicides to induce chromosome aberrations.

While chlorella is not a weed in the ordinary sense, it has been used in many studies on photosynthesis and its use for cytological studies on the effects of atrazine on the chloroplasts and their activity seems justified. Ashton et al. (1966) used *Chlorella vulgaris* to study these effects. They used cell samples taken 0, 24, 48 and 72 hours after atrazine treatment, and they measured chlorophyll content and packed cell volume. They found that atrazine prevented normal increase in cell volume and also chlorophyll development; inclusion of glucose in the culture medium counteracted the atrazine. *Chlorella* utilized both the endogenous products of photosynthesis and an exogenous supply of glucose as well. The atrazine-treated cells did not contain starch; control cells, atrazine-glucose, and glucose-treated cells developed starch grains. Atrazine did not cause any noticeable abnormalities in cell organelles.

Pyrazon, another Hill reaction inhibitor that, like the urea and triazine herbicides, is absorbed from soil by roots and translocated via the xylem, has been found to affect the anatomy of leaves. Rodebush and Anderson (1970) treated bean plants through the culture medium with pyrazon at 0, 5, 10, 25, 50, and 100 ppm. Seedlings exposed to pyrazon concentrations less than 25 ppm did not develop injury symptoms. Abnormalities were first noted 5 days after emergence in seedlings treated with 50 and 100 ppm pyrazon. The first symptoms were chlorosis of the margins of the unifoliate leaves; these failed to expand, chlorosis became severe, leaf margins became necrotic and rolled ab-axially. Seedlings grown in pyrazon treated vermiculite did not develop trifoliate leaves; the unifoliate leaves became senescent, the midribs and petioles became chlorotic and the plants died within 3 weeks after emergence. Internally the chloroplasts of chlorotic leaves failed to synthesize starch, and they became clumped in some cases. Later the epidermis collapsed, and the mesophyll became necrotic as the seedlings approached death.

In a subsequent paper Anderson and Schaelling (1970) describe ultrastructural effects of pyrazon on bean plants. With plants treated as in the above experiments they found that changes in the chloroplasts preceded visual symptoms. All chloroplasts in treated plant leaves whether from green or chlorotic tissues were devoid of starch; photosynthesis was evidently inhibited. Ultrastructural changes resembled those reported for atrazine by Ashton et al. (1963b). Chloroplasts became round and swollen; membrane permeability changes allowed the chloroplasts to become turgid. Chloroplasts became clumped; the grana contained fewer thylakoids than those of normal plants; ribosome numbers were not altered. In chlorotic leaf-margin cells chloroplast structure was severely affected. No grana were found; few thylakoids had formed; separations developed until the chloroplasts became highly vacuolated; lipid globules and osmiophylic bodies appeared. If pyrazon inhibits the light reaction of photosynthesis, production of ATP and $NADPH_2$ may cease hence structural protein synthesis may fail. This may result in disrupted and thylakoid structure according to Anderson and Schaelling (1970).

An interesting corollary to the above observations is that paraquat, a strictly contact herbicide, after destroying the plasmolemma membranes of treated leaf cells, ruptures chloroplast membranes, and prevents starch formation (Bauer et al., 1969). There were no differences in these effects between honey mesquite (*Prosopis juliflora*) plants treated and sampled in light, and those treated and sampled in dark. Other organelles in the cells were unchanged. The authors postulate that paraquat action in living tissues may be a two-phase process. Upon absorption the paraquat is rapidly reduced and re-oxidized forming organic radicals that disrupt membrane systems. The second phase involves blocking of anabolic processes in the chloroplast. Such inhibition could result from herbicide reduction by $NADPH_2$, a re-routing of endogenous reducing power for reduction of the herbicide rather than its utilization in normal redox reactions.

The vulnerability of leaf cells to 2,4-D depends upon their stage of development; mature leaves show no formative effects; leaves in the differentiating stages may be severely deformed; leaves just emerging from the shoot apex at time of treatment may be hardly recognizable (Sasaki and Kozlowski, 1967); with limited dosage a plant may slowly recover and again produce normal leaves. Gifford (1953) describes in detail the changes taking place in cotton leaf development following 2,4-D administration. Gorter and van der Zweep (1964) have reviewed the morphogenetic effects of herbicides on plants from the standpoint of normal anatomy and morphologic development. They describe abnormalities of roots, stems, and leaves that result from herbicides. While most of these follow the use of hormone-type compounds, they also describe abnormalities of roots caused by CDAA, amitrole, 2,6-dichlorobenzonitrile, and MH; loss of geotropism of roots due to naptalam; leaf aberrations caused by dalapon and TCA; and inhibition of flowering by MH. The formation of tubular leaves on cereal plants by 2,4-D, dalapon and TCA is treated in detail as is the production of abnormal heads of wheat and barley.

In studies on seedlings of pine, treated by soaking the seeds in herbicide solutions for 24 hours and then planting in soil, Sasaki et al. (1968) found that EPTC, CDEC, CDAA, and 2,4-D caused marked morphogenic changes during seedling development. The most conspicuous effect of EPTC, CDEC and 2,4-D treatment was inhibition of cotyledon development. Plants whose seeds had been treated with CDEC or EPTC had fused cotyledons and those treated with 2,4-D had swollen stems and shrivelled chlorotic cotyledons. Seed coats of many plants treated with EPTC, CDEC or 2,4-D were not shed during the 34 day growth period.

Diallate and triallate are compounds used through the soil to inhibit growth of wild oats (*Avena fatua*) in wheat and other cereals. Banting (1970) germinated wild oat and wheat seeds in vapors of diallate and triallate and found, after 3 days, that shoot tissue of wild oat was damaged more severely than root tissue by diallate. The meristem at the base of leaf one was more vulnerable than stem apex tissue since it showed more cell divisions and a greater number of abnormalities; triallate was less damaging.

In wheat again, shoot tissue was more sensitive than root tissue. Wheat proved more tolerant to both toxicants than oats and diallate was more severe than triallate. Shoot inhibition in both wild oats and wheat occurred at concentrations that did not affect mitosis.

Bensulide is a benzene sulfonamide used to kill weedy grass seedlings in turf, particularly dichondra. Since growth of seedlings is inhibited without emergence it seems that the mode of action must involve early root growth. Cutter et al. (1968) studied the effects of bensulide on the growth and structure of roots of oat seedlings. They found inhibition of length growth, precocious vacuolation and lignification, presence of short pitted tracheary elements, outgrowth of root hairs close to the root apex characteristic of roots that have been retarded in their growth. While elongation was inhibited, mitosis was not stopped. Epidermal cells were elongated radially but adjacent cortical cells were normal. Apparently this disturbance of root growth stops development of the whole embryo and eventually the seedlings die. Peterson and Smith found similar changes in root structure of quackgrass plants treated with pronamide (Peterson and Smith, 1971).

Ashton et al. (1969) studied the growth and structural modification of oats induced by bromacil, a non-selective herbicide used to control weeds on non-cropped land. Bromacil is a known inhibitor of the Hill reaction. Total root elongation was inhibited and inhibition increased with concentration through the series $10^{-6}M, 10^{-5}M, 10^{-4}M, 5 \times 10^{-4}M$, and $10^{-3}M$; growth at the latter concentration was almost mil. Inhibition was almost entirely restricted to the terminal 0.5 mm, the region of greatest elongation in untreated roots. Treated roots were swollen or necrotic and collapsed with necrosis present in meristem, procambium and epidermis. Precocious vacuolation was seen in cells of the root tip and inhibition of cell wall formation resulted in multinucleate cells. In the leaves bromacil inhibited development of the chloroplast grana and fret system; the loculi of the grana and fret vesicles swelled progressively and the chloroplast envelopes were modified; there appeared to be loss of membrane integrity.

REFERENCES

Akobundu, I. O., D. E. Bayer, and O. A. Leonard. 1970. The effects of dichlobenil on assimilate transport in purple nutsedge. *Weed Sci.* **18**:403–407.

Anderson, J. L. and J. P. Schaelling. 1970. Effects of pyrazon on bean chloroplast ultrastructure. *Weed Sci.* **18**:455–459.

Ashton, F. M., T. Bisalputra, and E. B. Risley. 1966. Effect of atrazine on Chlorella vulgaris. *Amer. J. Bot.* **53**:217–219.

Ashton, F. M., E. G. Cutter, and D. Huffstutter. 1969. Growth and structural modifications of oats induced by bromacil. *Weed Res.* 9:198–204.

Ashton, F. M., E. M. Gifford, Jr., and T. Bisalputra. 1963a. Structural changes in Phaseolus vulgaris induced by atrazine. I. Histological changes. *Bot. Gaz.* 124:329–335.

Ashton, F. M., E. M. Gifford, Jr., and T. Bisalputra. 1963b. Structural changes in Phaseolus vulgaris induced by atrazine. II. Effects on fine structure of chloroplasts. *Bot. Gaz.* 124:336–343.

Banting, J. D. 1970. Effect of diallate and triallate on wild oat and wheat cells. *Weed Sci.* 18:80–84.

Baur, J. R., R. W. Bovey, P. S. Baur, and Zonab El-Seify. 1969. Effects of paraquat on the ultrastructure of mesquite mesophyll cells. *Weed Res.* 9:81–85.

Bingham, S. W. 1968. Effect of DCPA on anatomy and cytology of roots. *Weed Sci.* 16:449–452.

Canvin, D. T. and G. Friesen. 1959. Cytological effects of CDAA and IPC on germinating barley and peas. *Weeds* 7:153–156.

Cutter, E. G., F. M. Ashton, and D. Huffstutter. 1968. The effects of bensulide on the growth, morphology and anatomy of oat roots. *Weed Res.* 8:346–352.

Eames, A. J. 1949. Comparative effects of spray treatments with growth regulating substances on the nutgrass *Cyperus rotundus* L. and anatomical modifications following treatment with butyl 2,4-dichlorophenoxy acetate. *Amer. J. Bot.* 36:571–584.

Eames, A. J. 1950. Destruction of phloem in young bean plants after treatment with 2,4-D. *Amer. J. Bot.* 37:840–847.

Ennis, W. B. 1948. Responses of crop plants to O-isopropyl N-phenyl carbamate. *Bot. Gaz.* 109:473–493.

Gentner, W. A. and L. G. Burk. 1968. Gross morphological and cytological effects of nitralin on corn roots.

Geronimo, J. and J. W. Herr. 1970. Ultrastructural changes of tobacco chloroplasts induced by pyriclor. *Weed Sci.* 18:48–53.

Gifford, E. M. 1953. Effect of 2,4-D upon the development of the cotton leaf. *Hilgardia* 21:606–644.

Gorter, C. J. and W. Van der Zweep. 1964. Morphogenetic effects of herbicides. In: *The Physiology and Biochemistry of Herbicides*. pp. 237–275. Ed. L. J. Audus. Academic Press, New York.

Hacskaylo, J. and V. A. Amato. 1968. Effect of trifluralin on roots of corn and cotton. *Weed Sci.* 16:513–515.

Hill, E. R., E. C. Putala, and J. Vengris. 1968. Atrazine-induced ultrastructural changes in barnyardgrass chloroplasts. *Weed Sci.* 16:377–380.

Ivens, G. W. and G. E. Blackman. 1949. The effects of phenylcarbamates on the growth of higher plants. *Symposia Soc. Exptl. Biol.* 3:266–282.

Kaufman, P. B. 1953. Gross morphological responses of the rice plant to 2,4-D. *Weeds* 2:223–253.

Kiermayer, O. 1964. Growth responses to herbicides. In: *The Physiology and Biochemistry of Herbicides.* pp. 207–233. Ed. L. J. Audus. Academic Press, New York.

Kreps, L. B. and H. P. Alley. 1967. Histological abnormalities induced by picloram on Canada thistle root. *Weeds* **15**:56–59.

Liang, G. H. L., K. C. Feltner, and O. G. Russ. 1969. Meiotic and morphological response of grain sorghum to atrazine, 2,4-D, oil, and their combinations. *Weed Sci.* **17**:8–12.

Maun, M. A. and P. B. Cavers. 1969. Effects of 2,4-D on seed production and embryo development of curly dock. *Weed Sci.* **17**:533–536.

Martin, S. C., S. J. Shellhorn, and H. M. Hull. 1970. Emergence of fourwing saltbush after spraying shrubs with picloram. *Weed Sci.* **18**:389–392.

McIlrath, W. J. 1950. Response of the cotton plant to maleic hydrazide. *Amer. J. Bot.* **37**:816–819.

Norton, J. A., J. P. Walter, Jr., and J. B. Storey. 1970. The effect of herbicides on lateral roots and nut quality of pecans. *Weed Sci.* **18**:520–522.

Peterson, R. L. and L. W. Smith. 1971. Effects of *N*-(1,1-dimethylpropynyl)-3,5-dichlorobenzamide on the anatomy of *Agropyron repens* (L.) Beaux. *Weed Res.* **11**:84–87.

Rodebush, J. E. and J. L. Anderson. 1970. Morphological and anatomical effects of pyrazon on bean. *Weed Sci.* **18**:443–446.

Sasaki, S. and T. T. Kozlowski. 1967. Effects of herbicides on carbon dioxide uptake by pine seedlings. *Canad. J. Bot.* **45**:961–971.

Sasaki, S., T. T. Kozlowski, and J. H. Torrie. 1968. Effect of pretreatment of pine seeds with herbicides on seed germination and growth of young seedlings. *Canad. J. Bot.* **46**:255–262.

Schweizer, E. E. 1970. Abberations in sugarbeet roots as induced by trifluralin. *Weed Sci.* **18**:131–134.

Scrifres, C. J. and M. K. McCarty. 1968. Reaction of Western ironweed leaf tissue to picloram. *Weed Sci.* **16**:347–349.

Scott, M. A. and E. G. Struckmeyer. 1955. Morphology and root anatomy of squash and cucumber seedlings treated with isopropyl *N*-(3-chlorophenyl) carbamate (CIPC). *Bot. Gaz.* **117**:37–45.

Stanifer, L. C. and C. H. Thomas. 1965. Response of Johnsongrass to soil-incorporated trifluralin. *Weeds* **13**:302–306.

CHAPTER 4

Absorption and Translocation of Herbicides

The Cuticle Barrier

In order to survive in their terrestrial environment, land plants have developed cuticle, the waxy layer that separates the wet cell wall phase from the drying effects of air. Whereas the air surrounding plants may range in relative humidity from values well below 50% to 100%, intercellular air in the stomatal chambers and intercellular spaces in plants seldom drops below 99%; when it goes to 98% or below the plants wilt and suffer from water deficiency. Table 4-1 presents data showing the relation between relative humidity, vapor pressure, and diffusion pressure deficit of atmospheric moisture at 20°C (Crafts, 1961c).

The outer cell wall of foliar organs is made up of four distinct substances that vary widely in composition and distribution: cutin, cutin wax, pectin, and cellulose. Cutins are polymerized acids and alcohols of high molecular weight. Cutin is isotropic in polarized light, insoluble in ordinary reagents, stainable in basic lipid dyes, saponifiable with sodium hydroxide, and strongly absorbent of ultraviolet light. This property probably affords protection to the photosynthetic mechanism of plant cells; particularly in the case of alpine plants at high altitudes.

Cutin waxes are short-chain esters and alcohols of relatively low molecular weight and lacking reactive end groups. They are optically negative, soluble in pyridine, stainable in lipid dyes, and they melt above 220°C. They do not absorb ultraviolet radiation.

Pectins are amorphous and highly hydrophilic; they are formed of long

34

TABLE 4-1.   Relation between Relative Humidity, Vapor Pressure, and Diffusion Pressure Deficit at Atmospheric Moisture at 20°C (Crafts et al., 1949)

Relative humidity, %	Vapor pressure, mm Hg	Diffusion pressure deficit, atm.
100	17.54	00.00
99	17.36	13.43
98	17.19	26.89
97	17.01	40.53
96	16.84	54.28
94	16.49	82.38
92	16.14	110.99
90	15.79	140.33
80	14.03	297.24
70	12.28	475.04
60	10.52	680.39
50	8.77	923.31

chain polygalacturonic acid molecules having side carboxyl groups. These are capable of forming salts; they give pectins base exchange properties. Polygalacturonic acid is water soluble; its calcium salts are insoluble. Some of the carboxyl groups of pectin are methylated through oxonium oxygen; these methoxy groups do not lower the water solubility. Pectins are isotropic, soluble in picric acid and hydrogen peroxide, stainable in Ruthenium red; they break down upon hydrolysis, are disintegrated by pectic enzymes and they do not absorb ultraviolet light.

Cellulose is fibrillar in structure, hydrophilic, and, having great tensile strength, is responsible for the high resistance of cell walls to stretching. Cellulose is optically positive, soluble in Sweitzers reagent, stainable in iodine-zinc chloride and readily hydrolyzed. It does not absorb ultraviolet radiation. It is composed of long chain molecules that are relatively stable. These molecules are organized into micelles and the micelles are associated into microfibrils. Thus the structure of cellulose, because of its fibrillar organization, imparts tensile strength and elasticity. It is this property of cell walls that resist expansion and hence results in turgor.

Turgor in turn results in form, enabling the plant to grow erect against the force of gravity, to extend its roots into the soil and tap a large water and nutrient supply, and to maintain its foliar organs in positions favorable for maximum absorption of light and CO_2.

Stomata are openings in the cuticularized epidermis that allow gaseous exchange to go on so that CO_2 can diffuse into leaves; such openings are essential to the import of CO_2 and the export of O_2 in photosynthesis.

A necessary corollary to the $CO_2:O_2$ exchange in leaves is the loss of water vapor by transpiration. In order that water loss cannot go on to the detriment of the plant, the stomata may close thus carrying on a regulatory function; by opening the stomata permit $CO_2:O_2$ exchange; by closing they prevent excessive water loss. Open stomata constitute a port of entry for spray solutions; closed stomata make the cuticle barrier complete.

Currier and Dybing (1959) have discussed the role of stomatal entry in the uptake of herbicides. They pointed out the variable nature of stomatal opening, the requirement of low surface tension if a solution is to enter stomata, and the fact that stomata occur only on the under surface of the leaves of many species. When the stomata are open and the conditions are right, stomatal penetration of spray solution may be rapid as compared with uptake through the cuticle which is relatively slow. Evidence concerning the exact role of stomatal uptake of herbicides under field conditions is conflicting, most workers observing that it seldom occurs, and when it does it is unimportant. Obviously stomatal opening cannot be depended upon as a means of penetration of herbicides, and formulation practice in general ignores it. The possibility of bringing about stomatal opening by chemical means is challenging but to date has not been successful. Mason (1960) has reported that 2,3,6-TBA-treated bean plants may have a higher number of open stomata than untreated plants. That 2,4-D application may result in closing of stomata has been observed a number of times. Ferri and Rachid (1949) report on stomatal closure from soil application of 2,4-D and Bradbury and Ennis (1952) found application to the foliage to result in the same effect. The closure is proportional to the dose applied to the leaves over a range from 10 to 1000 ppm and seems to result from a direct physiological action on guard cells and other leaf tissues. It is reported as being associated with a decrease in phosphorylase and amylase activity of stomatal cells (Maciejewska–Potapezyk, 1955).

Mechanics of Absorption

Penetration and absorption are the first processes involved in herbicide physiology. Work with ^{14}C-labeled compounds indicates that there are two routes by which exogenous molecules may traverse the distance from the cuticle surface into the living inner cells, a lipoid route and an aqueous pathway (Crafts, 1956, 1961a). Compounds that penetrate the cuticle in the lipoid-soluble form such as dinitro and phenoxy compounds do so principally in the non-polar, undissociated form. Penetration of such compounds is enhanced by formulation as the parent weak acids, as esters or as salts of weak bases.

Compounds that enter via the aqueous route move in slowly, and their penetration is greatly benefited by a saturated atmosphere. Smith et al. (1959)

have shown that maleic hydrazide (MH) is one such compound. Work by Clor, Crafts and Yamaguchi (1962, 1963, 1964) has proved that penetration of both types of compounds is greatly enhanced by providing a saturated atmosphere around the leaves. The effect results from both stomatal penetration and a maintained aqueous medium in the leaf surface for uptake via the cell-wall phase of the leaf.

There are five possible fates of an applied herbicide with respect to penetration: (1) it may volatilize and be lost into the atmosphere; (2) it may remain on the outer leaf surface and dry down to the crystalline form, or it may concentrate to a viscous liquid depending upon its physical state or the formulation; (3) it may penetrate into the cuticle and remain there in solution in this lipoid layer; (4) it may penetrate the cuticle and then part into the aqueous phase of the apoplast system where it is free to diffuse into the inner leaf structure. If not phloem-mobile it will remain in the apoplast and move with the transpiration water throughout the acropetal portion of the leaf with a tendency to accumulate around the edge; (5) it may follow the latter route into the mesophyll, be absorbed into the symplast and thence move to the phloem and be exported from the leaf via the assimilate stream. These five fates are not mutually exclusive; combinations may occur. During the course of the export of phloem-mobile compounds some of the herbicide may be diverted into the vacuoles of parenchyma cells along the route.

Examples of the above fates are: (1) light esters of 2,4-D and 2,4,5-T. Usually only a small fraction of the total dose is lost having little significance for the herbicidal treatment but great significance with respect to injury to neighboring sensitive crops; (2) iron sulfate, sodium chlorate, the sodium salts of 2,4-D, endothall and other polar materials; these may be largely lost from the foliage of the weeds; (3) oils, both aliphatic and aromatic, materials that may be held in the cuticle and the outer protoplasmic membranes where they carry out their toxic action by solubilization; (4) monuron, simazine, and uracil herbicides; these move acropetally in the apoplast system but fail to translocate out of leaves; (5) amitrole, dalapon, picloram, the heavy esters and emulsifiable acid forms of 2,4-D and 2,4-5-T and similar compounds. These materials move freely to the phloem and are rapidly transported from leaves. The chlorophenoxy compounds tend to be accumulated and retained by active parenchyma cells and translocation may be distinctly limited (Crafts, 1961c); see Figure 4-1.

Little is known of the exact mechanism by which assimilates or herbicides are moved into the sieve tubes of the phloem. Active transport schemes have been proposed, but until the concentration of these materials in the cytoplasm of the symplast is known it is fruitless to postulate such mechanisms because it seems possible that sugar concentrations in mesophyll cytoplasm may be

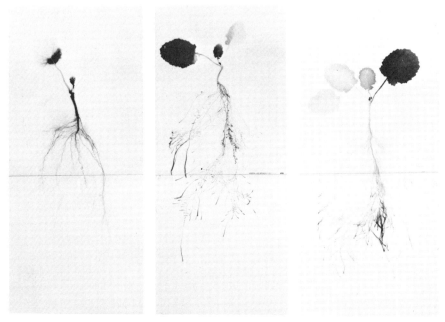

FIGURE 4-1.   Uptake and distribution of ^{14}C labeled 2,4-D (left), amitrol (center), and maleic hydrazide (right) by rape plants. Treatments were to leaves as 10 μl droplets, 4 droplets per leaf, containing a total of 0.2 μc. The specific activity of the tracers was standardized at 0.5 mc/mM. Treatment time was 2.5 days. (Crafts, 1961c.)

very high and movement into sieve tubes may be by diffusion accelerated within the cells by protoplasmic streaming (Crafts and Crisp, 1971). Assimilates move through the mesophyll and phloem parenchyma principally as hexose phosphates and amino acids. In the sieve tubes the hexoses are condensed and move principally as sucrose. Since the amount of inorganic phosphate in sieve tubes has no stoichiometric relation to the sucrose present it seems that the phosphorous may recycle within the mesophyll of the leaf (Ziegler, 1956). The way in which such diverse molecules as 2,4-D, amitrole, MH, dalapon, picloram, 2,3,6-TBA and dicamba are absorbed into the symplast, and moved into sieve tubes is unknown. Their very nature seems to preclude an active mechanism; presumably after they cross the plasmolemma barrier they are carried passively into the sieve tubes, along with assimilates.

In contrast with the above phloem-mobile compounds, the substituted ureas, uracils, symmetrical triazines, and certain diones (Jones and Foy, 1972) are apparently unable to surmount the plasmolemma barrier; they remain in

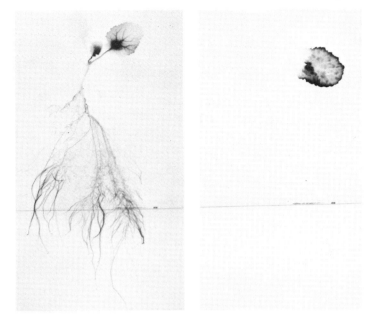

FIGURE 4-2. Uptake and distribution of ^{14}C-labeled urea (left) and monuron (right) by rape plants. Treatments as in Figure 4-1. (Crafts, 1961c.)

the apoplast and move only in an acropetal direction (Figure 4-2). Amitrole, MH, dalapon and many other compounds exhibit a combination of these effects; they may form typical apoplastic wedges and at the same time be transported via the phloem to assimilate sinks in roots, buds, and shoot tips (Figure 4-3).

Molecules that enter leaves via the lipoid route across the cuticle respond dramatically to buffering on the acid side of neutrality so that undissociated parent acid molecules are present in the spray solution for rapid movement. Table 4-2 shows the effect of pH of the applied solution on uptake and translocation of 2,4-D. Similar evidence for uptake of dinitro compounds was presented by Crafts and Reiber (1945). Buffering of the applied solution not only represses ionization of the herbicidal compounds but also of acid residues in the cuticle itself (van Overbeek, 1956).

In contrast with the above lipoid soluble compounds, it has been shown that adjustment of pH on the acid side has no effect on the penetration of MH, dalapon and picloram (Crafts, 1956; Crafts et al., 1958). Because these compounds penetrate most readily when the leaves are in a saturated atmosphere (Smith et al., 1959; Clor et al., 1962, 1963, 1964), it has been postulated by Crafts (1956) that they enter via an aqueous route. Electron micrographs of

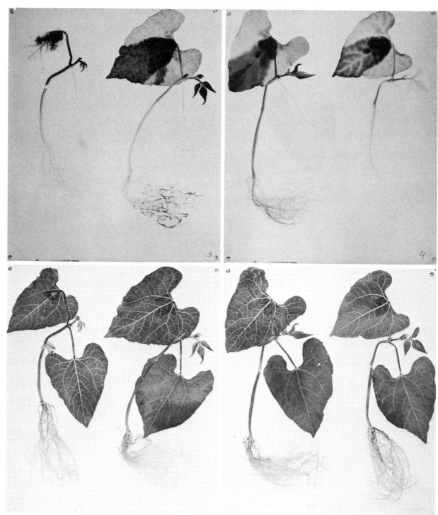

FIGURE 4-3.   Plants (below) and autoradiographs (above) of bean treated for 1 day with (left to right) 2,4-D, amitrole, MH, and dalapon. The 2,4-D treated plant shows symplastic movement only; the other three display both symplastic and apoplastic movement. (Crafts and Yamaguchi, 1964.)

cuticle indicate that it is porous in nature. The common phenomenon of cuticular transpiration also indicates that water molecules may escape from leaves with closed stomata. Work by Clor et al. (1962, 1963, 1964) gives convincing evidence for entry of herbicides into leaves via an aqueous route. When plants treated with ^{14}C-labeled molecules of 2,4-D, urea, or amitrole

TABLE 4-2.   Absorption and Translocation of 2,4-D as Shown by the Bean Bend Test (In degrees of bending; averages of 10 plants) (Crafts, 1961a)

	Hours after application of droplets							
pH	1	2	3	4	5	6	7	8
2	0	13	78	90	97	96	88	86
3	0	5	50	71	77	80	74	83
4	0	0	21	50	71	86	87	85
5	0	0	20	54	78	86	86	84
6	0	0	9	40	62	82	85	92
7	0	0	4	13	21	28	35	37
8	0	0	0	2	10	16	29	38
9	0	0	0	8	15	18	28	37
10	0	0	0	0	0	6	18	30

were enclosed in polyethylene bags so that they developed a super-saturated atmosphere from which moisture condensed on the treated leaves, much more chemical entered than in plants in the open greenhouse (Figure 4-4). Evidently all water soluble compounds may be induced to penetrate leaves by this method. Prasad et al. (1967) found dalapon to behave in this way. It is interesting that 2,4-D acid formulated in alcohol and surfactant for ready movement via the lipoid route moves even more readily along the aqueous route when the atmosphere is supersaturated. Since, under these conditions, the tracer moves into the opposite, untreated cotyledon and into all mature leaves it is evident that the treatment solution, containing surfactant, serves as a region for condensation and thus a source for movement of solution via the apoplast and xylem to all regions of the plant from which evaporation is taking place; transpiration under these conditions results from absorption of light by leaves which maintains them at a slightly elevated temperature and induces evaporization of water.

The most logical mechanism to explain this response is that the cuticle is a somewhat open sponge-like material made up of a lipoid framework interspersed with pectin strands, and possibly open pores. When the plant is in a completely saturated atmosphere the pores must be filled with water and the pectin highly hydrated. A droplet of spray solution falling on such a leaf would make instantaneous contact with the water continuum of the leaf and there is an open pathway for movement of solution into and along the apoplast. Open stomata would make free entry even more available. And materials diffusing along the apoplast would be available for movement into the symplast.

When the plant is under stress for water, the water in the pores of the

FIGURE 4-4. Autoradiographs (above) of plants (below) treated with ^{14}C-urea. The urea solution was applied to one cotyledon of each cotton plant and allowed to act for 30 hours. The two plants on the left were placed in a polyethylene bag immediately following the treatment; the two on the right remained in the green-house. The right-hand plant of each pair was steam-ringed just below the cotyledons before treatment. (Clor et al., 1962.)

cuticle recede, the pores fill with air, and a droplet falling on the leaf would entrap air bubbles in the pores. These would act as barriers to prevent contact with the water continuum and the aqueous route would be unavailable. However, spray solution absorbed through open stomata into the stomatal chambers would be in a nearly saturated environment; probability of aqueous contact would be much greater. This may be one of the principal advantages of stomatal uptake (Crafts, 1961b).

Under conditions of stress the lipoid route would still be available and it is under these conditions that the presence of non-dissociated parent acid enhances lipoid solubility and hence uptake by leaf tissues.

Two possibilities should be pointed out at this time. First, both the lipoid and the aqueous routes are available under most conditions within the diurnal cycle; the relative importance of either depends upon the potential of water in the plant (stressed or saturated), the nature of the molecules applied (lipoid vs. water soluble) and the formulation. If the spray solution, containing surfactant, dries down to a liquid of low surface tension compatible with the lipoid cuticle and hydrophilic pectin and cellulose, it would be advantageous. A low surface tension and high compatibility with cuticle pores should enable a well formulated spray solution to creep around the air blocks and establish contact with the aqueous system. This may be one of the principal advantages of spray formulations containing a high amount of surfactant (Leonard, 1958).

A second factor in the penetration of herbicides is the effect of the large amount of lipoid substance with which a lipoid-soluble toxicant may come in contact. On an acre of mesquite or live oak there are many pounds of cuticle which may serve to hold such a material in solution. Although there is evidence that the short chain esters of 2,4-D may hydrolyze during penetration (Crafts, 1960; Szabo, 1963) liberating the 2,4-D ion for movement via the aqueous route to the symplast, the matter of partition is still involved. Failure to kill sclerophyllous woody vegetation by low volume application of 2,4-D or 2,4,5-D by airplane may result from too little of the toxicant actually reaching the vascular channels in the plants.

Finally, contact injury to foliage must be considered. If severe injury occurs soon after application, further distribution may be prevented. Of the various factors that cause such injury, penetrability, concentration, and temperature are all important. One disadvantage of the short chain esters of 2,4-D and 2,4,5-T, especially when applied in a formulation containing much oil, is that rapid penetration causes contact injury, sometimes within minutes after application.

Considering concentration, there is evidence that with 2,4-D, 2,4,5-T, dalapon and picloram, concentration or dosage or both pass through an optimum. Dosages above optimum may cause rapid visible injury to foliage.

Temperature is also important, high temperature may increase the injury at a given dosage. Leonard and Yeates (1960) cite evidence that gorse (*Ulex europaeus*) leaves treated with ^{14}C-labeled amitrole remained uninjured even 15 days after treatment at 45°F; at 75°F they were injured after 7 days; at 95°F they were injured 1 day after treatment. Where injury was not serious translocation was markedly increased by a rise in temperature. However, because dosage passes through an optimum, whereas translocation increases uniformly with increasing temperature, there should be an optimum dosage for each temperature, the amount applied being the maximum that can be used without too rapid injury at the given temperature. In the trial cited above, Leonard could have applied a much higher dosage at 45°F than at 75°F; likewise the dosage at 75°F could exceed that at 95°F. Since translocation as well as contact injury is slow at the lower temperature, increased dosage should partially compensate for the reduced rate of movement so long as contact injury does not restrict uptake and movement by destroying the transport mechanism. Pallas (1960) found that uptake and translocation of labeled 2,4-D and benzoic acid were both increased by increased temperature and humidity.

Sutton et al. (1971), using the aquatic weed hydrilla (*Hydrilla verticillata*) found that certain herbicides (i.e., diquat, paraquat, ametryne, atrazine, terbutryne and 2,4-D) increased the uptake of copper ion from the culture medium. Copper was presented at 1.0 ppm and the herbicides at 1.0 and 10.0 ppm; treatment times were 2 days to 2 weeks. Herbicides that did not affect copper uptake were fenac, dichlobenil, diuron and endothall. Hydrilla treated with copper plus 0.1 ppm of paraquat contained 32.4% less phosphorus than did plants treated with paraquat alone. The phytotoxic effects of certain of these combinations may result from an increase in copper and a reduction in phosphorus content.

Absorption by Roots

With the discovery of the preemergence activity of 2,4-D (Anderson and Ahlgren, 1947; Anderson and Wolf, 1947) the use of the chlorophenoxy herbicides by soil application ushered in a new phase of chemical weed control. Annual weed control by chemicals in row crops soon displaced the mechanical cultivator. Studies since that time have shown that a great number of water-soluble compounds are readily absorbed by roots (Crafts, 1939, 1945; Sheets and Crafts, 1957; Sheets, 1958; Crafts and Yamaguchi, 1960, 1964). Comparative studies soon showed that plant roots absorb and accumulate some herbicides very rapidly (2,4-D, picloram, monuron, simazine) others more slowly (MH, dalapon, amitrole). Movement of these compounds into roots and upward into shoots occurs at varying rates. Substituted urea, uracil, and triazine compounds move into the xylem and upward in the transpiration stream very rapidly, amitrole, MH and dalapon move more slowly and 2,4-D may be retained in root cells if present at low concentration; at concentrations sufficient to cause rapid contact injury 2,4-D may move rapidly throughout the plant causing injury and death.

The mechanism of movement in some cases seems to involve absorption by root hairs and cortex parenchyma in the primary region behind the root tip. From these cells the molecules apparently migrate via the symplast into the stele where they leak from symplast to apoplast and ascend into the foliage via the transpiration stream. Possibly the leakage is conditioned by gradients of high O_2 and low CO_2 in the cortex shifting to low O_2 and high CO_2 in the stele (Crafts and Broyer, 1938; Crafts, 1961a).

In the case of the urea, uracil, and triazine, herbicides that apparently are unable to enter the symplast in leaves, molecules of these compounds apparently move inward with water along the apoplast system. At the endodermis the casparian strip is a watertight barrier separating cortex and stele, and here they must enter and pass symplastically or alternatively continue apoplastic migration by dissolving in and passing through the Casparian strip into the stele. This mechanism would seem to demand that these compounds be fat soluble enough to dissolve in the suberized layer and diffuse through it. The phenomenon of root pressure would seem to indicate that the polar water molecules cannot do this. Work of Minshall (1954) implies that the substituted urea compounds are dependent upon the transpiration stream for distribution in the foliar parts of plants; lateral accumulation may take place out of the xylem en route to the tops.

Nishimoto and Warren (1971b) have shown that diphenamid inhibits adventitious roots in the shoot zone of maize and sorghum; using this response in experiments they found that presence of adventitious roots increases the foliar injury from diuron, ametryne, or terbacil; DCPA chlorpropham, EPTC, and trifluralin injury was not affected by absence of adventitious roots.

Corbin et al. (1971) have shown that soil pH has profound effects upon uptake of herbicides by roots. Starting with highly organic soils and adjusting pH values of the cultures by addition of calcium and magnesium hydroxides they checked growth reduction as related to pH. Phytotoxicity increased as pH increased and peaked at pH 6.5 for dicamba, 2,4-D, prometone, and amitrole. Phytotoxicity increased as soil pH decreased with a maximum at pH 4.3 for dalapon, diquat, paraquat, and vernolate. Soil pH levels between 4.3 and 7.5 had no effect on phytotoxicity of chloramben, picloram, dichlobenil, isocil, diuron and nitralin.

So far absorption by leaves and roots has been considered. Work within the last decade has proved that some compounds may be absorbed from the soil by coleoptiles and young shoots as they develop and push upward through the soil following germination of seeds. Dawson (1963) studying the response of barnyardgrass (*Echinochloa crusgalli*) seedlings to soil-applied EPTC found that exposure of primary roots of the seedlings of this weed to EPTC in soil gave little or no response whereas exposure of the young shoot to concentrations above $\frac{1}{2}$ ppmw of EPTC resulted in severe injury, and in most instances death. For his 3 inch incorporation this represents about $\frac{1}{2}$ lb/A dosage. Dawson concluded that leaf tissue of barnyardgrass is the main site of EPTC uptake and also the prime site of injury.

Realizing the significance of Dawson's results, Appleby and Furtick (1965) devised a technique for controlled exposure of emerging grass seedlings to soil-active herbicides. This consisted of making small plastic envelopes to surround seeds with minute openings at the top for emergence of the coleoptile and at the bottom for emergence of the primary root. Where they wanted exposure of only coleoptiles they imposed a plastic barrier across the pot separating the soil of the root zone from that through which the first internode elongates. Thus the root can grow out of the envelope and into the soil beneath the plastic barrier, and the coleoptile can emerge through the soil above the plastic barrier. Table 4-3 gives the results of a test using oat seedlings in soils treated with diallate and propham.

TABLE 4-3. Dry Weight of Oat Seedlings Selectively Exposed to Diallate and IPC (Appleby and Furtick, 1965)

	Avg. dry wt/plant mgs	
Tissue exposed	Diallate	IPC
Coleoptile	0	6.1
Root	205.2	91.3
Coleoptile & root	0	0
Untreated control	303.8	302.8

Using the envelope method Appleby, Furtick, and Fang (1965) studied the results of soil placement with EPTC, diallate, and propham on oat seedlings. Using EPTC at 1.0 ppm, diallate at 2.0 ppm and propham at 3.0 ppm they found all three to be very toxic through coleoptile uptake, relatively innocuous when roots were exposed. Seeds allowed to imbibe for 36 hours in soil treated with EPTC at 1.0 ppm were unaffected; the seedlings grew normally.

Using radioactive EPTC Appleby et al. (1965) found that both roots and coleoptiles readily absorbed the herbicide from soil and detectable amounts moved both upward and downward. Under certain circumstances, as much unmetabolized EPTC could accumulate in roots as in shoots. This would indicate that the differential sensitivity between roots and shoots is not due to differences in uptake, translocation, or metabolic breakdown; it appears that the shoot is the major site of lethal action of EPTC in oats.

Parker (1966) developed a petri dish technique for bioassay of herbicides in soil or in sand culture. A modification of the technique allows the assessment of the relative importance of uptake by the root and by the shoot. He used a hybrid sorghum as his test plant and he used the herbicides listed in Table 4-4.

TABLE 4-4. The equieffective doses of herbicide required to cause a 50% inhibition of the extension of (a) roots of sorghum when the herbicide is in contact with roots only and (b) leaves when contact is with shoots only. The third column shows the ratio (Parker, 1966)

Compound	Equieffective concentrations ppm		
	Root	Leaf	Ratio
dichlobenil	0.055	1.25	1:23
SD 7961	0.085	1.8	1:21
trifluralin	0.065	2.7	1:42
CIPC	1.1	12.5	1:11
CP 31675[a]	8	8	1:1
EPTC	> 16	0.8	> 1:0.05
diallate	> 8	2.5	> 1:0.33
triallate	> 4	4	> 1:1
CDEC	15	0.8	1:0.053
CDAA	> 16	5.6	> 1:0.33

[a] Alphachloroaceto-2-methyl-6-*tert*-butylanilide.

As shown in this table, EPTC, diallate, CDEC, and possibly CDAA are dependent on uptake through the shoot prior to emergence. CIPC, dichlobenil, trifluralin, and SD 7961 depend not upon root absorption for their toxic effects. Linuron, dalapon, and isocil were inactive at 16 ppm within the 84 hour test period. Parker suggests that volatility may be a factor in shoot uptake of herbicides such as the thiocarbamates. If the action of the herbicide depends upon rapid uptake at an early stage, then volatility may be an advantage; obviously this would not apply to substituted urea, triazine, and uracil compounds.

The above effect is apparent in the results of Knake, Appleby, and Furtick (1967). Using the six herbicides atrazine, chloramben, EPTC, propachlor, linuron, and trifluralin they found the toxicity on tops of green foxtail (*Setaria viridis*) to be evident when the herbicide was placed in the shoot zone, but not when placed in the root zone, during the first two weeks after seeding. Shallow incorporation of trifluralin was beneficial at all moisture levels used. Shallow incorporation of amiben increased effectiveness under high moisture but decreased it under low moisture.

Speaking at the 1968 meetings of WSSA, Negi and Funderburk confirmed the fact that maize roots were little affected but shoots drastically injured by trifluralin; treatment in trifluralin vapor reproduced all of the typical symptoms.

By applying 11 herbicides to roots, seeds, and shoots of giant foxtail by exposure to 1-inch soil layers placed at different depths in cultures, Knake and Wax (1968) were able to show that all 11 compounds caused a reduction in dry weight when placed in the shoot zone; only chloramben caused a reduction when placed in the root zone. Only chloramben and naptalam were more effective in the seed zone than in the shoot zone. These tests were terminated at the end of a 2-week period from planting. Had the test run longer possibly atrazine, linuron, and siduron would have shown greater response from root treatment.

Oliver et al. (1968) continued studies on the site of root uptake and tolerance to EPTC of wheat, barley, sorghum, and giant foxtail (*Setaria faberi*). They were arranged so that (1) the upper shoot was exposed to treated soil; (2) the shoot and seed were exposed; (3) 2 to 4 mm of shoot and seed plus root were exposed; (4) the roots only were exposed and (5) the entire root and shoot were exposed. Results of this test are shown in Figure 4-5. Barley and wheat seedlings were quite tolerant to shoot exposure to EPTC; sorghum, oat, and giant foxtail were susceptible. The roots were the major site of uptake by barley; in oat root uptake gave more growth than any other treatment. These tests coupled with ^{14}C-EPTC studies indicated that differences in tolerance can be associated with the sites of uptake.

In experiments with EPTC, Prendeville et al. (1965) studied species differences in the site of shoot uptake and tolerance. Using pots with wax barriers to separate treated and untreated soil, they tested barley, wheat, oats, and sorghum. Wheat, barley, and oats were severely injured when treated at the coleoptilar internode; exposure of the shoot above this zone did not affect growth. Sorghum was severely injured regardless of the shoot zone exposed; these shoots absorbed double the amount of EPTC taken up by wheat. Evidently differences in tolerance to EPTC applied through soil depend upon stage of development at which treatment occurs. The shoot, even after emergence is still an important site of uptake and injury. Thus depth of soil incorporation of EPTC, as well as seed placement may be critical.

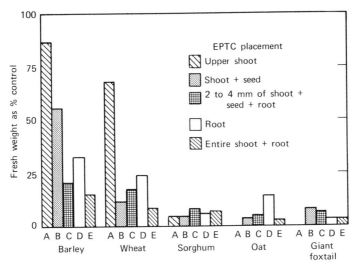

FIGURE 4-5. **Effect of EPTC placement on five grass species after 14 days. (Oliver et al., 1968.)**

In studies on the phytotoxicity of chloramben to a number of cucurbit species Ivany and Sweet (1971) found that when soil incorporated, chloramben was toxic to cucumber and watermelon, whereas muskmellon tolerated all seven of the analogues on test. Squashes had variable susceptibility; "Butternut" was least tolerant, "Table queen" was intermediate and "Boston Marrow" had the greatest tolerance. Using shoot vs. root exposure, cucumber was most susceptible to root exposure, but susceptible also to shoot exposure. Muskmelon and "Boston Marrow" squash were susceptible to root exposure but not to shoot exposure. Muskmelon was not susceptible to either hypocotyl or cotyledon exposure; cucumber was susceptible to both.

Barrentine and Warren (1971a) compared the phytotoxicity of trifluralin and nitralin on a number of plant species and found trifluralin to be more toxic than nitralin to shoots, the reverse was true of nitralin. Trifluralin prevented emergence of many species from seed; nitralin did not prevent emergence of any species. Through vapor activity trifluralin was much the more toxic of the two. In localized treatments with the above herbicides Barrentine and Warren (1971b) found that the most susceptible sites of uptake were the coleoptilar node of sorghum and hypocotyl hook of cucumber.

Studying phytotoxicity and loss of trifluralin vapors from soil, Swann and Behrens (1969) found that roots and shoots of both foxtail millet (*Setaria italica*) and proso millet (*Panicum miliaceum*) were inhibited by vapors arising from soil containing 5 ppm. Shoot exposure to this concentration of trifluralin in soil was lethal to both species; root exposure was not lethal at 20 ppm.

Vapor emission from soil increased with rising temperature and with increasing moisture. The authors concluded that vapors may be responsible for much of the herbicide activity of soil-applied trifluralin.

From this brief review it seems that the shoot of a seedling may be the major site of uptake of some herbicides, the roots may predominate in the case of others, and some may enter plants by either tissue. The coleoptilar node seems to be one important port of entry for EPTC and trifluralin in grass seedlings. At least some of the chemicals that enter and kill via the shoot are volatile, and they inhibit and kill the first leaves before they emerge from the coleoptile; in a sense they are contact toxicants; they are phytotoxic to young seedlings at an early stage.

In contrast to the above, those herbicides that kill as a response to light and inhibit the Hill reaction usually allow the plants to emerge and spread their first leaves; seedling death may take place only days or a week or so after emergence. Those tests reviewed above that lasted only one or two weeks may not have been run long enough to produce such results.

Examination of Table 2-2 shows that a large number of new soil-applied herbicides have appeared on the scene during the last decade. A few of these are general; most of them are selective, and many are specific for a relatively few crops. Their absorption characteristics and vascular mobilities will be considered in the chapters on chemical groups in which each herbicide is discussed.

Translocation

The translocation of assimilates in plants has long been a subject of controversy among plant physiologists. Research carried on within the past two decades has produced evidence that would seem to resolve the problems involved. Work on the anatomy of phloem using the electron microscope indicates clearly that sieve plates, the end walls in the sieve tubes, are open, enabling them to conduct the assimilate stream by mass flow; the plugging of the sieve plate pores by slime or cytoplasm so evident in histological preparations has been proved to be an artifact of the methods of preparation.

Work on phloem exudation from cut phloem and from mouth parts of aphids tapping the phloem indicate that the sieve tubes of the phloem constitute an open conduit system with osmotic pumps at the source to provide the driving force. This work further proves that the source-sink relation which determines the direction of flow may reverse when foliage is removed; phloem parenchyma of stems can shift from an importing system to an exporting one and maintain transport at a fairly normal rate. The sieve tube system acts like an inflated elastic osmometer and is subject to the normal laws of osmotic systems (Weatherley et al., 1959).

Work with labeled tracers in a wide variety of plants indicates that these move with assimilates in plants. Labeled sugars formed from $^{14}CO_2$ as applied directly, or from hydrolyzed urea, follow the normal source-sink pattern of distribution and exogenous phloem-mobile tracers apparently accompany sugars in their distribution; they go along together as in a stream. Many of the tracers used have been herbicides; 2,4-D, picloram, amitrole, dalapon, and 2,3,6-TBA to mention a few. It is difficult to visualize such compounds moving in a metabolically dependent conduit system; more likely the sieve tubes, though alive, play a passive role in the long-distance transport of such compounds. We recognize that 2,4-D may cause callose formation on sieve plates and plugging of the phloem within a few hours; this is not rapid enough to prevent movement into root systems of active plants (Crafts and Crisp, 1970).

Herbicides that enter and move freely in the symplast migrate to the phloem and translocate in the lumina of the sieve tubes in the assimilate stream (Crafts and Crisp, 1971). Release of molecules from the sieve-tube cytoplasm into the lumen may relate to the peculiar physiological state of the mature, functioning elements; they are enucleate and their tonoplasts disintegrate as do dictyosomes, mitochondria, and to some extent, endoplasmic reticulum. In a sense the open lumina of sieve tubes may be considered as an elaboration of the mesoplasm; that is, a highly specialized phase of the symplast adapted to rapid flow of the assimilate stream (Crafts, 1961a).

There is much evidence that transport is from regions of synthesis of foods, or hydrolysis of food reserves to regions of food utilization (Crafts and Yamaguchi, 1964). For this reason movement of assimilates from cotyledons and early leaves of seedlings is predominantly into roots; from lower leaves of mature plants it is the same (Lund-Höie and Bylterud, 1969), from later formed leaves it may be to both roots and shoot tips; from upper mature leaves it is predominantly into growing shoot tips, flowers, and fruits. Meanwhile mature leaves that are exporting assimilates are bypassed.

Comparative studies with labeled herbicides have shown that some compounds, for example 2,4-D, during migration across the mesophyll and along the sieve-tube conduits, are subject to export and accumulation within living cells as shown in Figures 4-2, and 4-6 (Crafts and Yamaguchi, 1964). Probably such accumulation takes place into the vacuoles of parenchyma cells and represents a type of storage.

Other compounds, for example amitrole, MH, picloram, dalapon, 2,3,6-TBA, and dicamba are not accumulated so avidly by living parenchyma; these move freely across the mesophyll, along the phloem and into young growing root tips. Some of these, for example MH, picloram, 2,3,6-TBA, and dalapon may migrate from phloem to xylem and thus circulate within the plant. And

FIGURE 4-6. (Upper) Autoradiographs of bean plants after 16-day treatments with ^{14}C-labeled 2,4-D, amitrole, MH, and dalapon. The amitrole treatment shows redistribution of the tracer from old to young leaves as the plant grows. (Lower) Redistribution of amitrole, MH, and dalapon if prominent; leakage is shown by the labeling of the opposite primary leaf. (Crafts and Yamaguchi, 1964.)

TABLE 4-5.   Radioactivity in Culture Solutions in Which Cotton Plants Treated with C^{14}-Labeled 2,4-D Were Growing (In counts per minute per plant) (Clor 1959)

Replication	Time intervals, days			
	0–2	2–4	4–8	8–16
1	2235	13165	2378	1254
2	1948	12944	3990	1358
3	1875	10760	7700	1120
4	2602	10892	2103	921
5	1928	8660	2728	845
6	2332	14524	3428	1304
7	2744	16635	4280	440
Average	2238	12511	3701	1035

some of them, particularly MH, picloram, and the benzoic acid compounds leak from roots into the culture medium. When 2,4-D does this it may be resorbed after a few days as shown in Tables 4-5 and 4-6.

Studies on the comparative mobility of labeled herbicides have shown that there exists a wide array of responses (Crafts, 1959; Crafts and Yamaguchi, 1964) (Table 4-7). Picloram, MH, and dalapon are freely mobile in the phloem, amitrole is less mobile, and 2,4-D moves even less freely. Urea, uracil, and triazine compounds are apparently unable to enter and move in the sieve tubes following application to foliage.

TABLE 4-6.   Radioactivity in Culture Solutions in Which Bean Plants Treated with C^{-14} Labeled 2,4-D Were Growing (In counts per minute per plant) (Clor, 1959)

Replication	Time intervals, days		
	0–2	2–6	6–14
1	220	1268	286
2	184	928	116
3	195	1451	366
4	312	1016	325
5	378	1384	223
Average	258	1209	263

TABLE 4-7. Mobility of tracers in plants. Compounds having no alternative designations were ^{14}C-labeled. Mobility varies between compounds; it may also vary between plants and between various treatments.

Free mobility			Limited mobility			Little or no mobility
In apoplast	In symplast	In both	In apoplast	In symplast	In both	
Atrazine	Amiben	Amino acids (some)	Barban	2,4-D	Naptalam	DCPA
Bioxone	Fenac	Amitrole	Dichlobenil	2,4-DP	Amino acids (some)	2,4-DB
Bromacil	Maleic hydrazide	Dalapon-^{14}C, dalapon-^{36}Cl	Diquat	MCPA	Ammonium thiocyanate	DNBP
Chlorpropham	Sodium acetate	Dicamba	Fluorodi-phen	2,4,5-T	$^{77}AsO_4^{\equiv}$	DNOC
Diuron	Sucrose	Picloram	Paraquat	TPA	Diallate	Endothall
Fluometuron		Pyriclor			Duraset	Nitrofen
Monuron		TBA			EPTC	PCP
Propham					Gibberellin	Trifluralin
Pyrazon					IAA	
Simazine					Ioxynil	
Sodium lauryl sulfate					Propanil	
T-1947					Sodium benzoate	
TCA						
Tetramine						
Tween 20						
Tween 80						

Note: Maleic hydrazide, picloram, TBA, sodium, rubidium, and cesium move from roots into the ambient culture medium. 2,4-D and many other compounds have been found to do this under special circumstances.

In contrast to these responses when the above compounds are applied to roots via the culture medium, compounds that are not phloem mobile are the first to penetrate to the xylem and move upward in the transpiration stream. Under comparable conditions the chlorophenoxy compounds are rapidly absorbed into the living root cells but they tend to be bound and move upward to the foliage relatively slowly. When applied at high concentration so that injury to the root cells results, 2,4-D moves rapidly to the tops and displays typical toxic effects. MH and amitrole accumulate somewhat more slowly than 2,4-D but they move more freely to the tops in the transpiration stream (Crafts and Yamaguchi, 1960). If 2,4-D treated roots also receive dinitrophenol, an uncoupler of oxidative phosphorylation, the 2,4-D moves more freely into the xylem and up the stem. Like 2,4-D, dalapon is accumulated into root tips of active root systems. In more mature portions of the roots however it is not strongly accumulated and it moves slowly but with comparative ease to the tops.

When the bipyridilium salts (diquat, paraquat) were first discovered (Brian et al., 1958) they were used as plant desiccants and considered not to be translocated. With further development and use as herbicides it became evident that diquat was moved in plants under certain conditions. Baldwin (1963) showed that when radioactive diquat was applied to a darkened leaf and left for 24 hours, return to the light resulted in distribution throughout the aerial portion of the plant within 5 hours. Baldwin concluded that movement took place in the xylem.

Smith and Davies (1965) made comparative translocation studies on dalapon, amitrole and paraquat in knotgrass (*Paspalum districhum*). ^{14}C-paraquat showed predominantly xylem movement with no apical accumulation and little basipetal transport. Smith and Sagar (1966) using autoradiography found that a period of darkness following diquat application was required for systemic distribution during a subsequent light exposure. The role of darkness was to allow penetration of the diquat; the light period produces the free radicals. Application in light results in such immediate contact action that penetration into the vascular system does not take place. Transport under the conditions described above is as activated diquat in free water in the xylem. Advance of the toxicant frees water en route and maintains the mobile stream. This is virtually a reversal of the transpiration stream as occurs in the use of the acid arsenical method (Kennedy and Crafts, 1927).

Slade and Bell (1966) obtained results with ^{14}C-labeled paraquat similar to those of Smith and Sagar described above. They used tomato, broadbean, and maize in their studies and concluded that paraquat transport takes place in the xylem.

Putnam and Ries (1968) found paraquat to be translocated in quackgrass (*Agropyron repens*); greater movement occurred under 6 hours of light than under 6 hours of darkness; 6 hours of darkness following application followed by a like period of light further enhanced transport from the treated leaf.

Taylor and Burrows (1968) found paraquat to move into underground rhizomes of young *Spartinia townsendii* plants when the youngest expanded leaf of a plant was immersed in a 0.6% aqueous solution. Killing of the untreated tissues was slow following a 24 hour exposure to the herbicide; some of the plants were dead 11 weeks after application; others were only partially desiccated and regrowth occurred. Under the conditions of treatment, translocation probably took place via xylem by reversal of the transpiration stream.

The herbicide 2,3,6-TBA is known to be freely mobile in both phloem and xylem of plants. Dicamba is also quite mobile. Chang and Vanden Born (1968) found that both foliage and roots of Canada thistle (*Cirsium arvense*) readily absorbed dicamba and that it was mobile in both xylem and phloem; it followed a source-to-sink pattern of distribution in the phloem following application to foliage. Following phloem transport to roots there was some exudation of the toxicant into the culture medium. After 54 days, 63.1% of dicamba applied to a leaf was still in the form of the unaltered compound; the remaining 36.9% was in unidentified compounds.

In later work Chang and Vanden Born (1971a) studied the translocation and metabolism of dicamba in Tartary buckwheat (*Fagopyrum tataricum*). Absorbed through either leaves or roots dicamba was translocated rapidly to meristems at the shoot apex and in leaf axils. Redistribution from mature to young tissues of the shoot continued for 20 days or more. Dicamba metabolism was very slow; products were 5-OH dicamba and dichlorosalicylic acid. Rapid entry and translocation, redistribution, and minimal detoxification explain the effectiveness of dicamba in controlling Tartary buckwheat.

Magalhaes et al. (1968) studied the translocation and fate of dicamba in purple nutsedge (*Cyperus rotundus*). This herbicide was slowly but appreciably translocated following application to leaves or roots. Foliarly applied dicamba transport proceeded both acropetally and basipetally, and the herbicide became widely distributed throughout the aerial parts and accumulated in meristems. Dicamba was barely detectable in the underground organs although it was excreted into the culture medium and passed through the rhizomes and tubers into daughter plants. Root-applied dicamba distribution was general except within the tubers and tips of the leaves. Dicamba was barely detectable within the tubers, although present in quantity at or near the surface; accumulation was apparent in tips of leaves. Translocation was greatest during the vegetative stage of development; after flowering translocation decreased; dicamba was not degraded by the plants during a 10-day period following treatment.

Binning et al. (1971) found that foliar applications of 2-chlorethylphosphonic acid (CEPA, Ethrel), a compound that breaks down to release ethylene, will increase basipetal translocation of dicamba. They propose that the ethylene release can alter the metabolic source-sink relationship so as to enhance subsequently applied dicamba.

Robertson and Kirkwood (1970) have published an excellent review on the mechanism of translocation of phenoxy-acid herbicides and factors influencing translocation, metabolism, and biochemical inhibition. They describe assimilate translocation and state that "it is generally accepted that movement of foliage-applied translocated herbicides takes place in the phloem along with photosynthates." In their summary of translocation they state, "Translocation within the sieve tubes is thought—to take place by mass-flow, though movement by a mechanism involving cytoplasmic streaming or electro-osmosis cannot be discounted." In this connection Crafts and Crisp (1971) point out that movement accelerated by cytoplasmic streaming, though probable in parenchyma cells of the mesophyll and cortex would be entirely inadequate physically since the volume of assimilate commonly moved in sieve tubes exceeds by a factor of three to five the volume of cytoplasm and that velocity may be 10 to 100 times the measured values of streaming. Furthermore movement by streaming would be indepenent for each molecular species and independent of the large volume of water known to be present in the lumina of sieve tubes. Such movement would take place for each solute along a gradient of that solute whereas translocation of endogenous tracers, including many herbicides, takes place from a source of assimilates to a sink for assimilates along a gradient of assimilates.

Movement by electro-osmosis would be physically impossible for many of the same reasons; conditions at the sieve plates simply are not right for such a mechanism to move assimilates at velocities up to 360 cm per hour. From the standpoint of phloem anatomy, of phloem exudation, and of tracer distribution, mass-flow appears to be the only possible mechanism that can provide rapid longitudinal transport of assimilates and attending herbicides in plants. For transport by any other mechanism, recommendations for use of foliage-applied translocated herbicides would be entirely different from what they are. As formulated these directions aim at applying the herbicide at such a time and in such a way that the herbicide, after entering the symplast, will be carried along in the assimilate stream by pressure derived by osmosis along a gradient of assimilates.

There have been many reports on translocation of individual herbicides in the literature within the past decade. These will be considered in detail in the chapters on chemicals or chemical groups that follow in later chapters.

REFERENCES

Anderson, J. C. and G. Ahlgren. 1947. Growing corn without cultivating. *Down to Earth* **3**:16.

Anderson, J. C. and D. E. Wolf. 1947. Pre-emergence control of weeds in corn with 2,4-D. *Amer. Soc. Agron. Jour.* **39**:341–342.

Appleby, A. P. and W. R. Furtick. 1965. A technique for controlled exposure of emerging grass seedlings to soil-active herbicides. *Weeds* **13**:172–173.

Appleby, A. P., W. R. Furtick, and S. C. Fang. 1965. Soil placement studies with EPTC and other carbamate herbicides on *Avena sativa*. *Weed Res.* **5**:115–122.

Baldwin. B. C. 1963. Translocation of diquat in plants. *Nature* **198**:872–873.

Barrentine, W. L. and G. F. Warren. 1971a. Differential phytotoxicity of trifluralin and nitralin. *Weed Sci.* **19**:31–37.

Barrentine, W. L. and G. F. Warren. 1971b. Shoot zone activity of trifluralin and nitralin. *Weed Sci.* **19**:37–41.

Binning, L. K., D. Penner, and W. F. Meggitt. 1971. The effect of 2-chloroethyl phosphonic acid on dicamba translocation in wild garlic. *Weed Sci.* **19**:73–75.

Bradbury, D. and W. B. Ennis, Jr. 1952. Stomatal closure in kidney bean plants treated with ammonium 2,4-dichlorophenoxyacetate. *Am. J. Bot.* **39**:324–328.

Brian, R. C., R. F. Homer, J. Stubbs, and R. L. Jones. 1958. A new herbicide. 1:1-ethylene-2:2-dipyridilium dibromide. *Nature* **181**:446–447.

Chang, F. Y. and W. H. Vanden Born. 1968. Translocation of dicamba in Canada thistle. *Weed Sci.* **16**:176–181.

Chang, F. Y. and W. H. Vanden Born. 1971a. Translocation and metabolism of dicamba in Tartary buckwheat. *Weed Sci.* **19**:107–112.

Chang, F. Y. and W. H. Vanden Born. 1971b. Dicamba uptake, translocation, metabolism, and selectivity. *Weed Sci.* **19**:113–117.

Clor, M. A., A. S. Crafts, and S. Yamaguchi. 1962. Effects of high humidity on translocation of foliar applied labeled compounds in plants. I. *Plant Physiol.* **37**:609–617.

Clor, M. A., A. S. Crafts, and S. Yamaguchi. 1963. Effects of high humidity on translocation of foliar-applied labeled compounds in plants. II. Translocation from starved leaves. *Plant Physiol.* **38**:501–507.

Clor, M. A., A. S. Crafts, and S. Yamaguchi. 1964. Translocation of C^{14}-labeled compounds in cotton and oaks. *Weeds* **12**:194–200.

Corbin, F. T., R. P. Upchurch, and F. L. Selman. 1971. Influence of *p*H on the phytotoxicity of herbicides in soil. *Weed Sci.* **19**:233–239.

Crafts, A. S. 1939. The relation of nutrients to toxicity of arsenic, borax, and chlorate in soils. *Jour. Agr. Res.* **58**:637–671.

Crafts, A. S. 1945. Toxicity of certain herbicides in soils. *Hilgardia* **16**:459–483.

Crafts, A. S. 1956. Weed control: applied botany. *Am. J. Bot.* **43**:548–556.

Crafts, A. S. 1959. Further studies on comparative mobility of labeled herbicides. *Plant Physiol.* **34**:613–620.

Crafts, A. S. 1960. Evidence for hydrolysis of esters of 2,4-D during absorption by plants. *Weeds* 8:19–25.

Crafts, A. S. 1961a. *Translocation in Plants.* Holt, Rinehart and Winston. New York. 182 pp.

Crafts, A. S. 1961b. Absorption and migration of synthetic auxins and homologous compounds. *Encyclopedia of Plant Physiology.* 14:1044–1054. Springer Verlag, Berlin.

Crafts, A. S. 1961c. *The Chemistry and Mode of Action of Herbicides.* Interscience Publishers. New York. 269 pp.

Crafts, A. S. and T. C. Broyer. 1938. The migration of solutes and water into the xylem of the roots of higher plants. *Amer. J. Bot.* 25:525–535.

Crafts, A. S. and C. E. Crisp. 1971. *Phloem Transport in Plants.* W. H. Freeman and Co., San Francisco. 481 pp.

Crafts, A. S., H. B. Currier, and H. E. Drever. 1958. Some studies on the herbicidal properties of maleic hydrazide. *Hilgardia* 27:723–757.

Crafts, A. S. and H. G. Reiber. 1945. Studies of the activation of herbicides. *Hilgardia* 16:487–500.

Crafts, A. S. and S. Yamaguchi. 1960. Absorption of herbicides by roots. *Amer. J. Bot.* 47:248–255.

Crafts, A. S., H. B. Currier, and C. R. Stocking. 1949. *Water in the Physiology of Plants.* Chronica Botanica Co., Waltham, Mass. 240 pp.

Crafts, A. S. and S. Yamaguchi. 1964. The autoradiography of plant materials. *Agr. Extension Serv. Manual* 35. Univ. of Calif., Berkeley, Calif. 143 pp.

Currier, H. B. and C. D. Dybing. 1959. Foliar penetration of herbicides— Review and present status. *Weeds* 7:195–213.

Dawson, J. H. 1963. Development of barnyardgrass seedlings and their response to EPTC. *Weeds* 11:60–67.

Ferri, M. G. and M. Rachid. 1949. Further information on the stomatal behavior as influenced by treatment with hormon-like substances. *Anais. Acad. Brazil Ciene.* 21:155–166.

Ivany, J. A. and R. D. Sweet. 1971. Response of cucurbits to certain analogues of chloramben. *Weed Sci.* 19:491–495.

Jones, D. W. and C. L. Foy. 1972. Absorption and translocation of bioxone in cotton. *Weed Sci.* 20:116–120.

Kennedy, P. B. and A. S. Crafts. 1927. The application of physiological methods to weed control. *Plant Physiol.* 2:503–506.

Knake, E. L., A. P. Appleby, and W. R. Furtick. 1967. Soil incorporation and site of uptake of preemergence herbicides. *Weeds* 15:228–232.

Knake, E. L. and L. M. Wax. 1968. The importance of the shoot of giant foxtail for uptake of preemergence herbicides. *Weed Sci.* 16:393–395.

Leonard, O. A. 1958. Studies on the absorption and translocation of 2,4-D in bean plants. *Hilgardia* 28:115–159.

Leonard, O. A. and J. S. Yeates. 1960. Absorption and translocation of radio-active herbicides in gorse and broom seedlings, and in Juncus. *Abstr. WSA*, p. 38.

Lund-Höie, K. and A. Bylterud. 1969. Translocation of amino triazole and dalapon in *Agropyron repens* (L.). Beauv. *Weed Res.* **9**:205–210.

Maciejewska-Potapezyk, W. 1955. The action of 2,4-D on some of the enzymes of the stomatal cells. *Acta Soc. Bot. Polon.* **24**:639–645.

Magalhaes, A. C., F. M. Ashton, and C. L. Foy. 1968. Translocation and fate of dicamba in purple nutsedge. *Weed Sci.* **16**:240–245.

Mason, G. W. 1960. The absorption, translocation and metabolism of 2,3,6-trichlorobenzoic acid in plants. Doctorate dissertation, Univ. of Calif., Davis, Calif. 139 pp.

Minshall, W. H. 1954. Translocation path and place of action of 3-(4-chloro-phenyl)-1:1-dimethyl urea. *Can. J. Bot.* **32**:795–798.

Negi, N. S. and H. H. Funderburk. 1968. Effect of solutions and vapors of trifluralin on growth of roots and shoots. *WSSA Abstr.*, pp. 37–38.

Nishimoto, R. K. and G. F. Warren. 1971a. Site of uptake, movement and activity of DCPA. *Weed Sci.* **19**:152–155.

Nishimoto, R. K. and G. F. Warren. 1971b. Shoot zone uptake and transloca-tion of soil-applied herbicides. *Weed Sci.* **19**:156–161.

Oliver, L. R., G. N. Prendeville, and M. M. Schreiber. 1968. Species differences in site of root uptake and tolerance to EPTC. *Weed Sci.* **16**:534–537.

Overbeek, J. van. 1956. Absorption and translocation of plant regulators. *Ann. Rev. Plant Physiol.* **7**:355–372.

Pallas, J. E., Jr. 1960. Effects of temperature and humidity on foliar absorption and translocation of 2,4-dichlorophenoxy acetic acid and benzoic acid. *Plant Physiol.* **35**:575–580.

Parker, C. 1966. The importance of shoot entry in the action of herbicides applied to the soil. *Weeds* **14**:117–121.

Prasad, R., C. L. Foy, and A. S. Crafts. 1967. Effect of relative humidity on absorption and translocation of foliarly applied dalapon. *Weeds* **15**:149–156.

Prendeville, G. N., L. R. Oliver, and M. M. Schreiber. 1968. Species differences in site of shoot uptake and tolerance to EPTC. *Weed Sci.* **16**:538–540.

Putnam, A. R. and S. K. Ries. 1968. Factors influencing the phytotoxicity and movement of paraquat in quackgrass. *Weed Sci.* **16**:80–83.

Robertson, M. M. and R. C. Kirkwood. 1970. The mode of action of foliage-applied translocated herbicides with particular reference to the phenoxyacid compounds. II. The mechanism and factors influencing translocation, metabolism and biochemical inhibition. *Weed Res.* **10**:94–120.

Sheets, T. J. 1958. The comparative toxicities of four phenylurea herbicides in several soil types. *Weeds* **6**:413–424.

Sheets, T. J. and A. S. Crafts. 1957. The phytotoxicity of four phenylurea herbicides. *Weeds* **5**:93–101.

Slade, P. and E. G. Bell. 1968. The movement of paraquat in plants. *Weed Res.* **6**:267–274.

Smith, L. W. and P. J. Davies. 1965. The translocation and distribution of three labeled herbicides in *Paspalum distichum* L. *Weed Res.* **5**:343–347.

Smith, J. M. and G. R. Sagar. 1966. A re-examination of the influence of light and darkness on the long-distance transport of diquat in *Lycopersicon esculentum.* Mill. *Weed Res.* **6**:314–321.

Smith, A. E., J. W. Zukel, G. M. Stone, and J. A. Riddell. 1959. Factors affecting the performance of maleic hydrazide. *J. Agr. and Food Chem.* **7**:341–344.

Sutton, D. L., R. D. Blackburn, and K. K. Steward. 1971. Influence of herbicides on the uptake of copper in hydrilla. *Weed Res.* **11**:99–105.

Swann, C. W. and R. Behrens. 1969. Phytotoxicity and loss of trifluralin vapors from soil. *WSSA Abstr.* No. 222.

Szabo, S. S. 1963. The hydrolysis of 2,4-D esters by bean and corn plants. *Weeds* **11**:292–294.

Taylor, M. C. and E. M. Burrows. 1968. Chemical control of fertile *Spartinia townsendii* (s.l.) on the Cheshire shore of Dee Estuary. II. Response of *Spartinia* to treatment with Paraquat. *Weed Res.* **8**:185–195.

Weatherley, P. E., A. J. Peel, and G. P. Hill. 1959. The physiology of the sieve tube. Preliminary experiments using aphid mouth parts. *J. Exptl. Bot.* **10**:1–16.

Ziegler, H. 1956. Untersuchungen über die Leitung und Sekretion der Assimilate. *Planta* **47**:447–500.

CHAPTER 5

Molecular Fate of Herbicides in Higher Plants

Degradation of Herbicides

Research on the physical and molecular fate of herbicides has received increasing emphasis during recent years. This is largely due to the increasing awareness of the public concerning the quality of the environment, and this is reflected by a greater concern with environmental pollution by all branches of the government at the federal, state, and local levels. Although certain specific pesticides have contributed to environmental pollution, the benefits of pesticides to mankind have far outweighed their detrimental effects. However, these facts point out the necessity for exhaustive research to determine the ultimate fate as well as the transitional fate of all pesticides and their degradation products. Those materials which are known to accumulate in the environment should not be marketed except in the most extreme emergency and even then only for limited periods of time and on restricted areas.

Casida and Lykken (1969) reviewed the metabolism of organic pesticides in higher plants and the book edited by Kearney and Kaufman (1969) reported on the degradation of herbicides in higher plant, soil, and animal systems. The following presentation is limited to the degradation of herbicides in higher plants.

Although the molecular modification of most herbicides results in a less phytotoxic compound there are exceptions to this. The β-oxidation of 2,4-DB to 2,4-D results in an increased phytotoxicity. Frequently the substitution of a single atom for another on a herbicide molecule will result in the formation of an almost totally non-phytotoxic compound or a compound of markedly different selectivity. The dechlorination of simazine and the concurrent

62

substitution of a hydroxyl group in place of the chlorine atom to form the hydroxysimazine results in a molecule which is in the order one thousand times less phytotoxic. Hydroxysimazine is not useful as a herbicide.

Higher plants have been shown to alter the molecular configuration of herbicides by a wide variety of chemical reactions. Most of these are probably catalyzed by specific enzymes; however some appear to be non-enzymatic. In most cases the specific enzyme(s) involved have not been isolated and characterized. The following types of reactions have been shown to be involved in herbicide degradation in higher plants: oxidation, decarboxylation, deamination, dehalogenation, dethioation, dealkylation, dealkyoxylation, dealkylthiolation, hydrolysis, hydroxylation and conjugation with normally occurring plant constituents. Representative examples of these reactions are discussed below, however a more complete presentation of the stepwise degradation of the various herbicides is presented in the chapters concerned with the specific herbicides.

Oxidation.   The phenoxy herbicides have been selected as examples of oxidation reactions since three different types of oxidation have been reported to occur. These have been classified as α-, β-, and ω-oxidation and involve oxidation at three different sites on the side chain. Fawcett et al. (1955, 1958) demonstrated that the ω-(2,4-dichlorophenoxy) alkane nitriles undergo α-oxidation, Figure 5-1.

$$R\!-\!O\!-\!(CH_2)_n\!-\!CN \longrightarrow R\!-\!O\!-\!(CH_2)_{n-1}\!-\!COOH$$

FIGURE 5-1. α-Oxidation of a model of phenoxy alkane nitrile.

The β-oxidation of the ω-phenoxyalkanoic acids has been studied by several investigators. Synerholm and Zimmerman (1947) postulated that only the 2,4-dichlorophenoxy acids with an even (I) number of carbon atoms in the side chain were active because when they were degraded by β-oxidation to 2,4-D whereas those with an odd (II) number of carbon atoms were degraded to an unstable intermediate which was converted to carbon dioxide and 2,4-dichlorophenol, Figure 5-2. The latter is not active.

$$R\!-\!O\!-\!(CH_2)_{n(even)}\!-\!CH_2COOH \xrightarrow{-(CH_2CH_2)n} R\!-\!O\!-\!CH_2COOH \quad (I)$$

$$R\!-\!O\!-\!(CH_2)_{n(odd)}\!-\!CH_2COOH \xrightarrow{-(CH_2CH_2)n}$$
$$[R\!-\!O\!-\!COOH] \longrightarrow R\!-\!OH + CO_2 \quad (II)$$

FIGURE 5-2. β-Oxidation of model phenoxyalkanoic acids.

The ω-oxidation of 10-phenoxy-*n*-decanoic acid was suggested by Fawcett et al. (1954) to explain the large amounts of phenol found as a degradation product in flax, Figure 5-3.

$$R-O-CH_2-(CH_2)_8-COOH \longrightarrow [R-O-\overset{\overset{\displaystyle O}{\|}}{C}-(CH_2)_8-COOH] \longrightarrow$$
$$R-OH + CH_3-(CH_2)_8-COOH$$

FIGURE 5-3.   ω-Oxidation of a model phenoxy-*n*-decanoic acid.

Decarboxylation.   Several herbicides including the phenoxy, benzoic, and urea derivatives have been shown to undergo decarboxylation, Figures 5-4 and 5-5. In the latter example with a urea herbicide the decarboxylation is accompanied by a simultaneous deaminization which follows demethylation reaction, Figure 5-5. The reaction also requires one molecule of water and therefore could be considered a hydrolysis.

$$R-COOH \longrightarrow RH + CO_2$$

FIGURE 5-4.   Decarboxylation of a model organic acid herbicide.

$$R-\overset{\overset{\displaystyle H}{|}}{N}-\overset{\overset{\displaystyle O}{\|}}{C}-NH_2 \xrightarrow{+H_2O} RNH_2 + CO_2 + NH_3$$

FIGURE 5-5.   Decarboxylation-deamination-hydrolysis of an aniline degradation product of a model urea herbicide.

Hydroxylation.   Hydroxylation of the ring of herbicide molecules has been demonstrated in higher plants. Some of these herbicides are derivatives of the phenoxy, benzoic, and triazine classes. *N*-hydroxy derivative formation of the side chain of propham has been suggested but the hydroxy compound was not isolated. With the phenoxy herbicides, ring hydroxylation may be accompanied by a shift in the position of a chlorine atom on the ring, Figure 5.6. However, with the benzoic acids hydroxylation occurs without a chlorine atom shift, Figure 5-7.

In the case with the triazine herbicides, ring hydroxylation involves dechlorination, demethoxylation, or demethylthioation. Figure 5-8 shows the hydroxylation of a 2-chlorotriazine, a similar reaction occurs with the 2-methoxy ($-OCH_3$) and 2-methylthio ($-SCH_3$) derivatives.

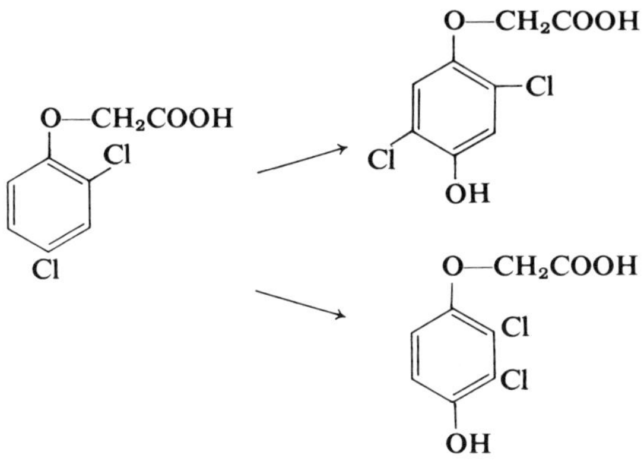

FIGURE 5-6.   Ring hydroxylation of 2,4-D in the 4 position with the chlorine atom shifting to the 3 or 5 position.

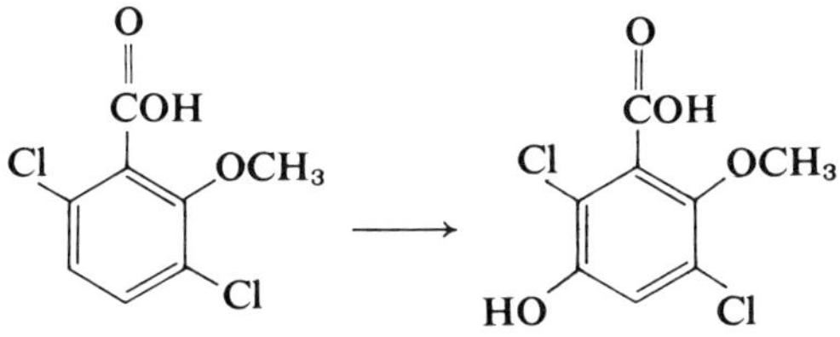

FIGURE 5-7.   Ring hydroxylation of dicamba.

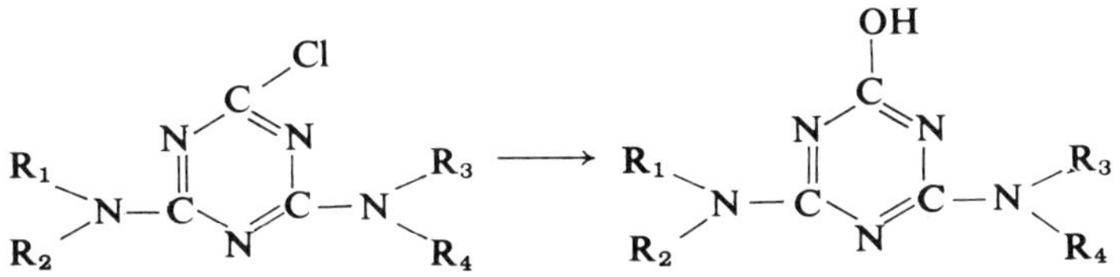

FIGURE 5-8.   Hydroxylation of a 2-chlorotriazine molecule accompanied by dechlorination.

Hydrolysis.   The degradation of herbicides in higher plants by hydrolysis is a common phenomenon. It is involved in the degradation of various formulations, i.e., phenoxy esters, as well as the basic molecules of several classes of herbicides. Some of these other classes are carbamates, thiocarbamates, triazines, and ureas. Hydrolysis of a herbicide molecule usually causes a major split in the molecule yielding two relatively large fragments which are non-phytotoxic. These two fragments are also often subject to additional degradation. The degradation of a model carbamate molecule is given as an example of hydrolysis, Figure 5-9.

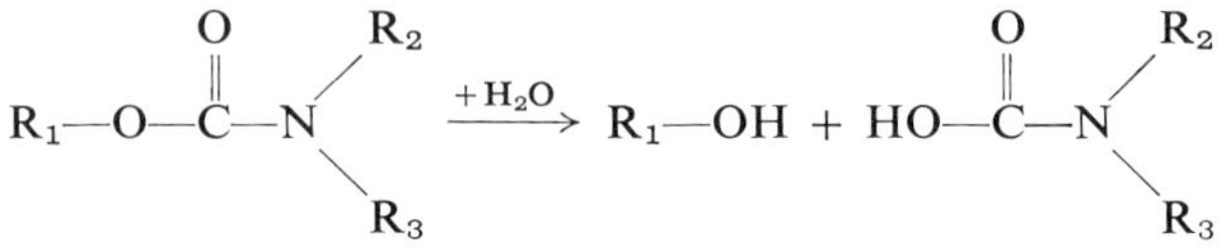

$$R_1-O-\overset{\overset{\textstyle O}{\|}}{C}-N\overset{\nearrow R_2}{\underset{\searrow R_3}{}} \quad \xrightarrow{+H_2O} \quad R_1-OH + HO-\overset{\overset{\textstyle O}{\|}}{C}-N\overset{\nearrow R_2}{\underset{\searrow R_3}{}}$$

FIGURE 5-9.   Hydrolysis of a model carbamate molecule.

Dealkylation.   The substitution of various alkyl groups on a basic structure of a given class of herbicide molecules has not only altered their absolute toxicity but it has also resulted in varied selectivity to different plant species. This latter fact is particularly evident with the triazine and urea type herbicides. Some of the classes of herbicides which have been shown to undergo dealkylation are the triazines, ureas, carbamates, thiocarbamates, and dinitroanilines. Certain herbicide molecules have alkyloxy substitutions rather than merely alkyl substitutions. They seem to undergo dealkyoxylation almost as readily as dealkylation occurs. Figure 5-10 illustrates the stepwise dealkylation and dealkyoxylation of a model urea herbicide.

$$R-HN-\overset{\overset{\textstyle O}{\|}}{C}-N\overset{\nearrow CH_3}{\underset{\searrow OCH_3}{}} \xrightarrow{\text{dealkylation}} R-NH-\overset{\overset{\textstyle O}{\|}}{C}-N\overset{\nearrow H}{\underset{\searrow OCH_3}{}} \xrightarrow{\text{dealkyoxylation}}$$

$$R-NH-\overset{\overset{\textstyle O}{\|}}{C}-N\overset{\nearrow H}{\underset{\searrow H}{}}$$

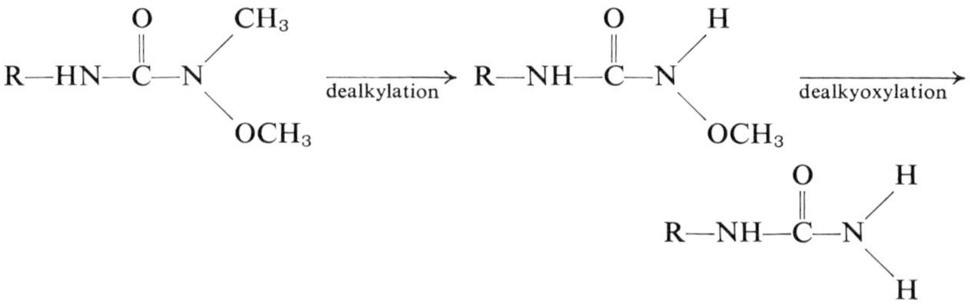

FIGURE 5-10.   Stepwise dealkylation and dealkyoxylation of a model urea herbicide.

Conjugation.   The conjugation of a herbicide or its degradation products with endogenous plant constituents have been frequently reported in the literature. The commonly observed types of conjugates involve sugars or amino acids and less frequently macromolecules such as protein or lignin. In the latter case they are initially suspected when the yield of soluble radioactivity from the applied radioactive herbicide is low and radioactivity in the insoluble residue is detected. Mild hydrolysis of this insoluble residue may release the intact herbicide molecule. Simple conjugates of the applied herbicide with a sugar or an amino acid are usually soluble in the extracting solvent. Two examples of such simple conjugates are given in Figure 5-11 (chloramben-glucose) and Figure 5-12 (amitrole-serine).

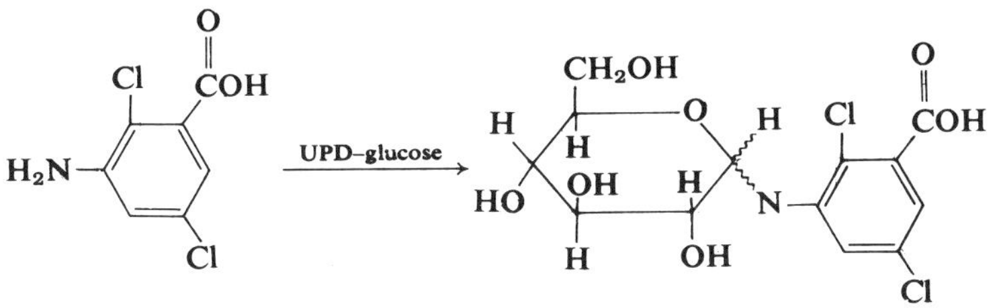

FIGURE 5-11. Conjugation of chloramben with glucose.

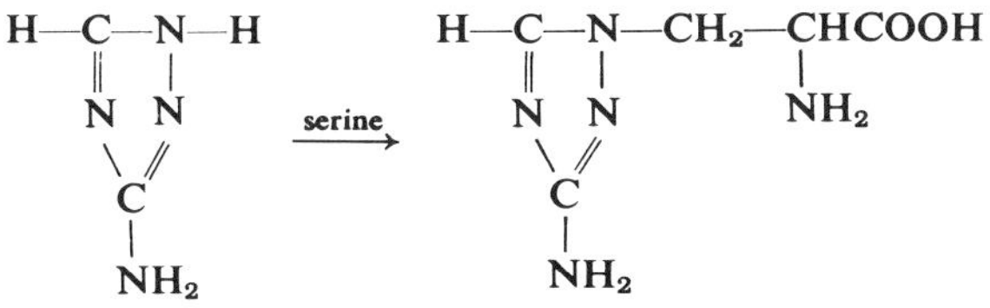

FIGURE 5-12. Conjugation of amitrole with serine.

Ring Cleavage. Apparently the splitting of the aromatic and heterocyclic ring structure contained in many herbicides proceeds very slowly in higher plants. Although the various substitutions on the ring are usually cleaved off, the ring itself may persist as a nontoxic compound throughout the life of the plant. Frequently the rings are bound to insoluble residues which complicates the research on them. Although some investigators have reported the release of radioactive $^{14}CO_2$ from ring-labeled herbicides in higher plants, the total amount and rate of $^{14}CO_2$ release has usually been quite low. Frear and Shimabukuro (1970) state, "There is no unambiguous evidence indicating that plants are capable of completely degrading these ring structures at a significant rate."

Although research on the pathway of degradation of herbicides in higher plants is continuing at an accelerated rate, the current trend appears to be placing an increasing emphasis on the isolation and characterization of enzymes responsible for the specific reactions.

REFERENCES

Casida, J. E. and L. Lykken. 1969. Metabolism of organic pesticide chemicals in higher plants. *Ann. Rev. Plant Physiol.* **20**:607–636.

Fawcett, C. H., J. M. A. Ingram, and R. L. Wain. 1954. The β-oxidation of ω-phenoxyalkylcarboxylic acids in the flax plant in relation to their growth-regulating activity. *Proc. Roy. Soc.* (London) **B142**: 60–72.

Fawcett, C. H., H. F. Taylor, R. L. Wain, and F. Wightman. 1958. The metabolism of certain acids, amides and nitriles within plant tissues. *Proc. Roy. Soc.* (London) **B148**: 543–570.

Fawcett, C. H., R. C. Seeley, H. F. Taylor, R. L. Wain, and F. Wightman. 1955. Alpha-oxidation of omega-(2:4-dichlorophenoxy) alkane nitriles and 3-indolyl-acetonitrile within plant tissues. *Nature* **176**:1026–1028.

Frear, D. S. and R. H. Shimabukuro. 1970. Metabolism and effects of herbicides in plants. *Technical Papers of the FAO Internat. Conf. Weed Control.* pp. 560–578. Davis, Calif. June 22–July 1, 1970.

Kearney, P. C. and D. D. Kaufman. 1969. *Degradation of Herbicides.* 349 pp. Marcel Dekker, Inc. New York, N.Y.

Synerholm, M. E. and P. W. Zimmerman. 1947. Preparation of a series of ω-(2,4-dichlorophenoxy) aliphatic acids and some related compounds with a consideration of their biochemical role as plant-growth regulators. *Contrib. Boyce Thompson Inst.* **14**:369–382.

CHAPTER 6

Biochemical Responses to Herbicides

When herbicides have penetrated the apoplast and come into contact with the living protoplasm, biochemical reactions of a great variety and number occur, depending upon the particular herbicide, the plant species, and the factors of formulation, method of application, temperature, humidity, etc. Most of the early salts and acids used were contact materials; they brought about rapid destruction of the delicate protoplasmic structures by virtue of their high acidity, osmotic concentration, and protein precipitating power. Little is known of the exact nature and sequence of the chemical reactions involved.

Oils, still widely used in weed control, destroy the semipermeable nature of living membranes by solubilization, the interpolation of oil molecules into the protein layer of the membrane with loss of bonding, disconfiguration, and leakage (van Overbeek and Blondeau 1954; Crafts and Robbins, 1962).

Many research papers and reviews covering the topic of the mechanism of action of herbicides describe a multitude of biochemical responses to the newer herbicides (Wort, 1964; van Overbeek, 1964; Moreland, 1967; Anon., 1969). Carbohydrate depletion, an early criterion of herbicidal action is still referred to in some situations (Weldon and Blackburn, 1969).

Moreland (1967) covers biochemical responses to herbicides under three headings: (1) Respiration and mitochondrial electron transport, (2) Photosynthesis and the Hill reaction, (3) Nucleic acid metabolism and protein synthesis. These three topics cover most of the biochemical reactions involved in plant response to modern herbicides. Much of this work has come from his own laboratory.

Respiration and Mitochondrial Electron Transport

Evidence from many sources indicates that mitochondria are the cellular organelles in which respiration takes place. Chemicals that alter the mitochondrial function may (1) uncouple the reactions responsible for synthesis of ATP, or (2) interfere with electron transport and energy transfer. Dinitrophenol (DNP) is a well-known uncoupler; dinosam, dinoseb, PCP, bromoxynil, and ioxynil are additional examples; propanil and chlorpropham have been observed to interfere with oxygen uptake and oxidative phosphorylation (Moreland et al. 1969). It has been suggested that uncouplers act on membranes of the mitochondria in which phosphorylation takes place; electric current (electrons) leaks through the membranes so that the charges that they normally separate are lost; energy accumulation in the form of ATP fails. In addition to preventing ATP synthesis, uncouplers may (1) stimulate respiration of isolated mitochondria suspended in a phosphate or phosphate acceptor deficient medium, (2) promote the hydrolysis of ATP in the medium, or (3) inhibit exchange reactions normally catalyzed by mitochondria in the absence of inorganic phosphate, ADP, ATP, and H_2O. A number of benzonitriles and benzimidazoles have recently been found to have uncoupling properties. Studies on ioxynil and its analogues (Kerr and Wain, 1964) have shown a correlation of uncoupling capacity and herbicidal activity; the reactivity of halogen substituents was I > Br > Cl. Ioxynil was more reactive than DNP.

Dichlobenil has been reported to stimulate oxygen utilization and to inhibit esterification of phosphorus with both succinate and α-ketoglutarate as substrates (Foy and Penner, 1965). Oxygen uptake and P/O ratios were both reduced by trifluralin and nitralin, considerably in soybean, somewhat less in maize and sorghum (Negi et al. 1968).

Lotlikar et al. (1968) found in cabbage mitochondria that carbamates, phenoxy acids, DMTT, naptalam 2,3,6-TBA, monuron, and CDAA inhibit phosphorylation to a greater extent than oxygen uptake. Respiration was stimulated by DNP or ADP and was inhibited by 2,4-D; the results suggest that 2,4-D may inhibit respiration in cabbage mitochondria by an effect on a reaction involved in coupling phosphorylation with electron transport. Table 6-1 presents some of their results.

While a good many other known herbicides including picloram, substituted benzoic acids and chlorophenoxy compounds have been reported to bring about uncoupling of oxidative phosphorylation (Table 6-1), it seems most probable that the growth regulating properties of these compounds are much more closely allied to their herbicidal potentials.

Inhibitors of electron transport act presumably by combining with electron carriers. Energy-transfer inhibitors may combine with intermediates of the

TABLE 6-1. The Effects of Herbicides on Oxidative Phosphorylation in Cabbage Mitochondria. (Lotlikar et al., 1968).

| Herbicide | Substrate | Herbicidal concentration | % Inhibition of | | P:O ratio[a] |
			O_2 uptake	Pi esterified	
		CARBAMATES			
IPC	Citrate	0	0	0	1.90
		5×10^{-4}M	13	53	1.03
		1×10^{-3}M	35	93	0.21
		3×10^{-3}M	72	100	0
		5×10^{-3}M	73	100	0
CIPC	Mixture	0	0	0	1.74
		1×10^{-3}M	85	91	1.03
		1×10^{-2}M	90	97	0.50
EPTC	Mixture	0	0	0	1.74
		1×10^{-3}M	77	100	0
		PHENOXYACETIC AND γ-BUTYRIC ACIDS			
2,4-D[b]	Mixture	0	0	0	1.81
		1×10^{-4}M	6	11	1.73
		5×10^{-4}M	27	55	1.12
		2.5×10^{-3}M	68	100	0
2,4-D[b]	Citrate	0	0	0	1.92
		1×10^{-3}M	54	83	0.71
		5×10^{-3}M	63	100	0
2,4,5-T	Mixture	0	0	0	1.29
		1×10^{-4}M	9	5	1.35
		5×10^{-4}M	17	53	0.73
2,4,6-T	Mixture	0	0	0	1.29
		1×10^{-4}M	10	21	1.13
		5×10^{-4}M	15	47	0.81
4-CPA[b]	Citrate	0	0	0	1.92
		1×10^{-3}M	18	36	1.49
		5×10^{-3}M	64	100	0
MCPA[b]	Mixture	0	0	0	1.81
		1×10^{-4}M	6	6	1.81
		5×10^{-4}M	12	12	1.80
		2.5×10^{-3}M	36	70	0.86
2-CPA[b]	Citrate	0	0	0	1.92
		1×10^{-3}M	17	25	1.75
		5×10^{-3}M	48	80	0.81

TABLE 6-1. The Effects of Herbicides on Oxidative Phosphorylation in Cabbage Mitochondria. (Lotlikar et al., 1968) (*continued*).

Herbicide	Substrate	Herbicidal concentration	O_2 uptake	Pi esterified	P:O ratio[a]
			\% Inhibition of		
2,4-DB[b]	Mixture	0	0	0	1.93
		5×10^{-4}M	49	88	0.44
		2.5×10^{-3}M	48	100	0
MCPB[b]	Mixture	0	0	0	1.93
		5×10^{-4}M	40	81	0.63
		2.5×10^{-3}M	73	94	0.43
MISCELLANEOUS HERBICIDES[c]					
Dalapon	Citrate	0	0	0	1.91
		1×10^{-3}M	3	0	1.94
		3×10^{-3}M	11	2	2.16
		6×10^{-3}M	9	14	1.81
		9×10^{-3}M	8	11	1.84
		1.2×10^{-2}M	14	15	1.90
2,2,3-TPA	Mixture	0	0	0	1.74
		1×10^{-3}M	1	1	1.75
		1×10^{-2}M	11	20	1.58
MH	Citrate	0	0	0	1.68
		1×10^{-3}M	1	5	1.61
		3×10^{-3}M	1	29	1.20
		6×10^{-3}M	7	16	1.50
		9×10^{-3}M	2	21	1.34
		1.2×10^{-2}M	18	29	1.45
Sodium chlorate	Citrate	0	0	0	2.14
		1×10^{-5}M	2	9	1.99
		1×10^{-4}M	1	4	2.08
		1×10^{-3}M	5	11	2.01
		1×10^{-2}M	9	12	2.09
		1×10^{-1}M	45	64	1.38
Amitrole	Citrate	0	0	0	1.93
		1×10^{-3}M	2	0	2.00
		3×10^{-3}M	5	0	2.11
		6×10^{-3}M	4	8	1.83
		9×10^{-3}M	3	7	1.86
		1.2×10^{-2}M	4	4	1.92

TABLE 6-1.   The Effects of Herbicides on Oxidative Phosphorylation in Cabbage
Mitochondria. (Lotlikar et al., 1968) (*continued*).

Herbicide	Substrate	Herbicidal concentration	% Inhibition of		P:O ratio[a]
			O_2 uptake	Pi esterified	
DMTT	Citrate	0	0	0	2.17
		1×10^{-3}M	26	43	1.69
		3×10^{-3}M	53	62	1.75
		6×10^{-3}M	52	82	0.81
		9×10^{-3}M	58	93	0.39
Naptalam	Citrate	0	0	0	2.04
		1×10^{-3}M	20	32	1.74
		3×10^{-3}M	52	80	0.84
		6×10^{-3}M	77	97	0.29
		9×10^{-3}M	87	100	0
		1.2×10^{-2}M	92	100	0
2,3,6-TBA	Citrate	0	0	0	2.57
		1×10^{-3}M	16	21	2.44
		3×10^{-3}M	44	47	2.43
		6×10^{-3}M	53	77	1.27
		9×10^{-3}M	69	91	0.72
		1.2×10^{-2}M	82	100	0
Monuron	Citrate	0	0	0	1.91
		5×10^{-4}M	20	57	1.03
		1×10^{-3}M	15	48	1.19
		3×10^{-3}M	25	63	0.96
CDAA	Citrate	0	0	0	2.00
		1×10^{-3}M	10	21	1.78
		3×10^{-3}M	23	40	1.58
		6×10^{-3}M	33	59	1.24
		9×10^{-3}M	59	78	1.07
Simazine	Citrate	0	0	0	2.30
		1×10^{-5}M	10	16	2.13
		3×10^{-5}M	8	13	2.14
		6×10^{-5}M	6	21	1.94

[a] P:O ratio, micromoles of inorganic orthophosphate esterified per microgram atom of
oxygen consumed.
[b] These herbicides were added as potassium salts in water solutions.
[c] Monuron and 2,3,6-TBA were dissolved in ethyl alcohol, and DMTT and simazine
were dissolved in acetone, for addition to the flasks. The other herbicides were dissolved
in water.

energy-coupling chain and thereby block the high-energy phosphorylation process (Moreland, 1967). Interference with mitochondrial electron transport and ATP synthesis may provide a critical mechanism for regulating growth. Oxidation and phosphorylation are subject to control by endogenous regulators. Changes in respiratory behavior often accompany various stages of growth and development; fruit ripening and dormancy are examples. While the action of a great number of growth regulators, stimulants, and depressants have been attributed to uncoupling action, it seems unlikely that any of the aforementioned herbicides bring about the type of phytotoxic changes that are responsible for lethal effects.

Recent studies on the effect of herbicides on oxygen uptake by mitochondria have utilized the polarographic-oxygen electrode method. This method in conjunction with variations in substrate, ADP and oxygen concentrations allow different steady-state conditions to be obtained with mitochondria. These steady-state conditions have been defined by Chance and Williams (1955), Table 6-2. An understanding of these steady-state conditions are necessary for the following discussion.

TABLE 6-2.   Steady-state Conditions in Mitochondria (adapted from Chance and Williams, 1955).

	State 1	State 2	State 3	State 4	State 5
ADP level	low	high	high	low	high
Substrate level	low	low	high	high	high
Oxygen level	high	high	high	high	zero
Respiration rate	slow	slow	fast	slow	zero
Limiting component	phosphate acceptor	substrate	respiratory chain	phosphate acceptor	oxygen

McDaniel and Frans (1969) reported that prometryne caused a decrease in second state 3 oxidation of both malate and succinate, but the decrease was overcome by DNP (a response similar to that of oligomycin). Oligomycin is considered to inhibit respiration by blocking the formation of a high-energy intermediate required for ATP production (Slater, 1963 and Lardy et al. 1964). Fluometuron caused an increase in second state 4 oxidation of both malate and succinate and overcame the inhibition of oxidative phosphorylation caused by oligomycin (an uncoupling response similar to DNP). Fluometuron was also shown to reverse the inhibition of oxidative phosphorylation caused by prometryne. Pyriclor completely blocked second state 3 oxidation but had no effect on second state 4 oxidation, suggesting that it inhibits the oxidation associated with ATP formation (Killion and Frans 1969). Additional studies by these workers, suggest that pyriclor either acts on a nonphosphorylating intermediate close to the electron carrier chain or with a component of the electron carrier chain or both. Moreland and Blackmon (1970) studied the effect of 3,5-dibromo-4-hydroxybenzaldehyde O-(2,4-dinitrophenyl)-oxime (C-9122), 3,5-dibromo-4-hydroxybenzaldoxime (bromoxime), bromoxynil and DNP on state 3, state 4 and oligomycin-inhibited respiration in mitochondria. All four compounds stimulated ADP-limited oxygen utilization (state 4 respiration), inhibited non-ADP-limited oxygen uptake (state 3 respiration) and relieved oliomycin-inhibited oxygen uptake. In a similar investigation, Moreland et al. (1970) found that the diphenylether herbicides (including nitrofen) also inhibit electron transport but the actions of this class of compounds are quite complex and will require additional studies before the types and sites of action can be comprehended completely. All of the aforementioned compounds interfere with ATP formation in mitochondria and it is conceivable that this could be a major, but not necessarily the only, herbicidal mechanism. Several of these compounds also interfere with ATP formation in chloroplasts via photophosphorylation.

Photosynthesis and the Hill Reaction

Soon after the introduction of the substituted urea herbicides it became evident that these compounds were effective by root treatment and that, following absorption, they were translocated into the tops of plants via the transpiration stream. It follows from this that their effectiveness should be related to the transpiration rate and Minshall (1954, 1957a) proved this to be true. Subsequent studies on the effects of light intensity (Minshall 1957a; Ashton, 1965) showed that light has profound effects upon the amount of the herbicides that reach the leaves, and hence upon their phytotoxicity; high temperatures and low humidity also act in the same way as high light intensity (Sheets, 1961).

A number of reports on the role of transpiration in regulating the effectiveness of the Hill-reaction inhibiting herbicides have appeared since these pioneering efforts. Van Oorschot has used the Hill-reaction inhibition in studies on herbicidal selectivity and physiology (1965, 1968). He has also studied the effects of transpiration of bean plants on photosynthesis inhibition by a number of root-applied compounds; the plants were treated with simazine, simetone, lenacil, isocil, diuron, and fenuron in nutrient media (van Oorschot, 1970). He found the decrease in photosynthesis to be more rapid at high transpiration rates. The total transpiration of plants up to the 50% inhibition level proved to be almost constant, even at different transpiration rates. Treatments with lenacil and diuron resulted in a lower degree of photosynthesis inhibition than with the other herbicides. Soil factors of water-holding capacity, adsorption potential, and organic matter content all tend to complicate uptake of these herbicides.

In 1959 Moreland et al. (1959) studied the action of simazine on barley plants and barley chloroplasts. Figure 6-1 from their paper shows simazine

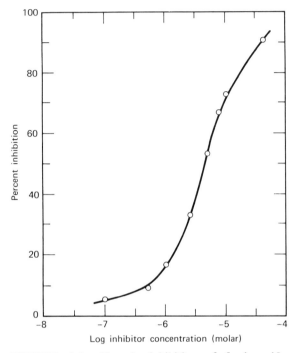

FIGURE 6-1. Simazin inhibition of ferricyanide reduction (Hill reaction) by isolated chloroplasts of barley. (Moreland et al., 1959.)

inhibition of ferricyanide reduction (Hill reaction) by isolated chloroplasts of barley. They found that simazine reduced the photochemical activity of isolated chloroplasts by 50% at 4.6×10^{-6}M. Glucose applied through open vascular bundles of severed leaf tips protected barley plants from the lethal effects of simazine.

In a subsequent paper, Moreland and Hill (1959) reported the action of alkyl-N-phenylcarbamates on the photolytic activity of isolated turnip chloroplasts. Since the carbamates can be removed from chloroplasts by washing and their photolytic activity subsequently restored, Moreland and Hill suggest that binding forces involved may be weak, possibly van der Waals or hydrogen bonds; the imino hydrogen appears to play an important role. They found ortho substitution on the benzene ring to negate inhibiting action; chlorine in the meta or para positions were effective; if IPC were assigned a rating of 1, CIPC had a relative inhibitory action of 4 and the 3,4-dichlorophenyl compound a rating of 36. Comparing toxicity tests in the greenhouse with inhibitory power, maximum activity was obtained with propyl and butyl esters, and from halogenation in the meta position on the ring.

Moreland and Hill continued their tests; in 1962 they studied the Hill reaction inhibition of phenylureas, dinitrophenol, chlorinated phenoxy acids, and chlorinated benzoic acids. From these and other studies they presented the results shown in Table 6-3.

TABLE 6-3. Molar Concentrations of 7 Herbicides Required to Inhibit the Hill Reaction of Chloroplasts isolated from Turnip Greens, Corn, and Soybeans by 50% (Moreland and Hill, 1962).

Herbicide	I_{50} concentrations for chloroplasts from		
	Turnip greens	Corn	Soybeans
Diuron	3.3×10^{-7}	3.9×10^{-7}	4.2×10^{-7}
Simazine	6.0×10^{-6}	7.8×10^{-6}	8.4×10^{-6}
DNBP	1.3×10^{-5}	1.2×10^{-5}	1.7×10^{-5}
3,4-DCIPC	3.3×10^{-5}	1.9×10^{-5}	2.5×10^{-5}
CIPC	2.9×10^{-4}	1.4×10^{-4}	2.3×10^{-4}
2,4,5-T	5.1×10^{-4}	3.3×10^{-4}	7.0×10^{-4}
3,4,5-TBA	2.0×10^{-3}	1.7×10^{-3}	3.6×10^{-3}

When turnip-green chloroplasts were treated with inhibitory concentrations of diuron, simazine, DNBP, 3,4-DCIPC, CIPC, 2,4,5-T, and 3,4,5-TBA in the dark, the herbicides could be removed by washing without seriously impairing their photolytic activity. However, irreversible changes occurred when the chloroplasts were illuminated while in contact with the herbicides, that is, photolytic activity was not completely restored except in the case of simazine.

Moreland and his associates continued their work studying acylanilides (Moreland and Hill, 1963a) where inhibition of the Hill reaction reached 50% at 7.8×10^{-6} in the case of N-(3-chlorophenyl)-2-methyl pentanamide, Kinetic studies with the 3,4-dichlorophenyl compound proved that both light and dark reactions were inhibited; the light reaction was the more sensitive. Moreland and Hill proposed that the inhibition measured involved hydrogen bonding between the imino hydrogen and the carbonyl oxygen of the acylanilides with appropriate receptors, possibly polypeptide chains of proteins or carbons 9 and 10 of the cyclopentane ring of chlorophyll.

Another study (Moreland and Hill, 1963b) involved a number of polycyclic ureas; norea was the strongest inhibiting the Hill reaction 50% at 4.0 × 10^{-6}M. In this paper Moreland and Hill suggested that chemicals possessing the amide moiety

$$\begin{matrix} H & O \\ | & || \\ -N & -C- \end{matrix}$$

(—N—C—) such as phenyl ureas, polycyclic ureas, N-phenyl carbamates, and acylanilides may undergo enolketo tantomerization of the following type.

$$\begin{matrix} H & O & & OH \\ | & || & & | \\ R-N & -C-R' \rightleftharpoons & R-N= & C-R' \end{matrix}$$

Either form might be stabilized and favored by hydrogen bonding, resonance, steric hindrance, and solvent effects. Since diuron has a Hill reaction inhibition of 3.3 × 10^{-7}, Moreland and Hill attribute the somewhat lower inhibitory power of the polycyclic ureas to increased lipophilicity, unfavorable spacial configurations and lack of resonance interaction between the ring system and the imino nitrogen.

In 1965 Camper and Moreland (1965) reported on work involving correlations between acidity of substituted phenylamides and inhibition of the Hill reaction. They found that the relative acidities of these substituted amides as determined by potentiometric titration using 0.1N tetrabutylammonium hydroxide methoxide in n-butylamine were correlated with inhibitory potency against the Hill reaction. Of the phenylamide families studied, for example the phenylureas, N-phenylcarbamates, and acylanilides, the phenylureas are the most acidic.

Within each family, increased acidity was correlated with increased inhibition for the unchlorinated, the 3- or 4-monochloro-, and 3,4-dichloro derivatives, respectively. A peak was found in the activity-acidity curve for various meta-substituted derivatives of 2-propyl N-phenyl carbamate which may be related to an optimum charge on the imino nitrogen required for high inhibitory activity. Variations in the length of the side chain did not markedly affect the acidity level. Substitution of a hydroxyl group for the imino hydrogen, replacement of the carbonyl oxygen with a sulfur atom and chlorination or saturation in the alkyl group increased acidity, but in general decreased inhibitory activity. A correlation between charge on the imino nitrogen and inhibition of the Hill reaction is apparent if comparisons are restricted to derivatives in which electronic rather than steric influences predominate. Steric influences are considered to control the fit, as well as the ease with which an inhibitor approaches the critical site. Electronic influences may control chemical reactivity and binding of the inhibitor to the active site.

In a paper on the structure-activity relations of inhibitors of chloroplast

TABLE 6-4. Inhibition of the Hill Reaction by Aryl and Alicyclic Dimethylureas (Moreland, 1969)

$$R-\underset{\overset{|}{H}}{N}-\underset{\overset{\|}{O}}{C}-N(CH_3)_2$$

Designation	R	$pI_{50}{}^a$
Chloroxuron		6.80
Diuron		6.75
Norea		5.80
Cycluron		5.25
Fenuron		5.00

[a] pI_{50} = the log of the reciprocal of the molar concentration causing 50% inhibition.

electron transport, Moreland (1969) reported Hill reaction inhibition by aryl and alicyclic dimethylureas, Table 6-4. Since all of these compounds are herbicides it is interesting to note the correlation shown between herbicidal activity and inhibitory potency. Table 6-5 gives similar data on a series of 3,4-dichlorophenylamides, all of which are herbicides.

Moreland also reported results comparing oxon and thion derivatives of some six herbicides; he reported on various ring substitutions in uracils and triazines; he presented data on ring substitution in seven positions on the benzimidazole ring system, in 5 positions in the imidazole ring and 3 positions on the ring of halogenated benzonitriles.

In his conclusions on this paper Moreland stated the obvious fact that not all herbicides are equally efficient in inhibiting the Hill reaction. His results presented in Table 6-6 confirm this. When one considers that DNBP is a

TABLE 6-5.   Inhibition of the Hill Reaction by 3,4-dichlorophenylamides (Moreland, 1969)

Designation	R	$pI_{50}{}^{a}$
Neburon	$-N-C_4H$ $\quad CH_3$	6.85
Diuron	$-N-CH_3$ $\quad CH_3$	6.75
Linuron	$-N-OCH_3$ $\quad CH_3$	6.70
—	(cyclopropyl with CH_3)	6.66
Dicryl	$-C=CH_2$ $\quad CH_3$	6.50
Karsil	$-CH(CH_2)_2CH_3$ $\quad CH_3$	6.48
Cypromid	(cyclopropyl)	6.24
Propanil	$-C_2H_5$	6.11
—	$-CH(CH_3)_2$	5.80
—	(furan ring)	5.68
—	$-N$ (morpholine) O	5.55
Swep	$-OCH_3$	5.42
3,4-DCIPC	$-OCH(CH_3)_2$	4.92

[a] pI_{50} = log of the reciprocal of the molar concentration causing 50% inhibition.

TABLE 6-6. Comparison of pI_{50} Values for the Most Active Members of Several Chemical Families (Moreland, 1969)

Designation	$pI_{50}{}^a$
Diuron	6.75
Karsil	6.48
Bromacil	6.31
Ioxynil	6.13
Simazine	5.66
DNBP	5.10
3,4-DCIPC	4.92
CIPC	3.92
2,4,5-T	3.61

^{a} pI_{50} = log of the reciprocal of the molar concentration causing 50% inhibition.

general contact foliar acting material, recognized as an uncoupler of oxidative phosphorylation, and that 2,4,5-T is a translocated foliar-acting growth regulator that injures plants as a result of growth aberrations which block phloem transport, it seems evident that one should take into consideration the nature of the herbicidal action involved in assaying herbicidal potency. If one selects materials that express their herbicidal action when absorbed by roots and translocated to tops the correlation of Hill reaction inhibition with herbicidal potential is more exact. Even here, however one notes that CIPC apparently acts as a mitotic poison on root tip meristems and so, while inhibiting O_2 release in tops, this is purely coincidental with its herbicidal activity.

Moreland stresses that the inhibition coefficients found in the literature do not provide an estimate of the concentration of the inhibitor actually present inside the chloroplast. Inhibition of the Hill reaction is achieved at concentrations much below the quantity of chlorophyll present. By dividing the chlorophyll concentration by the concentration of herbicide required to bring about 50% inhibition, chlorophyll to herbicide ratios of 100 to 1 have been estimated.

Good and Izawa (1964) and Izawa and Good (1965) using partition analysis and inhibition kinetics studied absorption and distribution of monuron, diuron, and atrazine in spinach chloroplasts; correcting for ineffective sites and partitioning they concluded that phenylureas and s-triazines act at the same site or on the same functional unit and that the number of inhibitor-sensitive sites is the same for both classes of inhibitors; one site serves for approximately 2,500 chlorophyll molecules. It may be pertinent that this size has approximately the dimensions of the unit of oxygen production found in research on photosynthesis (Good and Izawa, 1964).

Monaco and Moreland (1966) reported that for simazine, prometryne, and propanil in spinach chloroplasts there is one inhibitor-sensitive site for several hundred chlorophyll molecules. These results give a clue to the reason that these compounds of relatively low water solubility, acting through the tremendous diluting power of the soil solution, still kill plants effectively. Crafts and Drever (1960) found that monuron reduced growth of Kanota oats to a value below 50% at a concentration of 0.4 ppmw in the soil in one experiment, 0.8 ppmw in a second. Tests of this type, many of which occur in the literature, indicate that these very effective Hill reaction inhibitors are performing their herbicidal function in the chlorenchyma tissues at extremely low concentrations.

Moreland (1969) concludes that Hill reaction inhibition by herbicides based on structure-function relations can best be explained by multipoint attachment of the herbicide to appropriate receptors, at or near the active centers in the chloroplasts, possibly through hydrogen bonds. The requirement for a free and sterically unhindered amide or imino hydrogen has been demonstrated. If the amide or imino hydrogen is replaced with another group such as carbethoxy only a limited loss in inhibition occurs. Changes in molecular structure designed to decrease hydrogen bonding behavior result in compounds of decreased inhibitory activity. Free reversibility of inhibition indicates weak bonds such as hydrogen bonds. Substituents of herbicide molecules that participate in hydrogen bond formation include amide hydrogens, imino hydrogens, carbonyl oxygens, ester oxygens, and azomethine nitrogens in the s-triazines, benzimidazoles, and imidazoles. Hydrogen bonds involved in these same groupings are responsible for maintaining the functional structure of proteins and nucleic acids in biological systems. However, as Good and Izawa (1964) point out, there seems to be no consistent correlation between the bonding potentials of the imino hydrogens and inhibitory potency; the imino-hydrogens of the triazines and chloroacetyl-anilides form bonds with carbonyl oxygens very little if at all, yet these are some of the best inhibitors. Thioureas though unlike their oxygen counterparts, are poor inhibitors yet they form hydrogen bonds readily.

A compound having an intermediate though significant inhibitory activity on CO_2 assimilation, O_2 evolution and non-cyclic phosphorylation is pyriclor Meikle (1970) studied the effects of this compound on photochemical reactions that affect assimilatory power-generation of ATP and $NADHP_2$ at the expense of radiant energy by spinach chloroplasts. CO_2 assimilation and cyclic phosphorylation catalyzed by FMN were both inhibited at rather high pyriclor levels. O_2 evolution and non-cyclic phosphorylation were strongly inhibited. Meikle concluded that the O_2-evolving system and/or pigment system 2 of the photosynthetic apparatus are possible loci of attack.

While the Hill reaction has provided a convenient method by which potential phytotoxicity of herbicidal groups may be found *in vitro*, there are many commercial herbicides which do not inhibit the Hill reaction. And conversely there are many good inhibitors that are not phytotoxic. As Moreland (1969) observes, from the variety of compounds that inhibit the Hill reaction it is unlikely that all interfere at a common site through a common mechanism. Thus by identifying such compounds, sites of inhibition may be located and defined in terms of chemical architecture and chemical and physical properties; participating intermediates and components of the oxygen evolution pathway may eventually be identified. As far as discovery of new herbicides is concerned, it seems obvious that preliminary tests that prove that new compounds are uncouplers, growth regulators, mitotic poisons, or direct contact toxicants should indicate the futility of Hill reaction studies in search of the mode of action. When a chemical in the soil kills seedlings before they emerge into the light, some more direct mechanism must exist. When a compound causes severe growth aberrations, excess callus growth and crushing of the vascular tissues, obviously Hill reaction inhibition if present is secondary in its function. And when a compound interferes with normal root development as occurs with trifluralin and nitralin, Hill reaction studies are not indicated. While such studies have contributed much to our understanding of the phytotoxicity of the ureas, uracils, and triazines, and have speeded our search for new herbicides, it is well to understand the definite limitations of the method. Just as there are no universal herbicides there are no universal methods for testing herbicides.

Recently Moreland and Blackmon (1970) and Moreland et al. (1970) investigated the effect of several herbicides on various electron transport and phosphorylation reactions mediated by isolated chloroplasts. C-9122 [3,5-dibromo-4-hydroxybenzaldehyde O-(2,4-dinitrophenyl) oxime] inhibited both photoreduction and coupled phosphorylation, with water as the electron donor and ferricyanide and NADP as oxidants, and also cyclic photophosphorylation (Moreland and Blackmon, 1970). With ascorbate-2,6-dichlorophenolindophenol (ascorbate-DPIP) as the electron donor, phosphorylation coupled to NADP was inhibited, but the reduction of NADP was not inhibited. Bromoxime (3,5-dibromo-4-hydroxybenzaldoxime) and DNP behaved similarly to C-9122, but required higher concentrations. Bromoxynil differed from C-9122 in that it was a relatively poor inhibitor of cyclic photophosphorylation, and of the photophosphorylation coupled with electron transport when ascorbate-DPIP was the electron donor and NADP was the

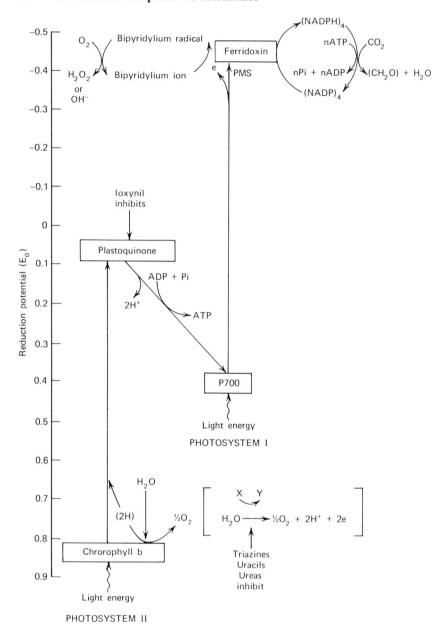

FIGURE 6-2. Simplified photosynthetic model and sites of action of several herbicides.

oxidant. The bromoxynil effects were very similar to those reported for ioxynil (Gromet-Elhanan, 1968). The effects of C-9122 on chloroplast reactions can be explained by action at two different sites: one near photosystem II and the oxygen evolution pathway (similar to diuron) and the second associated with energy transfer and formation of ATP, see Figure 6-2. Moreland et al. (1970) suggested that the diphenylether herbicides function as inhibitors of chloroplast electron transport. The main site of action appeared to be associated with light reaction II and the oxygen evolution pathway. The following decreasing order of inhibitory effectiveness was reported, 2,4,6-trichlorophenyl-4'-nitrophenyl ether (ML-1478) $\geq$ 2,4'-dinitro-4-trifluoromethyl-diphenyl ether (C-6989) > nitrofen. At high concentrations, they also marginally inhibited ascorbate-DPIP mediated transport. Therefore, inhibition at a site beyond light reaction II also was suggested. Interference with electron transport at this second site was also reported for C-9122, bromoxynil, bromoxime, ioxynil, and DNP.

Most of the herbicides discussed in the previous paragraph also interfere with ATP formation in isolated mitochondria, therefore the herbicidal mechanism of these compounds is, at least in part, associated with the inhibition of ATP formation in both mitochondria and chloroplasts. The reader seriously interested in the molecular basis of herbicide toxicology is encouraged to study these two papers of Moreland and associates in detail, since we are only able to present them in abbreviated form here. For more information on the mode of action of nitrofen, see Pereira (1971).

Nucleic Acid Metabolism and Protein Synthesis

Early work on the physiology of the herbicidal action of 2,4-D indicated that this powerful growth regulator was translocated in plants along with foods in the assimilate stream. And contemporary study of the anatomical responses of plants showed that 2,4-D soon inhibited translocation by bringing about callosing of sieve plates (Eames, 1949, 1950). A little later Muni (1959) found that despite the great differences in susceptibility of different plant species to 2,4-D, if a lethal dose was applied to either susceptible or resistant species, crushing of phloem and eventually xylem led to complete blocking of the vascular channels. Thus the physical mode of action of 2,4-D proved to be a loss of the ability of the vascular tissues to transport foods, minerals, and other essential materials in plants.

Remaining was the question as to what biochemically was the trigger that brought about the cell expansion, cell division and overgrowth that crushed the vascular tissues. Search for critical enzyme systems proved ineffective (Wort, 1964). Since the effective physical mechanism involves growth, it follows from this situation because growth is undoubtedly regulated by many enzymes no one of which would prove critical.

Reviews on the subject of the physiology of 2,4-D action (Penner and Ashton, 1966; Wort, 1964) cover a multitude of factors including the effects on many plant constituents. By the mid-fifties papers began to appear connecting auxin action with nucleic acids and in 1960 West et al. reported that 2,4-D in herbicidal concentrations increased the RNA and protein content of cucumber stem tissue; the microsomal fraction exhibited marked increases in RNA. However, low concentrations of 2,4-D tended to promote the loss of RNA and protein which accompanied endogenous growth. Basler and Nakazawa (1961) treated cotton seedling cotyledons with various amounts of 2,4-D and within 2 days found increases in protein and RNA.

Key (1959) found, in soybean seedlings, increased proteins, nucleic acids and acid soluble nucleotides from treatment with 5×10^{-4}M 2,4-D. Key and Hanson (1961) found, from 2,4-D treatment of soybean seedlings, increases in cytidine mono- and triphosphates and uridine mono-, di-, and triphosphates. Since directly or indirectly RNA and protein synthesis in regions of rapid cell proliferation is necessarily preceded by increased nuclear activity and thus by DNA. Chrispeels and Hanson (1962) proposed that the primary site of action of 2,4-D might well be the nucleus. The cytochemical basis of 2,4-D action would then lie in renewed nuclear activity and reversion of the tissue to meristem metabolism.

Since these early studies, there has been a great amount of research on the effects of herbicides on nucleic acid metabolism and protein synthesis. Key and Shannon (1964) found that the increase in RNA from 2,4-D treatment occurred primarily in the ribosomal fraction and that a net transfer of RNA from the nuclear to the ribosomal fraction takes place. A DNA-like RNA in excised soybean tissue seems to be essential to growth; this RNA is similar to messenger RNA of bacterial systems. Messenger RNA, in addition to ribosomal RNA, increases with 2,4-D treatment and there appears to be a general correlation between growth potential and the proportion of DNA-like RNA in both excised and intact soybean tissue.

Shannon et al. (1964) postulated that excessive nucleic acid and protein synthesis induced by 2,4-D would preclude normal cell function, and hence might prove to be the biochemical basis for the herbicidal action of 2,4-D.

Shröder et al. (1970) studying biosynthesis of DNA and RNA in *Neurospora crassa* as affected by culture solutions containing 2,4-D, amitrole, atrazine, chlorpropham, and chlorflurenol found growth of this fungus to be completely inhibited by chlorpropham at 10^{-3}M. The greatest effect on nucleic acid biosynthesis was given by 2,4-D at 10^{-3}M; synthesis of RNA was increased to 140% of the control. No other measurable effects on RNA or DNA synthesis were found.

Okubo and Switzer (1965) reported increased RNase activity and microsomal RNA following treatment of pea seedlings with herbicidal concentrations of mecoprop, 2,4-D and MCPA. Malhotra and Hanson (1966) reported that herbicidal activity of picloram was like that of 2,4-D; picloram enhanced DNA and RNA synthesis in soybean and cucumber but not in barley and wheat: the latter species are tolerant of picloram.

Chen et al. (1972) using roots of wheat and cucumber seedlings studied nucleic acid and protein changes induced by 2,4-D, 2,4,5-T, dicamba, and picloram. All four herbicides increased the DNA and protein in roots of the 4-day old seedlings; the increase was greatest in the cucumber. The greatest difference between the tolerant wheat and susceptible cucumber was on RNA levels; as concentrations of all four herbicides increased a progressive decrease in RNA occurred in wheat; in cucumber at 10 and 100 ppm, RNA levels increased over 200%. When protein levels in both species were compared on a per unit RNA basis the protein/RNA ratio in wheat was higher than controls, in cucumber lower. This indicates that the RNA produced in wheat is the normal type capable of translating for protein synthesis. In cucumber the auxin-like herbicides make more of the DNA template available for transcription resulting in more RNA; thus even though protein levels in cucumber are increased the protein/RNA ratios decrease with concentration. If some of the increased RNA consists of aberrant species, the resultant proteins may be abnormal. Chen et al. suggest that a differential alteration of RNA species and interference with protein synthesis may be the basis for the selectivity between wheat and cucumber.

Zukel (1963) summarized work indicating nuclear labeling and binding to RNA by MH. Evans and Scott (1964) found that MH treatment increased the time required for broad bean (*Vicia faba*) root tip cells to complete a mitotic cycle and that chromosome aberrations occurred. Temperli et al. (1966) found that with an arginine, thymine and uracil requiring strain of *E. coli*, prometryne and cyanuric acid or a derivative may partially replace uracil in RNA and thymine in DNA.

Amitrole has been found to inhibit the incorporation of labeled precursors into DNA, RNA and acid soluble nucleotides of wheat seedlings (Bartels and Wolf, 1965); it caused a slight decrease in the microsomal RNA of light-grown plants. Schweizer and Rogers (1964) reported that amitrole decreased the content of acid-soluble nucleotides in maize roots and etiolated coleoptiles. Williams et al. (1965) found that *E. coli* formed an alanine conjugate with amitrole as it does in higher plants; he concluded that the conjugate was subsequently built into cellular protein.

Mann et al. (1965) used 23 herbicides in tests on the incorporation of leucine into protein (Table 6-7). Only CDAA, CIPC, endothall, ioxynil and PCP had a pronounced effect. Barban caused about 75% inhibition of ^{32}P incorporation into both RNA and DNA in excised *Sesbania* roots. Shokraii and Moreland

TABLE 6-7.   Inhibition by Various Herbicides of Leucine-1-C^{14} Incorporation into Protein (Mann et al., 1965)

	Barley			Sesbania		
Herbicide	2 ppm	5 ppm	Importance of inhibition	2 ppm	5 ppm	Importance of inhibition
None	0^a	0^a		0^b	0^b	
Ethanol	−1	8		−6	23	
Chloramben	−20	—	−	4	22	−
Amitrole	−13	−4	−	19	33	−
Atrazine	7	11	−	37	32	−
CDAA	51	70^c	+	58	87^c	+
CDEC	25	29	−	7	0	−
CIPC	26	84	+	32	98	+
Dacthal	18	34	−	15	21	−
Dalapon	7	8	−	4	8	−
DATC	34	18^c	−	22	8^c	−
2,4-D	15	—	−	4	—	−
Dichlobenil	26	0	−	24	14	−
Diphenamid	21	27	−	17	22	−
Endothall	21	24	−	63	72	+
EPTC	38	22	−	14	11	−
Hadacidin	12	15	−	18	1	−
Ioxynil	44	70	+	82	88	+
Maleic hydrazide	6	−1	−	0	−3	−
Monuron	32	19	−	24	24	−
Naptalam	−20	—	−	20	12	−
PCP	13	62	+	42	65	+
Propanil	—	8	−	−7	14	−
Pyrazon	25	29	−	35	26	−
Trifluralin	8	3	−	13	29	−

[a] Control values of protein synthesis by barley ranged, in 6 experiments, from 4405 to 5785 cpm, with a mean of 4850 cpm; total uptake of leucine was approximately 4-fold greater than incorporation into protein.
[b] Control values of protein synthesis by Sesbania ranged, in 5 experiments, from 17,785 to 24,150 cpm, with a mean of 19,980 cpm, compared with total leucine uptake of 80,000 cpm.
[c] For reasons of solubility, 4 ppm rather than 5 ppm used.

(1966) found that propanil, CIPC, and DNBP inhibited leucine incorporation into protein and ATP incorporation into nuclear, mitochondrial, ribosomal, and soluble RNA of excised soybean hypocotyls. Observed uncoupling by ioxynil, DNBP, and PCP may relate to lack of ATP rather than to a direct influence on nucleic acid metabolism and protein synthesis.

Ashton et al. (1968) studied the proteolytic activity of some 26 herbicides in an attempt to correlate this activity with herbicidal effects through inhibition of seed germination. Ioxynil, chlorpropham, CDEC, dichlobenil, bromoxynil, bensulide, picloram, endothall, and 2,4-D all showed strong proteolytic activity inhibition which correlated well with growth inhibition of squash seedlings.

Moreland et al. (1969) have studied the effects of some 22 herbicides on the synthesis of RNA and protein in excised maize mesocotyl and soybean hypocotyl sections (Table 6-8). The assays measured ATP and orotate incorporation into RNA, leucine incorporation into protein and gibberellin induced α-amylase. Average results of the four assays suggested that 14 of the herbicides inhibited RNA and protein synthesis *in vivo*. The most inhibitory compounds were ioxynil, dinoseb, propanil, pyriclor, and chlorpropham. Isocil, CDAA and picloram inhibited only the α-amylase assay; atrazine inhibited the ATP incorporation assay and MH stimulated orotic acid incorporation as did amitrole. Trifluralin and chloramben had no measurable effects in any of the assays; dichlobenil was also inactive.

Slightly less inhibitory than the above seven active compounds was 2,4,5-T in ATP, leucine, and orotate assays; it was a strong inhibitor of α-amylase formation. Diuron, the strong inhibitor of the Hill reaction was only moderately active in these tests. Fenac inhibited leucine incorporation and α-amylase formation; RNA synthesis was affected only slightly. Propanil was a stronger inhibitor of the assays than karsil which effectively inhibited only leucine incorporation. Propachlor inhibited orotate and leucine incorporation and α-amylase induction but failed to affect ATP incorporation. EPTC and CDEC inhibited in the α-amylase assay; the latter was the stronger inhibitor.

Jones and Foy (1971) have studied the inhibition of the activity of α-amylase in barley half seeds induced by exogenous gibberellic acid. This activity was almost completely inhibited by 5×10^{-4}M fenac, bromoxynil, and endothall. At the same concentration paraquat inhibited 72%, dalapon 42%; trifluralin, SD 15418, and 2% dimethylsulfoxide were only slightly inhibitory. Such amolylitic activity is undoubtedly involved in germination of many seeds, and the subsequent growth of the seedlings.

Because the effects of some compounds on RNA and protein synthesis are complex, interpretation of results may be difficult. Factors such as synthesis of precursors, interruption of energy balance, species selectivity, and specificity of enzymes make generalizations hazardous.

While the above cases of interference by herbicides in nucleic acid metabolism and protein synthesis imply disorders of a phytotoxic nature the actions in this direction are only implied; in no case has direct lethality been proved. Both chloroplasts and mitochondria contain DNA differing in base composition from nuclear DNA. DNA synthesis, a DNA-dependent synthesis of

TABLE 6-8.   Effect (Percent of Control Activity) of 22 Herbicides on RNA Synthesis, Protein Synthesis, and Induction of α-Amylase Activity (Moreland et al., 1969)

Herbicide	ATP incorpo- ration	Leucine incorpo- ration	Orotate incorpo- ration	α-amylase induction	Average response[a]	Relative impor- tance[b]
Dinoseb	20*[c]	2*	9*	26*	14	Strongly
Ioxynil	22*	3*	31*	4*	15	inhibitory
Pyriclor	41*	12*	23*	33*	27	
Propanil	36*	10*	32*	35*	28	
Chlorpropham	28*	11*	19*	77	34	
2,4,5-T	56*	33*	59*	21*	42	Inhibitory
Diuron	38*	58*	63*	37*	49	
Fenac	87	30*	75	6*	50	
Propachlor	86	56*	59*	45*	62	
CDEC	78	67*	87	32*	66	
Dicamba	86	76	101	23*	72	
Karsil	73	54*	81	83	73	
EPTC	58*	76	98	61*	73	
Dichlobenil	75	67*	79	82	78	
Isocil	78	90	122	45*	84	No effect
Trifluralin	96	79	98	68	85	
CDAA	119	88	114	53*	94	
Picloram	89	101	127	60*	94	
Atrazine	63*	86	148**	84	95	
Chloramben	114	97	123	83	104	
MH	98	100	143**	128	117	
Amitrole	101	103	149**	132	121	Stimulatory

[a] $LSD_{0.05}$ for comparison of treatment means with the untreated control = 19.3; $LSD_{0.05}$ for comparison of any two treatment means = $\sqrt{2} \times 19.3 = 27.3$.
[b] See text for method of determination.
[c] * = inhibition and ** = stimulation as determined by the Wilcoxon signed rank test. Mean difference from control at 0.05 level of significance = 33 ± 6.

RNA, and protein synthesis are also viewed as taking place in these organelles. Many of the herbicides which inhibit oxygen uptake by mitochondria and oxygen evolution by chloroplasts are also reported to interfere with nucleic acid metabolism and protein synthesis. Additional sites of reaction by these herbicides, not found in the short-term oxygen measurement studies may well contribute to the overall biochemical responses to the herbicides. Regulation of RNA and protein synthesis, suppression or stimulation of DNase and RNase activity, and uncoupling of oxidative phosphorylation have been associated with certain herbicides. And many herbicides form complexes with amino acids, carbohydrates, and other cellular constituents. While herbicides do not resemble structurally known endogenous regulators they do mimic many of their actions. The chlorophenoxy compounds for example bring about many of the same growth promoting and metabolic responses shown by IAA, but they are not so readily inactivated (van Overbeek, 1964).

Gruenhagen and Moreland (1971) have attempted to bring the effects of herbicides on ATP content, oxidative phosphorylation and RNA and protein synthesis into perspective in a paper describing results of tests on the effects of some 22 compounds on the ATP content of soybean hypocotyl tissue. Measuring ATP enzymatically using luciferin-luciferase to initiate the photo-chemical reaction with ATP, they measured the emitted light with a photo-multiplier tube and recorded it. Table 6-9 presents their results. Data on inhibition of oxidative phosphorylation are from the literature; effects on RNA and protein synthesis are from a previous report (Moreland et al., 1969) which has been cited (Table 6-8). As shown in Table 6-9 the compounds tested are listed in order of their decreasing effects on ATP contents of the soybean hypocotyls; eight gave significant inhibitions. Six additional compounds brought about inhibitions of doubtful significance.

Of the reported compounds that affect oxidative phosphorylation, 16 inhibit this reaction; only MH among the reported compounds does not. Dinoseb, ioxynil, propanil and chlorpropham reduced ATP levels from 88 to 90%; they also strongly inhibit RNA and protein synthesis and are reported to inhibit oxidative phosphorylation. Tissue ATP levels were reduced from 60 to 69%, and RNA and protein synthesis were also inhibited by pyriclor, propachlor, 2,4,5-T and fenac. Pyriclor and 2,4,5-T are known to inhibit oxidative phosphorylation.

Shorter time tests on the ATP content of excised soybean hypocotyls incubated with 0.2mM ioxynil and 0.6mM chlorpropham, propanil, and pyriclor are reported in Figure 6-3. Ioxynil and pyriclor reduced ATP rapidly during the first hour, propanil brought about an increase and later a decrease, chlorpropham had little effect during the first hour but lowered the ATP content rapidly during the 3 succeeding hours.

Uptake and transfer of solutes by cells and cellular organelles are processes driven by ATP energy. Thus a herbicide that inhibits ATP production *in vivo* may reduce precursor incorporation in two possible ways: (1) by limiting the amount of available precursor, and (2) by limiting the energy available for driving the biosynthetic reactions. Since protein synthesis in a cell may account for up to 90% of the ATP expended, small changes in the availability of ATP energy may have profound effects upon the life of a plant.

TABLE 6-9.   Effects of 22 Herbicides on ATP Content, Oxidative Phosphorylation, and RNA and Protein Syntheses (Gruenhagen and Moreland, 1971)

Herbicide	ATP content[a]		Inhibition of oxidative phosphorylation[b]	Effect on RNA and protein syntheses[c]
	% of control	Relative status		
Dinoseb	10	Reduced	Yes (5,7)	− −
Ioxynil	10		Yes (3,10)	− −
Propanil	10		Yes (8, 9)	− −
Chlorpropham	12		Yes (13)	− −
Pyriclor	31		Yes (11)	− −
Propachlor	35		Not reported	−
2,4,5-T	37		Yes (13, 20)	−
Fenac	40		Not reported	−
Amitrole	83	No effect	Yes/No (4, 13, 17)	+
Dicamba	94		Yes (4)	−
Chloramben	95		Yes (4)	0
Diuron	97		Yes/No (4, 6, 9, 13, 15)	−
NIA-4562	97		Yes[d]	−
MH	98		No (13)	0
Trifluralin	101		Yes (16)	0
EPTC	102		Yes (1, 13)	−
Isocil	105		Not reported	0
CDEC	111		Not reported	−
Dichlobenil	118		Yes (4)	−
Atrazine	121		Yes/No (2, 4, 6, 13)	0
Picloram	121		Yes (4)	0
CDAA	134	Increased	Yes (13)	0

[a] $LSD_{0.05} = 22\%$. See text for method of determination of relative status.
[b] Numbers in parentheses refer to literature citations. See original paper.
[c] As reported by Moreland et al., (14): − −, strongly inhibited; −, inhibited; 0, no effect; and + stimulated.
[d] Blackmon, W. J. 1970. The effects of 3′,4′-dichloro-2-methylvaleranilide and related acylanilides on reactions mediated by mung bean mitochondria. Ph.D. Thesis, North Carolina State University at Raleigh.

All herbicides that reduced ATP levels were strong inhibitors of RNA and protein synthesis. Apparently this strong correlation points to reduced protein synthesis as presumably the major mode of action for these compounds. Where this correlation did not exist, some other mechanism could be involved. Since all of the compounds tested are effective herbicides they are all phytotoxic; as such they would be expected to affect oxidative phosphorylation. Inhibitions of this reaction would seem to be less specific than the other two reactions considered here.

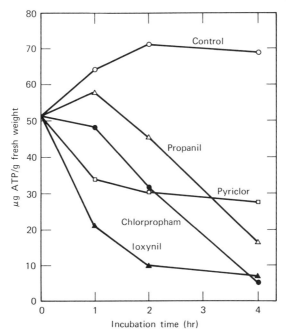

FIGURE 6-3. ATP content of excised soybean hypocrotyls incubated with 0.2 mM ioxynil and 0.6 mM chlorpropham, propanil, and pyriclor. (Gruenhagen and Moreland, 1971.)

Although beta-oxidation of the even numbered higher analogues of 2,4-D (2,4-DB for example) has been postulated as a mechanism of biochemical selectivity, Linscott and Hagin (1970) have found evidence in alfalfa for 2-carbon additions to the side chain of applied chlorophenoxy compounds, for example 4-(2,4-dichlorophenoxy) caproic acid following 2,4-D, and 4-(2,4-dichlorophenoxy) crotonic acid following 2,4-DB. They suggest the following possible reactions to explain their result 2,4-D $\rightleftharpoons$ intermediates $\rightleftharpoons$ 2,4-DB $\rightleftharpoons$ higher-chain analogues. The equilibrium may shift toward oxidation under adverse conditions.

In any growth involving synthesis of cytoplasm, membranes containing lipids are components. Mann and Pu (1968) using excised hypocotyls of hemp sesbania (*Sesbania exaltate*) investigated the incorporation of malonic acid-2-^{14}C into lipids. Inhibition of lipid synthesis was caused by dinoben, CDAA, CDEC, endothall, dichlobenil, ioxynil, and PCP. At a concentration of 1 mg per liter 2,4-D, 2,4,5-T, picloram, and chloramben stimulated incorporation of malonic acid into lipid; these latter compounds all have strong growth regulating properties.

With all of the growth regulating types of herbicides, and with many others low concentrations have been observed to have stimulatory effects, high concentrations, inhibitory. This is to be expected because most of these compounds have two basically different toxic effects on plants. For example in the common uses of 2,4-D in weed control sprays in the concentration range of 1 to 0.01% are used; 0.1 is a common average. At this dosage the foliage of the sprayed weeds usually dies and dries up in a matter of days; often a week or more may be necessary for the proliferative growth which, by crushing of vascular tissues, results in death of roots. Probably the high concentration applied to the foliage results in the inhibitory responses; reduction of oxygen utilization, uncoupling of phosphorylation; prevention of the Hill reaction etc; these occur in the treated leaves and stems. On the other hand the changes in nucleic acid and protein synthesis, the enlarged mitochondria, the formation of 2,4-D-aspartate conjugates, the 2,4-D promoted, enzymatically-catalyzed formation of aspartate complex involving the synthesis of new RNA and protein are all reactions requiring lower amounts of 2,4-D comparable to those that occur in roots, after translocation.

Thus the main biochemical reaction of 2,4-D would appear to involve a complex series of reactions initiated by the normal regulating synthesis of the enzyme RNase. The resultant synthesis of RNA and protein would be accompanied by massive cell proliferation in the presence of a proper auxin-cytokinin balance which would depend on 2,4-D dosage and translocation. If the newly-synthesized RNA and protein occur in the stem, proliferation may lead to swelling, appearance of gall-like protuberances, development of roots, etc. In roots the proliferation soon leads to crushing of the vascular tissues, death, and decay. Other mechanisms of action including inhibition of the Hill reaction and oxidative phosphorylation appear to be of secondary importance.

The complexity of plant responses to herbicides is well illustrated in the foregoing considerations. When one studies the interactions of combinations of herbicides the picture becomes even more complex. Hardcastle and Wilkinson (1970) have studied the interactions of some 8 herbicides used two at a time as they stimulated or inhibited root growth of rice. Of 25 possible combinations, 7 showed synergistic toxic action, 18 gave less than the expected response. In discussing the implications of the possible multiple modes of action the writers point out that combinations including 2,4-D indicated a possible auxin antagonism with a large number of herbicides including four substituted ureas, TCA, amitrole, and dalapon. They point out that the activity of these combinations indicates the presence of modes of action that have not been previously defined.

While many of the conclusions drawn from the type of research that has been under discussion must still be considered as only tentative with respect to herbicidal phytotoxicity, those concerning 2,4-D appear to be approaching a final answer. Much research must yet be done before the biochemical mechanisms of all of the multitude of compounds presently on the market can be satisfactorily explained.

REFERENCES

Anon. 1969. Physiological aspects of herbicidal action. In principles of plant and animal pest control. Vol. 2. *Weed Control* **2**:146–163. National Academy of Science, Washington, D.C.

Ashton, F. M. 1965. Relationship between light and toxicity symptoms caused by atrazine and monuron. *Weeds* **13**:164–168.

Ashton, F., D. Penner, and S. Hoffman. 1968. Effect of several herbicides on proteolytic activity of squash seedlings. *Weed Sci.* **16**:169–171.

Bartels, P. G. and F. T. Wolf. 1965. The effect of amitrole upon nucleic acid and protein metabolism of wheat seedlings. *Physiol. Plant.* **18**:805–812.

Basler, E. and K. Nakazawa. 1961. Some effects of 2,4-D on nucleic acids of cotton cotyledon tissue. *Bot. Gaz.* **122**:228–232.

Camper, N. D. and D. E. Moreland. 1965. Correlations between acidity of substituted phenylamides and inhibition of the Hill reaction. *Biochim. Biophys. Acta* **94**:383–393.

Chance, B. and G. R. Williams. 1955. Respiratory enzymes in oxidative phosphorylation. III. The steady state. *J. Biol. Chem.* **217**:409–427.

Chen, L. G., C. M. Switzer, and R. A. Fletcher. 1972. Nucleic acid and protein changes induced by auxin-like herbicides. *Weed Sci.* **20**:53–55.

Chrispeels, M. J. and J. B. Hanson. 1962. The increase in ribonucleic acid content of cytoplasmic particulates of soybean hypocotyl induced by 2,4-dichloro-phenoxy acetic acid. *Weeds* **10**:123–125.

Crafts, A. S. and H. Drever. 1960. Some experiments with herbicides in soils. *Weeds* **8**:12–17.

Crafts, A. S. and W. W. Robbins. 1962. *Weed Control, a Textbook and Manual.* McGraw-Hill Book Co., New York. 660 pp.

Eames, A. J. 1949. Comparative effects of spray treatments with growth-regulating substances on the nutgrass *Cyperus rotundus* L., and anatomical modifications following treatment with butyl 2,4-dichlorophenoxy acetate. *Amer. J. Bot.* **36**:571–584.

Eames, A. J. 1950. Destruction of phloem in young bean plants after treatment with 2,4-D. *Amer. J. Bot.* **37**:840–847.

Evans, H. J. and D. Scott. 1964. Influences of DNA synthesis on the production of chromatid aberrations by X-rays and maleic hydrazide in *Vicia faba. Genetics* **49**:17–38.

Foy, C. L. and D. Penner. 1965. Effect of inhibitors and herbicides on tricarboxylic acid cycle substrate oxidation by isolated cucumber mitochondria. *Weeds* **13**:226–231.

Good, N. E. and S. Izawa. 1964. Selective inhibitors of photosynthesis. *Record Chem. Progr.* **25**:225–236.

Gromet-Elhanan, Z. 1968. Energy-transfer photoreactions of chloroplasts by ioxynil. *Biochem. Biophys. Res. Commun.* **30**:28–31.

Gruenhagen, R. D. and D. E. Moreland. 1971. Effects of herbicides on ATP levels in excised soybean hypocotyls. *Weed Sci.* **19**:319–323.

Hardcastle, W. S. and R. E. Wilkinson. 1970. Bioassay of herbicide combinations with rice. *Weed Sci.* **18**:336–337.

Izawa, S. and N. E. Good. 1965. The number of sites sensitive to 3-(3,4-dichlorophenyl)-1,1-dimethylurea, 3-(4-chlorophenyl)-1,1-dimethylurea and 2-chloro-4-(2-propylamino)-6-ethylamino-*s*-triazine in isolated chloroplasts. *Biochim. Biophys. Acta* **102**:20–38.

Jones, D. W. and C. L. Foy. 1971. Herbicidal inhibition of GA-induced synthesis of α-amylase. *Weed Sci.* **19**:595–597.

Kerr, M. W. and R. L. Wain. 1964. Inhibition of the ferricyanide-Hill reaction of isolated bean leaf chloroplasts by 3,5-di-iodo-4-hydroxybenzonitrile (ioxynil) and related compounds. *Ann. Appl. Biol.* **54**:447–450.

Key, J. L. 1959. Biochemical effects of 2,4-dichlorophenoxy-acetic acid on plants. Ph.D. dissertation. Univ. of Illinois, Urbana, Ill.

Key, J. L. and J. B. Hanson. 1961. Effects of 2,4-dichlorophenoxy-acetic acid on soluble nucleotides and nucleic acid of soybean seedlings. *Plant Physiol.* **36**:145–152.

Key, J. L. and J. C. Shannon. 1964. Enhancement by auxin of ribonucleic acid synthesis in excised soybean hypocotyl tissue. *Plant Physiol.* **39**:360–364.

Killion, D. D. and R. E. Frans. 1969. Effect of pyriclor on mitochondrial oxidation. *Weed Sci.* **17**:468–470.

Lardy, H. A., J. L. Connelly, and D. Johnson. 1964. Antibiotics as tools for metabolic studies. II. Inhibition of phosphoryl transfer in mitochondria by oligomycin and aurovertin. *Biochem.* **3**:1961–1968.

Linscott, D. L. and R. D. Hagin. 1970. Additions to the aliphatic moiety of chlorophenoxy compounds. *Weed Sci.* **18**:197–198.

Lotlikar, P. D., L. F. Remmert, and V. H. Freed. 1968. Effects of 2,4-D and other herbicides on oxidative phosphorylation in mitochondria from cabbage. *Weed Sci.* **16**:161–165.

Malhotra, S. S. and J. B. Hanson. 1966. Nucleic acid synthesis in seedlings treated with an auxin-herbicide Tordon (4-amino-3,5,6-trichloropicolinic acid). *Plant Physiol. Abstr.* **41**:vi.

Mann. J. D., L. S. Jordon, and B. E. Day. 1965. A survey of herbicides for their effect upon protein synthesis. *Plant Physiol.* **40**:840–843.

Mann, J. D. and M. Pu. 1968. Inhibition of lipid synthesis by certain herbicides. *Weed Sci.* **16**:197–198.

McDaniel, J. L. and R. E. Frans. 1969. Soybean mitochondrial response to prometryne and fluometuron. *Weed Sci.* **17**:192–196.

Meikle, R. W. 1970. Inhibition of photosynthesis by pyriclor. *Weed Sci.* **18**:475–478.

Minshall, W. H. 1954. Translocation and place of action of 3-(4-chlorophenyl)-1,1-dimethylurea in bean and tomato. *Can. J. Bot.* **32**:795–798.

Minshall, W. H. 1957. Influence of light on the effect of 3-*p*-(chlorophenyl)-1,1-dimethylurea on plants. *Weeds* **5**:29–33.

Monaco, T. J. and D. E. Moreland. 1966. Partitioning and distribution of herbicides in spinach chloroplasts. *WSSA Abstr.* p. 46.

Moreland, D. E. 1967. Mechanisms of action of herbicides. *Ann. Rev. Plant Physiol.* **18**:365–386.

Moreland, D. E. 1969. Inhibitors of chloroplast electron transport: Structure-activity relations. *Progress in Photosynthesis Research.* **III**:1693–1711. Helmut Metzner, Ed.

Moreland, D. E. and W. J. Blackmon. 1970. Effects of 3,5-dibromo-4-hydroxy-benzaldehyde *O*-(2,4-dinitrophenyl)-oxime on reactions of mitochondria. *Weed Sci.* **18**:419–426.

Moreland, D. E., W. J. Blackmon, H. G. Todd, and F. S. Farmer. 1970. Effects of diphenylether herbicides on reactions of mitochondria and chloroplasts. *Weed Sci.* **18**:636–642.

Moreland, D. E., W. A. Gentner, J. L. Hilton, and K. L. Hill. 1959. Studies on the mechanism of herbicidal action of 2-chloro-4,6-bis(ethylamino)-3-triazine. *Plant Physiol.* **34**:432–435.

Moreland, D. E. and K. L. Hill. 1959. The action of alkyl *N*-phenylcarbamates on the photolytic activity of isolated chloroplasts. *J. Agr. and Food Chem.* **7**:832–837.

Moreland, D. E. and K. L. Hill. 1962. Interference of herbicides with the Hill reaction of isolated chloroplasts. *Weeds* **10**:229–236.

Moreland, D. E. and K. L. Hill. 1963a. Inhibition of photochemical activity of isolated chloroplasts by acylanilides. *Weeds* **11**:55–60.

Moreland, D. E. and K. L. Hill. 1963b. Inhibition of photochemical activity of isolated chloroplasts by polycyclic ureas. *Weeds* **11**:284–287.

Moreland, D. E., S. S. Malhotra, R. D. Gruenhagen, and E. H. Shokraii. 1969. Effects of herbicides on RNA and protein synthesis. *Weed Sci.* **17**:556–563.

Muni, A. P. 1959. Investigations on the anatomy of weeds and selective weed control. Ph.D. dissertation. Bose Inst., Univ. of Calcutta.

Negi, N. S., H. H. Funderburk, Jr., D. P. Schultz, and D. E. Davis. 1968. Effect of trifluralin and nitralin on mitochondrial activities. *Weed Sci.* **16**:83–85.

Okubo, C. K. and C. M. Switzer. 1965. Effects of various phenoxy herbicides on nucleic acid metabolism. *Plant Physiol. Abstr.* **40**:xiv.

Oorschot, J. L. P. van. 1965. Selectivity and physiological inactivation of some herbicides inhibiting photosynthesis. *Weed Res.* **5**:84–97.

Oorschot, J. L. P. van. 1968. The recovery from inhibition of photosynthesis by root-applied herbicides as an indication of herbicide inactivation. *Proc. 9th Brit. Weed Control Conf.* 624–632.

Oorschot, J. L. P. van. 1970. Effect of transpiration rate of bean plants on inhibition of photosynthesis by some root-applied herbicides. *Weed Res.* **10**:230–242.

Overbeek, J. van. 1964. Survey of mechanisms of herbicide action. *The Physiology and Biochemistry of Herbicides.* pp. 387–400. Editor L. J. Audus. Academic Press Inc., New York.

Overbeek, J. van and R. Blondeau. 1954. Mode of action of phytotoxic oils. *Weeds* **3**:55–65.

Penner, D., and F. M. Ashton. 1966. Biochemical and metabolic changes in plants induced by chlorophenoxy herbicides. *Residue Rev.* **14**:39–113.

Pereira, J. F. 1970. Some plant responses and the mechanism of selectivity of cabbage plants to nitrofen. Ph.D. dissertation. Univ. of Illinois, Urbana, Ill.

Schröder, I., M. Meyer, and D. Mücke. 1970. Die Wirkung der Herbicide 2,4-D, Amitrol, Atrazin, Chlorpropham und Chlorflurenol auf die Nucleinsaüre-Biosynthese des Ascomyceten *Neurospora crassa. Weed Res.* **10**:172–177.

Schweizer, E. E. and B. J. Rogers. 1964. Effects of amitrole on acid-soluble nucleotides and ribonucleic acid in corn tissue. *Weeds* **12**:310–311.

Shannon, J. C., J. B. Hanson, and C. M. Wilson. 1964. Ribonuclease levels in the mesocotyl tissue of *Zea mays* as a function of 2,4-dichlorophenoxy-acetic acid application. *Plant Physiol.* **39**:804–809.

Sheets, T. J. 1961. Uptake and distribution of simazine by oat and cotton seedlings. *Weeds* **9**:1–13.

Shokraii, E. H. and D. E. Moreland. 1966. Effect of herbicides on nucleic acid metabolism and protein synthesis P[32]. *WSSA Abstr.* pp. 46–47.

Slater, E. C. 1963. Uncouplers and inhibitors of oxidative phosphorylation, pp. 503–516. *In Metabolic Inhibitors.* R. M. Hochster and J. H. Quastel Ed. Vol. **2**. Academic Press, New York.

Temperli, A., H. Turler, and C. D. Ercegovich. 1966. Incorporation of s-triazines (cyanuric acid and prometryne) into bacterial nucleic acid. *Zeitschr. f. Naturforschung.* **21B**:903–904.

Weldon, L. W. and R. D. Blackburn. 1969. Herbicidal treatment effect on carbohydrate levels of alligator weed. *Weed Sci.* **17**:66–69.

West, S. H., J. B. Hanson, and J. L. Key. 1960. Effect of 2,4-dichlorophenoxy acetic acid on the nucleic acid and protein content of seedling tissues. *Weeds* **8**:333–340.

Williams, A. K., S. T. Cox., and R. G. Eagon. 1965. Conversion of 3-amino-1, 2,4-triazole into 3-amino-1,2,4-triazolyl alanine and its incorporation into protein by *Escherichia coli. Biochem. Biophys. Res. Comm.* **18**:250–254.

Wort, D. J. 1964. Effects of herbicides on plant composition and metabolism. pp. 291–334. In: *The Physiology and Biochemistry of Herbicides.* Ed. L. J. Audus. Academic Press, New York.

Zukel, J. W. 1963. A literature summary on Maleic Hydrazide. U.S. Rubber Co., Naugatuck, Conn. 111 pp.

CHAPTER 7

Mode of Action of Herbicides

The terms mode of action and mechanism of action mean different things to different people. Some have used the terms interchangeably and considered them to mean the same thing, namely the biochemical responses of plants to herbicides. Others have also used them interchangeably but have considered them to include physiological, biochemical, and other aspects of herbicidal action. Other usage has been for mechanism of action to refer to the biochemical responses and mode of action considered to be a more general term referring to all aspects of herbicidal action. The National Academy of Sciences book, entitled "Weed Control" (Anon., 1968) stated, "The term 'mode of action' refers to the entire sequence of events from introduction of a herbicide into the environment to the death of plants. 'Mechanism of action' refers to the primary biochemical or biophysical lesion leading to death."

In this book we are considering that "mode of action," comprises the sum total of anatomical, physiological, and biochemical responses that make up the total phytotoxic action of a chemical, as well as the physical (location) and molecular (degradation) fate of the chemical in the plant. The term "mechanism of action" will be restricted to the biochemical and biophysical responses of the plant that appear to be associated with the herbicidal action. These responses may not necessarily involve the primary biochemical lesion in a classical sense. The primary biochemical site of action (lesion) is the single enzyme or metabolic reaction that is affected at a concentration lower than any other enzyme or metabolic reaction, or the first reaction effected at a given low concentration. The primary biochemical lesion may only be part of the answer as to how a given herbicide is effective as a herbicide. The reason for this involves several factors. Perhaps foremost is the fact that the primary biochemical lesion may not be of sufficient importance in the overall metabolism of a plant that its inhibition will cause the drastic disturbance necessary to be considered herbicidal. It is also true that many herbicides increase in concentration at the subcellular level with time *in vivo*. This increasing

100

concentration could result in one or more less sensitive sites of action being involved in the herbicidal action.

We have to admit that at the present time, in this relatively new field, we have only fragmentary bits of knowledge concerning the great array of commonly used herbicides. However, even in the well established field of molecular pharmacology a similar but somewhat less extreme situation exists. Gehard Zbenden (1968) recently stated "The exact mode of action of many of our most valuable drugs is not known."

In an attempt to present the reader with a number of points of view of the mode of action of herbicides we have selected several papers published within the past decade dealing with the subject. They are presented in chronological order and therefore may represent the progressive thinking in this area. Since the earlier papers were written, the total amount of research information available has increased two to threefold and as a result some of the concepts developed would need reevaluation in light of recent findings. However, it is not our objective in this chapter to present latest ideas on the mode of action of a specific compound but rather present individual authoritative concepts of what the mode of action of herbicides involves. The last several pages of this chapter discuss a system for the localization of the primary biochemical site of action of herbicides.

Varying Concepts of Mode of Action

In 1964 van Overbeek (1964) stressed growth and photosynthesis as the major processes involved in the mechanism of herbicide action. He stated that out of some 71 named herbicides "three quarters of all chemical weed control is achieved with just three chemicals, their homologues, and their analogues. These chemicals are 2,4-D, by far the most widely used herbicide, simazine, and diuron. Just two fundamental physiological actions are involved by which these highly successful herbicides control weeds. 2,4-D causes abnormal plant growth which kills like cancer. Simazine and diuron upset photosynthesis." Some six additional chemicals "appear to act primarily on some phase or another of the growth process."

While the above statements would seem to represent a gross oversimplification of the concept of the mode of action, van Overbeek (1964) proceeds to an analysis of plant responses to a variety of herbicides. Starting with the action of herbicides on the growth process he describes research on 2,4-D, amitrole, dalapon, and dinoseb. He considers 2,4-D to have high auxin activity, and the herbicidal action results in various growth abnormalities. Amitrole he thinks ultimately disrupts protein synthesis and hence enzymes and structural proteins needed for growth are not provided. Dalapon action he attributes to abnormal protein metabolism accompanied by unusual accumulation of metabolites. Tolerant plants, he surmised, may be capable of detoxifying the breakdown products of protein rather than breakdown of dalapon itself. Dinoseb he explains is an uncoupling agent that unmeshes the gears of the mechanism responsible for generation of ATP. This process, he explains, may result from breaking mitochondrial membrane integrity enabling electric current to leak and thus to knock out the phosphorylation mechanism.

Attacking the problem of herbicidal action on photosynthesis, van Overbeek (1964) points out the nature of photosynthetic phosphorylation as a cyclic process that produces ATP at the expense of light. The urea herbicides apparently block the normal pathway by which electrons in chlorophyll are replenished; the chloroplasts are oxidized; they become bleached and the leaf is said to become chlorotic; it soon dies. The symmetrical triazines and the alkylamides apparently have the same basic action. As Good (1962) has said "All that we can say with confidence is this: the anilides (which include phenylureas), the alkylamides, and the triazines have similar modes of action and nothing that we know is inconsistent with the hypothesis that the mechanism for the oxidation of water to molecular oxygen is the process which is primarily affected."

Van Overbeek (1964) goes on to illustrate the fact that diuron, NIA-4562, and atrazine have the —NH— group in common; this group is known to form hydrogen bonds. Furthermore the phenylureas and acylanilides have the $=C=O$ group, and the triazines the $=C=N—$ group; these also have similar hydrogen bonding capacity. These compounds then may owe their herbicidal activity to their capacity to form hydrogen bonds with the protein of an enzyme involved in the oxidation of water. The dipyridylium herbicides may also act through the electron flow mechanism. Instead of blocking the return flow of electrons they apparently pick up single electrons and become free radicals. To confirm this relation van Overbeek points out the observation of Mees (1960) that monuron inhibits the herbicidal action of diquat.

In the 1963 volume of Annual Review of Plant Physiology, Hilton et al. (1963) have reviewed "Mechanisms of Herbicide Action." Grouping the majority of useful herbicides into 9 chemical categories, they stress the biochemical aspects of the mechanisms and point out the great interest in metabolic fate of herbicides in plants. They found evidence for the fact that a herbicide may be applied in one molecular form, translocated in a metabolized form and express its inhibition as still another chemical structure. Another herbicide may be absorbed, translocated, and accumulated at a given site as the original molecule. Most herbicides, they state, are rapidly and extensively metabolized; in some cases a relatively non-toxic chemical may be converted to a herbicidally active molecule; they cite the bipyridylium compounds and 2,4-DB as examples. In many other instances metabolism may result in detoxication through chemical degradation or chemical conjugation.

Hilton et al. (1963), while stressing the diversity of principles involved in herbicidal action, point to two major categories of compounds: (1) those that move only in the xylem with the transpiration stream and are absorbed by roots from the soil and accumulate in leaves—most of these compounds inhibit photosynthesis; (2) the other group consists of compounds that seem to move either apoplastically or symplastically—these become generally distributed throughout the plant and usually accumulate in the sinks of high growth activity. For such herbicides multiple sites and mechanisms of action must be considered; the most sensitive site may differ among species. They point out that it is difficult to substantiate the physiological significance of *in vitro* results as to what occurs under conditions of practical application in the field.

In his book, "Weeds of the World," King (1966) has a chapter on classification and mode of action of herbicides. In this chapter he discusses mode of action under six headings: contact herbicides, inhibitors of cell growth, auxin-type growth regulating chemicals, inhibitors of growth and of tropic responses, inhibitors of chlorophyll formation and of photosynthesis, and other translocated herbicides. Thus the concept of mode of action is broadening to cover the ever-expanding array of chemical weed killers. In his Appendix I, King lists 48 herbicides giving their properties and common uses. In Appendix II he lists 23 herbicides of recent introduction.

Moreland (1967), in a review paper, covered mechanisms of action of herbicides under three headings: (1) respiration and mitochondrial electron transport, (2) photosynthesis and the Hill reaction, and (3) nucleic acid metabolism and protein synthesis. Much of his report has been covered in Chapter 6.

In his concluding remarks Moreland (1967) stresses the need for identifying the sites through which plants express their reactions to herbicides at cellular and molecular levels. Criteria are needed for separating the primary from secondary or other effects of a given chemical. Because of the complex interrelations between cellular metabolism and growth and development, responses to chemicals may take place at sites removed from the region of application or the original site of action. Because of this it is difficult to evaluate experiments on intact plants when these are fractionated and assayed for particular biochemical changes; interpretation may be clouded by reactions that occur between the time of treatment and that of the assay.

Moreland (1967) goes on to point out that many herbicides seem to be metabolically nonspecific; instead of having a single site of action, there may be several sites and mechanisms involved in inhibition; different sites may vary in sensitivity to a given herbicide, and while the primary or most sensitive site should be affected first, this initial response may be difficult to single out and identify. As concentration of a chemical builds up inside the plant, additional sites may become involved. While the various sites affected by a given chemical may have a common molecular structure and show a common type of action, to pick the most important one may be impossible. In contrast to this situation, on some plant species, under a specified set of conditions, some herbicides are highly specific in their actions.

In a paper on the metabolism and effects of herbicides in plants Frear and Shimabukuro (1970) discuss the enzyme catalysis of herbicide biotransformations. They cite Stoller (1969) and Stoller and Wax (1968) as having shown that selectivity of chloramben is not the result of differences in the initial rates of absorption or binding but is directly correlated with differences in the concentration of chloramben necessary to saturate the enzyme responsible for the biosynthesis of N-glucosyl chloramben. This enzyme, arylamine N-glucosyltransferase is primarily associated with root tissues and has a broad specificity for substituted anilines. This enzyme system acts on chloramben, dinoben, pyrazon, and propanil.

Frear and Shimabukuro (1970) report further that arylacylamidase is an important factor in propanil selectivity in that it results in different rates of hydrolysis of propanil to 3,4-dichloroaniline in susceptable and resistant plants; several insecticidal carbamate and organophosphate compounds are potent inhibitors of rice arylacylamidase activity.

Selectivity of the triazines results from at least three different metabolic systems all of which appear to bring about detoxication. In highly resistant plants a pathway results from biosynthesis of a glutathione conjugate, glutathione S-transferase located in plant shoots; it is inhibited by halogenated aromatic and arylalkyl compounds; 2,3,6-trichlorophenylacetic acid is a synergist. Shimabukuro et al. (1970) have shown that the rate of conjugation in whole plants is directly correlated with the glutathione S-transferase activity levels. Rapid formation of the conjugate immobilizes atrazine at the site of application.

The primary site of action of triazines seems to be the chloroplast; chloroplasts from resistant or susceptible plants are inhibited to different degrees and recovery is correlated with rate of triazine metabolism.

Frear and Shimabukuro (1970) report that Kuratle et al. (1969) have shown that linuron selectivity is related not only to the rate of metabolism but to N-demethylated and N-demethoxylated metabolites which are more phytotoxic to susceptible than to resistant plants. A microsomal mixed function oxidase which catalyzes the N-methylation of 3-(phenyl)-1-methylurea herbicides has been isolated. Many of the reported biotransformations of herbicides in plants indicate the direct involvement of microsomal mixed function oxidase systems.

Hydrolysis, oxidation, and reduction of herbicides in plants often result from formation of substituted phenols, anilines, and heterocyclic compounds which are often conjugated and persistent in plants. Some such herbicides are propanil, 2,4-D, diuron, dichlobenil, dicamba, linuron, amitrole, pyrazon, and chloramben. For details and references the reader is referred to the original paper of Frear and Shimabukuro (1970).

This review is included here to support the concept of Frear and Shimabukuro (1970) that "a major factor responsible for herbicide selectivity is often a difference in the activities, specificities, and distribution of key enzyme systems directly associated with biotransformations in the plant." And "herbicide selectivity, as affected by specific plant enzymes, is obviously controlled by genetic factors."

Frear and Shimabukuro (1970) proceed with a discussion of "effects of herbicides" which is, in effect, a summary on mode of action. Listing atrazine, trifluralin, chloramben and analogues, and 2,4-D as having the greatest sales, and adding CDAA, propachlor, nitralin, 2,4,5-T, picloram, paraquat, and dicamba, they state that most of the herbicides act primarily by affecting one or more growth processes other than photosynthesis. They proceed with their discussion under the headings: morphological and cytological effects, effects on water and mineral uptake, nucleic acid and protein metabolism, effects on respiration, and effects on photosynthesis.

In their summary on effects of herbicides, Frear and Shimabukuro (1970) say "that herbicides may act on more than one biochemical or physiological site—the herbicide must reach the sensitive site or sites in its toxic form and at a concentration sufficient to cause severe disruption of normal growth." "Injury resulting from interaction at the most sensitive site may be severe enough to cause death in plants. This may be true of the photosynthetic inhibitors such as diuron and atrazine. With other herbicides, injury caused at individual sites may not be severe enough, but the sum of injury at several sites may reach a threshhold level where irreversible damage results and death becomes inevitable. This seems to be true of herbicides such as 2,4-D, amitrole, and the carbamates." These are only a few of the possible examples. It would be difficult to improve on this statement with respect to the current concepts of mode of action.

To return to 2,4-D as an example, the total array of effects include: (1) the growth response of epinasty, often evident within minutes after foliar application; (2) on a warm day the light aliphatic ester can cause wilting and collapse of mesophyll reminiscent of the loss of membrane integrity common to oil toxicity; (3) translocation via the symplastic system across mesophyll to the vascular tissues; (4) phloem transport to active metabolic sinks in shoot and root; (5) increase in nucleic acids resulting in expansion growth, often unordered; (6) development of callus-like tissues in cortex and pith leading to crushing and obliteration of phloem and xylem; (7) death and destruction by invading fungi and bacteria. When treatment is via the soil, the growing primary roots stop growth, swell as a result of unordered or lateral expansion of cells crushing the young vascular tissues and resulting in death of the seedlings. Soil treatment may result in development of symptoms of abnormal growth in the tops of perennial weeds but these weeds are not controlled by this manner of application.

In bringing about this array of responses undoubtedly many enzyme systems are disturbed, biosynthesis of many sorts are changed, and growth, of course, is altered and stopped. However, as noted by Crafts (1961) successful control of some weeds by the phenoxy-type herbicides may in the end not result in death but only in sufficient growth inhibition to limit in a practical way competition for water, light, and mineral nutrients. Weeds often controlled this way in cereal crops are Russian thistle (*Salsola kali*), yellow star thistle (*Centaurea solstitialis*), tar weeds (*Madia* sp.), and sometimes common lambsquarters (*Chenopodium album*) and pigweeds (*Amaranthus* sp.). Thus mode of action, in the modern sense, may cover a multitude of plant responses and often it is impossible to rank them with respect to importance. As pointed out by Frear and Shimabukuro (1970) control or death of a weed may result from one or a few key processes or, in contrast, it may require the sum of injury at several sites to attain a threshhold level where total injury is irreversible and complete loss of competitive ability or death results.

Localization of Primary Site of Action

Localization of the primary biochemical site of action (lesion) of a given herbicide is indeed a difficult task. Webb (1963) discusses the localization of the site of inhibition of metabolic inhibitors and states: "It is really only under fortuitous circumstances that a satisfactory localization can be readily made." Webb (1963) wrote an excellent chapter on this subject and it should be reviewed by the serious student of this type of research. Ashton (1967) speaking before the California Weed Conference discussed a program for the localization of the primary biochemical site of action of herbicides.

In biological systems there are a number of metabolic bypasses and hence the primary site of action may not coincide with the herbicidal site of action. Taking for example the reaction $A \rightarrow B \rightarrow C$, if formation of B from A is blocked by a herbicide it is possible that a bypass comes into action as follows.

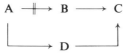

If the product D may be made from A by a pathway distinct from the reaction $A \rightarrow B$, then blocking this latter reaction may shift the pathway to C via D. Therefore the reaction $A \rightarrow B$ is not essential for the formation of C and C then can enter into subsequent essential reactions.

In research aimed at locating the primary site of action, Ashton (1967) presents 5 essential steps:

1. Location of sites of accumulation of a herbicide in the intact plant. This is most often accomplished by the use of labeled herbicides and autoradiography is a very useful tool.

2. Recognition of the symptoms of phytotoxicity. For example, if chlorosis is an early symptom one suspects chloroplast development and possibly chlorophyll synthesis has been blocked. Another symptom of herbicidal action is anthocyanin production, a sign that sugars are accumulating. Thus we may look for a block in sugar transport or possibly respiration which utilizes sugars. A loss of turgidity resulting in water-logging of mesophyll of leaves often indicates a loss of membrane integrity; such may result from oil toxicity; it also may be found in leaves of plants treated with simazine or monuron, an indication of a block in photosynthesis. The growth symptoms of the phenoxys is another aspect of herbicidal disturbance of growth; one that involves the endogenous hormone system and its disturbance.

3. Preliminary metabolic studies on intact plants to provide information on the essential pathways involved. A concentration relationship should be established over the total effective range from initiation of symptoms to death. Such information may be used in later *in vitro* studies on the various pathways that seem important. Next should follow time-course studies to follow the sequence of events that lead to death. A delay in the onset of inhibition may well indicate that the applied substance is not the inhibitor but must be metabolized to the inhibitor in order to function. Another aspect of time-course relation is the degree of reversibility as related to recovery. When a plant recovers it suggests a detoxication mechanism or that the herbicide is actually lost from the plant by leakage or volatilization.

4. Secondary metabolic studies using crude systems. Having blocked out the single major area of metabolism involved it is now possible to proceed in depth by studying the components of the affected metabolic process. For example, in photosynthesis, differentiation should be effected between light and dark reactions; CO_2 fixation or oxygen evolution should be measured and phosphorylation mechanism, cyclic or noncyclic should be determined. In respiration, distinction should be made between glycolysis and the TCA cycle.

5. Tertiary metabolic studies on purified systems. Such studies may involve single enzymes, single metabolic reactions or a single type of organelle. Using time and concentration series one should determine the critical

dosage-time relationships and compare these with the previously determined *in vivo* values. A close correlation of the *in vivo* and *in vitro* values increases the probability that one has indeed localized the primary site of action.

The above 5 processes represent a step-wise approach, by elimination, to arrive at a final result; such an approach is not fraught with the hazards of studying only one area of metabolism without previously determining whether this one area is more sensitive to the action of a specific herbicide than other areas of metabolism, or in not relating *in vivo* and *in vitro* results.

There are several biochemical techniques that are often useful in localization studies. Often when an inhibitor acts on a given enzyme the substrate acted upon by this enzyme tends to accumulate. Therefore, the accumulation of any intermediate suggests that the enzyme(s) acting upon this intermediate may be inhibited by the herbicide. Sometimes the addition of various substrates or intermediates of a given inhibited pathway may provide information as to the site of action. The addition of radioactively labeled substrates or intermediates to the inhibited system and the detection of the resulting alterations in the pathway induced by the herbicide, may be useful. The use of a second inhibitor whose site of action is known may be employed either to mimic the action of the herbicide or to isolate the inhibited metabolic pathway from the complexity of interfering pathways.

To indicate that the above program may not always lead to a discrete answer we find in Moreland's (1967) conclusions the following statement " For some herbicides, there is no documentation of interaction with any major metabolic pathway; whereas interference with several biochemical processes has been reported for other herbicides. The multitude of effects that can be imposed, and the interaction entered into, by a single herbicide may be illustrated by amitrole." Moreland proceeded to enumerate formation of complexes with amino acids and carbohydrates, interference with histidine biosynthesis, effects on purine and nucleic acid metabolism, and interactions with riboflavin as examples. He states that as of now, no investigator has been able to name a single mechanism through which the phytotoxicity of amitrole to higher plants can be explained.

As the absorption and translocation of herbicides, so essential to an understanding of herbicide phytotoxicity, are becoming well understood, more and more effort is being put into mechanism of action studies, particularly in their biochemical and biophysical aspects.

In reading the hundreds of papers required in the writing of this book, the authors are impressed with the mounting interest on the part of the research workers in mode of action studies. Although such studies may not lead directly to new and improved methods of weed control, they do establish a factual background upon which to base interpretation of the phytotoxic action of herbicides. Such interpretations become invaluable in trouble shooting where herbicides have failed. They also are valuable in the search for and synthesis of new herbicides.

REFERENCES

Anon. 1968. Principles of Plant and Animal Pest Control. Vol. 2. Weed Control. Publication 1597. National Academy of Science, Washington, D.C. 471 pp.

Ashton, F. M. 1967. Localization of primary site of action of herbicides. *Proc. 19th Ann. Calif. Weed Conf.* pp. 80–84.

Crafts, A. S. 1961. *The Chemistry and Mode of Action of Herbicides.* Interscience Publishers, New York. 269 pp.

Frear, D. S. and R. H. Shimabukuro. 1970. Metabolism and effects of herbicide in plants. *Technical Papers of the FAO International Conf. on Weed Control.* pp. 560–578. Davis, Calif.

Good, N. E. 1962. Inhibitors of the Hill Reaction. *Plant Physiol.* **36**:788–803.

Hilton, J. L., L. L. Jansen, and H. M. Hull. 1963. Mechanisms of herbicide action. *Ann. Rev. Plant Physiol.* **14**:353–384.

King, L. J. 1966. *Weeds of the World: biology and control.* Leonard Hill, London. Interscience Publ. Inc. New York, 526 pp.

Kuratle, H., E. M. Rohn, and C. W. Woodmansee. 1969. Basis for selectivity of linuron on carrot and common ragweed. *Weed Sci.* **17**:216–219.

Mees, G. C. 1960. Experiments on the herbicidal action of 1,1'-ethylene-2,2'-dipyridilium dibromide. *Ann. Appl. Biol.* **48**:601–612.

Moreland, D. E. 1967. Mechanisms of action of herbicides. *Ann. Rev. Plant Physiol.* **18**:365–386.

Overbeek, J. van. 1964. Survey of mechanisms of herbicide action. In: *The Physiology and Biochemistry of Herbicides.* pp. 387–400. L. J. Audus, Ed. Academic Press, Inc. New York.

Shimabukuro, R. H., D. S. Frear, H. R. Swanson, and W. Walsh. 1970. Glutathione conjugation. *WSSA Abstr.* No. 27.

Stoller, E. W. 1969. The kinetics of amiben absorption and metabolism as related to species sensitivity. *Plant Physiol.* **44**:854–860.

Stoller, E. W. and L. M. Wax. 1969. Amiben metabolism and selectivity. *Weed Sci.* **16**:283–288.

Webb, J. L. 1963. *Enzyme and Metabolic Inhibitors.* Volume 1. Academic Press, Inc, New York. 949 pp.

Zbenden, G. 1968. Mechanism of Toxic Drug Reaction, pp. 38–47. In P. E. Esieyler and J. H. Mayer III, *Pharmacologic Techniques in Drug Evaluation.* Year Book Medical Publishers, Inc., Chicago, Ill.

CHAPTER 8

Aliphatics

$$
\begin{array}{c}
\underset{\overset{|}{\underset{Cl}{\overset{|}{Cl}}}}{Cl-C-C-OH} \\
\overset{O}{\parallel}
\end{array}
\qquad
\begin{array}{c}
H\ Cl\ O \\
| \ \ | \ \ \parallel \\
H-C-C-C-OH \\
| \ \ | \\
H\ Cl
\end{array}
$$

TCA   Dalapon

(trichloroacetic acid)   (2,2-dichloropropionic acid)

Although this class of compounds is often referred to as the chlorinated or substituted aliphatic acids, they are almost exclusively used as the sodium salts. In most scientific publications they are referred as to TCA or dalapon when in fact the sodium salt has actually been used. This is in spite of the fact that the common name of TCA or dalapon, as designated by both the Weed Science Society of America and British Standards Institution, refers to the acids. Therefore, in keeping with common practice, TCA or dalapon will designate the corresponding sodium salt in the following presentation.

These two chlorinated aliphatic acids are widely used as herbicides. Although other chlorinated aliphatic acids have been synthesized and evaluated for herbicidal properties, they have not come into common usage. Both compounds are particularly effective against grasses, but they also control certain broad leaf weeds. TCA is registered for use in a limited number of crops; however, dalapon is registered for use in over twenty-five crops.

The most important application of this class of herbicides is the use of dalapon as a foliar treatment to control perennial grasses in certain crops and on non-cropped land. Although dalapon is one of the best herbicides available for this purpose, success depends upon a systematic retreatment program. For most perennial grass species, the best control is obtained when the first application is made relatively early in the growing season when leaves are mature and the seed heads have started to form. Subsequent retreatments are required about every five weeks throughout the growing season; the actual time between retreatments depends on the species and the environmental conditions; significant regrowth should have occurred. It may be necessary to make these repeated applications for more than one year for complete control. It must be emphasized that a program as outlined above is essential for the control of most perennial grasses; a single application is only of temporary value.

McWhorter (1971) found the effectiveness of dalapon to vary between wide limits in tests on some 55 morphologically distinct ecotypes of johnsongrass (*Sorghum halpense*); nitrogen fertilization of the plants enhanced control by dalapon.

Growth and Plant Structure

The phytotoxic symptoms of the chlorinated aliphatic acid herbicides are growth inhibition, leaf chlorosis and formative effects, especially at the shoot apex. Rapid foliar necrosis, and contact injury, may occur from TCA or with high concentrations of dalapon. Such injury would inhibit their systemic transport, by injuring the symplastic translocation system. Increased tillering in grasses is also a common response.

Mayer (1957a) reported that TCA inhibited the growth of both shoots and roots; however at very low concentrations root growth was stimulated and shoot growth was more sensitive than roots. Dalapon was found to be inhibitory to the elongation of the primary roots of maize and cucumbers, five times more dalapon was required for fifty percent inhibition of maize roots compared to cucumber roots (Ingle and Rogers, 1961). Root growth inhibition was first detectable four hours after dalapon treatment and growth ceased within twelve hours. Sublethal rates of dalapon progressively reduced the growth rates of *Lemna minor* and *Salvinia natans* (Prasad and Blackman, 1965a). This growth reduction was correlated with the rate of leaf or frond formation and mean leaf or frond area. Meyer and Buchholtz (1963) reported that TCA and dalapon inhibit the growth of shoots from isolated quackgrass (*Agropyron repens*) rhizomes and buds. An increase in tillering of barley from dalapon applications was reported by Hilton et al. (1959) and Wilkinson (1962). The growth of coleoptile sections of maize and peas was found to be inhibited by 10^{-3} and 10^{-2}M dalapon respectively, but oat coleoptiles were not affected from 10^{-4} to 10^{-2}M (Wilkinson, 1962). Funderburk and Davis (1960) noted that the reduction in height of maize plants following dalapon application was the result of a shortening of the internodes rather than a decrease in the number of nodes formed. The density of epidermal hairs on leaves of *Salvinia natans* was found to be reduced following the application of dalapon (Prasad and Blackman, 1965a).

Prasad and Blackman (1964) reported that the primary effect of dalapon on roots of several species studied was an interference with the meristematic activity of the root tip and that mitotic activity was arrested at prophase. Stem injury of *Elodea* caused by TCA was found to be associated with damage to epidermal, cortical, and vascular cells and the size of the nuclei in cells of the meristem was reduced 40% (Gooch and Erbe, 1967). Following TCA treatments, alteration of cell membranes and associated changes in permeability have been reported to be responsible for growth reductions (Mayer, 1957b) and for reduced wax excretions by leaves (Dewey et al., 1962).

Absorption and Translocation

TCA is usually applied to the soil and dalapon to the foliage of plants, therefore the uptake of TCA by the roots and the uptake of dalapon by leaves are more significant than the uptake of TCA by leaves or dalapon by roots.

[14]C-TCA applied to foliage was readily absorbed by maize but only small amounts were translocated from the treated leaf to other leaves, the shoot, the apex or the roots (Blanchard, 1954). Small amounts were present in the culture solution indicating leakage by the roots. Following root application of ^{14}C-TCA to maize and peas via the culture solution, TCA was readily absorbed and translocated throughout the plants including the leaves, stems, and shoot apex. The largest amounts were found in the older leaves; more even than those of the roots which were bathed in the herbicide solution. These results indicate that TCA is readily absorbed by both roots and leaves and is primarily translocated via the transpiration stream system, however small amounts are transported via the symplast system.

Although only a limited amount of information is available on the absorption and translocation of TCA, dalapon has been investigated in considerable detail. Most of this research utilized dalapon which was labeled with the radioactive atoms ^{14}C or [36]Cl. However, it is particularly interesting to note that the conclusions reached by Santelmann and Willard (1955), before the labeled compounds were available, were essentially correct. Using phytotoxicity symptoms as the criteria of absorption and translocation, they concluded that dalapon was readily absorbed by leaves and roots and was translocated throughout the plants. They also determined that dalapon was translocated with the photosynthate by symplastic movement, but that this was not the only means of translocation. This was determined by depleting the leaves of photosynthate, placing the plants in the dark, prior to the application of dalapon to the leaves, and finding a reduction in translocation.

However, the refined technique of using either ^{14}C or [36]Cl labeled dalapon in subsequent studies by numerous investigators, has not only confirmed the finding of Santelmann and Willard (1955) but has allowed increased confidence in their conclusions as well as extending them in several areas. For example,

Wilkinson (1956) and Foy (1961a) have shown that translocation of dalapon from a treated grass leaf is retarded by an active intercalary meristem at the leaf base.

Dalapon appears to be taken up by two different mechanisms. In studies with *Lemna minor*, Prasad and Blackman (1965a) reported that [36]C-dalapon entered both fronds and roots initially at a rapid rate and later at a slow but steady rate. The initial rapid rate was probably primarily an adsorption phenomenon. Foy (1962a) reported a small but significant amount of dalapon absorbed within 15 to 20 seconds by leaves of maize. Subsequent uptake of dalapon by *L. minor* was curvilinear and related to the herbicide concentration and temperature; it was inhibited by metabolic inhibitors such as dinitrophenol, arsenate, azide, iodoacetate, and phenyl mercuric acetate (Prasad and Blackman, 1965a). The inhibition by the latter was reversed by glutathione or cysteine and it was concluded that the slow continuous uptake was a metabolic process involving thiol groups.

Foy (1962a) found that the surfactant sodium dioctylsulfosuccinate (Vatsol T) at 0.10% enhanced absorption of dalapon placed on maize leaves up to 30 minutes but not at 3 hours. Dalapon entered the leaves of the hypostomatous plant *Tradescantia fluminensis* through both the cuticle and stomata, but stomatal absorption was erratic without a surfactant (Foy, 1962b). At 2 hours, Vatsol T increased cuticular and stomatal entry three to fourfold.

Several investigators have reported that dalapon enters the leaves and is distributed throughout the plant with the highest concentration in young or meristematic tissues; cotton (Foy, 1961a), sorghum (Foy, 1961a), maize (Foy 1962a, Blanchard, et al., 1960), sugarbeets (Andersen, et al., 1962), *Tradescantia fluminensis* (Foy, 1962a), yellow foxtail (*Setaria glauca*) (Andersen, et al., 1962) and dallisgrass (*Paspalum distichum*) (Smith and Davies, 1965). These data clearly show symplastic transport of dalapon via the phloem. However, as mentioned earlier, there also appears to be another important mechanism of dalapon translocation based on its acropetal distribution in a leaf from the treated spot and its distribution following root application, both apoplastic. Figure 8-1 (Wilkinson, 1956) shows both symplastic and apoplastic translocation of [36]C-dalapon from a treated spot on a barley leaf.

Detailed observations of radioautographs following ^{14}C-dalapon application to maize leaves showed that the first movement appeared as a diffusional pattern although subsequently the radioactivity was channelized into veinlets and larger vascular bundles (Foy, 1962a). The historadioautographic techniques used by Pickering (1965) showed that dalapon was present in both the xylem and phloem.

Studies showing both symplastic and apoplastic translocation patterns following root application of radioactive dalapon and subsequent radioautography include: cotton (Foy, 1961a, Smith and Dyer, 1961) sorghum (Foy, 1961a), maize (Blanchard, et al., 1960), soybean (Blanchard, et al.,

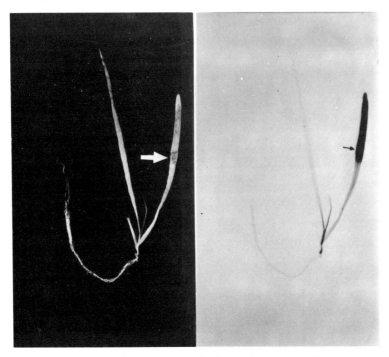

FIGURE 8-1. Radioautograph of the barley plant showing both symplastic and apoplastic translocation of dalapon; plant (left), autoradiograph (right); arrow shows point of application. (Wilkinson, 1956.)

1960), sugarbeet (Andersen, et al., 1962), and yellow foxtail (Andersen, et al., 1962).

The rate of absorption and translocation of dalapon is influenced by a number of factors including age of plant, surfactant, temperature, light intensity, and relative humidity. Wilkinson (1956) reported that leaves from barley plants up to 2 weeks of age exported dalapon whereas movement from leaves of 3-week old plants could not be detected. Studies on dalapon transport in quackgrass showed that shaded leaves exported more dalapon than unshaded leaves and the investigators suggested that this was the result of reduced water flow in the xylem which in turn reduced the removal of dalapon from the phloem to the xylem (Sagar, 1960). However, subsequent research by McIntyre (1962) with quackgrass showed no influence of a 50% reduction in light on transpiration, but placing plants in the dark following treatment accompanied by a 90% reduction in transpiration resulted in an increase in the amount of dalapon translocated to the roots and tillers. In addition, he placed various parts of a treated shoot in the darkness and from these results concluded that the influence of darkness on the translocation of dalapon was within the treated leaf itself and not an influence of transpiration as suggested by Sagar (1960). It was suggested that in the light a natural metabolite may be formed which complexes with dalapon and inhibits its export from the treated leaf. Prasad and Blackman (1965a) reported that the uptake of dalapon was not influenced by light intensity in the range of 300 to 900 foot-candles in studies with *Lemna minor* and *Salvinia natans*; however increases in temperature between 20° and 30°C did increase uptake. Relative humidity has also been implicated in the rate of dalapon absorption and translocation by leaves of barley, bean, zebrina, coleus, and nasturtium (Prasad, et al., 1967). They found that the amount of dalapon absorbed and translocated was greater at high relative humidities, 88%, than at the lower relative humidities of 60 or 28%. Pretreatment of the plants at high relative humidity (95%) also increased dalapon absorption and translocation, compared to low relative humidity (28%), when the plants were placed in the same relative humidity (95%) following treatment. They concluded that ontogeny and the degree of cuticle hydration are involved. The increased absorption at high relative humidities is also related to the decreased rate of drying of the applied herbicide drop. They also reported greater absorption and translocation of dalapon in bean leaves at 43°C than at 26°C.

In summary, dalapon is readily translocated via both the symplastic and apoplastic systems and undoubtedly moves from one system to the other rather freely with the rate depending on the concentration gradient and the relative retention and absorption potential of the two systems. The high degree of molecular stability of dalapon in the plant (see molecular fate section of this chapter), along with its intrinsic translocation pattern strongly suggests that dalapon actually circulates in the plant for a considerable period of time.

Molecular Fate

Foy (1969) recently reviewed the degradation of TCA and dalapon in plants, animals, and soils. In general, they are quite stable in higher plants and animals but degraded rapidly in soil.

When ^{14}C-TCA was added to the culture solutions containing maize or pea plants, and the plants harvested, extracted, and chromatographed after 15 days, only TCA was found to be present (Blanchard, 1954). TCA was also found to be unchanged in flax, radish, barley, maize, and common dandelion (*Taraxacum officinale*); however trichloromethyl compounds were detected in tomato and tobacco plants treated with TCA (Mayer, 1957a).

In relatively short-term studies using either ^{36}Cl-dalapon or ^{14}C-dalapon no degradation products were present in sugarbeets or yellow foxtail (Andersen, et al., 1962), cotton (Foy, 1961a, Smith and Dyer, 1961), sorghum (Foy, 1961a), maize (Foy, 1962a. Blanchard, et al., 1960), or soybean (Blanchard, et al., 1960).

DALAPON INDUCED
ABNORMALITIES IN
2ND-GENERATION
WHEAT

(a)

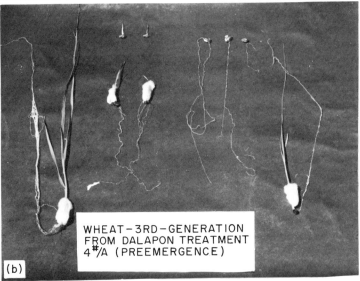

WHEAT – 3RD – GENERATION
FROM DALAPON TREATMENT
4#/A (PREEMERGENCE)

(b)

FIGURE 8-2.   Carry-over effect of characteristic dalapon symptoms in *a* (above) second generation, and *b* (below) third generation wheat following inhibition of first generation plants by pre-planting application of dalapon at 4 lb/A. Groups of heads in *a* are arranged left to right in order of occurrence on the same plant. In *b* the seeding at left is normal and others show inhibition or other anomalies. (Foy, 1961a.) Reprinted from P. C. Kearney and D. D. Kaufman, eds., *Degradation of Herbicides*, p. 214, by courtesy of Marcel Dekker, Inc.

In long term studies (9 to 10 weeks), Foy (1961b) determined that a small amount of non-extractable radioactivity was present in both cotton and sorghum. When mature cotton plants were treated with relatively high levels of ^{14}C-dalapon via severed petioles and the fruits dried, ground, extracted, and fractionated; 85 to 90% of the radioactivity was dalapon. The remainder was made up of an ether-soluble portion, the neutral and cationic fractions of the ethanol extract, and an insoluble plant residue. Additional fractionation or purification showed that small amounts of ^{14}C were associated with both the lipids and pigments (ether-soluble portion). A single non-dalapon compound was present in the neutral fraction; ^{14}C was present in cationic fractions indicating at least one more metabolite, and approximately 1% remained in the insoluble plant residue despite exhaustive extraction. Smith and Dyer (1961) also reported the presence of non-extractable radioactive residue in cotton following the application of radioactive dalapon but attributed this to occluded or trapped dalapon rather than evidence of degradation.

Additional evidence of the great stability of dalapon in higher plants has been provided by Foy (1961b) by a unique system in which the carry over of dalapon through three generations of wheat via the seed was demonstrated, Figure 8-2. However, Funderburk and Davis (1960) did not detect any carry-over of dalapon to the next generation in maize.

Although the evidence clearly showed that dalapon is extremely stable in higher plants, it also appears that a very slow degradation may occur. Because of the very slow rate of degradation the intermediates may not accumulate in

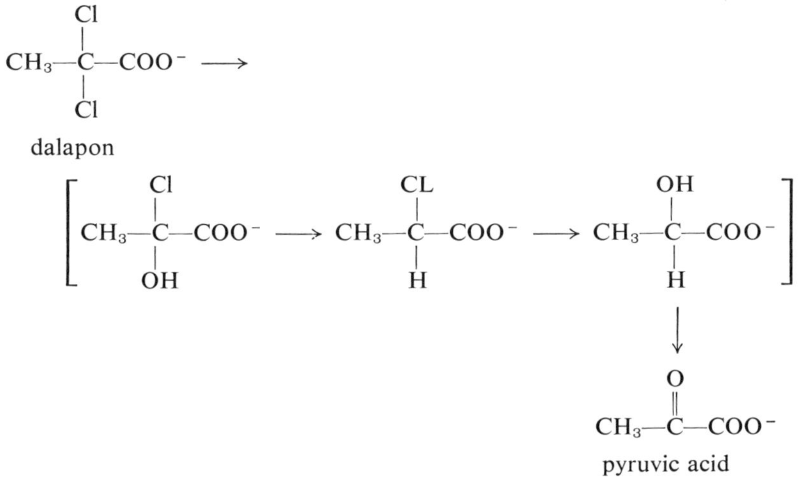

$$\underset{\text{dalapon}}{CH_3-\overset{\displaystyle Cl}{\underset{\displaystyle Cl}{C}}-COO^-} \longrightarrow$$

$$\left[CH_3-\overset{\displaystyle Cl}{\underset{\displaystyle OH}{C}}-COO^- \longrightarrow CH_3-\overset{\displaystyle CL}{\underset{\displaystyle H}{C}}-COO^- \longrightarrow CH_3-\overset{\displaystyle OH}{\underset{\displaystyle H}{C}}-COO^- \right]$$

$$\downarrow$$

$$CH_3-\overset{\displaystyle O}{\overset{\|}{C}}-COO^-$$

$$\text{pyruvic acid}$$

FIGURE 8-3.   Proposed pathway of metabolism of dalapon by higher plants.

sufficient quantity to be readily detected and therefore the metabolic pathway for the degradation of dalapon in higher plants has not been elucidated. However, it is probable that the pathway involves dechlorination, with concurrent hydroxylation followed by dehydroxylation for the removal of one of the chlorine atom and dechlorination with concurrent hydroxylation followed by dehydroxylation and concurrent oxidation for the removal of the other chlorine atom, yielding pyruvic acid, Figure 8-3. Pyruvic acid, a normal plant constituent, would be further metabolized via the usual pathways which would allow the labeling of many plant constituents following the application of radioactive dalapon.

Biochemical Responses

There are numerous reports in the literature of various biochemical events or metabolites which are influenced by TCA and dalapon. To date, these have not yielded a comprehensive understanding of how these herbicides kill higher plants at the molecular level. However, the authors will discuss selected possibilities for the mechanism of action of these compounds later.

Rebstock et al. (1953) showed that TCA applied through the soil increased the reducing-sugar content of wheat seedlings while non-reducing sugar decreased. Several reports suggest that dalapon has some influence on carbohydrate metabolism. McWhorter (1961) found that dalapon caused a general reduction of glucose with a corresponding increase in sucrose in johnsongrass (*Sorghum halpense*). Bourke et al. (1964) studied the effect of dalapon on glucose metabolism in peas and reported that it interfered with glycolysis. Jain et al. (1966) using barley shoot and root sections and Ross (1966) using bean leaf disks, having fed ^{14}C-1-glucose and ^{14}C-6-glucose, found no difference in the C_6/C_1 ratio or $^{14}CO_2$ release due to the dalapon treatment. They concluded that dalapon did not affect the total glucose utilization or shift the pathway of glucose metabolism via the pentose phosphate or Krebs cycle pathways. However, Jain et al. (1966) found a decrease in the 80% ethanol extract : 80% ethanol residue ratio whereas Ross (1966) reported them to be the same. Furthermore, Jain et al. (1966) reported that dalapon caused an increase in the labeling of sucrose, aspartic acid, glutamic acid, asparagine, and glutamine as well as a decrease in the labeling of alpha-ketoglutaric acid. Jain et al. (1966) concluded that it was not possible to demonstrate any specific site of blockage either at the initiation of the glycolytic-pathway or in the Krebs cycle, nor was it possible to propose any definite mechanism of action of dalapon. However, they did conclude: (a) the pentose phosphate pathway was not involved as a distinctly exclusive route of glucose utilization, (b) dalapon interfered with glucose utilization (ethanol extract:ethanol-insoluble residue ratio), and (c) partial inhibition may occur at initiation of the glycolytic pathway and within the Krebs cycle.

Lipid metabolism or deposition of wax in the cuticle is affected by dalapon. Dewey et al. (1956), Pfeiffer et al. (1957), and Juniper and Bradley (1968) reported that TCA and dalapon change the character of the surface wax of peas and maize making them more wettable by subsequent sprays. It is conceivable that modification of the cuticle is responsible for the increase in transpiration of TCA-treated bermudagrass reported by Corbadzijska (1962). Prasad and Blackman (1965a) observed that when *Salvinia natans* was treated with dalapon many leaves became submerged in the culture solution. Although they attributed this to a reduction in epidermal hairs, modification of the cuticle may also have been involved. Mashtakov et al. (1967a) found that TCA reduced both cuticle and lamina thickness in leaves of sensitive varieties of *Lupinus lutens*.

Nitrogen metabolism is also modified by the aliphatic acid herbicides. Mashtakov and Moshchuk (1967) reported that TCA increased the amount of asparagine and glutamine in a resistant variety of *Lupinus lutens*. They also noted a slight increase in protein and β-alanine and no change in the amount of free ammonia. However, in a sensitive variety, they reported a decrease in amides and β-alanine with an increase in free ammonia and protein. Similar results were also reported for dalapon (Mashtakov et al., 1967a). TCA increased the free ammonia content of the sensitive variety by 63% and dalapon by 30% after 48 hour herbicide treatment; corresponding figures for the resistant variety were 3 and 21%, respectively. They concluded that the action of these herbicides was an inhibition of the enzymes which are involved in the conversion of ammonia to amides, thereby allowing the accumulation of toxic levels of free ammonia. Andersen et al. (1962) observed that dalapon caused an increase in the degradation of protein to amino acids and an increase in amides. Further breakdown of the free amino acids with the liberation of ammonia was indicated. The amides appeared to act in ammonia detoxication by serving as storage sites for the released ammonia. In sugarbeets the amide was glutamine whereas in yellow foxtail, the amide was asparagine. After some period of time the amide and amino acids returned to normal levels in the resistant species sugarbeets, but this did not occur in the susceptible species, yellow foxtail. Jain et al. (1966) reported an increase in radioactive aspartic acid, glutamic acid, asparagine, and glutamine following ^{14}C-glucose feeding of dalapon treated plants. Mashtakov and Moshchuk (1967) reported that TCA inhibited proteinase activity especially in the resistant variety. However, Ashton et al. (1968) did not find any inhibition of proteinase activity in soybean or squash (*Cucurbita maxima*) seeds grown in 10^{-3}M dalapon.

TCA is the classical compound used by the biochemist for precipitating proteins in the test tube; however, concentrations of about 5% are usually used for this purpose and this is entirely out of the physiological range of herbicidal action. Dalapon and TCA were reported to be protein precipitants by Redemann and Hamaker (1957) and concentrations as low as about 200 ppm produced a visible precipitate from egg yolk and egg white protein. This is still a relatively high concentration. However, an aspect of the effect of the aliphatic acid herbicides which has not been investigated but warrants serious study is the effect of these compounds on conformational changes in proteins, including enzymes. Foy (1969) notes that halogenated acetates and propionates are theoretically able to alkylate the sulfhydryl or amino groups in enzymes. If these actually occur they could bring about conformational changes. Hydrogen bonding of dalapon to *N*-methylacetamide, a model protein, has been reported (Kemp et al., 1969). We have seen from the previous discussion that apparently many enzyme reactions are influenced by these herbicides, but none in a specific way that would explain its mechanism of action. Compounds which actually bring about precipitation of proteins at 200 ppm must produce some conformational changes at much lower concentration, perhaps in the physiological range. If such conformational changes do indeed occur, they could explain many of the results which have been reported.

Ingle and Rodgers (1961) concluded that dalapon does not interfere with the production of metabolic energy but rather with its utilization. Results of work by Jain et al. (1966), discussed earlier, can be interpreted in a similar manner. Ross (1966) reported that dalapon caused an increase in ATP and a reduction in an unidentified sugar phosphate. He concludes from this and from the fact that several other energy producing steps are not affected by dalapon, that if dalapon does have a specific site of action, this site is probably phosphorous metabolism and the build-up of ATP occurs because it is not utilized for phosphorylation of some compound needed in synthetic processes.

Several chlorinated aliphatic acids interfere with the biosynthesis of pantothenic acid in certain microorganisms (Hilton et al., 1958; van Oorschot and Hilton, 1963). Pantoic acid and β-alanine are precursors of pantothenic acid. Pantothenic acid is a precursor of the universally required coenzyme A. Since coenzyme A is required for several essential biochemical reactions in higher plants as well as microorganisms, this would appear to be a likely site of action of TCA and dalapon. Hilton et al. (1959) reported that the isolated pantothenate synthesizing enzyme from *Escherichia coli* was inhibited by dalapon competing with pantoate for a site on the enzymes and both pantoate and pantothenate caused significant reduction in the abnormal tiller development on barley plants which resulted from application of sublethal concentrations of dalapon. Pantothenate was more effective than pantoate, but even pantothenate was not effective in overcoming dalapon inhibition of total growth.

Ingle and Rogers (1961) did not obtain any significant reversal of dalapon root growth inhibition in cucumber or maize by pantothenate application.

Pretreatment of seeds with pantothenate was somewhat effective but a foliar spray of pantothenate was ineffective on certain species (Leasure, 1964). Pantothenate could only counteract the growth inhibition of dalapon on *Lemna minor* and *Salvinia natans* to a small degree (Prasad and Blackman, 1965b). Similar results were reported for wheat (Aberg and Johansson, 1966). These workers concluded that although growth responses to dalapon can usually be counteracted to some extent by pantothenate applications it is not plausible that this plays any essential part in its herbicidal action in higher plants. Mashtakov et al. (1967b) reported that TCA reduced the pantothenic acid content of forage lupins; however Andersen et al. (1962) showed that dalapon increased pantothenic acid in sugarbeets and yellow foxtail but it later returned to normal in the resistant species, sugarbeets. In wheat, dalapon caused an increase in fumaric acid and a decrease in aconitic acid and citric acid which was considered to be consistant with a reduced level of coenzyme A (Oyolu and Huffaker, 1964). Jain et al. (1966) and Ross (1966) concluded from their studies on the effect of dalapon on glycolysis, pentose phosphate pathway and the Krebs cycle that dalapon does not inhibit coenzyme A activity. Although some of these results have been interpreted as supporting the pantothenate hypothesis, others are in opposition to it; the current generally accepted position of most herbicide physiologists is that this is not the primary site of action of the chlorinated aliphatic acid herbicides in higher plants.

A number of other processes have been reported to be influenced by TCA or dalapon; however they are not considered to be particularly significant in relation to the localization of the primary site of action of these herbicides. TCA appears to increase respiration to a degree (Rai and Hammer, 1956; Corbadzijska, 1962), however dalapon does not affect this process (Ingle and Rogers, 1961; Ross, 1966; Jain et al., 1966). Photosynthesis is not inhibited by dalapon (Ross, 1966). Parshakova and Mashtakov (1967) reported that TCA caused an increase in bound IAA and a decrease in free IAA which he attributed to enzyme inactivation. Dalapon and TCA increased the auxin content of seed tissue (Mashtakov et al., 1967a). Dalapon appears to have weak anti-auxin properties (Wilkinson, 1962; Aberg and Johansson, 1966). Flanagan and Langille (1963) observed an increase in the concentration of phenol in rhizomes of quackgrass following dalapon treatment.

Mode of Action

The chlorinated aliphatic acid herbicides cause formative effects, growth inhibition, leaf chlorosis, leaf necrosis, and eventually death. They interfere with the meristematic activity of root tips and probably the apical meristems. Reduced wax formation on leaf surfaces and alteration of cell membranes have been reported.

These compounds are readily absorbed by both roots and shoots. Dalapon is readily translocated via both the symplastic and apoplastic systems and moves from one system to the other, however TCA appears to be almost exclusively transported via the apoplastic system.

These compounds are very resistant to degradation by higher plants. Perhaps TCA is essentially non-degradable by higher plants, however one report indicates that trichloromethyl compounds are formed. Dalapon apparently undergoes a very slow degradation yielding pyruvic acid through several steps of dechlorination, hydroxylation, and dehydrogenation.

Although dalapon has been reported to alter carbohydrate, lipid and nitrogen metabolism, most of these responses are probably of a secondary nature and reflect a more basic primary site of action common to all. It is suggested that this primary site of action is associated with modification of protein structure, including enzymes. At the concentrations of dalapon which occur *in vivo*, protein precipitation *per se* probably does not take place. However, it is likely that more subtle changes such as conformational changes do occur. Conformational changes of proteins could alter enzymes and change membrane permeability which could result in the many modifications of metabolic pathways which have been observed.

A rather specific effect of dalapon appears to be an increase in ammonia. This is most apparent in susceptible species, which do not seem to be able to form amides at a sufficient rate to prevent the accumulation of toxic levels of ammonia. Of course, this could be caused by conformational changes in asparagine synthetase and/or glutamine synthetase.

Unfortunately very little information is available on the mode of action of TCA. At the moment we can only assume the TCA probably acts in a manner similar to dalapon.

REFERENCES

Aberg, B. and I. Johansson. 1966. On the mechanism of action of dalapon. *LantbrHögsk. Annlr* **32**:245–254.

Andersen, R. N., A. J. Linck, and R. Behrens. 1962. Absorption, translocation and fate of dalapon in sugar beets and yellow foxtail. *Weeds* **10**:1–3.

Andersen, R. N., R. Behrens, and A. J. Linck. 1962. Effects of dalapon on some chemical constituents in sugar beets and yellow foxtail. *Weeds* **10**:4–9.

Ashton, F. M., D. Penner, and S. Hoffman. 1968. Effect of several herbicides on proteolytic activity of squash seedlings. *Weed Sci.* **16**:169–171.

Blanchard, F. A. 1954. Uptake, distribution, and metabolism of carbon-14 labeled TCA in corn and pea plants. *Weeds* **3**:274–278.

Blanchard, F. A., U. U. Muelder, and G. N. Smith. 1960. Herbicide uptake and distribution: Synthesis of carbon-14-labeled dalapon and trial applications to soybean and corn plants. *J. Agr and Food chem.* **8**:124–128.

Bourke, J. B., J. S. Butts, and A. C. Fang. 1964. Effect of various herbicides on glucose metabolism in root tissue of garden peas. II. Plant growth regulators and other herbicides. *Weeds* **12**:272–279.

Corbadzijska, B. 1962. The effect of the herbicide sodium trichloroacetate on the respiration, transpiration and vitality of the protoplasm in Bermuda grass (*Cynodon datylon*) and grape vine. Nauc. Trud. viss. selskostop. Inst. "V. Kolarov", *Agron. Fak.* **11**:87–98.

Dewey, O. R., P. Gregory, and R. K. Pfeiffer. 1956. Factors affecting the susceptibility of peas to selective dinitroherbicides. *Proc. Brit. Weed Control Conf.* 3rd Meeting. pp. 313–326.

Dewey, O. R., G. S. Hartley, and J. W. G. Mackaughlan. 1962. External leaf waxes and their modification by heat-treatment of plants with trichloroacetate. *Proc. Roy. Soc.* (B) **155**:432–450.

Flanagan, T. R. and A. R. Langille. 1963. Phenol in quackgrass associated with dalapon. *Science* **140**:179–180.

Foy, C. L. 1961a. Absorption, distribution, and metabolism of 2,2-dichloropropionic acid in relation to phytotoxicity. I. Penetration and translocation of [36]Cl- and ^{14}C-labeled dalapon. *Plant Physiol.* **36**:688–697.

Foy, C. L. 1961b. Absorption, distribution, and metabolism, of 2,2,-dichloropropionic acid in relation to phytotoxicity. II. Distribution and metabolic fate of dalapon in plants. *Plant Physiol.* **36**:698–709.

Foy, C. L. 1962a. Penetration and initial translocation of 2,2-dichloropropionic acid (dalapon) in individual leaves of Zea mays L. *Weeds* **10**:35–39.

Foy, C. L. 1962b. Absorption and translocation of dalapon-2-^{14}C and [36]Cl in *Tradescantia fluminensis. Weeds* **10**:97–100.

Foy, C. L. 1969. The chlorinated aliphatic acids. pp. 207–253. In P. C. Kearney and D. D. Kaufman, *Degradation of Herbicides.* Marcel Dekker, Inc., New York.

Funderburk, H. H. and D. E. Davis. 1960. Factors affecting the response of Zea mays and Sorgum halepense to sodium 2,2-dichloropropionate. *Weeds* **8**:6–11.

Gooch, F. S. and L. W. Erbe. 1967. The effects of monochloroacetic acid, dichloroacetic acid, and trichloroacetic acid on the stem tissues of Elodea. *Proc. 20th Southern Weed Conf.* p. 287.

Hilton, J. L., J. S. Ard, L. L. Jansen, and W. A. Gentner. 1959. The pantothenate synthesizing enzyme, a metabolic site in the herbicidal action of chlorinated aliphatic acids. *Weeds* **7**:381–396.

Hilton, J. L., L. L. Jansen, and W. A. Gentner. 1958. Beta-alanine protection of yeast growth against the inhibitory action of several chlorinated aliphatic acid herbicides. *Plant Physiol.* **33**:43–45.

Ingle, M. and B. J. Rogers. 1961. Some physiological effects of 2,2-dichloropropionic acid. *Weeds* **9**:264–272.

Jain, M. L., E. B. Kurtz, and K. C. Hamilton. 1966. Effect of dalapon on glucose utilization in the shoot and root of barley. *Weeds* 14:259–262.

Juniper, B. E. and D. E. Bradley. 1958. The carbon replica technique in the study of the ultrastructure of leaf surfaces. *J. Ultrastr. Res.* 2:16–27.

Kemp, T. R., L. P. Stoltz, N. W. Herron, and W. T. Smith. 1969. Hydrogen bonding of dalapon. *Weed Sci.* 17:444–446.

Leasure, J. K. 1964. The Halogenated Aliphatic Acids. *J. Agr. and Food Chem.* 12:40–43.

Mashtakov, S. M., V. P. Deeva, and A. P. Volynets. 1967a. Histochemical changes in germinating seeds under the influence of herbicides, pp. 47–59. In Mashtakov, S. M. *Physiological Effects of Herbicides on Varieties of Crop Plants.* Nauka: Tekhnika, Minsk.

Mashtakov, S. M., V. P. Deeva, and A. P. Volynets. 1967b. The physiological effect of grass-type herbicides on varieties of leguminous plants, pp. 124–146. In Mashtakov, S. M., *Physiological Effects of Herbicides on Varieties of Crop Plants.* Nauka: Tekhnika, Minsk.

Mashtakov, S. M. and P. A. Moshchuk. 1967. The effect of sodium trichloroacetate on the content of nitrogenous substances in varieties of lupin resistant and sensitive to herbicides. *Agrokhimiya* 9:80–89.

Mayer, F. 1957a. Reaction of trichloroacetic acid and other halogen acetates with sulfhydryl and amino groups and also vegetable matter. *Biochem. Z.* 328:433–442.

Mayer, F. 1957b. Effect of trichloroacetate on higher plants. *Z. Naturforsch.* 12B:336–346.

McIntyre, G. I. 1962. Studies on the translocation in Agropyron repens of ^{14}C-labeled 2,2-dichloropropionic acid. *Weed Res.* 2:165–176.

McWhorter, C. G. 1961. Carbohydrate metabolism of Johnsongrass as influenced by seasonal growth and herbicide treatment. *Weeds* 9:563–568.

McWhorter, C. G. 1971. Control of Johnsongrass ecotypes. *Weed Sci.* 19:229–233.

Meyer, R. E. and K. P. Buchholtz. 1963. Effect of chemicals on buds of quackgrass rhizomes. *Weeds* 11:4–7.

Oyolu, C. and R. C. Huffaker. 1964. Effects of 2,2-dichloropropionic acid (dalapon) on organic acid content of wheat (*Triticum vulgare*). *Crop Sci.* 4:95–96.

Parshakova, Z. P. and S. M. Mashtakov. 1967. The effect of TCA-sodium on the heteroauxin and tryptophan contents of leguminous plants. *Dokl. Akad. Nauk Belorussk. SSR,* 11:271–273.

Pfeiffer, R. K., O. R. Dewey, and R. T. Brunskill. 1957. *Proc. Intern. Congr. Crop Protection,* 4th Meeting. (Hamburg, Germany, Sept. 8–15.)

Pickering, E. R. 1965. Foliar penetration pathways of 2,4-D, monuron and dalapon as revealed by historadioautography. Ph.D. Dissertation, University of Calif. Davis. 186 pp.

Prasad, R. and G. E. Blackman. 1964. Studies in the physiological action of 2,2-dichloropropionic acid. I. Mechanisms controlling the inhibition of root elongation. *J. Exp. Bot.* **15**:48–66.

Prasad, R. and G. E. Blackman. 1965a. Studies on the physiological action of 2,2-dichloropropionic acid. II. The effects of light and temperature on the factors responsible for the inhibitions of growth. *J. Exp. Bot.* **16**:86–106.

Prasad, R. and G. E. Blackman. 1965b. Studies in the physiological action of 2,2-dichloropropionic acid. III. Factors affecting the level of accumulation and mode of action. *J. Exp. Bot.* **16**:545–568.

Prasad, R., C. L. Foy, and A. S. Crafts. 1967. Effects of relative humidity on absorption and translocation of foliarly applied dalapon. *Weeds* **15**:149–156.

Rai, G. S. and C. L. Hamner. 1956. Respiratory activity of certain plant species as affected by sodium trichloroacetate (TCA). *Mich. Agr. Expt. Sta. Q. B.* **38**:555–558.

Rebstock, T. L., C. L. Hamner, R. W. Lueeke, and H. M. Sell. 1953. The effect of sodium trichloroacetate upon the metabolism of wheat seedlings. (*Triticum vulgare* L.). *Plant Physiol.* **28**:437–442

Redemann, C. T. and J. Hamaker. 1957. Dalapon (2,2-dichloropropionic acid) as a protein precipitant. *Weeds* **3**:387–388.

Ross, M. A. 1966. Utilization of ^{14}C metabolites, incorporation of ^{32}P and changes in respiration occurring in leaves from dalapon treated beans. *WSSA Abstr.* p. 50.

Sagar, G. R. 1960. An important factor affecting the movement of 2,2-dichloropropionic acid (dalapon) in experimental systems of Agropyron repens. *Proc. 5th Brit. Weed Control Conf.*, pp. 271–278.

Santelmann, P. W. and C. J. Willard. 1955. The absorption and translocation of dalapon. *Proc. 9th Northeast Weed Control Conf.* pp. 21–29.

Smith, L. W. and P. J. Davies. 1965. The translocation and distribution of three labelled herbicides in *Paspalum distichum. Weed Res.* **5**:343–347.

Smith, G. N. and D. L. Dyer. 1961. Fate of 2,2-dichloropropionic acid (dalapon) in the cotton plant. *J. Agr. and Food Chem.* **9**:155–160.

Van Oorschot, J. L. P. and J. L. Hilton. 1963. Effects of chloro substitutions on aliphatic acid inhibitors of pantothenic metabolism in Escherichia coli. *Arch. Biochem. Biophys.* **100**:289–294.

Wilkinson, R. E. 1956. The physiological activity of 2,2-dichloropropionic acid. Ph.D. Dissertation, University of Calif. Davis. p. 148.

Wilkinson, R. E. 1962. Growth inhibitions by 2,2-dichloropropionic acid. *Weeds* **10**:275–281.

CHAPTER 9

Amides

_____

$$R_1-\overset{\overset{\displaystyle O}{\displaystyle \|}}{C}-N\overset{\displaystyle R_2}{\underset{\displaystyle R_3}{\diagdown}}$$

The amide herbicides comprise a diverse group of chemicals (Table 9-1). They range from such structures as N-substituted α-halo acetamides and α,α-diacyl acetamides through substituted aromatic anilides of aliphatic acids and cyclopropyl carboxylic acids to N-naphthalamic acids. Likewise, their biological properties and uses vary widely. They are almost exclusively used as selective herbicides in a variety of crops. Most of them are used as preemergence or preplant-soil-incorporated herbicides, i.e., alachlor, CDAA, CDEA, diphenamid, naptalam, delachlor, pronamide, propachlor, and R-7465. However cypromid, dicryl, solan, and propanil are applied to the foliage of the weeds to be controlled.

Growth and Plant Structure

CDAA inhibits root elongation, the inhibition increases with increasing concentrations and varies with different species. Canvin and Friesen (1959) investigated the effect of CDAA on cell division in root tips of barley and peas. The number of dividing cells in the meristematic region was determined at 0, 1, 10, and 1000 ppm of CDAA and in barley was 33.7, 30.4, 8.2, and 1.5, respectively; and in peas 76.3, 73.2, 68.7, and 62.4, respectively. The effect of CDAA on cell division of pea roots was much less pronounced than with barley. This is in agreement with the general concept that CDAA is a more effective herbicide for grasses than for many broadleaf species.

Propachlor, like CDAA, inhibits root elongation. Duke (1968) reported that propachlor inhibited the root growth of cucumbers and that the degree of inhibition was closely correlated with the inhibition of protein synthesis in root tips. He also noted that propachlor inhibited auxin (2,4-D) induced cell expansion in cucumber hypocotyl sections and postulated that this was due to the prevention of auxin induced enzyme formation.

Alachlor inhibits growth of shoots of both yellow and purple nutsedge; the yellow species proved to be the more susceptible (Keeley et al., 1972). Tests on cotton seedlings proved that alachlor inhibited shoot and root growth and lateral root development (Keeley et al., 1972).

Diphenamid appears to permit early germination of seeds but kills the seedling plant prior to emergence from the soil. At sub-lethal concentrations severe inhibition of root development occurs in many species. Preplant-soil-incorporation treatments of diphenamid at relatively high rates permit apparently normal germination of tomatoes but result in chlorosis of leaf margins followed by necrosis at the margins of the cotyledons and leaves of seedlings. Schubert (1966) reported that diphenamid reduced the fresh weight and number of primary roots of runner plants of strawberries, as well as inhibiting the formation of well branched secondary roots.

Deli and Warren (1971) used 57 weed species in a selectivity study of diphenamide; the 10 most sensitive were all grasses. In a moderately sensitive group of 8 species two were gramineae. Eight tolerant species were all broadleaf weeds. In general roots were more sensitive than shoots. The authors considered this to be evidence that the primary site of action is in the roots. Lynch and Sweet (1971) found that high relative humidity and low light intensity increased the activity of diphenamid against 4 crop species following foliar application.

Diphenamid affects germination and tube length of stimulated *Orobanche ramosa* seeds when applied in the germination fluid at concentrations of 100 ppm and above (Saghir and Abu-Shakra, 1971). Both tube length and germination were progressively reduced as concentration of diphenamid was increased.

Propanil, in contrast to these soil applied herbicides, is applied to the foliage. In susceptible species it causes chlorosis followed by necrosis. When propanil remains as discrete small droplets on leaves of susceptible species, as may happen in drift of the chemical, the symptoms may be observed as a speckled pattern, i.e., *Prunus* species. Hofstra and Switzer (1968) reported that propanil reduced the growth of tomatoes, inhibited radical elongation in tomatoes and retarded the auxin induced growth of *Avena* coleoptiles. They also observed that propanil destroyed the permeability of red beet root membranes and chloroplast membranes. Propanil concentrations of 10 to 1000 ppm stopped protoplasmic streaming in cells of *Hydrilla verticillata* leaves mounted in water; at 100 to 1000 ppm, the chloroplasts turned yellowish green and plasmolysis and granulation seemed to occur in some cells (Sierra and Vega, 1967). However, similar differences could not be reproduced in paradermal

TABLE 9-1. Common Name, Chemical Name and Chemical Structure of the Amide Herbicides

Common name	Chemical name	R_1	R_2	R_3
alachlor	2-chloro-2',6'-diethyl-N-(methoxymethyl)acetanilide	$Cl-CH_2-$	3,5-diethylphenyl (H_3CH_2C)	CH_3-O-CH_2-
CDAA	N,N-diallyl-2-chloroacetamide	$Cl-CH_2-$	$CH_2=CH-CH_2-$	$CH_2=CH-CH_2-$
CDEA	2-chloro-N,N-diethylacetamide	$Cl-CH_2-$	CH_3-CH_2-	CH_3-CH_2-
cypromid	3',4'-dichlorocyclopropane-carboxanilide	cyclopropyl	3,4-dichlorophenyl (Cl, Cl)	$H-$
delachlor	2-chloro-N-(isobutoxymethyl)-2',6'-acetoxylidide	$Cl-CH_2-$	2,6-dimethylphenyl (CH_3, CH_3)	$(CH_3)_2CH-CH_2-O-CH_2-$
dicryl	3',4'-dichloro-2-methyl-acrylanilide	$CH_2=C(CH_3)-$	3,4-dichlorophenyl (Cl, Cl)	$H-$
diphenamid	N,N-dimethyl-2,2-diphenyl-acetamide	$H-C(C_6H_5)_2-$	CH_3-	CH_3-

naptalam	N-1-naphthylphalamic acid	(2-COOH-phenyl); (1-naphthyl)	H—
pronamide	N-(1,1-dimethylpropynyl)-3,5-dichlorobenzamide	(3,5-dichlorophenyl); $HC{\equiv}C{-}C(CH_3)(CH_3){-}$	H—
propachlor	2-chloro-N-isoproylacetanilide	$Cl{-}CH_2{-}$; (phenyl)	iso—C_3H_7—
propanil	3′,4′-dichloropropionanilide	$C_2H_5{-}$; (3,4-dichlorophenyl)	H—
solan	3′-chloro-2-methyl-p-valerotoluidide	$CH_3{-}(CH_2)_2{-}CH(CH_3){-}$; (3-chloro-4-methylphenyl)	H—
R-7465	2-(α-naphthoxy)-N,N-diethylpropionamide	$CH_3{-}C(H){-}O{-}$(1-naphthyl); $C_2H_5{-}$	C_2H_5—

leaf sections of rice or barnyardgrass (*Echinochloa crusgalli*). They also reported that propanil had little adverse effect on the growth of rice but caused a reduction in dry weight of barnyardgrass which became more severe with increasing concentrations.

Naptalam has the unique property of acting as an antigeotropic agent although it is not clear that this is associated with its herbicidal propensities. Hoffman and Smith (1949) reported that at 0.1 ppm of naptalam leaf-rolling in tomatoes was induced; at 0.31 ppm, epinasty was observed; and at 20 ppm, stem swelling occurred. Mentzer and Netien (1950) showed that naptalam at 10^{-3} to 10^{-6}M caused apparent negative geotropic response of peas, lentils, crucifers, tomatoes, sunflowers, and cucumber seedlings. Treated tap roots were thicker and shorter than the controls. In a study of the effect of naptalam on germination at early stages of growth, Netien and Conillot (1951) observed that it retarded the germination of some species and inhibited the growth in most species tested. Grigsby et al. (1954) found that naptalam caused a 70% reduction in straight growth of peas and a complete loss of geotropic sensitivity. A disproportional inhibition of growth and curvature was found in naptalam treated roots and shoots of maize and pea, but no disproportionality was observed in oat coleoptiles (Tsou et al., 1956). Teas and Sheehan (1957) found that naptalam inhibited the geotropic response of flower stems of snapdragon which were placed in a horizontal position and that IAA and gibberellic acid partially reversed the naptalam inhibition. Mentzer and Netien (1950) suggested that naptalam also inhibited phototropic responses. Keith and Baker (1966) observed that naptalam was a strong inhibitor of basipetal polar auxin (IAA) transport in excised stem sections of bean. The relationship of this to the tropic responses of plants to naptalam is not clear but they may be related. Although the antitropic responses induced by naptalam are interesting and may be associated with the same basic process which inhibits seed germination, the latter is more likely to be the herbicidal effect.

Peterson and Smith (1971), studying the effects of pronamide on anatomy of quackgrass, observed cell enlargement, necrosis and increase in nuclear volume in treated rhizome apices. Abnormal metaxylem vessels (little lignification) and necrotic phloem characterized the vascular tissues. Roots showed precocious differentiation and maturation with eventual vacuolation of the apical meristem. Meristem cells of the inhibited roots were much enlarged and contained several nucleoli.

Solan and dicryl are foliar applied herbicides. The leaf symptoms require several days to become apparent, although shoot growth may cease soon after application. A few species have some resistance to the phytotoxic action of these herbicides. Bingham and Porter (1961a) reported dicryl reduced the rate of enlargement of both cotyledons and true leaves of cotton and also the elongation of stem internodes and roots. However these are probably secondary effects caused by a limited food supply.

Absorption and Translocation

CDAA (Jaworski, 1964) and propachlor (Jaworski and Porter, 1965) are rapidly taken up by maize and soybean roots and translocated to the upper parts of the plant. They are also readily absorbed by germinating seeds of several species but selectivity was not correlated with the amount of herbicide taken up (Smith et al., 1966). Nishimoto et al. (1967) and Knake and Wax (1968) demonstrated that propachlor is more readily taken up from soil by the shoot than by the roots of plants.

Diphenamid absorption and translocation was studied by Lemin (1966) and Golab et al. (1966). Lemin (1966) showed that seedling tomatoes removed about 60% of the applied diphenamid from the culture solution in 7 days. Golab et al. (1966) found that the radioactivity in strawberries grown for 40 days in soil treated with ^{14}C-diphenamid had the following percentage distribution; leaves and petioles 61, calyxes 18, crowns 13, roots 7, and fruits 1. This would indicate that diphenamid was absorbed relatively rapidly by roots and translocated upward with most of the herbicide accumulating in the leaves.

Devlin and Yaklich (1971) found that the presence of gibberellic acid in the culture medium around the roots of bean plants greatly enhanced the uptake and movement of naptalam. GA at 10^{-4} to 10^{-3}M increased the naptalam concentration of bean plants by almost 60%. This would seem to substantiate Devlin's (1967) observation that either IAA or GA increases the sensitivity of redtopgrass to simazine and the finding of Devlin et al. (1969) that these growth regulators enhance the herbicidal effect of silvex on poison ivy.

Propanil absorption rates by rice and barnyardgrass leaves were reported to be different by Takematsu and Yanagishima (1963) and Adachi et al. (1966). Their data showed that barnyardgrass absorbed propanil at a more rapid rate than rice and they suggested that this was the basis of selectivity between these two species. However, Yih et al. (1968) reported that during a 100 hour period propanil was absorbed at essentially the same rate by rice and barnyardgrass leaves. These differences are probably due to the fact that Takematsu and Yanagishima (1963) and Adachi et al. (1966) used bioassays to determine the amount of propanil absorbed while Yih et al. (1968) used ^{14}C-propanil in their studies. The use of the bioassay technique fails to take into account the degradation of the herbicide. Since it has been shown that propanil is degraded more rapidly in rice than barnyardgrass (Yih et al., 1968), one would expect to find more propanil in barnyardgrass than rice by a bioassay even if the absorption rate was essentially the same. The translocation of propanil from a foliar application was very limited (Yih et al., 1968).

Dicryl absorption and translocation has not been studied in detail. However, to exert its relatively rapid phytotoxic symptoms following foliar applications it must be absorbed quite rapidly. Indirect evidence suggests that it is not readily translocated. Porter et al. (1960) proposed that dicryl could be used as a directed spray under the cotyledons of young cotton without injury to the plant. Such an application would permit some dicryl to come in contact with the stems; apparently it is not translocated to the foliage or injury would result. Funderburk and Porter (1961) applied dicryl to roots of maize seedlings via culture solution and foliar symptoms did not develop, indicating that dicryl is not readily absorbed by roots and translocated to the leaves.

Solan absorption by leaves of tomato and egg plant was shown to occur within 10 minutes and it continued for at least 72 hours, although the most rapid rate of uptake was within the first 24 hours (Colby and Warren, 1965). The rate of foliar absorption was about the same for the two species even though tomato is resistant and egg plant susceptible to solan injury. Therefore, differential absorption was not the basis of selectivity between these two species to solan. Differential wetting does not appear to be the selective factor either, since the addition of a wetting agent to the spray solution did not change the relative margin of tolerance between the two species (Colby and Warren, 1962). However, the wetting agent did enhance the toxicity of solan to both species.

Molecular Fate

CDAA degradation pathway in higher plants has been reviewed recently by Jaworski (1969) and Casida and Lykken (1969). Jaworski (1964) studied the metabolism of ^{14}C-CDAA in which either the carbonyl carbon or the number 2 carbon of the allyl moiety was labeled. The roots of maize seedlings were treated via soil with the labeled CDAA and harvested 4, 7, 13, and 17 days after treatment. Extracts of the plants were subjected to paper chromatography. CDAA was rapidly degraded to glycolic acid and probably diallyamine (Figure 9-1). Glycolic acid was formed from the chloroacetyl moiety and was thought to be in equilibrium with glyoxylic acid (Jaworski, 1969). The degradation appears to start either with the hydrolysis of the amide or with dechlorination to form the hydroxy derivative of CDAA and continuing with dechlorination or hydrolysis, respectively. Since neither hydroxylated CDAA nor chloroacetic acid were detected, this detail of the pathway remains uncertain. Although free ^{14}C-diallylamine or ^{14}C-monoallylamine was not detected in plants treated with CDAA labeled in the allyl moiety, ^{14}CO$_2$ was evolved indicating that degradation of the allyl moiety did occur. Other possible degradation products such as acrolein, allyl alcohol, and acrylic acid were sought but not found. Glycolic acid and glyoxylic acid of the carbonyl moiety, and the products of the allyl moiety were further metabolized via normal metabolic pathways resulting in the incorporation of ^{14}C-atoms into many metabolites.

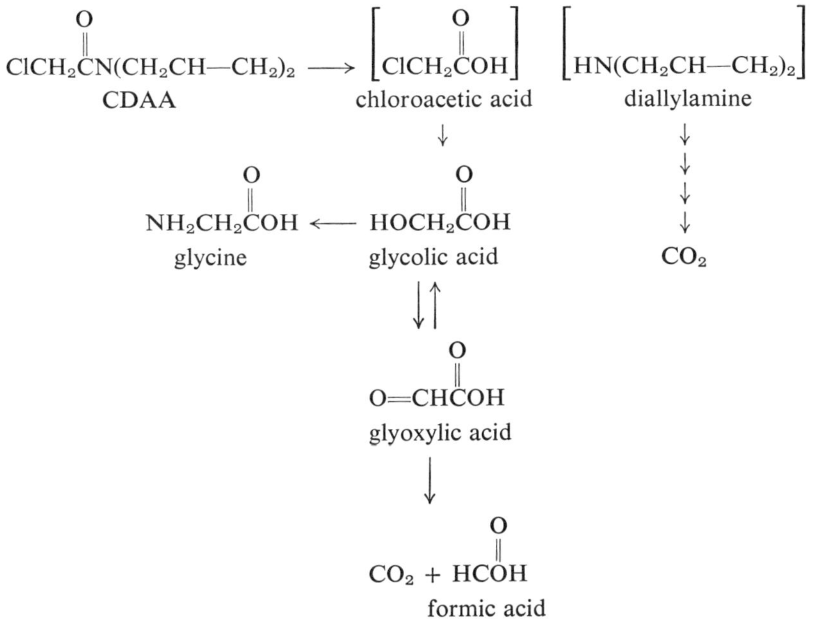

FIGURE 9-1.   Proposed degradation pathway of CDAA in higher plants (Modified from Jaworski, 1964).

Propachlor degradation in maize and soybeans has been reported (Jaworski and Porter, 1965; Porter and Jaworski, 1965; Jaworski, 1969). Uniformly ring labeled (^3H) propachlor was applied to the roots via soil. Plants were harvested at various periods of time after herbicide treatment and extracted with an acetone : water mixture (80 : 20). The extract was subjected to paper and thin-layer chromatography as well as ion-exchange fractionation. Degradation of propachlor in these two species was very rapid since no propachlor was detected even at the earliest harvest, 5 days after treatment. The only radioactive product detected was a water-soluble acidic metabolite. Although the structure of this metabolite has not been determined, basic hydrolysis and vapor phase chromatography indicated that it contains N-isopropylaniline. Acidic hydrolysis and vapor-phase chromatography yielded the 2-hydroxy-N-isopropylacetanilide suggesting that the chloroacetanilide was conjugated through its active chlorine to some natural product to form a glycosidic linkage. Therefore, the metabolite contains essentially the entire structure of the original herbicide molecule with the exception of the chloro group which appears to have been displaced, probably by some nucleophilic endogenous substrate. Apparently this metabolite is quite stable in the plant once it is formed and presumably relatively non-phytotoxic.

^{14}C-diphenamid degradation was investigated in tomatoes (Lemin, 1966) and in strawberries (Golab et al., 1966). In both studies the carbonyl carbon atom was labeled and the roots treated via the soil or nutrient solutions. In tomato seedlings degradation products were found as early as 12 hours; after 7 days 59% of the radioactivity was in diphenamid, 36% in diphenylacetamide, and a trace in diphenylacetic acid. After 21 days the major radioactive compound was N-methyl-2,2-diphenylacetamide, a small amount of diphenylacetamide was also present. Lemin (1966) proposed that the degradation mechanism involved direct hydroxylation of the N-alkyl group with the production of formaldehyde. Golab et al. (1966) treated strawberry plants and examined berries harvested 27 to 40 days after treatment, as well as leaves at the final harvest, for degradation products. The major radioactive metabolite found was N-methyl-2,2-diphenylacetamide. Diphenylacetamide, diphenylacetic acid, and p- and α-hydroxy-2,2-diphenylacetic acids were tentatively identified as minor breakdown products of diphenamid. The degradation pattern of diphenamid, as derived from these two research reports, is presented in Figure 9-2.

Propanil degradation in higher plants has been studied more thoroughly than any other amide herbicide. This work has been recently reviewed by Casida and Lykken (1969) and Matsunaka (1969). Several investigators have reported that rice plants can hydrolyze propanil to 3,4-dichloroaniline (McRae et al., 1964; Adachi et al., 1966; Ishizuka and Mitsui, 1966; Still and Kuzirian, 1967, and Yih et al., 1968). Although certain of these studies suggest that propionic acid was a degradation product, Yih et al. (1968) provide evidence that it was not. Furthermore these investigators showed that 3,4-dichloroacetanilide was an intermediate between propanil and 3,4-dichloroaniline. However, 3,4-dichloroacetanilide appears to be a transient intermediate in rice and is not isolatable under normal conditions. The formation of 3,4-dichloroaniline was reported to be catalyzed by a macromolecule (Still and Kuzirian, 1967) which was partially purified and characterized by Frear and Still (1968). They refer to it as an aryl acylamidase (aryl-acylamine amidohydrolase, EC 3.5.1a). The propionic moiety is rapidly degraded to CO_2 via β-oxidation (Still, 1968a). The 3,4-dichloroaniline moiety is converted to three complexes one of which is the glucosylamine (Still, 1968b). Yih et al.

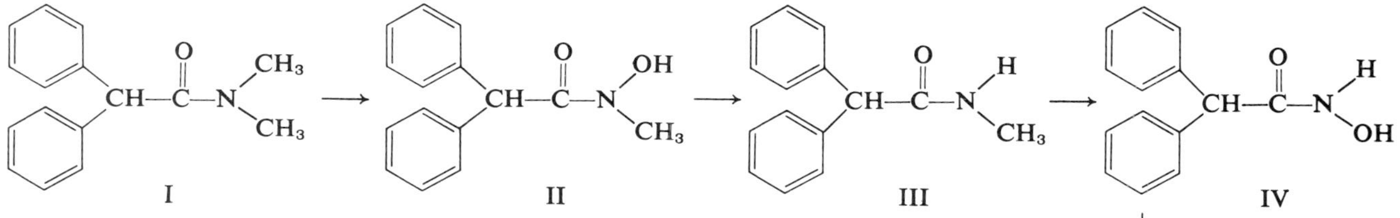

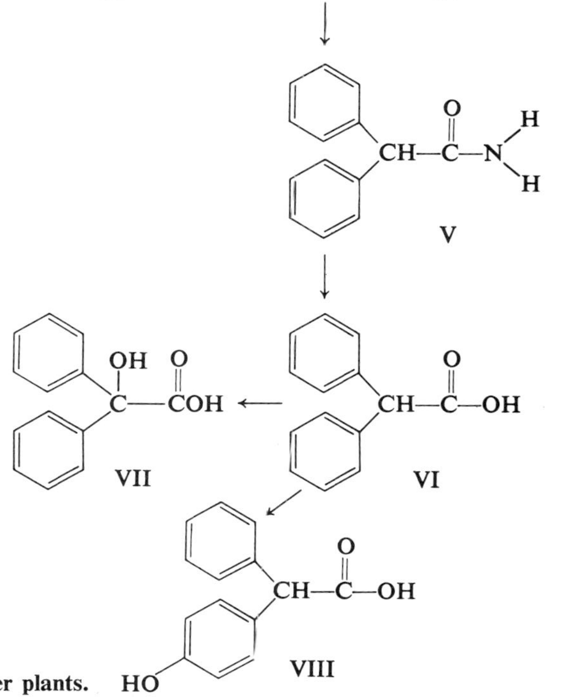

I. diphenamid
II. *N*-methyl-*N*-hydroxy-2,2-diphenylacetamide
III. *N*-methyl-2,2-diphenylacetamide
IV. *N*-hydroxy-2,2-diphenylacetamide
V. 2,2-diphenylacetamide
VI. 2,2-diphenylacetic acid
VII. α-hydroxy-2,2-diphenylacetic acid
VIII. *p*-hydroxy-2,2-diphenylacetic acid

FIGURE 9-2. Proposed degradation pathway of diphenamid in higher plants.

(1968a) also found that 3,4-dichloroaniline complexes with soluble carbohydrates and they isolated four complexes, (1) the *N*-(3,4-dichlorophenyl)-glucosylamine, (2) an unstable complex which readily decomposed to the glucosylamine, (3) a complex which contained glucose, xylose, and fructose, and (4) a sugar derivative of 3,4-dichloroaniline but not further identified. Frear (1968) isolated an enzyme from soybeans which catalyzes the biosynthesis of *N*-glucosylarylamines. It is specific for the nucleotide glucosyl donors, uridinediphosphate glucose (UDPG) and thymidinediphosphate glucose (TPDG) but has a broad specificity toward acceptor arylamines, including 3,4-dichloroaniline. Yih et al. (1968a) reported that relatively large amounts of the 3,4-dichloroaniline moiety also complexes with cell wall components namely cellulose, hemicellulose and lignin. The degradation pattern of propanil, as proposed by Yih et al. (1968) is presented in Figure 9-3.

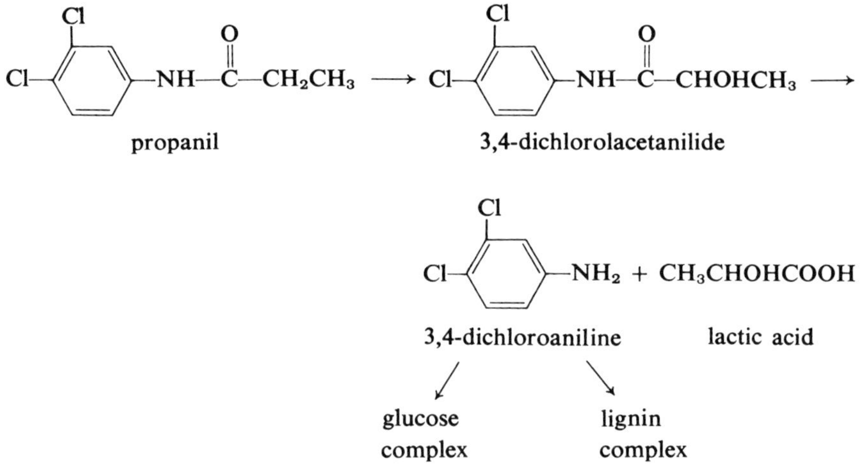

FIGURE 9-3. **Proposed degradation pathway of propanil in higher plants (Modified from Yih et al., 1968).**

   The resistance of rice to propanil has been attributed to the ability of this species to degrade propanil more rapidly than most weedy species. Adachi et al. (1966) reported the rate of inactivation of propanil by homogenates of several species (Table 9-2) and found that rice was about ten times as effective as barnyardgrass. Still and Kuzirian (1967) reported that rice was at least twenty times more effective than barnyardgrass in inactivating propanil. Frear and Still (1968) reported that the degradation of propanil takes place at a much faster rate in the leaves of rice than the roots but little degradation occurs in barnyardgrass leaves or roots (Table 9-3). Although these results suggest lower specific enzyme levels in the susceptible species, Ishizuka and

TABLE 9-2. Inactivation of Propanil by Homogenates of Various Plants (Adachi et al., 1966)

Source	% hydrolyzed in 24 hours
Monochoria	0.0
Smart weed	5.5
Barnyardgrass	7.0
Dayflower	30.0
Crabgrass	58.4
Rice plant	69.6

TABLE 9-3. Distribution of a Propanil Hydrolyzing Aryl Acylamidase in Rice and Barnyardgrass Tissue (Frear and Still, 1968)[a]

Tissue	Units[a]/gm (fresh wt.)	Total Units	Specific Activity
Rice leaves	314.4	6288	9.8
Barnyardgrass leaves	5.2	103	0.5
Rice roots	5.7	226	2.2
Barnyardgrass roots	5.8	230	2.1

[a] An enzyme unit is the amount of enzyme required to catalyze the hydrolyses of 1 mu mole of propanil per hour.

Mitsui (1966) proposed that the selectivity results from differences in the substrate specificity of the enzyme from different plants rather than the level of the enzyme.

Hodgson (1971) found that the metabolism of propanil in rice is quantitatively modified by temperature and daylength. Absorption and metabolism were most rapid under high (32°C) temperature and long (16 hr) day conditions. Greater quantities of 3,4-dichloroaniline and N-(3,4-dichlorophenyl) glucosamine were recovered from plants in the high temperature, long day environments, 47% of the propanil plus water-soluble metabolites was in the roots. High temperature and long days increased the percentage in the shoots but because of increased tissue mass due to growth, concentration of propanil metabolites in plant tissue decreased from 0.18 to 0.12 μmoles.

Field observations have shown that when certain insecticides were applied together with propanil or when these insecticides were used just before or after a propanil treatment rice plants were subject to increased leaf necrosis. Subsequent research by Bowling and Hodgins (1966), Adachi (1966) and

Adachi et al. (1966) has shown that the organophosphate and carbamate insecticides increase propanil injury to rice whereas chlorinated hydrocarbons
have no effect. Presumably these insecticides are interfering with the enzymatic
detoxication mechanism of rice.

Biochemical Responses

CDAA mechanism of action studies are quite limited despite the fact that it is
a relatively old compound which has been widely used in both maize and
soybeans. Jaworski (1956) reported that ryegrass seeds are more susceptible
to CDAA than wheat seeds. Furthermore, the difference in response of these
two species to CDAA was demonstrated at the biochemical level. Respiration
of ryegrass seeds was strongly inhibited by 10 ppm of CDAA, whereas wheat
seed respiration was only moderately affected. This was also demonstrated
by a marked decrease in respiratory quotient in ryegrass seeds and only slight
affect in wheat seeds. However, a strong growth inhibition of coleoptile elongation was caused by 10 ppm CDAA. The addition of glutathione, calcium,
pantothenate, or α-lipoic acid to the seed culture medium prevented the respiration and respiratory quotient inhibition of CDAA in both species, but did
not reverse the growth inhibition of the coleoptile. Jaworski concluded that
CDAA inhibits certain sulhydryl-containing enzymes that are involved in
respiration. However, it appears that CDAA affects a mechanism even more
intimately connected with growth. Sasaki and Kozlowski (1966) have observed
that CDAA inhibits respiration in pine seedlings.

Mann et al. (1965) and Moreland et al. (1969) have investigated the affect
of CDAA on protein synthesis in tissue segments by ^{14}C-leucine incorporation.
Mann et al. (1965) reported that CDAA inhibited protein synthesis in barley
and hemp sesbania (*Sesbania exaltata*) but Moreland et al. (1969) did not
observe any significant protein synthesis inhibition in soybean. In addition
to the inhibition of protein synthesis, an inhibition of ^{14}C-leucine uptake was
also reported by Mann et al. (1965). However, the degree of inhibition of
amino acid uptake was substantially less than that of protein synthesis, therefore a real inhibition of protein synthesis was observed. Perhaps the fact that
soybeans are resistant to CDAA injury could explain the opposite results of
these two investigations. However, it is more likely that some detail of experimental technique is involved. Moreland et al. (1969) reported that CDAA
does not inhibit RNA synthesis but does inhibit gibberellic acid induced α-
amylase synthesis in half-seeds of barley. Smith and Jaworski (unpublished)
have also shown that CDAA inhibits the gibberellic acid induced α-amylase
system. CDAA has been shown to inhibit the development of proteolytic
activity in squash cotyledons (Ashton et al., 1968). Whether the inhibition of
these hydrolytic enzymes by CDAA is caused by an interference with protein
synthesis or their hormonal control mechanisms is not yet clear. Since CDAA

has been reported to uncouple oxidative phosphorylation (Lotlikar et al., 1968) perhaps the inhibition of protein synthesis by CDAA is the result of a reduced level of ATP which is required for protein synthesis. However, the research of Duke et al. (1967) with propachlor indicates that this is not the case.

Propachlor mechanism of action studies have been conducted by Duke et al. (1967) and Smith and Jaworski (unpublished). Duke et al. (1967), found that propachlor inhibited both root and shoot growth but that an inhibition of protein synthesis occurred just prior to the cessation of growth. The inhibition of ^{14}C-leucine incorporation into protein was not accompanied by a decrease in respiration or oxidative phosphorylation. The meristematic zone of the root was found to be the most sensitive in regard to protein synthesis inhibition. They propose that propachlor inhibits at the level of nascent protein formation and this is probably due to the prevention of the transfer of an enoacyl-sRNA to the polypeptide chain. These results are consistent with the research of Smith and Jaworski (unpublished) which showed that propachlor inhibited gibberellic acid-induced α-amylase synthesis in half-seeds of barley. Penner (1970) reported that propachlor inhibited phytase development in squash. Although the research on the mechanism of action of propachlor and CDAA is limited, the parallelism of most of the research results would suggest that their mechanisms of action may be quite similar.

Diphenamid at 10 ppm was reported to completely inhibit RNA synthesis in oat roots (Briquet and Wiaux, 1967). However, this result is difficult to reconcile with other data which indicate that diphenamid does not inhibit protein synthesis. Mann et al. (1965) showed that diphenamid had only a slight inhibitory effect on the incorporation of ^{14}C-leucine into protein in barley coleoptile and hemp sesbania hypocotyl. Amylase synthesis in barley seeds was only slightly inhibited by diphenamid (Penner, 1968). The development of proteolytic activity in squash cotyledons was not inhibited by diphenamid (Ashton et al., 1968). However, it is possible that RNA synthesis in roots is more sensitive to diphenamid than it is in other plant organs or that there is a higher concentration of diphenamid at the site of action in the system of Briquet and Wiaux (1967) than in the others. Diphenamid reduced the uptake of magnesium, calcium, potassium, and phosphorus by cabbage from a nutrient solution (Nashed and Illnicki, 1968). Analysis of the tissue from this experiment showed that the shoots contained less of these elements, especially calcium, than control plants and the roots contained less magnesium but more calcium than control plants. It appears that diphenamid not only inhibits the uptake of inorganic ions but also influences the distribution of calcium within the plant.

Propanil mechanism of action studies indicate that it alters several biochemical reactions including photosynthesis. Camper and Moreland (1967) reported that propanil was bound to both animal protein (bovine serum albumin) and plant protein (arachin), but the binding was six times stronger to the plant protein than to the animal protein. Moreland et al. (1969) observed that propanil was a strong inhibitor of RNA and protein synthesis as well as an inhibitor of gibberellic acid induced α-amylase synthesis in half-seeds of barley. However, Mann et al. (1965) had reported that propanil did not inhibit protein synthesis. Investigations on the effect of propanil on respiration have been variable, although in general there appears to be inhibition. Sierra and Vega (1967) found that respiration increased one hour after propanil at 100 ppm was applied to barnyardgrass. However, respiration was reduced when the leaves were soaked in 1000 ppm propanil. Hofstra and Switzer (1968) reported that when propanil, 3.45×10^{-3}M was sprayed on the leaves of tomato and common lambsquarters (*Chenopodium album*), respiration was inhibited in both species initially but that in tomato it returned to near normal after 72 hours and in common lambsquarters respiration continued to decline. They also noted that oxygen uptake and phosphate esterification in soybean mitochondria were inhibited by propanil but that there was an uncoupling effect. The inhibition of the Hill reaction of photosynthesis by propanil has been reported by Good (1961) and Moreland and Hill (1963). More recently, Nishimura and Takamiya (1966) specified the site of inhibition of photosynthesis by propanil as the reduction of cytochrome 553 by System II. Photosynthesis was inhibited within 20 minutes following the application of propanil to tomatoes and lambsquarters but began to recover in tomatoes after 6 hours and completely recovered after 72 hours (Hofstra and Switzer, 1968). However, in common lambsquarters it did not recover. Monaco (1968) reported that the rapid absorption and accumulation of propanil by isolated spinach chloroplasts were related to the lipid content but independent of the protein content of the chloroplasts. Propanil also appears to destroy chloroplast membranes and increase the permeability of red beet membranes (Hofstra and Switzer, 1968). Transpiration has been reported to be reduced by propanil (Smith and Buchholtz, 1962).

Smith et al. (1971) have shown that pronamide markedly changed the levels of DNA, RNA, and protein in treated rhizomes. These changes were accompanied by a rise in levels of cellulase activity and gross anatomical aberrations.

Dicryl was reported to inhibit respiration in maize leaf tissue (Funderburk and Porter, 1961a) and alter the action of several enzymes. Although dicryl had little effect on catalase activity 2 days after treatment, 5 days after treatment its activity was reduced about 50% and similar results were noted on peroxidase and glycolic acid oxidase. Funderburk and Porter (1961b) also reported that 3 to 4 days after maize was treated with dicryl there was a sudden increase in ascorbate oxidation. Dicryl caused only a slight decrease in respiration in cotton, but reduced ascorbic acid oxidase in the cotyledons (Bingham and Porter, 1961a). It also prevented the normal increase in catalase and peroxidase but had no effect in glycolic acid oxidase (Bingham and Porter, 1961b).

Solan causes considerable injury to egg plant as a postemergence treatment; however tomatoes are resistant to similar treatments. Colby and Warren (1962) reported that the resistance of both of these species increased with age, but the resistance increased much more rapidly in tomato and that addition of glucose increased the resistance of tomato. This latter experiment suggests that photosynthesis inhibition may be involved. In more refined experiments ($^{14}CO_2$ fixation), Colby and Warren (1965) showed that solan inhibited photosynthesis in both species initially but that tomato recovered relatively rapidly whereas in egg plant photosynthesis continued to decline. Light appears to be a requirement of solan injury, as is true with several other herbicides which inhibit photosynthesis. These workers showed that no solan injury occurred on plants placed in darkness after spraying, regardless of plant species or rate of herbicide applied.

Mode of Action

In considering the mode of action of the amide herbicides, one must take into account that certain of these are applied to the soil and are active via the root system or seeds of plants while others are applied to the foliage. All of the soil applied herbicides classified as amides and studied for their effect on root growth have been reported to inhibit root elongation. These are alachlor, CDAA, propachlor, diphenamid, and naptalam. Although only naptalam has been reported to inhibit seed germination, all of the aforementioned herbicides are generally considered to inhibit seed germination or early seedling growth. If the susceptible seedlings do emerge from the soil they are either greatly stunted and/or malformed. These effects are due probably to an interference with cell division, as reported for CDAA, and/or cell enlargement, as reported for propachlor. The inhibition of stem, coleoptile, shoot, and/or leaf growth has been reported for one or more of these herbicides. Even propanil whose action is primarily in the leaves, has been reported to inhibit root and coleoptile growth when applied to these organs. Therefore the amide herbicides are general growth inhibitors and especially inhibitors of root elongation. In addition, naptalam has its unique antigeotropic property.

The initial injury of plants from the foliar applied amide herbicides is restricted to the leaves where they cause either localized or general necrosis. These are propanil, dicryl, solan, and cypromid. This injury may be associated with alteration of the cellular membranes as has been demonstrated with propanil. Later inhibition of growth in other plant parts occur but these are probably secondary effects.

CDAA, propachlor, and diphenamid are readily absorbed by roots and translocated to the shoot of plants. CDAA and propachlor are also absorbed by seeds. However, it has been observed that propachlor is more readily absorbed from the soil by the shoot than by the root.

Propanil, dicryl, solan, and cypromid are readily absorbed by leaves but it has been reported that transport of propanil and dicryl from the treated leaf is quite limited. The transport of solan and cypromid from the treated leaf is unknown.

The molecular fate of specific amide herbicides in higher plants has been presented previously in this chapter. Those which have been studied are metabolized and have their own particular pathway of degradation. One generalization which can be made is that propanil and propachlor, which have a substituted aniline as a degradation product form complexes with normally occurring compounds. A substituted aniline-glucose complex is most common but others have been reported.

Mechanism of action studies on the amide herbicides have not yielded a consistent pattern, therefore this aspect of their mode of action is presented in tabular form (see Table 9-4).

TABLE 9-4.   Mechanism of Action of Amide Herbicides[a]

	Photosynthesis	Respiration	RNA synthesis	Protein synthesis	Amylase	Proteinase	Dipeptidase	Other enzymes
CDAA	o	+	+	±	+	+	−	+
Propachlor	o	−	o	+	+	o	o	+
Diphenamid	o	o	+	−	−	−	o	o
Propanil	+	±	+	±	+	o	o	o
Dicryl	o	+	o	o	o	o	o	+
Solan	+	o	o	o	o	o	o	o

[a] + inhibited; − not inhibited; ± contradictory reports; o no information.

REFERENCES

Adachi, M. 1966. Selective herbicidal action of 3,4-dichloropropionanilide. II. Propanil degrading enzyme of rice plant. *Noyaku Seisan Gijyutsu* **15**:11–14.

Adachi, M., K. Tonegawa, and T. Uejima. 1966. Selective herbicidal action of

3′,4′-dichloropropionanilide. I. Penetration into plants and detoxication by their tissues. *Noyaku Seisan Gijutsu* **14**:19–22.

Ashton, F. M., D. Penner, and S. Hoffman. 1968. Effect of several herbicides on proteolytic activity of squash cotyledons. *Weed Sci.* **16**:169–171.

Bingham, S. W. and W. K. Porter, Jr. 1961a. The activities of certain enzymes from cotton treated in the cotyledon stage with *N*-(3,4-dichlorophenyl)methacrylamide. *Weeds* **9**:290–298.

Bingham, S. W. and W. K. Porter, Jr. 1961b. Glycolic acid oxidase, catalase and peroxidase of cotton tissue. *Weeds* **9**:299–306.

Bowling, C. C. and H. R. Hodgins. 1966. The effect of insecticides on the selectivity of propanil on rice. *Weeds* **14**:94–95.

Briquet, M. V. and Wiaux, A. L. 1967. Herbicides and RNA synthesis of *Pisum* and *Avena* roots. *Meded. Rijksfac. LandbWet. Gent.* **32**:1040–1049.

Camper, N. D. and D. E. Moreland. 1967. Sorption of diuron and propanil to proteins. *WSSA Abstr.* pp. 66–67.

Canvin, D. T. and G. Friesen. 1959. Cytological effects of CDAA and IPC on germinating barley and peas. *Weeds* **7**:153–156.

Casida, J. E. and L. Lykken. 1969. Metabolism of organic pesticide chemicals in higher plants. *An. Rev. Plant Physiol.* **20**:607–636.

Colby, S. R. and G. F. Warren. 1962. Selectivity and mode of action of *N*-(3-chloro-4-methylphenyl)-2-methylpentanamide. *Weeds* **10**:308–310.

Colby, S. R. and G. F. Warren. 1965. Selective action of solan on tomato and egg plant. *Weeds* **13**:257–263.

Deli, J. and G. F. Warren. 1971. Relative sensitivity of several plants to diphenamid. *Weed Sci.* **19**:70–72.

Devlin, R. M. 1967. Preliminary studies of the influence of indole-3-acetic acid and gibberellic acid on the uptake of simazine by *Agrostis alba* L. *Proc. 21st Northeast Weed Control Conf.* pp. 585–588.

Devlin, R. M., K. H. Deubert, and I. E. Demoranville. 1969. Poison ivy control on cranberry bogs. *Proc. 23rd Northeast Weed Control Conf.* pp. 58–62.

Devlin, R. M. and R. W. Yaklich. 1971. Influence of GA on uptake and accumulation of naptalam by bean plants. *Weed Sci.* **19**:135–137.

Duke, W. B. 1967. Ph.D. Thesis, U. Ill., 124 pp.

Duke, W. B. 1968. Effect of CP-31393 on auxin-induced growth. *Proc. 22nd Northeast Weed Control Conf.* pp. 504–511.

Frear, D. S. 1968. Herbicide metabolism in plants. I. Purification and properties of UDP glucose: arylamine *N*-glucosyl-transferase from soybean. *Phytochemistry* **7**:381–390.

Frear, D. S. and G. G. Still. 1968. The metabolism of 3,4-dichloropropionanilide in plants. Partial purification and properties of an aryl acylamidase from rice. *Phytochemistry* **7**:913–920.

Funderburk, H. H. and W. K. Porter, Jr. 1961a. Effects of *N*-(3,4-dichloro-phenyl)methacrylamide (DCMA) on growth and certain respiratory enzymes of corn. *Weeds* **9**:538–544.

Funderburk, H. H. and W. K. Porter, Jr. 1961b. Effect of *N*-(3,4-dichloro-phenyl)methacrylamide on the oxidation of ascorbic acid by corn. *Weeds* **9**:545–557.

Golab, T., R. J. Herberg, S. J. Parka, and J. B. Tepe. 1966. The metabolism of carbon-14 diphenamid in strawberry plants. *J. Agr. and Food Chem.* **14**:592–596.

Good, N. E. 1961. Inhibitors of the Hill reaction. *Plant Physiol.* **36**:788–810.

Grigsby, B. H., T. M. Tsou, and G. B. Wilson. 1954. Some effects of alanap (NP) and amizole on germinating peas. *Proc. 11th North Cent. Weed Control Conf.* p. 88.

Hodgson, R. H. 1971. Influence of environment on metabolism of propanil in rice. *Weed Sci.* **19**:501–507.

Hoffman, O. L. and A. E. Smith. 1949. A new group of plant growth regulators. *Science* **109**:588.

Hofstra, G. and C. M. Switzer. 1968. The phytotoxicity of propanil. *Weed Sci.* **16**:23–28.

Ishizuka, K. and S. Mitsui. 1966. Activation or inactivation mechanisms of biological active compounds in higher plants. II. On anilide degrading enzyme. *Abstr. Ann. Meeting Agr. Chem. Soc. Japan*, p. 62.

Jaworski, E. G. 1956. Biochemical action of CDAA, a new herbicide. *Science* **123**:847–848.

Jaworski, E. G. 1964. Metabolism of 2-chloro-*N,N*-diallylacetamide (CDAA) and 2-chloroallyl-*N,N*-diethyldithiocarbamate (CDEC) by plants. *J. Agr. and Food Chem.* **12**:33–37.

Jaworski, E. G. 1969. Chloroacetamides, pp. 165–185. In P. C. Kearney and D. D. Kaufman, *Degradation of Herbicides*. Marcel Dekker, Inc., New York.

Jaworski, E. G. and C. A. Porter. 1965. Uptake and metabolism of 2-chloro-*N*-isopropylacetanilide in plants. *Abstr. 149th Meeting*, Am. Chem. Soc. 21A.

Keeley, P. E., C. H. Carter, and J. H. Miller. 1972. Evaluation of the relative toxicity phytotoxicity of herbicides to cotton and nutsedge. *Weed Sci.* **20**:71–74.

Keith, G. W. and R. A. Baker. 1966. Auxin activity of substituted benzoic acids and their effect on polar auxin transport. *Plant Physiol.* **41**:1561–1569.

Knake, E. L. and L. M. Wax. 1968. The importance of the shoot of giant foxtail for uptake of preemergence herbicides. *Weed Sci.* **16**:393–395.

Lemin, A. J. 1966. Absorption, translocation, and metabolism of diphenamid-1-^{14}C by tomato seedlings. *J. Agr. and Food Chem.* **14**:109–111.

Lotikar, P. D., L. F. Remmert, and V. H. Freed. 1968. Effects of 2,4-D and other herbicides on oxidative phosphorylation on mitochondria from cabbage. *Weed Sci.* **16**:161–165.

Lynch, M. R. and R. D. Sweet. 1971. Effect of environment on the activity of diphenamid. *Weed Sci.* **19**:332–337.

Mann, J. D., L. S. Jordan, and B. E. Day. 1965. A survey of herbicides for their effect upon protein synthesis. *Plant Physiol.* **40**:840–843.

Matsunaka, S. 1969. Activation and inactivation of herbicides by higher plants. *Residue Rev.* **25**:45–58.

McRae, D. H., R. Y. Yih, and H. F. Wilson. 1964. A biochemical mechanism for the selective action of anilides. *WSSA Abstr.* p. 87.

Mentzer, C. and G. Neitien. 1950. Sur un procede permettant de troubler le geotropisme des racines. *Bull. Mens. Soc. Linneenne Lyon* **19**:102–104.

Monaco, T. J. 1968. The partitioning and distribution of simazine and propanil in spinach chloroplasts. Ph.D. Thesis North Carolina State University, Raleigh, 94 pp.

Moreland, D. E. and K. L. Hill. 1963. Inhibition of photochemical activity of isolated chloroplasts by acylanilides. *Weeds* **11**:55–60.

Moreland, D. E., S. S. Malhotra, R. D. Gruenhagen, and E. H. Shokraii. 1969. Effects of herbicides on RNA and protein synthesis. *Weed Sci.* **17**:556–563.

Nashed, R. B. and R. D. Ilnicki. 1968. The effect of diphenamid on the uptake and distribution of macronutrient elements in cabbage. *Proc. 22nd Northeast Weed Control Conf.* p. 500.

Netien, G. and R. Conillat. 1951. Action de lacide naphthyl-phtalamique sur la germination et les premiers stades de croissance des vegetaux. *Bull. Mens. Soc. Linneenne*, Lyon **20**:49.

Nishimoto, R. K., A. P. Appleby, and W. R. Furtick. 1967. Site of uptake of preemergence herbicides. *WSSA Abstr.* pp. 46–47.

Nishimura, M. and A. Takamiya. 1966. Energy and electron transfer systems in algal photosynthesis. I. Action of two photochemical systems in oxidation-reduction reactions of cytochrome in *Porphyra. Biochem. Biophys. Acta* **120**:45–56.

Penner, D. 1968. Herbicidal influence on amylase in barley and squash seedlings. *Weed Sci.* **16**:519–522.

Penner, D. 1970. Herbicide and inorganic phosphate on phytase in seedlings. *Weed Sci.* **18**:360–364.

Peterson, R. L. and L. W. Smith. 1971. Effects of *N*-(1,1-dimethylpropynyl)-3,5-dichlorobenzamide on the anatomy of *Agropyron repens* (L.) Beaux. *Weed Res.* **11**:84–87.

Porter, C. A. and E. G. Jaworski. 1965. Metabolism of tritiated 2-chloro-*N*-isopropylacetanilide in plants. *Plant Physiol. Abstr.* **40**:xiv–xv.

Porter, W. K., Jr., C. H. Thomas, W. L. Sloane, and D. R. Melville. 1960. A proposed method for post-emergence weed control in cotton. *Proc. 13th Southern Weed Conf.* p. 30.

Saghir, A. R. and S. Abu-Shakra. 1971. Effect of diphenamid and trifluralin on the germination of *Orobanche* seeds *in vitro*. *Weed Res.* **11**:74–76.

Sasaki, S. and T. T. Kozlowski. 1966. Influence of herbicides on respiration of young *Pinus* seedlings. *Nature* **210**:439–440.

Schubert, O. E. 1966. Effect of diphenamid, simazine, and atrazine on the weight and development of strawberry mother and runner plants. *Proc. West Va. Acad. Sci.* **38**:92–101.

Sierra, J. N. and M. R. Vega. 1967. The response of rice and barnyard grass to propanil. *Philipp. Agric.* **51**:438–452.

Smith, D. and K. P. Buchholtz. 1962. Transpiration reduction with herbicides and its effect on water requirement. *Proc. 19th North Cent. Weed Control Conf.* pp. 45–46.

Smith, G. R., C. A. Porter, and E. G. Jaworski. 1966. Uptake and metabolism of ^{14}C labeled α-chloroacetamides by germinating seeds. *Abstr. 152nd Meeting, Am. Chem. Soc.*, A-42.

Smith, L. W., R. L. Peterson, and R. F. Horton. 1971. Effects of a dimethylpropynyl benzamide herbicide on quackgrass rhizomes. *Weed Sci.* **19**:174–177.

Still, G. G. 1968a. Metabolic fate of 3,4-dichloropropionanilide in plants: the metabolism of the propionic moiety. *Plant Physiol.* **43**:543–546.

Still, G. G. 1968b. Metabolism of 3,4-dichloropropionanilide in plants: the metabolic fate of the 3,4-dichloroaniline moiety. *Science* **159**:992–993.

Still, G. G. and O. Kuzirian. 1967. Enzyme detoxication of 3′,4′-dichloropropionanilide in rice and barnyard grass, a factor in herbicide selectivity. *Nature* **216**:799–800.

Takematsu, T. and S. Yanagishima. 1963. Studies on the herbicidal phenomenon and the function of dichloropropionanilide, a weed killer having a selective herbicidal effect on genera of Gramineae. *Il Riso* **12**:37–70.

Teas, H. J. and T. J. Sheehan. 1957. Chemical modification of ageotropic bending in snapdragon. *Plant Physiol. Abstr.* **32**:*xlii*.

Tsou, T. M., R. H. Hamilton, and R. S. Bandurski. 1956. Selective inhibition of the geotropic response by *n*-1-naphthylphthalamic acid. *Physiol. Plant.* **9**:546–548.

Yih, R. Y., D. H. McRae, and H. F. Wilson. 1968a. Mechanism of selective action of 3′,4′-dichloropropionanilide. *Plant Physiol.* **43**:1291–1296.

Yih, R. Y., D. M. McRae, and H. F. Wilson. 1968b. Metabolism of 3,4-dichloropropionanilide: 3,4-dichloroaniline-lignin complex in rice plants. *Science* **161**:376–378.

CHAPTER 10

Arsenicals

Arsenic has been used since ancient times as a poison, a stimulant, and a conditioner for animals. In sufficient quantities arsenic is an acute poison to both plants and animals. At low dosage it is used as a stimulant by athletes and mountain climbers. As a conditioner is produces smooth shiny coats on animals. Fowler's solution containing potassium arsenite was used for years in human and veterinary medicine and arsenic is still used in mixed poultry feeds. Arsenic is a very common element present in nearly all soils and hence in plant and animal products. At 1 ppm and below there is no known harmful effect of arsenic. Pentavalent arsenic in arsonic acid feed additives may provide up to 100 ppm of arsenic at recommended feeding levels (Frost, 1968, 1969). Unlike lead which accumulates, arsenic is readily excreted and it does not accumulate in the body.

Arsenic is amphoteric; arsenic trichloride (arsenous chloride, $AsCl_3$) was the active ingredient in KMG (Kill Morning Glory), an acid-arsenical herbicide used in the period 1920–1940 as a translocated herbicide to kill bindweed and other perennials. Arsenic trioxide is the form in which arsenic is recovered in the process of smelting copper. With water, arsenic trioxide (arsenous oxide, As_2O_3) forms arsenious acid.

$$As_2O_3 + 3H_2O \longrightarrow 2H_3AsO_3$$

When the hydrogens are replaced by sodium, the weed killer sodium arsenite is formed; it may contain one, two, or three sodium ions and the alkalinity increases through the series; Na_3AsO_3 is quite caustic.

Sodium arsenite was the weed killer of commerce for many years; it was used in large quantities on railroad rights-of-way in the USA, in sugar cane, and rubber plantations in tropical countries. An acidified solution of dilute sodium arsenite was used as an acid-arsenical translocated spray against perennial weed for several years (Crafts, 1933, 1937).

Sodium arsenite is an extremely toxic compound and accidental fatalities occurred wherever it was used. Applied to mixed vegetation, sodium arsenite sprays rapidly release aromatic compounds which are attractive to animals; literally thousands of cattle and sheep poisoning incidents occurred so long as it was used on the railroads. Such use is now prohibited by law in most countries and other less hazardous materials are used on plantations.

Used for years around the home to kill weeds and sterilize soil on driveways, tennis courts, and sidewalks, sodium arsenite was a constant hazard. Innumerable cases of poisoning occurred when the poisons were stored in garages and tool sheds; frequently they were placed in soft drink bottles and children would find these and drink the contents by mistake. Home use is now prohibited in nearly all advanced countries.

The pentavalent form of arsenic as sodium arsenate found some use in the control of prickly pear cactus (*Opuntia* spp.) in Australia in the years before the discovery of biological control by the moth borer, *Cactoblastis cactorum*. Pentavalent arsenic is only one half or less toxic than the trivalent form; in inorganic form it was not widely used in weed control.

More recently organic arsenic compounds have been found useful as general contact herbicides to control weeds, particularly grasses. Cacodylic acid, hydroxydimethylarsine oxide, was one of the first of the organic

$$CH_3\!-\!\underset{\underset{CH_3}{|}}{\overset{\overset{O}{\|}}{As}}\!-\!OH$$

Cacodylic acid

arsenicals to be introduced. It has an oral LD_{50} of 1350 mg/kg, and an oral toxicity rating of 4 in a class rating 1 to 6 where 4 is described as slightly toxic. A probable lethal dose to man would be one ounce or more. It has little or no dermal toxicity.

Cacodylic acid and its sodium salt are used in forest management to control weedy or unwanted trees, in lawn renovation, and as a general contact herbicide in non-crop areas, in orchard and vineyard weed control, and by directed spraying in cotton.

More recently a group of salts and esters of monomethyl arsinic acid have been introduced and these are apparently translocated herbicides. The mono-

$$CH_3\!-\!\underset{\underset{OH}{|}}{\overset{\overset{O}{\|}}{As}}\!-\!OH$$

MMA

sodium acid salt (MSMA), the disodium salt (DSMA), the monoammonium acid salt (MAMA), and certain amine salts have been tested and are available as formulated materials, some with surfactants, and some with 2,4-D or other supplementary herbicides added. These compounds are used principally against johnsongrass (*Sorghum halpense*) in orchards and vineyards and in cotton. They are used against quackgrass (*Agropyron repens*), dallisgrass (*Paspalum dilatum*), nutsedge (*Cyperus* spp.), goosegrass (*Eleusine indica*), barnyardgrass (*Echinochloa crusgalli*), sandbur (*Cenchrus* spp.), cocklebur (*Xanthium* spp.), ragweed (*Ambrosia* spp.), and puncture vine (*Tribulus terrestris*) in various crops or in non-crop areas. In turf they are used to control any of the above weeds plus bahia grass (*Paspalum notatum*), common chickweed (*Stellaria media*), and wood sorrel (*Oxalis* spp.). The triethanol-amine salt is recommended against the above list of weeds plus common dandelion (*Tararacum officinale*), plantain (*Plantago* spp.), and other broad-leaf weeds. Because these organic arsonates have proved effective against a number of perennial weeds it is evident that they are absorbed and trans-located into the underground organs—rhizomes, tubers, or roots. Chemical analyses have proved this to be true and work with [77]As-arsonate by auto-radiography indicates that these compounds move in plants in much the same way as phosphate does. Use on cotton in the early to mid-bloom stage results in relatively large amounts of arsenic in the seeds (Baker et al., 1969).

In order to broaden the spectrum of weed control and to combine the desirable properties of two arsenical herbicides, a mixture of MSMA, sodium cacodylate and cacodylic acid has been formulated. This proprietary mixture is recommended as a general, broad spectrum, postemergence herbicide. It provides a quick contact burn on vegetation while retaining the systemic effects necessary to control deep-rooted perennial weeds. Many of our most troublesome weeds are listed as susceptible to this formulation (Anon., 1970a).

Many of the formulations of organic arsenicals that have been available for years are now being offered in new, highly concentrated forms. These are more convenient and economical than the older formulations (Anon., 1970b).

Growth and Plant Structure

The organoarsenical herbicides are not growth regulators in the sense of plant hormones; they apparently act through enzyme systems to inhibit growth. They kill relatively slowly and the first symptoms are usually chlorosis, cessation of growth, and gradual browning followed by dehydration and death. Rhizomes and tubers may show browning of the storage tissues; buds fail to sprout and the whole structures eventually decompose.

When resprouting of tubers or rhizomes does occur treatment should be repeated when some of the leaves have reached full size; treatment before this time will not result in translocation because movement of the assimilate stream into the underground organs is necessary to carry the toxicant to the proper sites of action.

Rumberg et al. (1960) found chlorosis from DSMA treatment to increase with increasing temperatures as shown in Table 10-1; translocation also increased.

TABLE 10-1. The Effect of Temperature After Treatment on the Degree of Chlorosis Induced by DSMA as Indicated by Ratings on Crabgrass 5 and 10 Days After Application (Rumberg et al., 1960)

| DMSA lb/acre | Chlorosis rating[a] | | | | | |
| | 60°F | | 75°F | | 85°F | |
	5 days	10 days	5 days	10 days	5 days	10 days
2	0.2	2.7	3.2	2.2	1.8	2.0
6	0.2	3.3	6.8	6.0	5.1	7.1
18	0.3	2.7	7.1	7.1	5.4	7.6
Mean	0.2	2.9	5.7	5.1	4.1	5.6

[a] 0 to 10 (0 = no chlorosis and 10 = complete chlorosis).

Absorption and Translocation

Sodium arsenite, long used as a general contact herbicide, was not considered to be translocated, probably because, as a contact treatment it was used at a concentration that rapidly destroyed the foliage. This prevented assimilate movement. At lower concentrations it was combined with acid to hasten penetration; this also rapidly killed the foliage. Rumberg et al. (1960) compared DSMA at 100 mg per ml with sodium arsenite ($4As_2O_3:3NaOH$) at 50 mg per ml; both were labeled with ^{76}As. Using soybean plants they found evidence for translocation of sodium arsenite applied in the radioactive form.

Table 10-1 shows the effect of temperature on translocation of DSMA in large crabgrass (*Digitaria sanguinalis*) as indicated by chlorosis.

Table 10-2 presents the data on translocation of activity from sodium arsenite in soybean 8 hours after treatment at 85°F and Table 10-3 shows transport of ^{76}As-DSMA in soybean 6 hours after treatment at 60 and 85°F. The organic form of arsenic proved to be the more mobile of the two. Rumberg et al. (1960) suggests that the rapid injury from sodium arsenite treatment may be responsible for the lesser transport; the arsenite usually

TABLE 10-2. Activity Found in Sections of Soybean Plant 8 Hours After Treatment with 2.5 mg of Sodium Arsenite—[76]As at 85°F (Rumberg et al., 1960)

Section of plant	Mean activity, CPM	% recovered activity
Trifoliate + one primary leaf	169	0.10
Stem below the treated leaf	960	0.57
Treated leaf	166,000	99.32

TABLE 10-3. Activity Found in Sections of Soybean Plant 6 Hours After Treatment with 5 mg of DSMA—[76]As at 85 and 60°F (Rumberg et al., 1960)

	60°F		85°F	
Section of plant	Activity cpm	% activity recovered	Activity cpm	% activity recovered
Trifoliate + one primary leaf	132	0.19	419	0.74
Stem below the treated leaf	428	0.61	618	1.07
Treated leaf	70,000	99.2	57,000	98.2

shows injury symptoms within hours after treatment whereas DSMA requires many hours or even days before the chlorosis appears. Sodium arsenite injury appears, first as loss of turgidity indicating possible effects on membrane integrity whereas the organic arsenicals usually cause slowly developing chlorosis with little or no wilting; possibly different enzyme systems are affected.

Cacodylic acid is considered to be a general contact toxicant. It is applied in solution to cuts around the base of trees, or as solution on foliage. Apparently its only translocation is apoplastic. DSMA and an amine methylarsonate (equal parts of octyl and dodecyl ammonium methyl arsonates) were shown by Long and Holt (1959) to be effective for controlling purple nutsedge (*Cyperus rotundus*) in bermudagrass (*Cynodon dactylon*) turf. This result can only be explained by translocation from foliage to tubers, a movement with the assimilate stream via the phloem. In a later paper Long, Allen, and Holt (1962) showed that the above organic arsenicals will kill nutsedge tubers if they are applied several times over a two-year period. Long and Holt (1959) showed amine methyl arsonate to be somewhat superior to DSMA at equivalent rates of application, a result that seems logical in view of the mechanism of herbicide activation described by Crafts and Reiber (1945).

In further studies on purple nutsedge Holt et al. (1967) using single and repeated applications of amine methyl arsonate to shoots of single tubers and shoots of terminal tubers on chains of tubers, found that arsenic was translocated laterally into the tubers, separated from the treated shoot by up to four tubers. The tuber at the opposite end of the chain from the treated shoot tended to be higher in arsenic content than the tubers between; translocated arsenic tended to be higher in tubers from which active growth was taking place. This follows from the source-to-sink nature of food movement; actively growing tubers are active sinks.

There was no apparent relationship between arsenic content of tubers and their ability to produce new shoots and there was a tendency for tubers to produce more than one shoot in the regrowth following treatment. Evidently the arsenic treatment altered the apical dominance in tubers receiving arsenic. The writers concluded that death of tubers, following repeated treatments, was due to depletion of food reserves rather than to the level of arsenic in the tubers. Both the variability in arsenic content in killed tubers and the varying number of treatments required to kill tubers suggest that failure to sprout is not related to the overall arsenic content of the tuber; in some cases viable tubers contained more arsenic than some dead ones. Interruption of normal oxidative phosphorylation and exhaustion of the food supply resulting from increased sprouting may also have been involved in ultimate death of tubers. In the repeated application tests, the arsenic level decreased in tubers from which new rhizomes, tubers, and shoots developed. This indicates retranslocation, a property common to phosphate, amitrole, and dalapon in plants. The high arsenic content of terminal tubers and the appearance of chlorosis in untreated shoots confirm this interpretation. Roots and newly developing shoots were not analyzed and these might account for the loss of arsenic following the initial influx into a sprouting tuber.

Duble, Holt, and McBee (1968) studied the translocation of two organic arsenicals in purple nutsedge. DSMA and amine methane arsonate (referred to as AMA) were the forms used and tracers containing ^{14}C were applied to greenhouse grown plants. Chromatographic tests on extracts from ^{14}C-DSMA treated plants indicated that the compound was not readily degraded; the ^{14}C—As bond appeared to remain intact although some ^{14}CO$_2$ was found several days after treatment. Comparison of the Rf values of plant extract-DSMA with that of standard DSMA-^{14}C suggested that a plant extract-DSMA conjugate might have been formed; the values for extract and standard solution were 0.59 and 0.66 respectively; only one spot was found in each case. Over 85% of the material applied to the plant remained in the treated shoots. DSMA moved both acropetally and basipetally in single leaves and such movement was not influenced by relative age of the leaf. The writers report that both DSMA and AMA are moved symplastically and apoplastically, a property they share with amitrole. An autoradiograph of ^{14}C distribution in an untreated shoot (their figure 1) appears to be very similar to some of those reported by Anderson (1958) for amitrole distribution in the same plant.

It seems apparent from the above results that translocation in nutsedge follows a source : sink pattern and that the amount of arsenic moved into a tuber depends very much upon the sink activity of that tuber. Duble et al. (1968) found that actively growing terminal tubers in a chain accumulated arsenic whereas intermediate and dormant tubers did not. Thus, as noted above, the arsenic level in a tuber may not serve as an index of the lethality of a treatment; the effects of the initial import of arsenic upon subsequent growth activity may be the critical factor in lethality and continuing transport of arsenic to growing roots, other tubers, and shoots. This may mask the effects of the arsenic content as determined by analysis at any one time.

McWhorter (1966) found no evidence for translocation of DSMA in johnsongrass in field and greenhouse tests. Sckerl and Frans (1969), by contrast, found that ^{14}C-MAA was both xylem- and phloem-mobile in johnsongrass and cotton. Root uptake of this labeled arsenical by johnsongrass from nutrient solution was rapid and translocation into all portions of the plant took place within 4 hours. Apoplastic movement in both plants was more rapid than symplastic movement; in johnsongrass, symplastic movement was more rapid than in cotton. As Duble et al. (1968) found, chromatography of extracts from johnsongrass revealed values that differed from those of treated plants and standard ^{14}C-MAA solutions; Sckerl and Frans suggest possible complexing with sugars, organic acids, or both. When amino acid fractions were prepared from methanol extracts of both plants an MAA metabolite with a positive ninhydrin reaction was found in the johnsongrass fraction. Comparing Rf values the authors suggest a possible complex with histidine or one of its analogues. Amino acid accumulation was noted in johnsongrass as a result of MAA treatment and the authors suggest that the MAA metabolite may block protein synthesis or some other biosynthetic pathway.

Wilkinson and Hardcastle (1969) produced radioactive arsenic (^{76}As) by neutron activation and determined concentrations of this radioisotope in new (unsprayed), and old (sprayed) cotton leaves, and in the soil from beneath the plants; treatment rate was 2.24 kg/ha of MSMA. By using long counting they were able to detect 5 ng (1×10^{-9} μg) of their ^{76}As in the cotton leaf samples. Table 10-4 presents the results of the analyses.

Kempen (1970) has made a study of MSMA in plants, principally johnsongrass. He used both detached leaves and whole plants in his studies. He found that relatively high temperature and light (35°C and 2800 ft-c) produced 50% necrosis of rhizomatous johnsongrass foliage in less than one day whereas lower temperature and light (15°C and 320 ft-c) required 12 days to give the

TABLE 10-4.   Arsenic Content of Leaf and Soil Samples taken from Field-Grown Cotton Treated with 2.24 kg/ha MSMA (Wilkinson and Hardcastle, 1969)

Number of treatments	New leaves	Old leaves	As μg per g soil
0	0.05	0.17	3.75
1	0.13	0.45	8.25
2	0.95	1.05	8.35
3	0.23	0.90	7.50
4	3.15	17.30	10.15
5	4.80	41.10	4.00
6	10.30	30.25	11.20

same results. Regrowth from rhizomes showed similar trends, indicating that the arsenic had translocated into the rhizomes. Translocation of the label from leaf-applied ^{14}C-MSMA was primarily acropetal in the xylem, but small amounts also moved basipetally proving that this organic arsenical is phloem mobile. Within a week the arsenic was transported from a treated mature leaf into the leaf base and sheath, to meristematic regions and to roots and rhizomes; this indicates symplastic movement. Presence of the methane arsonate was detected by autoradiography and by counting.

Sachs and Michael (1971) found cacodylic acid and MSMA to be transported about equally from the leaves to the terminal buds and expanding leaves of bean plants. There was no indication that either was demethylated or reduced to a trivalent compound. Using ^{14}C-MSMA Sachs and Michael found that about 40% of the ^{14}C and arsenic recovered was bound to another molecule to form a ninhydrin-positive complex.

In studies using ^{14}C-labeled MSMA, MAA, and DSMA, Keeley and Thullen (1971a) found little translocation of these herbicides from cotyledons to developing leaves of cotton seedlings. An exception was noted where ^{14}C-MSMA was applied at 13°C; there was little contact injury of the treated cotyledons and the terminal leaves were well labeled.

Using these same three herbicides in studies in purple and yellow nutsedge, Keeley and Thullen (1971b) found the yellow species to be more susceptible; the yellow species absorbed and translocated more of the ^{14}C-labeled tracers than did the purple. In chromatography of plant extracts and standards there was less than 5% variation; nutsedge plants did not readily metabolize these arsenicals in 72 hours.

Surfactants have been found to increase the penetration and translocation of the organic arsenical herbicides. The principal manufacturer of the methane arsonates formulates mixtures containing tested surfactants. Apparently anionic and non-ionic surfactants are satisfactory with these materials.

Molecular Fate

Arsenic, being a metal, is indestructible, and hence, any arsenical compound applied to the soil, or applied to a plant, the residues of which return to the soil, will become a soil residue upon decomposition of the original molecule in which it is applied. Conceivably, under strong reducing conditions in the soil, arsenic might be reduced to arsine, AsH_3, which is volatile. However, this is unlikely to happen under most conditions and the most probable form in which it would occur would be arsenious oxide (As_2O_3) or arsenic pentoxide (As_2O_5). In moist soil or in moist organic residues a certain proportion of each of these would exist as arsenious acid (H_3AsO_3) or arsenic acid ($HAsO_3$). Possibly since the As—C bond of the organic arsenical compounds mentioned above is quite stable, some of the arsenic might occur as methane substituted oxides or acids.

Arsenic is known to be bound tightly to soils and to resist leaching in moving water (Crafts, 1935; Arnott and Leaf, 1967). For this reason residues resulting from frequent repeated use of arsenicals are common; they are found in plantation soils where sodium arsenite has been used for weed control, in cotton soils where calcium arsenate was long used for boll worm control, and in orchard soils following extensive use of lead arsenate for codling moth control. However, arsenic is tightly bound and relatively unavailable to plants, hence these residues have not usually resulted in decreased yields. An exception is the apple orchards in the Yakima Valley in Washington, where lead arsenate residues have limited growth of cover crop plants and hence upset the normal orchard practices.

Concerning the modern organic arsenical herbicides, there have been no reports of soil residues to date but with frequent repeated use residues are bound to accumulate. For this reason it seems obvious that users of these materials should adopt an integrated program, using combinations or rotations of herbicides in such a way that harmful residues of arsenic do not accumulate. The principal reason for using these herbicides is that for particular circumstances, they are the most effective materials available. But since, when used properly, they give high levels of control, they should be used infrequently, with other methods employed to complete the eradication program. Such supplementary materials might be dalapon which undergoes degradation in soils, or carbon bisulfide which volatilizes or leaves harmless residues. Hamilton (1969), using MSMA and DSMA, found six to ten applications using 3 to 6 lb aihg (active ingredient per 100 gal) necessary to control mature johnsongrass plants in Arizona. Such treatments, totaling

24 to 60 lb/A, would seem to be approaching the level of tolerance for such use of arsenic. Surely a repeat of such treatments for several successive years would result in excessive arsenic residues in the soil. Millhollon (1969) found five treatments with MSMA at 3.6 lb/A proved effective to control mature johnsongrass on drainage ditch banks in Louisiana. Similar treatments with dalapon at 7.4 lb/A were also effective—fenac, bromacil, and TCA proved useful as preemergence treatments on seedlings. McBee, Johnson, and Holt (1967) reported the presence of high concentrations of arsenic in bermuda-grass forage harvested soon after application of organic arsenicals for weed control. Stage of plant maturity, rainfall between treatment and harvest, rate of application and the amount of growth occurring between treatment and harvest all influenced the arsenic residue levels. Regrowth, following harvest of treated forage, had lower arsenic content. Removal of treated growth following application of arsenical herbicides was the most effective means of reducing arsenic residues.

Kleifeld (1970) in Israel and Hamilton and Arle (1970) in Arizona found that directed applications of MSMA and DSMA were effective in controlling emerging weeds in irrigated cotton. Preplant applications of trifluralin reduced the number of directed postemergence applications needed to give satisfactory weed control. The directed-spray method of application reduces to a minimum the problem of soil residues of arsenic.

Little is known of the metabolism of organic arsenicals in plants. Duble, Holt, and McBee (1968) obtained evidence for a conjugate of DSMA with some plant component. Although such conjugation might serve to retain the arsenic within the plant, ultimately, with the complete degradation of the plant residues within the soil, the arsenic would be freed; even the CH_3—As bond would be broken since the methane group is a potential source of energy for soil microorganisms. Under most soil conditions the ultimate form of the arsenic would be as the pentoxide (As_2O_5) and this in moist soil would hydrolyze to arsenic acid. While this would tend to bond with soil organic matter, it should ultimately leach to lower and lower depths. Considering the mass of soil and water involved it should have an insignificant effect upon ground water.

Biochemical Responses

When arsenic in solution has penetrated the cuticle and entered the apoplast system it bathes the external surface of the plasmolemma of the symplast. Here are at least some of the enzymes that make up the active system responsible for the living condition of the plant. One of the first symptoms of injury by sodium arsenite is wilting caused by loss of turgor and this immediately suggests an alteration in membrane integrity. It has long been recognized that trivalent arsenic is an uncoupling agent; that is, it inhibits the transformation of energy to ATP in the process of oxidative phosphorylation to a lesser degree than oxygen consumption. This reaction is responsible for providing the energy from respiration. This energy is essential for those biosynthetic processes that enable the plant to accumulate nutrients, to store foods, and to grow. When uncoupling takes place the oxidative process goes on, foods are utilized, but the energy is dissipated as heat and is unavailable for living processes. This could well explain the effects of trivalent arsenic in membrane degradation, injury, and eventually death.

Trivalent arsenic reacts readily with SH groups as follows:

$$2R\text{—}SH + O=As\text{—}R^1 \rightleftharpoons \begin{matrix} RS \\ \diagdown \\ \diagup \\ RS \end{matrix} As\text{—}R^1 + H_2O \tag{1}$$

$$\begin{matrix} CH_2\text{—}SH \\ | \\ CH_2 \\ | \\ CH\text{—}SH \\ | \\ (CH_2)_4\text{—}COOH \end{matrix} + O=As\text{—}R^1 \longrightarrow \begin{matrix} CH_2\text{—}S \\ | \quad \diagdown \\ CH_2 \quad As\text{—}R^1 + H_2O \\ | \quad \diagup \\ CH\text{—}S \\ | \\ (CH_2)_4\text{—}COOH \end{matrix} \tag{2}$$

The first of these reactions (1) may be reversed by adding an excess of a thiol compound such as cysteine. With dithiol compounds such as reduced lipoic acid (2) the arsenical reagents form stable ring structures and such reactions are not reversible (Dixon and Webb, 1958).

In general, arsenates are less toxic than arsenites. The arsenate symptoms involve chlorosis but not rapid loss of turgor, at least through the early expression of toxicity. Arsenate contact action is more subtle than that of arsenites.

Arsenate is known to uncouple phosphorylation. For example, normal ATP formation may result in the following way:

$$\begin{matrix} CH_2 \cdot O \cdot H_2PO_3 \\ | \\ CHOH \\ | \\ C=O \\ | \\ S \cdot Enzyme \end{matrix} + H_3PO_4 \rightleftharpoons \begin{matrix} CH_2 \cdot O \cdot H_2PO_3 \\ | \\ CHOH \\ | \\ CO \cdot O \cdot H_2PO_3 \end{matrix} + HS \text{ enzyme} \tag{3}$$

$$\begin{matrix} CH_2 \cdot O \cdot H_2PO_3 \\ | \\ CHOH \\ | \\ CO \cdot O \cdot H_2PO_3 \end{matrix} + ADP \rightleftharpoons \begin{matrix} CH_2 \cdot O \cdot H_2PO_3 \\ | \\ CHOH \\ | \\ COOH \end{matrix} + ATP \tag{4}$$

In the presence of arsenate the reactions go as follows:

$$
\begin{array}{l}
CH_2 \cdot O \cdot H_2PO_3 \\
| \\
CHOH \\
| \\
CO + H_3AsO_4 \\
| \\
S \cdot Enzyme
\end{array}
\rightleftharpoons
\begin{array}{l}
CH_2 \cdot O \cdot H_2PO_3 \\
| \\
CHOH \\
| \\
CO \cdot O \cdot H_2AsO_3 + HS \cdot Enzyme
\end{array}
\tag{5}
$$

$$
\begin{array}{l}
CH_2 \cdot O \cdot H_2PO_3 \\
| \\
CHOH \\
| \\
CO \cdot O \cdot H_2AsO_3
\end{array}
\rightleftharpoons
\begin{array}{l}
CH_2 \cdot O \cdot H_2PO_3 \\
| \\
CHOH \\
| \\
COOH + H_3AsO_4
\end{array}
\tag{6}
$$

In this way the coupled phosphorylation of ADP is abolished; the energy of ATP is not available; the plant must slowly succumb (Dixon and Webb, 1958).

Arsenate has other profound effects on plant systems. For example, Figure 10-1 shows the relative effects of arsenite and arsenate upon the activation of the enzyme fumarase. Fumaric acid is a common constituent of all plants being involved in the citric acid cycle. Fumarase carries on a hydrolysis of fumarate to L-malate.

The above examples typify the role played by the organic arsenical herbicides in plant metabolism. When one considers the number of reactions in plants involving SH groups and phosphorus it is easy to appreciate the ways in which arsonates in particular may upset plant metabolism and interfere with normal growth. Probably the ability of arsonate to enter into reactions in place of phosphate is the most important way in which arsenic serves as a toxicant. Not only does it substitute for phosphate in a number of ways; work with labeled arsonates indicates that these compounds are absorbed and translocated much as phosphates are. It is difficult to visualize a more effective way in which an herbicide might kill a plant.

Duble and Holt (1970) studied the effects of AMA on synthesis and utilization of food reserves in the tubers of purple nutsedge. Their tests indicated that repeated applications (3-week intervals) of AMA accelerated starch hydrolysis but had little effect on the fat and protein contents. The loss of starch was correlated with increased arsenic in the tubers. In leaf tissue arsenic treatment significantly reduced the sugar content and increased proteins. Tracer studies on the rate of $^{14}CO_2$ fixation and evolution suggested that AMA-treated plants had a higher rate of utilization of the assimilates than did untreated plants. Respiration of treated leaf discs from AMA treated plants was higher than that of control discs for a period of 8 days after treatment. While Duble and Holt (1970) did not measure photosynthesis they found that CO_2 utilization was high in AMA-treated plants. Thus untreated plants accumulated more photosynthates than AMA-treated plants.

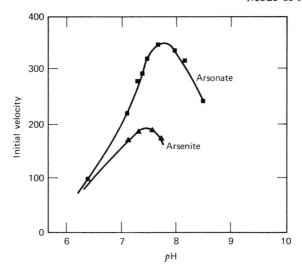

FIGURE 10-1.   Activation of fumarase by arsonate and arsentite ions as a function of *p*H. (Dixon and Webb, 1958.)

Keeley and Thullen (1970), using tubers from treated and control green-house cultures of yellow nutgrass (*Cyperus esculentus*), found that DSMA-and MSMA-treated plants produced tubers having significantly larger quantities of arsenic (4 to 33 ppm) than did control tubers (1 ppm). Tubers from treated plants sprouted in fewer numbers; small tubers from treated plants contained more arsenic (23 to 33 ppm) than did large tubers (4 to 12 ppm). The treatments reduced the vitality of small tubers but had little or no effect on sprouting of large tubers.

Mode of Action

Considering the overall action of arsenicals as herbicides in plants it seems important that they are able to penetrate the cuticle and enter into the apoplast phase of the plant system. Here they may move with transpiration water and bathe the cells of the foliar organs to which they have been applied. At low concentration it seems possible that arsenites may be absorbed into the symplast, and translocated at least for short distances. Under most condi-tions in which these compounds have been used in the field, their concentra-tions have been such that rapid contact injury has precluded any extensive translocation. This is related, partly at least, with their rapid effect in mem-brane degradation.

The arsonates, in contrast, have much lower contact toxicity; they are absorbed and translocated, at least in those species that have succumbed to treatment; johnsongrass and nutsedges are examples. In these susceptible perennial weeds the great virtue of MSMA and DSMA has been their ability to penetrate into and cause destruction of underground tubers and rhizomes. Thus from a few repeated applications these arsenicals have controlled two of the most serious perennial weed species that we have; species that have resisted control by any other means.

As topical sprays these compounds are inactivated almost instantaneously upon contact with the soil and hence they may be used with impunity in many row crops; cotton is one of the more important of these. While arsenicals in ordinary herbicidal dosages are rapidly rendered unavailable to plants in the soil, and although most soils have a very great capacity to inactivate and hold arsenic, eventually arsenic residues in soils may become troublesome. For this reason, in any weed control activity involving arsenical herbicides, integrated programs using herbicide rotation should be employed. In this way occasional use of the organic arsenicals in the particular roles in which they are highly effective may not result in soil residues of any significance.

As for the chemical mechanisms by which the organic arsonates kill plants, their relatively slow action involving translocation and producing chlorosis as a primary symptom would seem to implicate disturbance of phosphorous metabolism as was indicated on pages 157–158. Not only are they absorbed and translocated much like phosphates by plants; in the cells they affect many organelles including the chloroplasts, in all of which phosphorus plays important roles.

This interpretation is further strengthened by evidence of Schweizer (1967) that the addition of phosphorus to two silt loam soils increased the toxicity of DSMA to cotton possibly by saturating sites in these soils upon which both arsenite and phosphorus are fixed. As early as 1934, Albert reported that residues of calcium arsenate became more toxic to several crops where heavy applications of phosphate fertilizer were made. Substantial evidence indicates that phosphates and arsonates tend to replace each other chemically but that arsenic cannot serve the many essential roles of phosphorus in plants.

Possibly the uncoupling of oxidative phosphorylation and the formation of complexes with sulfhydryl containing enzymes may also enter the picture of arsenic phytotoxicity. However, trivalent arsenic is the form commonly associated with these effects and this would implicate arsenites rather than arsenates or arsonates.

REFERENCES

Albert, W. B. 1934. Arsenic solubility in soils. *South Carolina Agr. Expt. Sta. Ann. Rept.* **47**:45–46.

Anderson, O. 1958. Studies on the absorption and translocation of Amitrol (3-amino-1,2,4-triazole) by nutgrass. *Weeds* 6:370–385.

Anon. 1970a. *Weeds Controlled by Broadside.* The Ansul Company. Marinette, Wis. Form No. C7091.

Anon. 1970b. *Ansar 170 H.C.; Ansar 529 H.C.; Ansar 8100; Broadside.* The Ansul Company. Marinette, Wis.

Arnott, J. T. and A. L. Leaf. 1967. The determination and distribution of toxic levels of arsenic in a silt loam soil. *Weeds* 15:121–124.

Baker, R. S., H. F. Arle, J. H. Miller, and J. T. Holstun, Jr. 1969. Effects of organic arsenical herbicides on cotton response and chemical residues. *Weed Sci.* 17:37–40.

Crafts, A. S. 1933. The use of arsenical compounds in the control of deep-rooted perennial weeds. *Hilgardia* 7:361–372.

Crafts, A. S. 1935. The toxicity of sodium arsenite and sodium chlorate in four California soils. *Hilgardia* 9:461–498.

Crafts, A. S. 1937. The acid-arsenical method in weed control. *Jour. Amer. Soc. Agron.* 29:934–943.

Crafts, A. S. and H. G. Reiber. 1945. Studies on the activation of herbicides. *Hilgardia* 16:487–500.

Dixon, M. and E. C. Webb. 1958. *Enzymes.* Longman Group Ltd., Harlow, England. 782 pp.

Duble, R. L. and E. C. Holt. 1970. Effect of AMA on synthesis and utilization of food reserves in purple nutsedge. *Weed Sci.* 18:174–179.

Duble, R. L., E. C. Holt, and G. G. McBee. 1968. The translocation of two organic arsenicals in purple nutsedge. *Weed Sci.* 16:421–424.

Frost, D. V. 1968. Arsenic: Science or superstition. *Food and Nutrition News* (Oct. 1968).

Frost, D. V. 1969. Arsenic: Milestones in history. *Food and Nutrition News* (May 1969).

Hamilton, K. C. 1969. Repeated foliar applications of herbicides on johnson-grass. *Weed Sci.* 17:245–250.

Hamilton, K. C. and H. F. Arle. 1970. Directed applications of herbicides in irrigated cotton. *Weed Sci.* 18:85–88.

Holt, E. C., J. L. Faubian, W. W. Allen, and G. G. McBee. 1967. Arsenic translocation in nutsedge tuber systems and its effect on tuber viability. *Weeds* 15:13–15.

Keeley, P. E. and R. J. Thullen. 1970. Vitality of tubers of yellow nutsedge treated with arsenical herbicides. *Weed Sci.* 18:437–439.

Keeley, P. E. and R. J. Thullen. 1971a. Cotton response to temperature and organic arsenicals. *Weed Sci.* 19:297–300.

Keeley, P. E. and R. J. Thullen. 1971b. Control of nutsedge with organic arsenical herbicides. *Weed Sci.* 19:601–606.

Kempen, H. M. 1970. A study of monosodium methane-arsonate in plants. Masters dissertation, University of Calif. Davis. pp. 129.

Kleifeld, Y. 1970. Combined effect of trifluralin and MSMA on johnsongrass control in cotton. *Weed Sci.* **18**:16–18.

Long, J. A. and E. C. Holt. 1959. Selective and non-selective performance of several herbicides for the control of southern nutgrass (*Cyperus rotundus*). *12th Southern Weed Control Conf.* p. 195.

Long, J. A., W. W. Allen, and E. C. Holt. 1962. Control of nutsedge in bermudagrass turf. *Weeds* **10**:285–287.

McBee, G. G., P. R. Johnson, and E. C. Holt. 1967. Arsenic residue studies on coastal bermudagrass. *Weeds* **15**:77–79.

McWhorter, C. G. 1966. Toxicity of DSMA to johnsongrass. *Weeds* **14**:191–194.

Millhollon, R. W. 1969. Control of johnsongrass on drainage ditchbanks in sugarcane. *Weed Res.* **17**:370–373.

Rumburg, C. B., R. E. Engel, and W. F. Meggitt. 1960. Effect of temperature on the herbicidal activity and translocation of arsenicals. *Weeds* **8**:582–588.

Sachs, R. M. and J. L. Michael. 1971. Comparative toxicity among four arsenical herbicides. *Weed Sci.* **19**:558–564.

Schweizer, E. E. 1967. Toxicity of DSMA soil residues to cotton and rotational crops. *Weeds* **15**:72–76.

Sckerl, M. M. and R. E. Frans. 1969. Translocation and metabolism of MAA-^{14}C in johnsongrass and cotton. *Weed Sci.* **17**:421–427.

Wilkinson, R. E. and W. S. Hardcastle. 1969. Plant and soil arsenic analyses. *Weed Sci.* **17**:536–537.

CHAPTER 11

Benzoics

Many substituted benzoic acids have been investigated for their possible use as herbicides. However, only those listed in Table 11-1 have been developed into commercial products.

TABLE 11-1. Common Name, Chemical Name, and Chemical Structure of Benzoic Acid Herbicides

Common name	Chemical name	—2	—3	—4	—5	—6
2,3,6-TBA	2,3,6-trichlorobenzoic acid	—Cl	—Cl	—H	—H	—Cl
dicamba	3,6-dichloro-*o*-anisic acid	—OCH$_3$	—Cl	—H	—H	—Cl
tricamba	3,5,6-trichloro-*o*-anisic acid	—OCH$_3$	—Cl	—H	—Cl	—Cl
chloramben	3-amino-2,5-dichlorobenzoic acid	—Cl	—NH$_2$	—H	—Cl	—H
PCA	mixture: trichloro- and tetra-chlorobenzoic acids predominate, but other chlorinated benzoic acids also present					

Zimmerman and Hitchcock (1942) pointed out the growth regulating properties of the substituted benzoic acids. As early as 1948 the Jealott's Hill

Experiment Station in England evaluated the herbicidal properties of 2,3,6-TBA under field conditions. Similar investigations were also conducted in the United States at about this same time. PCA and 2,3,6-TBA are primarily used to control perennial broadleaf weeds and selected woody plants. These two materials are resistant to degradation in soils and therefore persist for some period of time.

Acceptable selectivity to several crop species has been obtained with certain chlorinated benzoic acids by substituting a methoxy (—OCH_3) or an amino (—NH_2) group for one of the chlorine atoms. Dicamba is used in barley, corn (maize), grasses (pasture and rangeland), oats, sorghum, and wheat. Chloramben is used in asparagus, beans, corn (maize), peanuts, peppers, pumpkins, soybeans, squash, sunflower, sweet potatoes, and tomatoes. Dicamba has been shown to reduce the intensity of witchweed infestation of corn plants but could not be recommended for this purpose because of yield decreases (Sand et al., 1971). The greatest use of any of the substituted benzoic acid herbicides is that of chloramben in soybeans.

Growth and Plant Structure

Trichlorobenzoic acid (2,3,6-TBA) promoted cell elongation, proliferation of tissue, and the induction of adventitious roots in studies by Zimmerman and Hitchcock (1951). Epinasty of young shoots following treatment with 2,3,6-TBA is often observed. In addition to these effects, 2,3,6-TBA often completely inhibits the apical meristems of dicotyledons and, although the plant does not die, it is incapable of further growth. However, in monocots it primarily affects the nodal meristems causing a weakness that results in lodging. Buchholtz (1958) reported that 2,3,6-TBA repressed the regrowth of quackgrass (*Agropyron repens*) to a greater degree than the 2,3,5-TBA or other mono-, di-, or tetra-chlorinated benzoic acids. Keitt (1960) showed that 2,3,6-TBA not only inhibited root elongation but also inhibited geotropic curvature of the roots. Inhibition of the typical geotropic response of stems by 2,3,6-TBA has been reported by Jones et al. (1954) for *Avena* coleoptile. Vander Beek (1959) observed that upright shoot growth of oats, barley, and *Cucumis sativus* was inhibited by 2,3,6-TBA and to a lesser extent by 2,6-dichlorobenzoic acid. Growth of isolated coleoptile segments of *Avena* coleoptile was stimulated by 10^{-3} and 10^{-4}M 2,3,6-TBA but the geotropic response was inhibited (Schrank, 1960; Schrank and Rumsey, 1961; and Schrank, 1964). The analog 2,6-dichlorobenzoic acid was only about one-half as active in inhibiting the geotropic response. Vander Beek (1967) showed that 2,3,6-TBA and 2,6-dichlorobenzoic acid interfered with the phototropic response as well as the geotropic response. The inhibition of the geotropic response of the chlorinated benzoic acids has been suggested to be caused by a depression of the gravity perceiving mechanism (Schrank, 1964).

Dicamba affects plant growth in much the same way as 2,3,6-TBA. Rogerson and Foy (1968) observed that dicamba caused proliferative growth and consequent increase in diameter of stems in nearly all zones, especially in the first and second nodes above the cotyledons in Black Valentine beans. In each zone where stem swelling occurred there was a very marked decrease in anthocyanin content. Pate et al. (1965) reported that dicamba caused destruction of the tissue within the nodes of alligatorweed (*Alternanthera philoxeroides*). This was associated with dissolution of phloem and parenchyma cell walls with apparent multinucleate coagulated protoplasts. There was no collapse of cell walls. In transverse sections, areas of destroyed tissue were observed including cortical parenchyma adjacent to the bundles. Increased permeability of cell membranes of purple nutsedge (*Cyperus rotundus*) leaves induced by dicamba has been noted (Magalhaes and Ashton, 1969). Wuu and Grant (1966) reported that dicamba inhibited mitosis in barley.

Keitt and Baker (1966) reported that 2,3,6-TBA, dicamba, and chloramben promoted growth with cell division in the presence of kinetin as well as cell enlargement. This information, along with that mentioned above, shows that these herbicides have auxin-like properties.

Absorption and Translocation

In a series of papers, Venis and Blackman (1966a, 1966b, 1966c) reported on the uptake of 2,3,6-TBA by stem segments of a number of species. They found that 2,3,6-TBA was absorbed relatively rapidly but after 2 to 6 hours the rate of loss from the stem segments into the external medium exceeded the rate of uptake, resulting in a reduction in the amount of the herbicide present. They found that this "leakage" of the herbicide from the stem tissue was prevented by streptomycin and other compounds containing a cationic nitrogen group. They proposed that the uptake of 2,3,6-TBA was governed by an unstable accumulatory system (type I). The type I system involves absorption by some membrane system through an interaction between carbonyl anions and the quarternary ammonium group of the choline moiety of α-lecithin. The hydrolysis of lecithin by phospholipase-D destroys the type I binding while cationic nitrogen compounds maintain a positive herbicide uptake by competing with the choline quaternary ammonium group of lecithin for the anionic site of phospholipase-D.

Linder et al. (1958) observed that when 2,3,6-TBA was applied to the leaves or stems of young bean plants it was absorbed and translocated to the roots. It was then excreted (leaked) into the surrounding soil. Untreated bean plants growing in the soil then absorbed the 2,3,6-TBA and translocated it to the above ground parts of the plant where it induced growth modifications characteristic of the herbicide. Subsequent quantitative studies by Linder et al. (1964) showed that when 63 μg of ^{14}C-2,3,6-TBA was applied over 1 cm^2 of a bean leaf in Tween-20-lanolin mixture, 1.5 μg per day was exuded from the roots during the 2 days immediately following the treatment. It would be interesting to know whether the "leakage" of 2,3,6-TBA from roots could be inhibited by compounds containing a cationic nitrogen group as was reported by Venis and Blackman (1966b, 1966c) for 2,3,6-TBA "leakage" from stem segments.

The translocation of 2,3,6-TBA from a host plant, maize or sorghum, to thè parasitic plant, witchweed (*Striga lutea*), has been reported (Egley and Kust, 1964).

Mason (1960), using tritiated 2,3,6-TBA, proved that translocation out of the primary leaf of bean occurred within 3 hours and that the chemical was rapidly accumulated in the developing trifoliate leaf bud. After 54 hours more than 50% of the activity absorbed by the treated leaf was concentrated in this bud. The strong morphological symptoms which developed in the trifoliate leaf system after 27 hours were associated with the progressive accumulation of activity in the bud. From the pattern of distribution of the radioactivity, it was concluded that movement of 2,3,6-TBA from the leaf took place via the living cells of the phloem.

In contrast to bean, the more resistant corn plant translocates 2,3,6-TBA from the treated leaf only slowly. Accumulation occurs not in the apical bud but in the more mature parts of the untreated leaves. Some movement from the treated spot toward the leaf tip apparently takes place in the non-living apoplast. Root application of 2,3,6-TBA to bean results in rapid uptake and translocation into the tops via the xylem. There is little accumulation in the root cells. Accumulation of activity in the trifoliate leaf bud by secondary movement in the phloem takes place within 9 hours. Accumulation at sites of meristematic activity in bean is associated with movement in the assimilate stream.

Root uptake of 2,3,6-TBA by corn is slow, and translocation from roots is limited. Accumulation is not in regions of meristematic activity of corn, such as the apical bud and intercalary region, but in marginal and terminal areas of the leaves. Primary movement of 2,3,6-TBA from roots was via the transpiration stream; secondary movement in the assimilate stream was much less prominent than in bean. Guttate from treated barley seedlings exposed to 2,3,6-TBA in the culture solution contained the original labeled unmetabolized compound. This indicates that monocotyledons may retain 2,3,6-TBA in the xylem and transfer much less to the phloem than do susceptible dicotyledons.

Apparently, accumulation of 2,3,6-TBA in active meristems is a major factor determining susceptibility. The rate of accumulation is determined by the relative rates of absorption, translocation, and retention within the site of action. There was no evidence that corn had greater ability than bean to degrade 2,3,6-TBA; no metabolites were identified from either plant whether uptake was by leaf or root. The labeled compound assayed in the distribution study in both species was shown by co-chromatography and radiochromatogram scanning to be unmetabolized ^{3}H-2,3,6-TBA. On retreatment of bean with this extracted material, characteristic morphological symptoms were again produced in the developing trifoliate leaf. Activity that could not be extracted from plant tissue by standard methods was shown to be associated with protein. The labeled compound was freed by denaturation of the precipitated protein. Apparently the high residual character of 2,3,6-TBA found in greenhouse and field studies is due to its great resistance to biological breakdown.

The above results show that 2,3,6-TBA is readily absorbed by leaves, stems, and roots; furthermore, it is translocated via both the symplastic and apoplastic systems with accumulation in areas of high metabolic activity, i.e., active meristems.

A review by Zick and Castro (1966) has summarized much of the research conducted on dicamba.

^{14}C-dicamba absorption by leaf sections of wheat and wild buckwheat (*Polygonum convolvulus*) was investigated by Quimby and Nalewaja (1968). A sigmoidal uptake pattern was obtained with both species when the amount of dicamba absorbed was plotted arithmetically against the logarithm of time from 0.75 to 48 hours. The Q_{10} of absorption was positive. Boiled leaf sections absorbed about twice as much dicamba. The addition of a 100-fold concentration of unlabeled dicamba increased the absorption of ^{14}C-dicamba. These workers suggested that the absorption was associated with binding to proteins and that the increased uptake with high dicamba concentrations and boiling of leaf sections was related to an increase in the number of binding sites on the adsorptive proteins. However, it is also possible that these variables may have altered membrane permeability and the increased uptake could also have been associated with this phenomenon.

Leonard and Glenn (1968) studied the absorption and translocation of ^{14}C-dicamba in detached primary leaves of bean. The herbicide was absorbed by the lamina and transported basipetally into the petiole with concentration in the basal position. However, pretreatment of the leaves for 5 to 10 minutes with unlabeled dicamba prior to the application of ^{14}C-dicamba caused a more uniform labeling of the petiole and somewhat greater total amount.

The absorption and translocation of dicamba from leaf applications has been studied by many workers. As early as 1964, Linder et al. (1964) showed that dicamba was absorbed by leaves of beans, translocated basipetally and exuded from the roots into the culture solution; 1% of the applied dicamba was present in the culture solution after 24 hours. They used bioassays in this study. Egley and Kust (1964) demonstrated that dicamba was absorbed and translocated basipetally from leaves of maize or sorghum and translocated acropetally in the attached parasitic witchweed (*Striga lutea*) plant. Abnormal development of the terminal and lateral buds of witchweed was used to detect this movement. Although these investigations showed that dicamba was absorbed and translocated from foliar applications, more detailed studies were possible when radioactive ^{14}C-dicamba became available. Many researchers have shown that dicamba is absorbed by leaves and exported to other parts of the plant in several species; beans (Hurtt and Foy, 1965a and 1965b); Canada thistle (*Cirsium arvense*) (Vanden Born, 1966; Chang and Vanden Born, 1968); purple nutsedge (Magalhaes et al., 1968; Ray and Wilcox, 1969); skeleton weed (*Chondrilla juncea*) (Greenham, 1968); Tartary buckwheat (*Fagopyrum tataricum*) (Vanden Born, 1966; Chang and Vanden Born, 1971); wild buckwheat (Quimby and Nalewaja, 1966); quackgrass and Kentucky bluegrass (*Poa pratenis*) (Vanden Born, 1966); wheat (Quimby and Nalewaja, 1966); maize and cucumber (Hahn et al., 1969). Several of these investigators reported an accumulation of dicamba in meristems or young developing plant parts indicating symplastic translocation. Leakage of dicamba into the culture solution or soil was also observed in several of these studies. The translocation of dicamba appears to be somewhat slower from treated leaves of grasses (Vanden Born, 1966) and sedges (Magalhaes et al., 1968) than from leaves of many broadleaf species. Dicamba has been shown to be translocated to some extent from "mother plants" to "daughter plants" of purple nutsedge (Magalhaes et al., 1968; Ray and Wilcox, 1969) and from main culm to tillers of wheat (Quimby and Nalewaja, 1966). Magalhaes et al. (1968) found that translocation and distribution of dicamba in purple nutsedge from a treated leaf was reduced after the plant had flowered.

Dicamba has also been shown to be readily absorbed by roots and translocated throughout the plant in several species: grapes (Leonard et al., 1966); purple nutsedge (Magalhaes et al., 1968); maize and cucumber (Hahn et al., 1969); Tartary buckwheat, wild mustard, barley, and wheat (Chang and Vanden Born, 1971). In purple nutsedge, where dicamba translocation from foliar and root applications was investigated, root applications resulted in an accumulation of dicamba at the tips of mature leaves (Magalhaes et al., 1968). This suggests that at least in certain species the rate of apoplastic transport of dicamba exceeds that of symplastic transport.

These results show that dicamba is readily absorbed by leaves and roots, furthermore it is readily translocated via both the symplastic and apoplastic systems.

Chloramben absorption by seeds has probably been investigated more extensively than the uptake of any other herbicide by seeds. These types of studies are especially relevant for soil-applied herbicides which act on seed germination or early seedling growth rather than mature plants. Unfortunately such studies are lacking for many herbicides of this type. Haskell and Rogers

(1960) presoaked seeds of maize, soybean, and jimsonweed (*Datura stramonium*) for various periods from 12 to 72 hours and then placed them into solutions of radioactive chloramben for 3 hours. The distribution of the herbicide in the seed was determined by microautoradiography. Chloramben did not penetrate the seeds of maize within the 12-hour presoaking period but was present in the embryo axis within the 24-hour presoaking period and was distributed throughout the seed within the 72-hour presoaking period. In soybean, chloramben was found throughout the seed even at the minimal presoaking time, 12 hours. However, with the dormant seed of jimsonweed, chloramben did not enter the seed during the 3-hour herbicide treatment period at any of the presoaking times. Rieder et al. (1970) showed that the uptake of chloramben by dry seeds of soybean was directly proportional to the concentration; uptake increased with increasing temperatures from 10° to 30°C; uptake was reversible and uptake by dead seeds was similar to that of living seeds. Furthermore, they found that the rate of uptake was not associated with the rate of water uptake and concluded that the uptake was largely a physical process. Swan and Slife (1965) added another dimension to the study of seed uptake of chloramben by removing soybean seeds from the radioactive herbicide solution after the initial uptake period, placing them in sand culture and later solution culture, and observed the subsequent distribution of radioactivity with time. The cotyledons retained most of the radioactivity through the 32-day experiment indicating very little translocation or redistribution. Most of the small amount of radioactivity which was translocated moved into the roots and only a trace was found in the aerial portions. These studies show that the absorption of chloramben by seeds is influenced by plant species, moisture status of the seed, and temperature. The uptake appears to be a physical phenomenon and redistribution of the herbicide from the seed to other plant parts during subsequent development is quite limited.

Chloramben has been shown to be absorbed by the roots of a number of species; however, subsequent translocation is quite limited in most species: soybean (Swan and Slife, 1965); carrots (Ashton, 1966); barley, cucumber, and soybean (Hodgson, 1967); cucumber and squash (Baker and Warren, 1962); tomatoes (Colby et al., 1964); beans and barley (Crafts and Yamaguchi, 1964); and squash, soybean, cucumber, morning glory (*Ipomoea hederocea*), and velvetleaf (*Abutilon theophrasti*) (Stoller and Wax, 1968). However, in contrast to the above generalization, Stoller and Wax (1968) reported that 29% of the root absorbed chloramben was transported to the shoot of giant foxtail (*Setaria faverii*) in 24 hours. In some of these studies the plant tolerance relationship to transport was examined and in general there was little if any correlation.

Many of the reports mentioned in the previous paragraph suggest that chloramben is tightly bound in the roots of plants and therefore is not available for transport to the aerial portions of the plant. In carrots, it has been

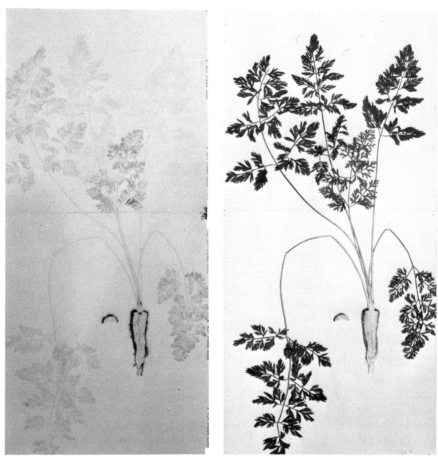

Figure 11-1. Carrot plant grown in soil treated with ^{14}C amiben at 6 lb/A 191 days after planting. Plant right; radioautograph left. (Ashton, 1966.)

further shown that the radioactivity is bound very close to the surface of the root and remains there even 191 days after chloramben treatment, Figure 11-1 (Ashton, 1966). This radioactivity may be in the form of N-glucosyl chloramben rather than unaltered chloramben, since chloramben is rapidly converted to a sugar conjugate in carrot roots.

Stoller (1969) used excised roots of the chloramben-sensitive velvetleaf and chloramben-tolerant morning glory in a kinetic study of chloramben absorption. The initial rates of absorption and binding were identically concentration dependent for both species for external concentrations from 0.02 to 500 mg/liter.

Moody et al. (1970) investigated the release of chloramben from soybean roots of intact plants following a 1-hour uptake period from solutions containing 0.04, 0.08, and 0.16mM chloramben. The release of chloramben from the roots after 30 minutes in water was 93, 70, and 55%, respectively. After 2 hours these values were 59, 54, and 50%, respectively. The reduced values after 2 hours were the result of reabsorption of the released chloramben. They suggest that the herbicide released after a 1-hour uptake was that diffusing out of "free space" rather than that released from uptake or adsorptive sites.

Crafts and Yamaguchi (1964) reported that chloramben uptake and movement by a bean leaf is symplastic, with medium-strong transport to roots and no retransport via the xylem. In a barley leaf, uptake is medium with appreciable movement to the roots. However, Slife (1963) suggested in a review that the movement of chloramben from leaves appears to be almost imperceptibly slow.

Taylor and Warren (1968) studied the translocation of chloramben through vertical 6 mm-long bean petiole sections into agar blocks; although several herbicides did move into the agar block there was no movement of the absorbed chloramben. However, when the sections were treated with dinoseb, chloramben was transported but moved primarily as N-glucosyl chloramben. Sublethal concentrations of dinoseb with chloramben, applied to whole plants in nutrient solution, did not appreciably influence the uptake and movement of chloramben by bean, cucumber, or squash. Higher dinoseb concentrations decreased uptake and subsequent translocation. It appears that dinoseb stimulated translocation of chloramben through the bean petiole sections by increasing membrane permeability.

Molecular Fate

Swanson (1969) recently reviewed the degradation of the benzoic acid herbicides in plants and soils.

Trichlorobenzoic acid (2,3,6-TBA) appears to be very stable in higher plants. Mason (1960) was unable to detect any degradation product of 2,3,6-TBA from maize, barley, or bean plants. However, about 10% of the radioactivity was unextractable and was found to be 2,3,6-TBA which was associated with protein. Straw and seed heads of wheat were found to contain 2,3,6-TBA at harvest following root or foliage treatments (Balagannis et al., 1965). It was detected by gas-liquid chromatography and other peaks were detected that could have been degradation products of 2,3,6-TBA. However, since they were not identified, it is difficult to know whether the herbicide was actually degraded. There appears to be no clear evidence that 2,3,6-TBA is actually degraded by higher plants, although one or more conjugates may be formed.

Dicamba degradation rate in higher plants varies greatly with species. In purple nutsedge no metabolism of dicamba was detected after 10 days (Magalhaes et al., 1968; Ray and Wilcox, 1969), whereas in wheat dicamba was completely metabolized in 18 days although a small amount persisted as a conjugate after 29 days (Broadhurst et al., 1966). These later investigators also found that the major metabolite was 5-hydroxy-2-methoxy-3,6-dichloro-benzoic acid and a minor metabolite was identified as 3,6-dichlorosalicylic acid. The major metabolite accounted for 90% of the radioactivity while the minor metabolite contained 5%. The remaining 5% was dicamba. These identifications were made after hydrolysis, suggesting that they may have been present as conjugates. They obtained similar results with Kentucky bluegrass. Quimby and Nalewaja (1966) also found that dicamba was degraded in wheat, but considered their major metabolite to perhaps be 3,6-dichlorosalicylic acid and not the 5-hydroxy derivative of dicamba. They also reported that more of the metabolite was formed in wheat than in wild buckwheat. Ray and Wilcox (1967) reported that in roots of maize dicamba was converted to the 5-hydroxy derivative as the major metabolite and 3,6-dichlorosalicylic acid as the minor metabolite. However, in barley roots, they suggest that both the 5-hydroxy derivative and 3,6-dichlorogentisic acid are major metabolites and 3,6-dichlorosalicylic acid is a minor metabolite.

Hull and Weisenberg (1967) observed that while dicamba was unchanged in beans after 5 days, in johnsongrass about 50% had been converted to at least two metabolites. Canada thistle, Tartary buckwheat, and quackgrass contained 85% dicamba up to 6 weeks after treatment (Vanden Born, 1966). Apparently the conjugation of the metabolites of dicamba also varies with species since Chang and Vanden Born (1968) found only 1.0 to 1.7% of the total radioactivity of the tissue was released from the ethanol-insoluble fraction by acid hydrolysis in Canada thistle 54 days after treatment. They also found that about 20% of the total activity was released as $^{14}CO_2$, suggesting the decarboxylation of dicamba or a metabolic product.

Chang and Vanden Born (1971a) found that detoxification of dicamba occurred in Tartary buckwheat, wild mustard, barley, and wheat, but not at equal rates. Figure 11-2 shows the time-course of dicamba metabolism in experiments using ^{14}C-dicamba (Chang and Vanden Born, 1971b). Wheat and barley obviously alter the dicamba molecule more extensively than do Tartary buckwheat and wild mustard. A non-toxic metabolite was 5OH-dicamba. Decarboxylation did occur but at a very slow rate.

Quimby and Nalewaja (1971) used ^{14}C-dicamba in studies on the selectivity of this herbicide between wild buckwheat, a susceptible species, and wheat, a resistant one. Selectivity could not be explained on the basis of differences in uptake by leaf sections. ^{14}C-dicamba accumulated in meristems of wild buckwheat but not in the youngest tillers of wheat; the main culms of wheat

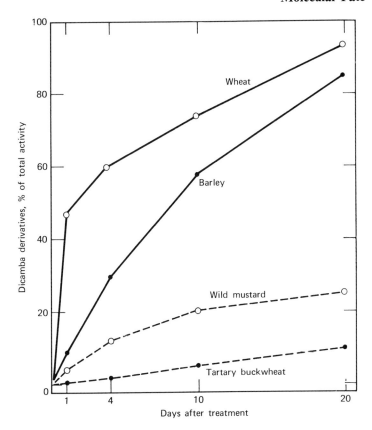

FIGURE 11-2. Time-course of dicamba metabolism in Tartary buckwheat, wild mustard, barley, and wheat, following application of 0.1 μc of ^{14}C-dicamba to a single leaf of each plant. Radioactivity in dicamba derivatives is shown as a percentage of the total radioactivity recovered. Data plotted are means for two replicates of four (Tartary buckwheat and wild mustard) or ten (barley and wheat) plants each. (Chang and Vanden Born, 1971b.)

conjugated or metabolized ^{14}C-dicamba more quickly than buckwheat meristems. The writers suggest that selectivity of dicamba is related to interspecific differences in translocation and metabolism.

A proposed pathway of dicamba degradation in higher plants is given in Figure 11-3.

Chloramben molecular fate in higher plants is largely confined to its complexing with endogenous compounds although small amounts of unidentified degradation products have been isolated and some CO_2 may be evolved from

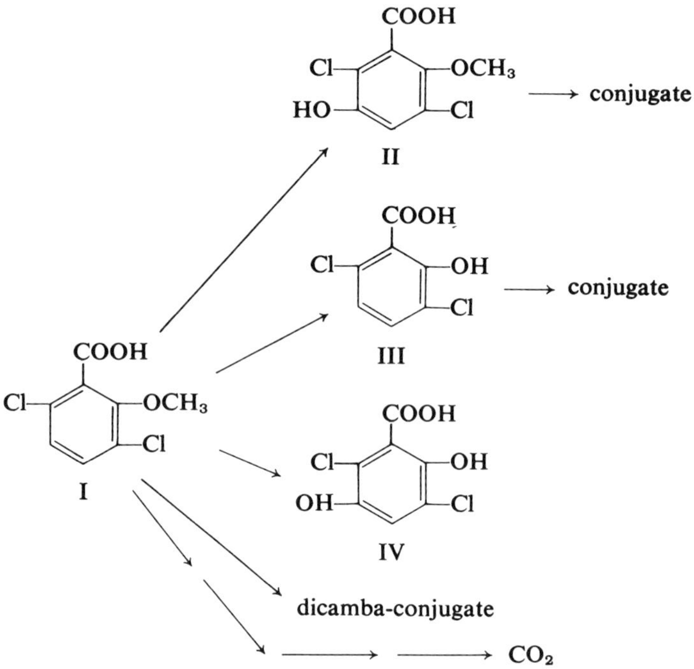

FIGURE 11-3. Proposed pathways for dicamba degradation in higher plants. Specific pathways apparently vary with species and in some species degradation has not been detected: (I) dicamba; (II) 5-hydroxy-2-methoxy-3,5-dichlorobenzoic acid; (III) 2-hydroxy-3,6-dichlorobenzoic acid (3,6-dichlorosalicylic acid); (IV) 2,5-dihydroxy-3,6-dichlorobenzoic acid (2,5-dihydroxygentisic acid).

chloramben in some species. Sutherland (1961) observed that chloramben formed a conjugate in soybean plants and that chloramben could be released from this conjugate by alkaline hydrolysis. Subsequently numerous papers have been published confirming the formation of a conjugate in a variety of species. Researchers have also found a chloramben complex in soybean (Colby, 1965; Colby, 1966; Swanson et al., 1966a). A chloramben complex has also been reported in carrots (Ashton, 1966), cucumbers (Baker and Warren, 1962), tomato (Colby et al., 1964), barley (Colby, 1966), sugarbeets (Swanson et al., 1966a), peas (Swanson et al., 1966b), as well as several other species. The major conjugate was isolated from soybean and shown to be chromatographically identical to synthetic N-glycosyl chloramben by Colby (1965). Additional structural proof of N-glycosyl chloramben, N-(3-carboxy-2,5-dichlorophenyl)-glycosylamine, was provided by Swanson et al. (1966a) using thin-layer and gas chromatography and confirmed by infrared spectroscopy.

Frear et al. (1967) reported that N-glucosyl chloramben biosynthesis in soybean hypocotyl sections was heat labile, sensitive to freezing and thawing, unaffected under anaerobic conditions, dependent on chloramben concentration, and proportional to tissue section concentration. Its formation was maximal between pH 6.3 and 7.3 and was inhibited by several metabolic inhibitors. In addition to the soybean hypocotyl section studies, they also investigated N-glucosyl chloramben formation in various organs of tomato, barley, sugarbeets, cucumber, and soybeans. They found little correlation between the ability of a plant to form N-glucosyl chloramben and its resistance to the herbicide. A direct comparison of root tissue with leaf tissue from barley and soybean indicated that relatively little N-glucosyl chloramben formation occurs in leaf tissue.

Stoller and Wax (1968) investigated the absorption, translocation, and metabolism of chloramben in five species of plants in an attempt to determine their bases of selectivity. In intact plant, neither the amount of chloramben absorbed, the transport to the shoots, the concentration of radioactivity in a soluble methanol fraction, nor the distribution of radioactivity among chloramben, N-glucosyl chloramben, and chloramben-X (an unknown conjugate) was associated with species sensitivity. However, using excised shoot or root seedling tissue, they found that the percentage composition of each of the three compounds was significantly correlated with plant sensitivity. There was an equilibrium established between the three compounds within 24 hours after treatment. As plant tolerance increased, the distribution of radioactivity shifted toward a smaller percentage of chloramben-X and chloramben with proportionately more in N-glucosyl chloramben. The relationship of this finding to selectivity is still open to question since even in excised seedling tissue the concentration or total radioactivity or content of these three compounds was not correlated with sensitivity. It has been speculated that N-glucosyl chloramben is relatively non-phytotoxic and Stoller (1968) showed that chloramben-X is relatively non-phytotoxic. Additional research by Stoller (1969) has suggested that bases of selectivity may be that the tolerant plant possesses the capacity to (1) sustain higher internal concentrations of free chloramben for a parallel expression of the herbicidal effect and to (2) conjugate the absorbed chloramben more rapidly and to a greater extent than the susceptible plant. Freed et al. (1961) reported that a measurable amount of $^{14}CO_2$ was released from carboxyl-labeled chloramben fed to intact soybean plants and showed that this decarboxylation occurred in the roots but not the shoots. The lack of decarboxylation of chloramben by soybean shoots was also shown by Swan and Slife (1965). Ashton (1966) observed that carrot root disks also decarboxylated chloramben to a small degree, 4%, after 6 days. He also reported that 64% was present as a conjugate containing a sugar moiety, probably N-glucosyl chloramben, 19% as chloramben, and 12% was distributed among four additional unknown compounds. These results suggest that decarboxylation of chloramben is not a major degradation pathway but that it does occur in the root of at least some species. A proposed pathway of chloramben metabolism in higher plants is given in Figure 11-4.

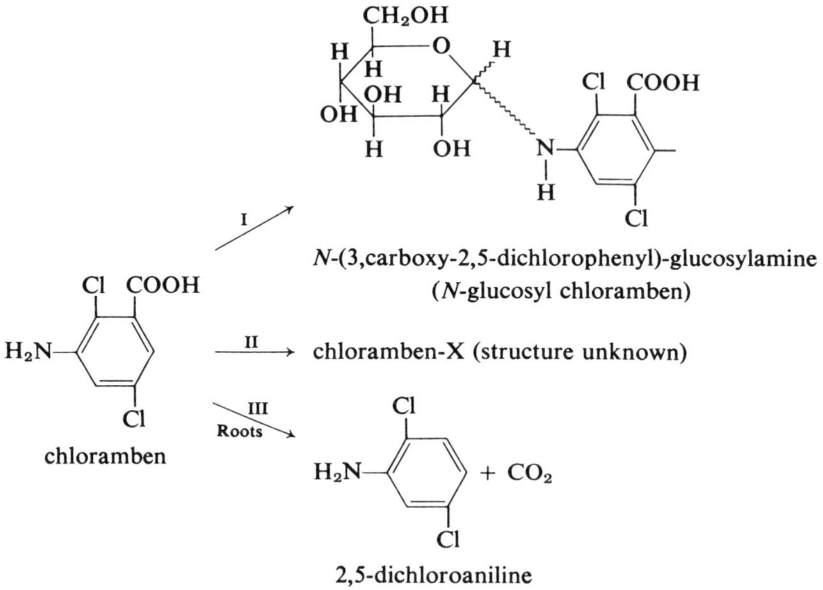

N-(3,carboxy-2,5-dichlorophenyl)-glucosylamine
(N-glucosyl chloramben)

chloramben-X (structure unknown)

2,5-dichloroaniline

FIGURE 11-4.   A proposed pathway of chloramben degradation in higher plants. Reaction I is the major pathway.

Biochemical Responses

Information on the biochemical changes induced in higher plants by the benzoic acid-type herbicides is quite limited. Some data have been derived from survey studies on the effect of a number of herbicides on a given process. However, in these cases the necessary detailed experimentation with the benzoic acid herbicides has usually not been conducted. Therefore, the following tends to be a compilation of the research which has been conducted in this area and any attempt to develop a mechanism of action would be purely speculative. This also suggests that this should be a fertile area for research.

The injury symptoms of the benzoic acid herbicides suggest that they may be acting as auxins or interfering with endogenous auxin transport. Keitt and Baker (1966) reported that chloramben, dicamba, and 2,3,6-TBA inhibited basipetal polar auxin (IAA) transport in excised stem sections of beans, although the concentrations required for severe transport inhibition were generally toxic to the tissue. Napthlam and 2,3,5-triiodobenzoic acid were at least 300 times more active in inhibiting auxin transport than chloramben, dicamba, or 2,3,6-TBA. Therefore, it appears that the inherent auxin-like properties of the benzoic herbicides are associated with their mechanism of action rather than their effect on auxin transport.

Pretreatment of segments of etiolated pea stems with 2,3,6-TBA induced the formation of 3-indolacetyl-L-aspartate and 1-naphthaleneacetyl-L-asparate from indolacetic acid and naphthalene acetic acids, respectively (Sudi, 1966). Enzyme induction was thought to have been involved.

Moreland and Hill (1962) reported that the relatively high concentration of $2 \times 10^{-3}M$ was required for a 50% inhibition of the Hill reaction of photosynthesis by 3,4,5-trichlorobenzoic acid. Several other chlorinated benzoic acids, including 2,3,6-TBA, did not inhibit the Hill reaction at this concentration. These results show that Hill reaction inhibition is not the mechanism of action of 2,3,6-TBA.

The effect of 2,3,6-TBA on oxidative phosphorylation by isolated mitochondria has been studied by Lotlikar et al. (1968) using citrate as the substrate. They found that at $1 \times 10^{-3}M$ of 2,3,6-TBA, oxygen uptake was inhibited 16% and phosphorylation was inhibited 22%. These inhibitions increased with increasing concentrations until at $1.2 \times 10^{-2}M$, oxygen uptake was inhibited 82% and phosphorylation 100%. Foy and Penner (1965) observed that 2,3,6-TBA inhibited oxygen uptake by isolated mitochondria 30, 43, 75, and 100% at 10^{-5}, 10^{-4}, 10^{-3}, and $10^{-2}M$, respectively, using succinate as the substrate. Alpha-ketoglutarate as the substrate yielded somewhat greater inhibitions. Although 2,3,6-TBA inhibits both oxygen uptake and phosphorylation, as well as uncoupling the reaction to a degree, the concentrations required are relatively high leading to some question as to the role of oxidative phosphorylation in the mechanism of action of the 2,3,6-TBA.

Chernyshev (1968) studied the effect of annual treatments of 2,3,6-TBA on carbohydrate metabolism in roots of Russian knapweed (*Centaurea repens*) under field conditions over a three-year period; the inulin content of the shallow roots progressively decreased. By the third year, no inulin was found on the roots from the 20- to 40-cm horizon but in the 40- to 80-cm horizon some was found; roots in the 80- to 100-cm horizon contained only one-half as much as the controls. The disaccharide content of the roots was also progressively reduced over the three-year period. The biochemical basis for the reduction in carbohydrate content of roots of Russian knapweed is not apparent from this study; however, Zhirmunskaya (1966) previously suggested that they are utilized as substrates in increased nitrogen metabolism.

He sampled roots of Canada thistle and perennial sowthistle (*Sonchus arvensis*) from 3 to 24 months following 2,3,6-TBA treatment and reported that there was a 2- to 3-fold increase in total nitrogen; this was accompanied by an intensive synthesis of protein and accumulation of non-protein forms of nitrogen at the expense of the amides. He further showed that this increase was not due to translocation from the aerial portions of the plants.

The adsorption or binding of 2,3,6-TBA to a model protein, bovine serum albumin, has been shown by equilibrium dialysis, Sephadex gel filtration, and a partitioning technique (Camper and Moreland, 1966).

Dicamba has been shown to inhibit the oxygen uptake of leaves of purple nutsedge as much as 40% following a foliar spray of 10^{-3} or 10^{-5}M (Magalhaes and Ashton, 1969). Foy and Penner (1965) reported that dicamba inhibited the oxygen uptake of isolated mitochondria 5, 51, and 68% at 10^{-5}, 10^{-4}, and 10^{-3}M, respectively, when succinate was used as the substrate. However, when α-ketoglutarate was used as the substrate only a slight inhibition was observed.

Quimby (1967) reported that dicamba influenced the DNA-precipitating properties of histone and that it may thereby interfere with the normal function of the genetic mechanism. Moreland et al. (1969) reported that dicamba inhibited gibberellic acid-induced α-amylase biosynthesis in distal halves of barley seeds but that it had little if any effect on RNA or protein synthesis. Wild buckwheat and wheat were used by Arnold and Nalewaja (1971) to study the effect of dicamba on RNA and protein synthesis. Dicamba increased RNA and protein content in wild buckwheat at the 5–8 cm growth stage and when flowering; these constituents were increased in wheat at the boot stage. Dicamba affected the transition temperature and precipitation of reconstituted nucleohistone; it did not affect the uncombined nucleic acid nor histone *in vitro*. Possibly a DNA-histone-dicamba complex was involved. Binding of dicamba to protein varied with different proteins. Ashton et al. (1968) reported that dicamba at 10^{-3}M had only a slight inhibitory effect on the development of proteolytic activity in the cotyledons of germinating squash seeds although it had a marked effect on seedling growth. Dicamba at 10^{-4}M did not significantly affect the normal increase in phytase activity in germinating barley seeds (Penner, 1970). All of the aforementioned enzymes are considered to be synthesized *de novo* and increase during germination. Therefore it does not appear that dicamba interferes with enzyme synthesis to a sufficient degree to explain its phytotoxic effect. However, the results of Moreland et al. (1969) seem to be an exception to this general statement.

Dicamba has also been reported to reduce the anthocyanin content of bean stems (Rogerson and Foy, 1968) and the transpiration of barley and Tartary buckwheat (Friesen and Dew, 1967).

Chloramben has been shown to inhibit oxygen uptake by isolated mitochondria 20, 35, and 75% at 10^{-5}, 10^{-4}, and 10^{-3}M, respectively, when succinate was used as the substrate (Foy and Penner, 1965). However, when α-ketoglutarate was the substrate no inhibition was detected at these concentrations. The reason for this difference is not apparent but with the relatively high herbicide concentration required for a marked inhibition with succinate as the substrate and lack of inhibition with α-ketoglutarate it is doubtful respiration *per se* is an essential part of the mechanism of action of chloramben.

Protein synthesis is not inhibited by chloramben (Mann et al., 1965; Moreland et al., 1969). Moreland et al. (1969) also reported that chloramben does not inhibit RNA synthesis or gibberellic acid induced α-amylase formation in deembryonated barley half-seeds. This latter effect was previously reported by Penner (1968); in addition, he reported that chloramben inhibited α-amylase development in intact seeds and the addition of gibberellic acid could not overcome this block. This evidence suggests that the embryo may exert a more complex control on amylase synthesis in the aleurone cells than merely supplying them with gibberellic acid and chloramben interferes with some aspect of this more complex control mechanism. Chloramben increased the phytase level in the embryonic axis of 2-day-old squash seedlings by about 50% (Penner, 1970).

Mann and Pu (1968) reported that chloramben stimulated the incorporation of radioactivity from malonic acid-2-^{14}C into lipid by excised hypocotyls of hemp sesbania (*Sesbania exaltata*). This stimulation of lipid synthesis was also observed for other auxin-like herbicides, whereas other non-auxin herbicides were inhibitory. This provides additional evidence of the auxin-like properties of the benzoic acid-type herbicides.

Mode of Action

The auxin-like properties of the benzoic acid-type herbicides has been well documented but the available research data are insufficient to propose a mechanism of action. It might very well be that it is similar to the phenoxy-type herbicides, which have been shown to interfere with nucleic acid metabolism.

Dicamba and 2,3,6-TBA appear to have similar absorption-translocation patterns. They are readily absorbed by leaves and roots and translocated via both the symplastic and apoplastic systems with accumulation in areas of high metabolic activity. Furthermore, they are excreted or they leak from the roots into the surrounding medium. However, most reports have shown that chloramben, although absorbed by roots and leaves, is translocated very little from either. The translocation of chloramben from seeds into the developing seedling is also very limited.

Trichlorobenzoic acid (2,3,6-TBA) appears to be very stable in higher plants. Dicamba is also relatively stable in higher plants but it undergoes hydroxylation or demethoxylation with hydroxylation. These products may then be conjugated with some endogenous metabolite. Other minor metabolites have also been isolated. The rate and pathway of dicamba degradation appears to vary significantly in different species. Chloramben is conjugated with glucose to form the N-glucosyl chloramben fairly rapidly in higher plants; other minor metabolites have also been reported. In roots, 2,5-dichloroaniline has been reported as a degradation product as well as the N-glucosyl chloramben.

Reports on the effect of the benzoic acid-type herbicides on the biochemistry of plants is fragmentary, therefore any attempt to development of a mechanism of action would be speculative.

REFERENCES

Arnold, W. E. and J. D. Nalewaja. 1971. Effect of dicamba on RNA and protein. *Weed Sci.* **19**:301–305.

Ashton, F. M. 1966. Fate of ^{14}C-amiben in carrots. *Weeds* **14**:55–57.

Ashton, F. M., D. Penner, and S. Hoffman. 1968. Effect of several herbicides on proteolytic activity of squash seedlings. *Weed Sci.* **16**:169–171.

Baker, R. S. and G. F. Warren. 1962. Selective herbicidal action of amiben on cucumber and squash. *Weeds* **10**:219–224.

Balagannis, P. G., M. S. Smith, and R. L. Wain. 1965. Studies on plant growth regulating substances. XIX. Stability of 2,3,6-trichlorobenzoic acid in wheat plants and in the rabbit and mouse. *Ann. Appl. Biol.* **55**:149–157.

Broadhurst, N. A., M. L. Montgomery, and V. H. Freed. 1966. Metabolism of 2-methoxy-3,6-dichlorobenzoic acid (dicamba) by wheat and bluegrass plants. *J. Agr. and Food Chem.* **14**:585–588.

Buchholtz, K. P. 1958. The sensitivity of quackgrass to various chlorinated benzoic acids. *WSSA Abstr.* pp. 33–34.

Camper, N. D. and D. E. Moreland. 1966. Adsorption of herbicides to proteins. *WSSA Abstr.* p. 32.

Chang, F. Y. and W. H. Vanden Born. 1971b. Dicamba uptake, translocation, Canada thistle. *Weed Sci.* **16**:176–181.

Chang, F. Y. and with Vanden Born. 1971a. Translocation and metabolism of dicamba in tartary buckwheat. *Weed Sci.* **19**:107–112.

Chang, F. Y. and W. H. Vanden Born. 1971. Dicamba uptake, translocation, metabolism, and selectivity. *Weed Sci.* **19**:113–117.

Chernyshev, I. D. 1968. Carbohydrate metabolism in the roots of Russian knapweed (*Centaurea repens*) as an index of herbicidal efficiency. *Biol. Nauki* **11**:75–79.

Colby, S. R. 1965. *N*-glycoside of amiben isolated from soybean plants. *Science* **150**:619–620.

Colby, S. R. 1966. The mechanism of selectivity of amiben. *Weeds* **14**:197–201.

Colby, S. R., G. F. Warren, and R. S. Baker. 1964. Fate of amiben in tomato plants. *J. Agr. and Food Chem.* **12**:320–321.

Crafts, A. S. and S. H. Yamaguchi. 1964. The autoradiography of plant materials. *Calif. Agr. Exp. Sta. Manual*, No. 35, 143 pp.

Egley, G. H. and G. A. Kust. 1964. Translocation of herbicides to witchweed (*Striga asiatica*). *WSSA Abstr.*, p. 80.

Foy, C. L. and D. Penner. 1965. Effect of inhibitors and herbicides on tricarboxylic acid cycle substrate oxidation by isolated cucumber mitochrondria. *Weeds* **13**:226–231.

Frear, D. S., C. R. Swanson, and R. E. Kadunce. 1967. The biosynthesis of *N*-(3-carboxy-2,5-dichlorophenyl) glucosylamine in plant tissue sections. *Weeds* **15**:101–104.

Freed, V. H., M. Montgomery, and M. Kief. 1961. The metabolism of certain herbicides by plants—a factor in their activity. *Proc. 15th Northeast Weed Control Conf.* pp. 6–16.

Friesen, H. A. and D. A. Dew. 1967. The effect of two herbicides, bromoxynil and dicamba, on transpiration of tartary buckwheat and barley. *Can. J. Pl. Sci.* **47**:533–537.

Greenham, C. G. (1968). Studies on herbicide contents in roots of skeleton weed (*Chondrilla juncea*) following leaf applications. *Weed Res.* **8**:272–282.

Hahn, R. R., O. C. Burnside, and T. L. Lang. 1969. Dissipation and phytotoxicity of dicamba. *Weed Sci.* **17**:3–8.

Haskell, D. A. and B. J. Rogers. 1960. The entry of herbicides into seeds. *Proc. 17th North Central Weed Control Conf.* p. 39.

Hodgson, R. H. 1967. Some effects of environment and species on absorption and metabolism of amiben in plants. *WSSA Abstr.*, pp. 64–65.

Hull, R. J. and M. R. Weisenberg. 1967. Translocation and metabolism of dicamba in *Sorghum halepense* and *Phaseolus vulgaris*. *Plant Physiol. Abstr.* **42**:S–49.

Hurtt, W. and C. L. Foy. 1965a. Excretion of foliar-applied dicamba and picloram from roots of Black Valentine beans grown in soil, sand and culture solution. *Proc. 19th Northeast Weed Control Conf.* p. 602.

Hurtt, W. and C. L. Foy. 1965b. Some factors influencing the excretion of foliar-applied dicamba and picloram from roots of Black Valentine beans. *Plant Physiol. Abstr.* **40**:xlviii.

Jones, R. L., T. P. Metcalfe, and W. A. Sexton. 1954. The relationship between the constitution and the effect of chemical compounds on plant growth. IV. Derivatives and analogues of 2-benzoylbenzoic acid. *J. Sci. Food Agric.* **5**:32–43.

Keitt, G. W., Jr. 1960. Effects of certain growth substances on elongation and geotropic curvature of wheat roots. *Botan. Gaz.* **122**:51–62.

Keitt, G. W. and R. A. Baker. 1966. Auxin activity of substituted benzoic acids and their effect on polar auxin transport. *Plant Physiol.* **41**:1561–1569.

Leonard, O. A., L. A. Lider, and R. K. Glenn. 1966. Absorption and translocation of herbicides by Thompson Seedless (Sultanina) grape, *Vitis vinifera. Weed Res.* **6**:37–49.

Leonard, O. A. and R. K. Glenn. 1968. Translocation of herbicides in detached bean leaves. *Weed Sci.* **16**:352–356.

Lindner, P. J., J. C. Craig, Jr., F. E. Cooker, and J. W. Mitchell. 1958. Movement of 2,3,6-trichlorobenzoic acid from one plant to another through their root systems. *J. Agr. and Food Chem.* **6**:356–357.

Lindner, P. J., J. W. Mitchell, and G. D. Freeman. 1964. Persistence and translocation of exogenous regulating compounds that exude from roots. *J. Agr. and Food Chem.* **12**:437–438.

Lotlikar, P. D., L. F. Remmert, and V. H. Freed. 1968. Effects of 2,4-D and other herbicides on oxidative phosphorylation in mitochondria from cabbage. *Weed Sci.* **16**:161–165.

Magalhaes, A. C. and F. M. Ashton. 1969. Effect of dicamba on oxygen uptake and cell membrane permeability in leaf tissue of *Cyperus rotundus. Weed Res.* **9**:48–52.

Magalhaes, A. C., F. M. Ashton, and C. L. Foy. 1968. Translocation and fate of dicamba in purple nutsedge. *Weed Sci.* **16**:240–245.

Mann, J. D., L. S. Jordan, and B. E. Day. 1965. A survey of herbicides for their effect upon protein synthesis. *Plant Physiol.* **40**:840–843.

Mann, J. D. and M. Pu. 1968. Inhibition of lipid synthesis by certain herbicides. *Weed Sci.* **16**:197–198.

Mason, G. W. 1960. The absorption, translocation, and metabolism of 2,3,6-trichlorobenzoic acid in plants. Ph.D. dissertation, University of Calif., Davis, 139 pp.

Moody, K., C. A. Kust, and K. P. Buchholtz. 1970. Release of herbicides by soybean roots in culture solutions. *Weed Sci.* **18**:214–218.

Moreland, D. E. and K. L. Hill. 1962. Interference of herbicides with the Hill reaction of isolated chloroplasts. *Weed Sci.* **10**:229–236.

Moreland, D. E., S. S. Malhotra, R. D. Gruenhagen, and E. H. Shokraii. 1969. Effect of herbicides on RNA and protein synthesis. *Weed Sci.* **17**:556–562.

Pate, D. A., H. H. Funderburk, J. M. Lawrence, and D. E. Davis. 1965. The effect of dichlobenil and dicamba on nodal tissues of alligatorweed. *Weeds* **13**:208–210.

Penner, D. 1968. Herbicidal influence on amylase in barley and squash seedlings. *Weed Sci.* **16**:519–522.

Penner, D. 1970. Herbicide and inorganic phosphate influence on phytase in seedlings. *Weed Sci.* **18**:360–364.

Quimby, P. C. 1967. Studies relating to the selectivity of dicamba for wild buckwheat (*Polygonum convolvulus* L.) vs. Selkirk wheat (*Triticum aestivum* L.) and a possible mode of action. Ph.D. thesis, N. Dakota State University, 87 pp.

Quimby, P. C., Jr. and J. D. Nalewaja. 1971. Selectivity of dicamba in wheat and wild buckwheat. *Weed Sci.* **19**:598–601.

Quimby, P. C. and J. D. Nalewaja. 1966. The uptake, translocation, and fate of dicamba in wheat and wild buckwheat. *Proc. 23rd North Central Weed Control Conf.*, pp. 22–23.

Quimby, P. C. and J. D. Nalewaja. 1968. Uptake of ^{14}C-dicamba by leaf sections of wheat and wild buckwheat. *WSSA Abstr.*, p. 15.

Ray, B. R. and M. Wilcox. 1967. Metabolism of dicamba in Zea and Hordeum. *Abstr. 154th Meeting Am. Chem. Soc.*, p. A28.

Ray, B. and M. Wilcox. 1969. Translocation of the herbicide dicamba in purple nutsedge, *Cyperus rotundus. Physiol. Plant* **22**:503–505.

Rieder, G., K. P. Buchholtz, and C. A. Kust. 1970. Uptake of herbicides by soybean seed. *Weed Sci.* **18**:101–105.

Rogerson, A. B. and C. L. Foy. 1968. Effect of dicamba and picloram on growth and anthocyanin content in Black Valentine beans under different environmental conditions. *Proc. 21st Southern Weed Conf.*, p. 347.

Sand, P. F., G. H. Egley, W. L. Gould, and C. A. Kust. 1971. Witchweed control by herbicides translocated through host plants. *Weed Sci.* **19**:240–244.

Schrank, A. R. 1960. Differential effect of 2,3,6-trichlorobenzoic acid on growth and geotropic curvature of *Avena* coleoptiles. *Plant Physiol.* **35**:735–741.

Schrank, A. R. 1964. Growth and geotropic curvature of *Avena* coleoptiles in the presence of 2,6-dichlorobenzoic acid. *Physiol. Plant* **17**:346–351.

Schrank, A. R. and A. F. Rumsey. 1961. Effects of 2,3,6-trichlorobenzoic acid on growth and geotropic curvature of *Avena* coleoptiles. *Physiol. Plant* **14**:231–241.

Slife, E. W. 1963. The translocation of amiben in plants. *Hormolog* **4**:11–12.

Stoller, E. W. 1968. Differential phytotoxicity of an amiben metabolite. *Weed Sci.* **16**:384–386.

Stoller, E. W. 1969. The kinetics of amiben absorption and metabolism as related to species sensitivity. *Plant Physiol.* **44**:854–860.

Stoller, E. W. and L. M. Wax. 1968. Amiben metabolism and selectivity. *Weed Sci.* **16**:283–288.

Sudi, J. 1966. Increases in the capacity of pea tissue to form acylaspartic acids specifically induced by auxins. *New Phytol.* **65**:9–21.

Sutherland, M. L. 1961. ^{14}C-amiben pre-emergent tracer study in soybean. *Tech. Report.* Amchem Products Inc., Ambler, Pa.

Swan, D. G. and F. W. Slife. 1965. The absorption, translocation, and fate of amiben in soybeans. *Weeds* **13**:133–138.

Swanson, C. R. 1969. The benzoic acid herbicides, pp. 299–320. In: Kearney, P. C. and D. D. Kaufman, *Degradation of Herbicides.* Marcel Dekker, Inc., New York.

Swanson, C. R., R. E. Kadunce, R. A. Hodgson, and D. S. Frear. 1966a. Amiben metabolism in plants. I. Isolation and identification of an *N*-glucosyl complex. *Weeds* **14**:319–323.

Swanson, C. R., R. A. Hodgson, R. E. Kadunce, and H. R. Swanson. 1966b. Amiben metabolism in plants. II. Physiological factors in *N*-glucosyl amiben formation. *Weeds* **14**:323–327.

Taylor, T. D. and G. F. Warren. 1968. Herbicide transport as influenced by certain metabolic inhibitors. *WSSA Abstr.* pp. 11–12.

Vanden Born, W. H. 1966. Translocation and persistence of dicamba in Canada thistle, Tartary buckwheat, quackgrass, and Kentucky bluegrass. *WSSA Abstr.* p. 62.

Vander Beek, L. C. 1959. Effect of 2,3,6-trichlorobenzoic acid and 2,6-dichlorobenzoic acid on the geotropic and phototropic responses of seedlings of various species. *Plant Physiol.* **34**:61–65.

Vander Beek, L. C. 1967. 2,3,6-trichlorobenzoic and 2,6-dichlorobenzoic acid interference with phototropic and geotropic responses of seedlings. *Ann. N.Y. Acad. Sci.* **144**:374–382.

Venis, M. A. and G. E. Blackman. 1966a. The uptake of growth substances. 7. The accumulation of chlorinated benzoic acids by stem tissues of different species. *J. Exp. Bot.* **17**:270–282.

Venis, M. A. and G. E. Blackman. 1966b. The uptake of growth substances. 8. Accumulation of chlorinated benzoic acids by *Avena* segments: A possible mechanism for the transient phase of accumulation. *J. Exp. Bot.* **17**:771–789.

Venis, M. A. and G. E. Blackman. 1966c. The uptake of growth substances. 9. Further studies of the mechanism of uptake of 2,3,6-trichlorobenzoic acid by *Avena* segments. *J. Exp. Bot.* **77**:790–808.

Wuu, K. D. and W. F. Grant. 1966. Morphological and somatic chromosomal aberrations induced by pesticides. *Can. J. Genet. Cytol.* **8**:481–501.

Zhirmunskaya, N. M. 1966. The effect of herbicides on nitrogen metabolism in roots of *Cirsium arvense* and *Sonchus arvensis*. *Khimiya sel'. Khoz.* **4**:46–51.

Zick, W. A. and T. R. Castro. 1966. Dicamba-dissipation in and on living plants. *Proc. 8th Br. Weed Cont. Conf.*, 265–265d.

Zimmerman, M. H. and A. E. Hitchcock. 1942. Substituted phenoxy and benzoic acid growth substances and the relation of structure to physiological activity. *Contr. Boyce Thompson Inst.* **12**:321–343.

Zimmerman, M. H. and A. E. Hitchcock. 1951. Growth-regulating effect of chlorosubstituted derivatives of benzoic acid. *Contr. Boyce Thompson Inst.* **16**:209–213.

CHAPTER 12

Bipyridyliums

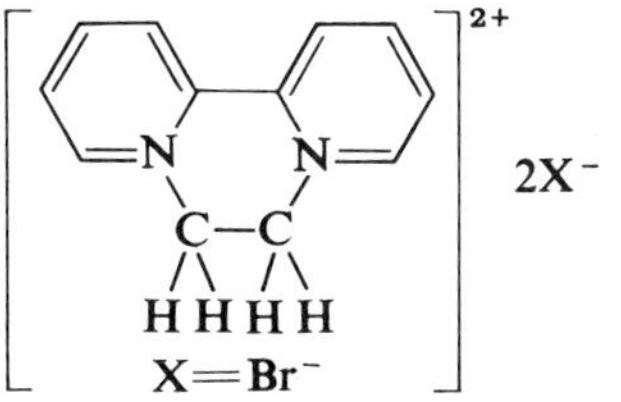

Diquat

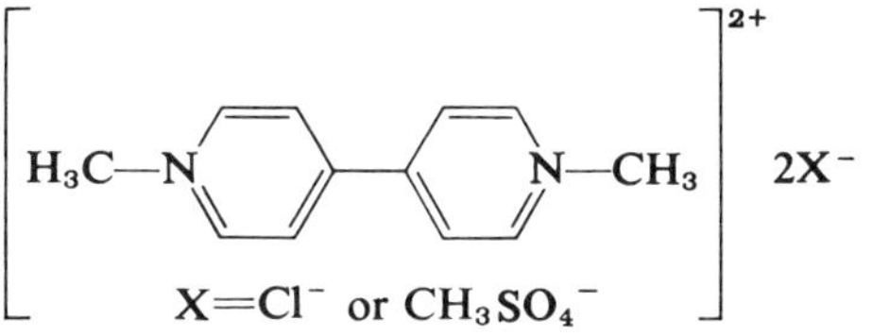

Paraquat

These two heterocyclic organic compounds are members of the bipyridylium (or dipyridylium) quaternary ammonium class. They are also referred to as viologens, especially when used as redox indicators. Although their chemistry has been known for some time, the herbicidal properties of diquat (6,7-dihydrodipyrido[1,2-a:2′,1′-c]pyrazinediium ion) were not discovered until the mid-1950's. The herbicidal activity of paraquat (1,1′-dimethyl-4,4′-bipyridinium ion) was found shortly thereafter. The herbicidal properties of these compounds are associated with their redox potential (Calderbank, 1971). Only those compounds with a redox potential less negative than -800mV have herbicidal properties. The most active herbicides are those having redox potential values in the region of -300 to -500mV. Diquat has a redox potential of -349mV and paraquat -446mV (Homer et al., 1960).

These herbicides are usually used as general contact weed control agents and are applied to the foliage. They have also been used selectively for the control of annual weeds in herbaceous perennial crops where perennial species will recover from a contact spray whereas the annual species will be killed, i.e., pasture renovation. They are bound so tenaciously by soil components by means of base exchange that they are considered to be essentially biologically inactive in most soils. Therefore, they may be used selectively in treatment providing direct contact with the crop is avoided, i.e., preplant, preemergence, or directed-postemergence. Paraquat is used for these types of

185

applications in many crops; whereas diquat is used for preharvest desiccation or aquatic weed control. Both compounds are also used as non-selective contact herbicides. The contact action of these herbicides is very rapid, often visible within a few hours.

Several excellent reviews of the mode of action of bipyridylium herbicides have been published and the reader is referred to these as additional sources of information (Cronshey, 1961; Funderburk and Bozarth, 1967; Calderbank, 1968; Akhavein and Linscott, 1968; Funderburk, 1969; Zweig, 1969; Zweig et al., 1969; Caseley, 1970; Calderbank, 1971).

Growth and Plant Structure

The bipyridylium-type herbicides cause wilting and rapid desiccation of the foliage to which they are applied. Mees (1960) showed that light increased the rate of development of the phytotoxic symptoms but was not essential for herbicidal action. He treated broad bean (*Vicia faba*) plants for 24 hours in the dark by dipping one leaflet into a tube containing 15 ml of 3×10^{-3}M diquat. The treated leaflet was removed and the plants placed in the light. Within 10 minutes the leaves adjacent to the treated one began to blacken and after 2 hours they began to wilt. Experiments in which the treated leaflets were not removed from the plants began to wilt after 48 to 72 hours and were dead in 5 days. Green and etiolated wheat seedling leaves were dipped into a 10^{-2}M diquat solution for 15 seconds and placed in the light. The green seedlings began to show necrosis after 24 hours and were almost dead after 4 days, while the etiolated seedlings required 40 hours for the initial symptom, wilting. These experiments show that both light and the presence of chlorophyll are necessary for maximal diquat injury. The slow injury which occurs in the dark may involve an entirely different mechanism. Following this report, a relatively large number of papers were published on the effect of light on the action of the bipyridylium herbicides; however, since they do not contribute to knowledge of the effect of these herbicides on plant structure they will be discussed later.

The effect of paraquat on the ultrastructure of mesophyll cells of honey mesquite (*Prosopis juliflora* var. *glandulosa*) was investigated by Baur et al. (1969). They treated a leaflet of the second fully-expanded leaf from the apex with 20 μl of an aqueous solution of 0.01M paraquat containing 0.1% X-77 surfactant. Within 5 minutes, 39% of the cells showed plasmalemma disintegration and after 40 minutes disruption of the chloroplast membrane became evident, see Figure 12-1a and 12-1b. These ultrastructural observations are consistent with the gross morphological symptoms, the disruption of membrane integrity should result in wilting, necrosis, and ultimate death of the leaves.

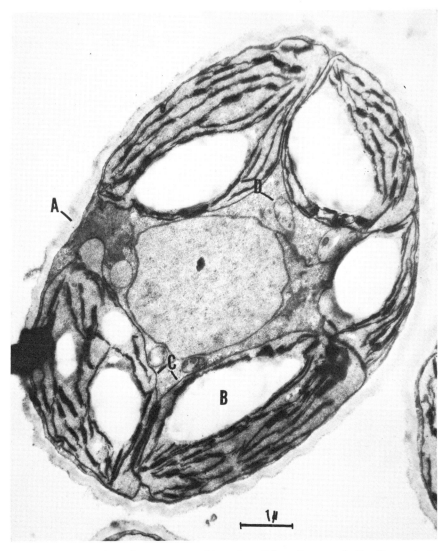

Figure 12-1a.   Effect of paraquat on chloroplast. Mesquite mesophyll cells, control. Preconditioned under light bank, 15-hr daylength. (A) plasmalemma; (B) starch deposits; (C) chloroplast membrane; (D) mitochondria. (Baur et al., 1969.)

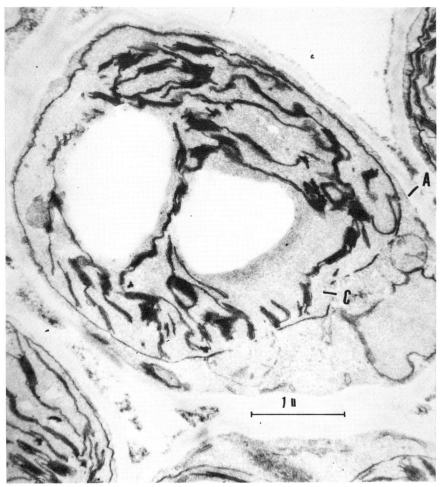

Figure 12-1b. Effect of paraquat on chloroplasts. Mesquite mesophyll cell, 40 min after treatment with 0.01 M paraquat. Preconditioned under light bank, 15-hr day-length. Note partial disintegration of plasmalemma (A) and breakage of chloroplast membrane (C). (Baur et al., 1969.)

Absorption and Translocation

The interpretation of the absorption-translocation research into a working hypothesis is possibly more difficult with the bipyridylium herbicides than any other class. This is primarily due to their very rapid action in the foliage and the pronounced influence of light. Not only does the quantity, quality, and period of light influence their action; when light is considered in relation to time of treatment, it is more important than with most herbicides. Other environmental factors such as relative humidity, temperature, soil moisture, and even oxygen play a role. However, the latter is probably never limiting to the foliage of terrestrial plants under normal conditions.

Shortly after the introduction of the bipyridylium herbicides, field observations indicated that these compounds were more effective when applied in the late afternoon or evening rather than in the morning or midday. Preliminary experiments by Brian (unpublished) showed that plants treated with diquat on one leaf only and kept under normal daylight suffered only localized damage, while if kept in the dark for a period after treatment they were rapidly and completely killed when brought into the light. Using ^{14}C-diquat and autoradiographic techniques, Mees (unpublished) found that the herbicide moved very little from the treated leaf in the light. In a series of refined experiments, Baldwin (1963) showed that ^{14}C-diquat did not move out of the treated leaf of tomato when the plants were kept in darkness for 24 hours but when placed in the light for 5 hours following 24 hours of darkness diquat was found throughout the aerial portion of the plant. On the other hand, if the plants were placed in the light immediately after treatment there was essentially no movement out of the treated leaf. Steam ringing of the leaf petiole did not prevent the movement of diquat out of the treated leaf. He suggested that these results were compatible with the concept that diquat was transported in the xylem. When the tissue was damaged, as occurred in the light with diquat, movement was from this area to other transpiring surfaces and thus distributed throughout the aerial portions of the plant. Although this did not explain why diquat was not translocated from the treated leaf when it was placed in the light immediately after treatment, he suggested that this was associated with rapid localized damage which prevented the herbicide from reaching the conducting elements. Similar experiments were conducted on sugarbeets, cabbage, French beans, and broad beans with the same results. Paraquat behaved in the same manner.

Wood and Gosnell (1965) reported that the translocation of paraquat from leaves of purple nutsedge (*Cyperus rotundus*) and yellow nutsedge (*Cyperus esculentus*) was much greater when the plants were in darkness for 24 hours following treatment than when they were immediately placed in the light. They also noted that the greatest movement was toward the tip of the treated leaf which suggests apoplastic translocation; however, some did move downward with accumulations in the new growth of rapidly growing plants, suggesting some symplastic movement. Slade and Bell (1966) showed that paraquat was translocated from young developing leaves as well as mature leaves of tomato and that steam ringing did not affect the movement, suggesting xylem transport. When the plants were placed in the dark for 24 hours following treatment with paraquat, transport was increased relative to being placed in light immediately after treatment. Once absorbed, light induced damage was required to permit significant movement through the rest of the plant; however, this damage then inhibited further entry of paraquat into the xylem. There was no further movement of the radioactive paraquat from the treated leaf once it was killed with non-radioactive paraquat whether the treated plant was placed in the light or darkness. Smith and Sagar (1966) also studied the influence of light and darkness on diquat translocation. In tomato they found that a period of darkness after diquat application was necessary for reliable systemic action during subsequent exposure to light. However, darkness was necessary only in the area of application and then only to allow adequate penetration and this was its only role. After penetration, the desiccation following death allowed the transfer of free water containing diquat to the other leaves. They concluded that this was the primary force of long distance transport and that light was only essential for the rapid toxic action and not directly related to translocation. This was confirmed by applying ^{32}P-orthophosphate or ^{14}C-urea with non-radioactive diquat and finding distributions similar to ^{14}C-diquat alone. They concluded that phloem translocation was not involved. Smith and Davies (1965) also concluded that paraquat was translocated via the xylem in dallisgrass (*Paspalum distichum*). Funderburk and Lawrence (1963) reported that there was only slight translocation of paraquat or diquat upward from root application or downward from foliar application in the aquatic species water stargrass (*Heteranthera dubia*) or alligator weed (*Alternanthera philoseroides*). Several workers have applied these principles to the use of the bipyridylium herbicides to obtain more effective control of weeds (Brian and Headford, 1968; Putnam and Ries, 1968; Brian, 1969; Akhavein and Linscott, 1970).

Although it was stated in the introduction of this chapter that the bipyridylium herbicides are essentially biologically inactive in most soils, root uptake can occur from culture solution or even from soil if it is leached into the root zone of plants. Coats et al. (1965) reported appreciable uptake of paraquat and diquat from preemergence applications to sand or loamy sand when the chemical is leached to depths greater than 0.5 inch. Weber and Scott (1966) showed that cucumber seedlings could absorb paraquat which was tightly bound to montmorillonite or kaolinite clay particles. Damonakis et al. (1970) found that paraquat which was tightly bound to peat soil could be absorbed by a number of species; however, the inhibition occurred primarily in the roots and although shoot growth was also inhibited this was considered a secondary effect since typical paraquat symptoms did not appear in the shoots. Apparently the lack of activity of the bipyridylium herbicides via the soil results from lack of leaching into the root zone and limited translocation from roots to shoots.

The absorption of the bipyridylium herbicides by leaves is influenced by their cationic and polar properties. Their cationic property effect has been observed by Brian (1967) as well as Davies and Seaman (1968). Brian (1967) showed that approximately 33% of the total amount of diquat applied to a beet leaf was adsorbed within 30 seconds and could not be removed by washing. He concluded that this was a physical adsorption involving an ionic attraction between cations of the herbicide and negative charges on the leaf surface. Davies and Seaman (1968) reported that diquat uptake by the aquatic plant *Elodea canadensis* appeared to consist of a rapid initial passive adsorption phase followed by a slower accumulation phase. The rapid initial phase was not reversible. Their polar-property influence on absorption is probably associated with the non-polar cuticle and perhaps of the effect of the relative humidity of the air on the absorption of these herbicides. Thrower et al. (1965) concluded that the entry of diquat into the leaf is directly related to the amount of cuticular wax on the leaves of various species of plants. However, this conclusion was apparently based on the degree of wetting of the various species as correlated with amount and characteristic of the cuticular wax rather than the entry of diquat *per se*.

The increased effectiveness of the bipyridylium herbicides when applied in the late afternoon or evening may involve relative humidity as well as light. Brian (1966) showed that diquat and paraquat were absorbed 2 to 5 times as much at 93% relative humidity as compared to a relative humidity of 55 to 65%. Furthermore, he reported that high humidity for 8 or more hours before the herbicide treatment was more effective than increased humidity after treatment; and that 16 hours of low humidity following the high humidity treatment, but before the herbicide treatment, eliminated the high humidity effect. This suggests that hydration of the cuticle by the high humidity treatment is involved with the increased absorption. Brian and Ward (1967) reported a two-fold increase in diquat uptake by potato stems when relative humidity was increased from 50 to 94%. Thrower et al. (1965) also noted increased uptake of diquat at high relative humidities but associated this with slowness of drying of the applied droplets.

The information presented above shows that the bipyridylium herbicides are almost exclusively translocated via the apoplastic system. However, the fact that they are poorly translocated from the leaf surface to the xylem elements in the light, due to rapid herbicidal injury, and that they have been shown to accumulate in the new growth of rapidly growing plants, suggests some symplastic movement. The reason for only limited transport from the roots to the shoots once they have been absorbed remains obscure. Perhaps they are bound too tightly to cellular components of the cortical parenchyma and do not enter the xylem.

Molecular Fate

Although the amount of research is very limited, apparently the bipyridylium herbicides are not degraded by higher plants. Funderburk and Lawrence (1964), using ^{14}C-labeled diquat and paraquat, reported that neither herbicide was degraded by alligator weed or beans. However, these compounds do undergo photodecomposition (Crosby and Li, 1969). Slade (1966) applied radioactive paraquat to leaves of maize, tomato, and broad beans and concluded that it was not degraded in the plants but undergoes photodecomposition on the leaf surface. Paraquat was not degraded in the dark. A dark control is rarely used by workers studying the molecular fate of herbicides in plants. Such a dark treatment would be a desirable control for others to use making similar studies with other herbicides since many other herbicides are known to undergo photodecomposition. The molecular fate of the bipyridylium herbicides has been reviewed by Funderburk and Bozarth (1967).

Paraquat and diquat form free radicals in plants and this could be considered a molecular change. However, the reaction is reversible and these herbicides act essentially as a catalyst without any consumption of the compound in this reaction. Therefore these herbicides are not degraded in our usual understanding of the term. This will be discussed in greater detail in the next section since it is intimately related to the mechanism of action of these herbicides.

Biochemical Responses

The key to the mechanism of action of the bipyridylium herbicides is their ability to form free radicals by reduction and subsequent autooxidation to yield the original ion, as illustrated in Figure 12-2. The photosynthetic apparatus, light and oxygen, are required. The reader is referred to Figure 6-2 for a presentation of selected biochemical reactions of photosynthesis and sites of inhibition by herbicides for orientation into the following discussion.

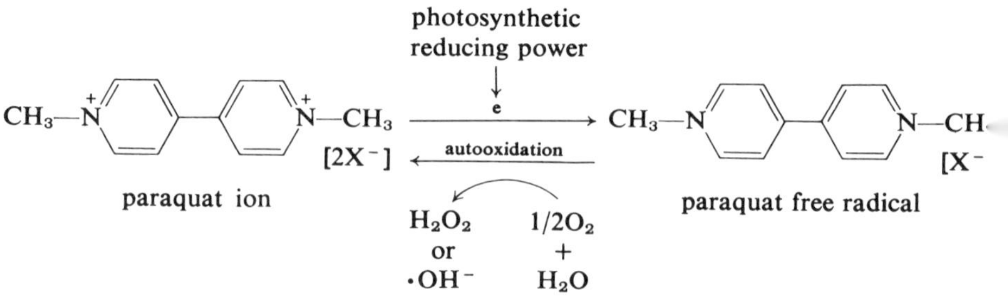

FIGURE 12-2.   Free radical formation from paraquat ion and autooxidation of free radical yielding H_2O_2 or $\cdot OH^-$.

The physiological studies of Mees (1960) which showed that diquat action in plants was much faster in the light than in the dark and that the herbicide was absorbed in the dark but not activated until brought into the light, as well as subsequent research with light as a variable by others, suggested that photosynthesis was involved in the mechanism of action of the bipyridylium herbicides. Mees (1960) also showed that the presence of oxygen was necessary for the development of the phytotoxic symptoms.

The light reaction of photosynthesis involves the absorption of light by chlorophyll or accessory pigments and transformation of this light energy into chemical energy via a series of electron transfer reactions leading to the formation of reduced pyridine nucleotide (NADPH) and adenosine triphosphate (ATP). Both of these compounds are required for the dark reactions of photosynthesis which result in the fixation of carbon dioxide into simple carbon compounds and eventually into sugars. Since light was required for the development of the phytotoxic symptoms of the bipyridyliums, the light reactions were the obvious ones to investigate. It was later shown that carbon dioxide fixation was also inhibited (Couch and Davis, 1966; van Oorschot, 1964; Zweig, 1969). However, inhibition of carbon dioxide fixation is the result of the lack of reducing power from the light reactions of photosynthesis and therefore a secondary effect. Furthermore, the bipyridylium herbicides are relatively poor inhibitors of carbon dioxide fixation when compared to the herbicides which inhibit the Hill reaction, i.e., diuron.

In retrospect, one wonders if the fact that Horowitz (1952) provided indirect evidence that the viologens (4,4'-bipyridyliums) were reduced in a photosynthetic reaction fostered their examination by Imperial Chemical Industries Ltd. as potential herbicides, or whether their discovery was by empirical screening. Viologens have been shown to catalyze ATP formation from ADP and inorganic phosphate in isolated chloroplasts (Jagendorf and Avron, 1958; Hill and Walker, 1959). Wessels (1959) and Kandler (1960) noted that the viologens were unique in stimulating photosynthetic phosphorylation because of their low redox potentials and because the reduced form was autooxidizable, in contrast to other reducible electron carriers such as phenazine methosulfate and vitamin K. Later it was shown that diquat could also act as an electron carrier (Davenport, 1963; Zweig and Avron, 1965). Wessels (1959) suggested that viologens are reduced by the same mechanism as NADP in photosynthesis and Davenport (1963) has shown diquat competitively inhibits NADP reduction by isolated chloroplasts. The reduction of paraquat has also been reported by Kok (1963), Zweig and Avron (1965), Zweig et al. (1965), Black (1965), and Black and Meyers (1966).

Mees (1960) showed that pretreatment of leaves with monuron delayed the appearance of phytotoxic symptoms of diquat. This suggests that the site of action of diquat occurs somewhat later in the photosynthetic system than the

Hill reaction since monuron inhibits the Hill reaction. Kok et al. (1965) suggested that the photoreduction of methyl viologen by chloroplasts can be explained by the primary reductant, X, of photosystem I and the substrate, thus placing the site of action in photosystem I. Zweig et al. (1965) observed that diquat competitively inhibits photophosphorylation catalyzed by phenazine methosulfate (PMS) and concluded that diquat acts at the same site as PMS in the electron-transfer path of chloroplasts. As evident from their proposed site of action, diquat can thus shunt electrons from ferredoxin to form the diquat free radical and prevent the normal reduction of NADP to NADPH (see Figure 6-2).

The reduction of diquat and paraquat can also occur in the electron transport system of respiration to form the free radical and presumably undergo autooxidation in the presence of oxygen (Bozarth et al., 1965). This work was carried out with yeast and bacteria, but presumably would also occur in higher plants. However, it is assumed that the reduction proceeds at a much slower rate in respiration than in photosynthesis since the phytotoxic symptoms in higher plants develop much more rapidly in the light than in the dark. This may explain why Mees (1960) reported that light increased the rate of development of the phytotoxic symptoms in broad beans but was not essential for herbicidal action. Perhaps the following examples of non-photosynthetic related injury caused by paraquat also involve free radical formation via the respiratory electron transport system—inhibition of radical elongation of dark-grown honey mesquite seedlings (Merkle et al., 1965), inhibition of dark-grown tobacco callus tissue (Jordan et al., 1966), inhibition of percent germination of several species of grasses (Appleby and Brenchley, 1968), and injury of white leaves of *Hibiscus rosa sinensis* (Bovey and Miller, 1968). Most other workers have suggested that light is essential; it is possible in these later studies that observations were not continued long enough in the dark for the phytotoxic symptoms to develop. Nevertheless under normal environmental conditions light would be present during the day and the formation of the free radical via photosynthesis would proceed so rapidly that the contribution of free radicals via respiration to the herbicidal action would be of minor significance.

The autooxidation or reoxidation of the free radical of the bipyridylium herbicides has been demonstrated by Kok et al. (1965), Zweig et al. (1965), and Black (1965, 1966). They showed reduction of the bipyridylium ion in the light to the free radical and complete reoxidation to the ion when the light was replaced by darkness in the presence of oxygen. In the absence of oxygen and presence of light, reduction occurred rapidly but oxidation in the dark was relatively small. This demonstrates the catalytic nature of the bipyridylium herbicides.

Therefore, the bipyridylium herbicides remove electrons from the electron–

transport system of photosystem I, thus inhibiting the reduction of NADP to NADPH. The latter is required for carbon dioxide fixation into sugars through a series of dark reactions. However, such a mechanism would result in a relatively slow death by starvation which is not in agreement with the rapid development of the phytotoxic symptoms.

Several workers have suggested that hydrogen peroxide produced during reoxidation of the bipyridylium free radical is the toxicant causing the rapid injury to plants (Mees, 1960; Calderbank and Crowdey, 1962; Black and Meyers, 1966; Calderbank, 1968). Calderbank (1968) also proposed that reactive hydroxyl radicals may also be involved. Baldwin (unpublished) has demonstrated hydrogen peroxide formation during air oxidation of chemically reduced paraquat. Davenport (1963) showed the formation of a metmyoglobin-peroxide complex in a chloroplast-metmyoglobin system with catalytic amounts of diquat. Davenport (1963) and Zweig et al. (1965) reported that the photoreduction of diquat proceeds more rapidly in the presence of a catalase-ethanol peroxide trap, which suggests the production of hydrogen peroxide. In similar studies, Good and Hill (1955) indirectly demonstrated peroxide formation in a benzyl-viologen chloroplast system. Kok et al. (1965) also suggested hydrogen peroxide formation in a chloroplast system using several viologens. Since catalases and peroxidases are widely distributed in higher plants one must assume that the bipyridylium herbicides produce sufficient hydrogen peroxide to exceed the capacity of these enzymes to destroy it, or accept Calderbank's hypothesis (1968) that the reactive hydroxyl radical is the toxicant.

The rejection of the hypothesis that the bipyridylium free radical itself is the actual toxicant is based on the fact that oxygen is required for maximal phytotoxic symptoms and in the absence of oxygen the free radical must accumulate since it would not be subject to autooxidation to the ion.

Mode of Action

Diquat and paraquat are bipyridylium quaternary ammonium salts which are used as general contact herbicides. Selectivity may be obtained with certain crops by avoiding contact with the desired species. They are highly water soluble and tenaciously bound to soil components.

These herbicides cause rapid desiccation of the foliage to which they are applied. Wilting is an early symptom of this desiccation. This is followed by necrosis and ultimate death of the entire leaf. At the cellular level they cause loss of integrity of the membranes of the cells and the chloroplasts. Light, molecular oxygen, and chlorophyll are required for the maximum development of these phytotoxic symptoms.

Their translocation following a foliar application appears to be almost exclusively via the apoplastic system. However, following the loss of membrane integrity they do move into untreated leaves presumably along with the flow of some "freed" cellular contents. They are poorly translocated from root applications to the shoots of plants, apparently because they are tightly bound to cellular components of the cortical parenchyma. There is evidence of some symplastic movement, i.e., accumulation in new growth of rapidly growing plants and the restriction of foliar transport by rapid injury.

These herbicides are not degraded in higher plants in the usual sense. However, they are reversibly converted from the ion form to the free radical form.

The mechanism of action of these herbicides involves the formation of the free radical by reduction of the ion and subsequent autooxidation to yield the original ion. The free radical itself does not appear to be the primary toxicant but rather the OH^- radical or H_2O_2 which is formed during the autooxidation of the free radical to the ion. The photosynthetic apparatus, light, and molecular oxygen are required co-factors for these reactions.

REFERENCES

Akhavein, A. A. and D. L. Linscott. 1968. The dipyridylium herbicides, paraquat and diquat. *Residue Rev.* **23**:97–145.

Akhavein, A. A. and D. L. Linscott. 1970. Effects of paraquat and light regime on quackgrass growth. *Weed Sci.* **18**:378–382.

Appleby, A. P. and R. G. Brenchley. 1968. Influence of paraquat on seed germination. *Weed Sci.* **16**:484–485.

Baldwin, B. C. 1963. Translocation of diquat in plants. *Nature* **198**:872–873.

Baur, J. R., R. W. Bovey, P. S. Baur, and Z. El-Seify. 1969. Effects of paraquat on the ultrastructure of mesquite mesophyll cells. *Weed Res.* **9**:81–85.

Black, C. C. 1965. Reduction of trimethylene dipyridyl with illuminated chloroplasts. *Science* **149**:62–63.

Black, C. C., Jr. and L. Meyers. 1966. Some biochemical aspects of the mechanism of herbicidal activity. *Weeds* **14**:331–338.

Bovey, R. W. and F. R. Miller. 1968. Phytotoxicity of paraquat on white and green hibiscus, sorghum, and alpinia leaves. *Weed Res.* **8**:128–135.

Bozarth, G. A., H. H. Funderburk, E. A. Curl, and D. E. Davis. 1965. Preliminary studies on degradation of paraquat by soil microorganisms. *Proc. 18th Southern Weed Conf.* p. 615.

Brian, R. C. 1966. The bipyridylium quaternary salts. The effect of atmospheric and soil humidity on the uptake and movement of diquat and paraquat in plants. *Weed Res.* **6**:292–303.

Brian, R. C. 1967. The uptake and adsorption of diquat and paraquat by tomato, sugar-beet, and cocksfoot. *Ann. Appl. Biol.* **59**:91–99.

Brian, R. C. 1969. The influence of darkness on the uptake and movement of diquat and paraquat in tomatoes, sugar-beet, and potatoes. *Ann. Appl. Biol.* **63**:117–126.

Brian, R. C. and D. W. R. Headford. 1968. The effect of environment on the activity of bipyridylium herbicides. *Proc. 9th Br. Weed Control Conf.* pp. 108–114.

Brian, R. C. and J. Ward. 1967. The influence of environment on potato haulm killed by diquat and its residue in the tubers. *Weed Res.* **7**:117–130.

Calderbank, A. 1968. The bipyridylium herbicides. *Adv. in Pest Control Res.* **8**:127–235.

Calderbank, A. 1971. Chemical structure and biological activity of the bipyridylium herbicides. *Internat. Conf. Pest. Chem.* (in press).

Calderbank, A. and S. H. Crowdy. 1962. Bipyridylium herbicides. *Rept. Progress Applied Chem.* **47**:536.

Caseley, J. 1970. Herbicide activity involving light. *Pestic. Sci.* **1**:28–32.

Coats, G. E., H. H. Funderburk, J. M. Lawrence, and D. E. Davis. 1965. Studies on translocation, degradation, and factors affecting the persistence of diquat and paraquat. *Proc. 18th Southern Weed Control Conf.* p. 614.

Couch, R. W. and D. E. Davis. 1966. Effects of atrazine, bromacil, and diquat on $^{14}CO_2$-fixation in corn, cotton, and soybeans. *Weeds* **14**:251–255.

Cronshey, J. F. H. 1961. A review of experimental work with diquat and related compounds. *Weed Res.* **1**:68–77.

Crosby, D. G. and M. Y. Li. 1969. Herbicide photodecomposition, pp. 321–363. In P. C. Kearney and D. D. Kaufman, *Degradation of Herbicides.* Marcel Dekker, Inc., New York.

Damonakis, M., D. S. H. Drennan, J. D. Fryer, and K. Holly. 1970. The toxicity of paraquat to a range of species following uptake by roots. *Weed Res.* **10**:278–283.

Davenport, H. E. 1963. The mechanism of cyclic phosphorylation by illuminated chloroplasts. *Proc. Roy. Soc.* **157B**:332–345.

Davies, P. J. and D. E. Seaman. 1968. Uptake and translocation of diquat in *Elodea. Weed Sci.* **16**:293–295.

Funderburk, H. H., Jr. 1969. Diquat and paraquat, pp. 283–298. In P. C. Kearney and D. D. Kaufman, *Degradation of Herbicides.* Marcel Dekker, Inc., New York.

Funderburk, H. H. and G. A. Bozarth. 1967. Review of the metabolism and decomposition of diquat and paraquat. *J. Agr. Food Chem.* **15**:563–576.

Funderburk, H. H. and J. M. Lawrence. 1963. Absorption and translocation of radioactive herbicides in submersed and emerged aquatic weeds. *Weed Res.* **3**:304–311.

Funderburk, H. H. and J. M. Lawrence. 1964. Mode of action and metabolism of diquat and paraquat. *Weeds* **12**:259–264.

Good, N. E. and R. Hill. 1955. Photochemical reduction of oxygen in chloroplast preparations. II. Mechanisms of the reaction with oxygen. *Arch. Biochem. Biophys.* **57**:355–366.

Hill, R. and D. A. Walker. 1959. Pyocyanine and phosphorylations with chloroplasts. *Plant Physiol.* **34**:240–245.

Homer, R. F., G. C. Mees, and T. E. Thomblinson. 1960. Mode of action of dipyridyl quaternary salts as herbicides. *J. Sci. Food Agric.* **11**:309–315.

Horowitz, L. 1952. Investigations on Hill reactions of isolated chloroplasts and whole cells and their relation to photosynthesis. Ph.D. dissertation, University of Minnesota, 135 pp.

Jagendorf, A. T. and M. J. Avron. 1958. Cofactors and roles of photosynthetic phosphorylation by spinach chloroplasts. *J. Biol. Chem.* **231**:277–290.

Jordan, L. S., T. Murashige, J. D. Mann, and B. E. Day. 1966. Effect of photosynthesis-inhibiting herbicides on non-photosynthetic tobacco callus tissue. *Weeds* **14**:134–136.

Kandler, O. 1960. Energy transfer through phosphorylation mechanisms in photosynthesis. *Ann. Rev. Plant Physiol.* **11**:37–54.

Kok, B. 1963. Significance of P_{700} as an intermediate in photosynthesis. *Proc. 5th Intern. Congr. Biochem.* **6**:73–81.

Kok, B., R. J. Rurainski, and O. V. H. Owens. 1965. The reducing power generated in photoact I of photosynthesis. *Biochem. Biophys. Acta* **109**:347–356.

Mees, G. C. 1960. Experiments on the herbicidal action of 1,1'-ethylene-2,2'-dipyridylium dibromide. *Ann. Appl. Biol.* **48**:601–612.

Merkle, M. G., C. L. Leinweber, and R. W. Bovey. 1965. The influence of light, oxygen, and temperature on the herbicidal properties of paraquat. *Plant Physiol.* **40**:832–835.

Putnam, A. R. and S. K. Ries. 1968. Factors influencing the phytotoxicity and movement of paraquat in quackgrass. *Weed Sci.* **16**:80–83.

Slade, P. 1966. The fate of paraquat applied to plants. *Weed Res.* **6**:158–167.

Slade, P. and E. G. Bell. 1966. The movement of paraquat in plants. *Weed Res.* **6**:267–274.

Smith, J. M. and G. R. Sagar. 1966. A re-examination of the influence of light and darkness on the long-distance transport of diquat in *Lycopersicon esculentum*. *Weed Res.* **6**:314–321.

Smith, L. W. and P. J. Davies. 1965. The translocation and distribution of three labeled herbicides in *Paspalum distichhum* L. *Weed Res.* **5**:343–347.

Thrower, S. L., N. D. Hallam, and L. B. Thrower. 1965. Movement of diquat (1,1'-ethylene-2,2'-dipyridylium) dibromide in leguminous plants. *Ann. Appl. Biol.* **55**:253–260.

van Oorschot, J. L. P. 1964. Some effects of diquat and simetone on CO_2-uptake and translocation of *Phaseolus vulgaris*. *Proc. 7th Br. Weed Control Conf.* pp. 321–324.

Weber, J. B. and D. C. Scott. 1966. Availability of a cationic herbicide adsorbed on clay minerals to cucumber seedlings. *Science* **152**:1400–1402.

Wessels, J. S. C. 1959. Studies on photosynthetic phosphorylation. III. Relation between photosynthetic phosphorylation and reduction of triphosphopyridine nucleotide by chloroplasts. *Biochem. Biophys. Acta* **35**:53–64.

Wood, G. H. and J. M. Gosnell. 1965. Some factors affecting the translocation of radioactive paraquat in *Cyperus* species. *Proc. S. Afr. Sugar Technol. Ass.* p. 7.

Zweig, G. 1969. Mode of action of photosynthesis inhibitor herbicides. *Res. Rev.* **25**:69–79.

Zweig, G. and M. Avron. 1965. On the oxidation-reduction potential of the photoproduced reductant of isolated chloroplasts. *Biochem. Biophys. Res. Commun.* **19**:397–400.

Zweig, G., J. E. Hitt, and D. H. Cho. 1969. Mode of action of dipyridyls and certain quinone herbicides. *J. Agr. Food Chem.* **17**:176–181.

Zweig, G., N. Shavit, and M. Avron. 1965. Diquat (1,1′-ethylene-2,2′-dipyridylium dibromide) in photoreactions of isolated chloroplasts. *Biochem. Biophys. Acta* **109**:332–346.

CHAPTER 13

Carbamates

General Properties

The carbamate herbicides derive their basic structure from carbamic acid (NH_2COOH). The common name, chemical name, and chemical structure are given in Table 13-1.

TABLE 13-1. Common Name, Chemical Name, and Chemical Structure of the Carbamate Herbicides

$$\underset{R_1-N-C-O-R_2}{\overset{\overset{\displaystyle H \quad O}{\displaystyle | \quad \|}}{}}$$

Common name	Chemical name	R_1	R_2
barban	4-chloro-2-butynyl *m*-chlorocarbanilate		$-CH_2C{\equiv}CCH_2Cl$
chlorpropham	isopropyl *m*-chloro-carbanilate		
dichlormate	3,4-dichlorobenzyl methylcarbamate	$-CH_3$	

200

Common name	Chemical name	R₁	R₂
karbutilate	m-(3,3-dimethylureido) phenyl-tert-butylcarbamate[a]	CH₃ —C—CH₃ CH₃	(structure: phenyl—N—C(=O)—N(CH₃)₂)
phenmedipham	methyl m-hydroxy-carbanilate m-methyl-carbanilate	(phenyl)—CH₃	(phenyl—N(H)—C(=O)—O—CH₃)
propham	isopropyl carbanilate	(phenyl)	—C(H)(CH₃)—CH₃
swep	methyl 3,4-dichloro-carbanilate	(phenyl)—Cl, —Cl	—CH₃
terbutol	2,6-di-tert-butyl-p-tolyl methylcarbamate	—CH₃	(phenyl with two —C(CH₃)₃ groups and —CH₃)

Also referred to as: tetra-bulycarbamic acid ester with 3-(m-hydroxyphenyl)-1,1-dimethylurea.

In 1945, Templeman and Sexton (1945) described the herbicidal properties of propham. It has been widely used to control grasses in tolerant crops: sugar beets, soybeans, onions, garlic, peas, flax, sunflower, rape, and mustard to name a few. Propham has proved toxic to oats, barley, wheat, quackgrass (*Agropyron repens*), red rice (*Oryza sativa*), fescue (*Festuca* spp.), maize, and timothy. Ineffective when applied to foliage of most species this herbicide is almost exclusively applied through the soil.

Freed (1951) pointed out that one should differentiate between application of propham on annual weeds and on perennials. Against annuals it should be applied during their very early seedling stages or, preferably preemergence. Against perennial grasses Freed recommended application to mature foliage in an oil carrier to be followed by tillage. In this way there is a combined contact action on the foliage followed by exposure to the roots following the tillage; for this treatment the soil should be moist. Quackgrass will succumb to this treatment when conditions are right.

Propham is rapidly broken down by microorganisms in the soil; such degradation is promoted by warmth and moisture. For this reason propham has proved most useful in cool season crops. In Britain propham is recommended only for replanting use; it has controlled wild oats (*Avena fatua*) in peas and sugar beets when applied before final seedbed preparation.

Soon after the discovery of the herbicidal properties of propham, chlorpropham was found to have even greater toxicity against grassy weeds; it also had somewhat different selectivities; its residual action in the soil proved longer. Chlorpropham has largely replaced propham as a herbicide in field crops.

Chlorpropham is lower in volatility than propham and more persistent in soils especially during the warmer part of the year. Danielson (1959) has reported on the mode and rate of release of chlorpropham from granules made of several materials. Neither propham nor chlorpropham translocate symplastically in plants. They are readily fixed in soils and so, diluted upon mixture; for this reason they are applied to the soil surface but are not incorporated before planting. They tend to remain fixed in the top 1 inch of soil.

These carbamate herbicides are available as emulsifiable concentrates, wettable powders, and granular formulations. They may move from granules into the surface soil in the vapor form.

Barban is a carbamate that has a high selectivity; it is commonly used to control wild oats in spring wheat, durum wheat, barley, flax, peas, sugar beets, safflower, mustard as a seed crop, and soybeans. Application of barban is by ground or aerial spray, postemergence to the crop and when wild oats are in the $1\frac{1}{2}$ to the $2\frac{1}{2}$ leaf stage. The time of application is critical; beyond the $2\frac{1}{2}$ leaf stage control of wild oats diminishes rapidly.

Barban is also effective for the control of several species of *Rumex* and *Polygonum*. Reed canarygrass (*Phalaris arundinacea*) is very sensitive. All small grains with the exception of oats, buckwheat, and some rye varieties show a high level of tolerance. Studies have shown that effective treatment results in swelling of the apical growing point, distortion of the meristems and failure of axillary buds to develop. Oats show these effects to a much greater extent than does wheat. Results of barban application are best when the crop and weeds are growing actively; use of fertilizer to promote growth materially increased effectiveness in greenhouse trials. In dry, cold weather use of barban may be ineffective.

Under favorable conditions yields of wheat have been increased as much as 60% by controlling wild oats with barban; barley and flax have responded with increases of 40%. In flax, $\frac{1}{4}$ to $\frac{1}{2}$ lb/A is used in the 8 to 10 leaf stage. Barban and MCPA are compatible and they may be used in combination where both wild oats and broadleaf weeds infest a crop.

Phenmedipham which is sold under the trade name of Betanal contains two carbamate radicals in a single molecule. It is a postemergence herbicide particularly effective in sugar beets. It is recommended for use against common lambsquarters (*Chenopodium album*), shepherds purse (*Capsella bursapastoris*), dogfennel (*Eupatorium capillifolium*), wild turnip (*Brassica campestris*), common chickweed (*Stellaria media*), wild radish (*Raphanus raphanistrum*), ragweed (*Ambrosia* spp.), kochia (*Kochia scoparia*), wild buckwheat (*Polygonum convolvulus*), green foxtail (*Setaria viridis*), nightshade (*Solanum* spp.), and field pennycress (*Thlaspi arvense*).

Phenmedipham acts through foliar absorption, and rain falling immediately after application may reduce its effectiveness. It breaks down, soon after application, by hydrolysis, and it has a half-life of approximately 25 days. It is used at rates of 1 to $1\frac{1}{2}$ lb/A broadcast in the 2 to 4 leaf stage of the weeds. Diluted with water it may be used at 20 gal/A broadcast. It is also useful by band application, tillage is then used to control weeds between the bands.

Norris (1972) has found that crystallization of phenmedipham in the spray tank depends upon temperature and dilution. When temperature of the spray solution is below 40°F and when the dilution is greater than about 1 to 25, the phenmedipham forms crystals which clog screens and prevent effective spraying.

Dichlormate, developed under the trade name of Rowmate and Sirmate is a preemergence herbicide recommended for controlling annual grass and broadleaf weeds in established alfalfa, asparagus, bush berries, cabbage, castor bean, cotton, field peas, and a variety of field and vegetable crops. It can be used in orchards, vineyards, and ornamental plantings. Selectivity is greatest from preemergence application; postemergence activity is general, but some selectivity may be obtained by directed spray application. This herbicide is available as an emulsifiable concentrate or in granules and it is recommended at 4 to 8 lb/A depending upon the soil type; the lighter dosage is used in the light soils. Water is used as carrier for the emulsifiable concentrate.

Dichlormate inhibits pigment formation in leaves. It has no effect on seed germination or early seedling growth. Treated weeds turn chlorotic, wither, and die.

Swep is used preemergence on large-seeded legumes and for pre- or postemergence weed control in rice. In the latter crop swep has proved effective against barnyardgrass (*Echinochloa crusgalli*), crabgrass (*Digitaria* ssp.), bittercress (*Cardamine* ssp.), smartweed (*Polygonum* ssp.), and lambsquarters. It is also effective in the control of spike rush (*Eleocharis acicularis*), a perennial weed that is a serious pest in rice culture.

Swep has been a popular herbicide in Japan where rice is a critical crop requiring a considerable amount of hand labor. The use of swep allows the Japanese to plant dry-field, direct-seeded rice which eliminates the rice nursery crop and the very laborious transplanting by hand. In addition in conserves water at a time when the supply is critical for weed control in the transplanted paddys. By using swep, it is possible to work the soil and prepare the seedbed as in planting other cereal crops; flooding is avoided. The herbicide is applied directly to the soil during the period from seeding until emergence of the rice plants. Where swep is used the very laborious and time consuming process of hand replanting is eliminated. Thus, as shown in Table 1-8 many millions of dollars in planting and weeding costs may be saved each year by the use of herbicides.

Swep is prepared as a wettable powder, in granular form and mixed in granules with MCPA. The wettable powder may be applied as a suspension by hand sprayer, power equipment or aerial application. The granules may be applied by ground rig or by air. Where the rice is planted under water, as in California, aerial application becomes the only available method.

Tandex (karbutilate) is a general preemergence or postemergence herbicide useful in the control of annual and broadleaf weeds and grasses on railroad rights-of-way, highway, and utility areas, industrial sites and other non-cropped areas. It has activity against woody species if used at proper dosage.

The chemical name and structure of karbutilate, Table 13-1, shows that it could be considered a urea type herbicide as well as a carbamate. Although physiological and herbicidal properties appear to be more like the urea-type herbicides than the carbamate-type herbicides, we have included it with the carbamates because the chemical name ends with carbamate.

Karbutilate is absorbed through the roots and translocated apoplastically. The 80% wettable powder can be applied in either water or herbicidal oil; a surfactant aids the contact action of the aqueous spray. Recommended dosages are 2 to 4 lb/A for annual weeds; 4 to 8 lb/A for perennial weeds; and 8 to 20 lb/A for woody species.

Terbutol, sold under the trade name of Azak is a preemergence herbicide recommended for the control of crabgrass in established turf. Application at from 5 to 15 lb/A is required for season-long control; seedlings of most grasses are susceptible to terbutol. This herbicide may be applied as a wettable powder suspended in water or as granules; the latter may be mixed with fertilizer.

Terbutol is absorbed from the soil by roots. It inhibits growth of roots and rhizomes at the terminal meristems. In the soil terbutol leaches slowly because of its low solubility; it is not strongly fixed on soil colloids.

Growth and Plant Structure

The carbamates considered in this chapter constitute too diverse a group of compounds, both in terms of molecular structure and in symptology, to allow any broad generalizations. Propham and chlorpropham were early demonstrated to be mitotic poisons that killed roots by inhibiting cell division. Ennis (1948) reported propham blocked cell division in roots of onion at metaphase and that there was no spindle formation. He also found that chlorpropham acted similarly (Ennis, 1949). Concurrent research by Ivens and Blackman (1949) showed that the responses of plants to ethyl phenylcarbamate are similar to those of colchicine; they consist of disruption of mitoses in roots with an increase in arrested metaphase figures. In barley roots there was an increase in cell size, many polyploid nuclei, varying degrees of chromosome contraction, production of multinucleate cells, and a great increase in metaphase nuclei compared with anaphase and telophase. The writers proposed that the carbamate esters associate with lipophilic side chains of the spindle proteins leading to intramolecular precipitation, folding of protein chains and disintegration of the spindle. Scott and Struckmeyer (1955) found that concentrations of chlorpropham of 5 ppm and above produced severe distortion of root tissues. Endodermal and pericyclic cells were radially elongated, accounting for the swelled appearances of the roots. Many meristematic cells were hypertrophied; some enlarged cells have abnormal nucleoli; thickenings of the cell walls of the endodermis were found, cessation of growth resulted in maturation of tissues up to within a short distance of the root apex.

Canvin and Friesen (1959) studying the effects of propham found disturbance of the orderly polarized arrangement of cells. Endopolyploidy was common. Chromosomes were scattered and appeared to be in extended pseudometaphase, chromosome clumping was common and the presence of tripolar and polypolar anaphases showed abnormal spindle behavior; failure of cell wall formation resulted in multinucleate cells. Storey et al. (1968) demonstrated that propham, chlorpropham, dichlōrmate, 2,3-dichlorobenzyl methylcarbamate, methyl-4-methoxycarbonylaminobenzosulphonyl-carbamate and benzyladenine produced rapid inhibition of cell activity and contraction of chromosomes, probably at all stages of cell division. Propham has also been shown to alter the orientation of spindle microtubules of dividing endosperm cells of *Haemanthus katherinae* (Helper and Jackson, 1969).

Bartels and Pegelow (1967, 1968) reported that dichlormate reduced growth and resulted in chlorotic leaves of wheat. Ultrastructural studies showed that the chloroplasts lacked normal grana-fret membranes and chloroplast ribosomes. In contrast, the cytoplasmic ribosomes and endoplasmic reticulum were abundantly present. Organelles other than chloroplasts were morphologically normal.

Absorption and Translocation

All of the early work on propham and chlorpropham indicated that these compounds, having low water solubility, are readily absorbed from the soil but that they have little phytotoxicity when applied to foliage. In recommending propham for control of perennial grasses Freed (1951) advocated using oil as a carrier to aid in penetration; he also suggested cultivation following application to bring about contact of the chemical with the rhizomes and stolons. Recognizing propham as a mitotic poison Freed stated that it is most effective used against germinating seeds and very young seedlings; adequate soil moisture was instrumental in assuring intimate contact of the chemical with the plant.

One of the first studies on the absorption and translocation of the carbamate herbicides was that of Baldwin et al. (1954). They reported that propham was absorbed through cut surfaces of leaves, cut surfaces of roots, and intact roots; these are listed in descending order of the absorption rate. Intact leaf surfaces did not absorb an appreciable amount of propham. These results suggest that plant leaf surfaces are a barrier to the absorption of propham. They also noted that propham was absorbed to a greater degree by roots of maize than oats.

Barrentine and Warren (1970) found that an isoparaffinic oil used as a carrier increased the penetration of chlorpropham as much as 8-fold in ivyleaf morning glory (*Ipomoea hederacea*) and 4-fold in giant foxtail (*Setaria faberi*) as compared with water as a carrier. This isoparaffinic oil had no phytotoxicity on onions, carrots, peppermint, or spearmint. It evidently was completely free of aromatic compounds. The fact that the chlorpropham in the isoparaffinic oil gave 100% control of giant foxtail indicates that the low phytotoxicity of this herbicide applied in water is largely a matter of penetration.

Prendeville et al. (1968) applied sublethal concentrations of chlorpropham labeled with ^{14}C in the ring or on the side chain to foliage and to roots of redroot pigweed (*Amaranthus retroflexus*), pale smartweed (*Polygonum lapathifolium*), and parsnip. They found that the labeled herbicide did not move out of the treated leaves of the pigweed and smartweed; there was evidence for slight movement in parsnip in 21 days. In root treatments of three days duration the herbicide moved to all plant parts and movement was similar in all three species. These results indicate that chlorpropham undergoes apoplastic distribution but is not moved symplastically. From the similarity in response to foliage and root applications of propham it seems probable that this compound moves in the same way in plants. Prendeville et al. (1968) concluded that differences in absorption and translocation were not sufficient to account for the different susceptibilities of these three plant species.

Knake and Wax (1968) studied the phytotoxicity of a number of herbicides including chlorpropham when various plant parts (root zone, shoot zone, root zone plus shoot zone, seed zone) were exposed to treatment via the soil. Chlorpropham proved to be relatively ineffective in controlling seedlings of giant foxtail when only the roots were exposed; exposure of the lower shoot to chlorpropham in the top one inch of soil was the most effective treatment. In contrast to general recommendations, Jordan et al. (1968) found best control of weeds with chlorpropham following rotary tiller incorporation into preirrigated soil.

New light has been thrown onto the way in which chlorpropham may kill plants. Slater et al. (1969) have found that dodder, a weed that has responded dramatically to treatment with granular chlorpropham, may be killed in the seedling stage by vapors given off by this herbicide in contact with most soil. There is a definite time factor involved in this response; dodder seedlings must be exposed to the vapors for 16 hours or more in order that wrappings of the dodder stem on the host plant does not take place. Table 13-2 gives data on the effect of chlorpropham vapors on dodder seedlings exposed at different stages of development.

TABLE 13-2. **Effect of Chlorpropham Vapors on Dodder Seedlings Exposed at Different Stages of Development as Indicated by the Number that Wrapped and became Attached to Alfalfa (Slater et al., 1969)**

Dodder stage	Number of pairs transplanted	Herbicide	% wrapped	% attached
Hook, 3–20 mm long	20	Chlorpropham	10	0
		none	100	75
Bent, 5–22 mm long	8	Chlorpropham	50	0
		none	100	75
Straight, 13–30 mm long	20	Chlorpropham	45	0
		none	95	85
Straight, 31–50 mm long	15	Chlorpropham	73	0
		none	100	68

When chlorpropham exposure to alfalfa plants in the field was tested it was found that of 110 seedlings set in untreated soil 64 (58%) wrapped on the host plants; of 225 set in chlorpropham-treated soil only 8 (4%) wrapped. Control of dodder was evident throughout a 19-day period after the application. The alfalfa foliage evidently restricts air movement of the herbicide sufficiently to provide for vapor toxicity to the dodder seedlings.

Dawson (1969), who had done much of the pioneering work on dodder control with chlorpropham, tested some 28 different herbicides for longevity of dodder control on alfalfa. Chlorpropham at 6 lb/A usually controls dodder completely for from 3 to 5 weeks. Of the 27 additional compounds tested seven controlled dodder completely for from 6 to 18 weeks; nine were comparable to chlorpropham; eleven controlled the pest to a lesser extent than chlorpropham. The period of dodder control by chlorpropham was doubled when p-chlorophenyl N-methyl-carbamate (PCMC) at 1.5 lb/A was included with it. PCMC is a microbial inhibitor which presumably slows the microbiological degradation of chlorpropham in the soil.

The absorption and translocation of chlorpropham by germinating seeds has been investigated (Ashton and Helfgott, 1966; Helfgott and Ashton, 1966; and Helfgott, 1969). Seeds of soybean, maize, peanut, and castor bean were used in these studies. When air-dry seeds were placed in a solution of ^{14}C-chlorpropham the herbicide entered the seed more rapidly than the water. Hydration was not necessary since absorption took place with fully imbibed seeds. Furthermore, when air-dry seeds were exposed to chlorpropham vapors in a closed system the seeds accumulated substantial amounts of the herbicide rapidly. The absorption of chlorpropham by seeds killed with methyl bromide was similar to living seeds. The temperature coefficient (Q_{10}) for the process was about 1.4 and a direct proportional relationship was found between absorption and concentration. The addition of sodium azide or dinitrophenol to the herbicide solution did not alter the absorption. Therefore, it was concluded that the absorption of chlorpropham during the initial stages of germination was a physical accumulation process which was more rapid and independent of water uptake. The chemical composition of the various species of seeds used appeared to have little, if any, influence in the absorption of chlorpropham. All the soybean and corn seed parts were penetrated by the chemical. However, there was no radioactivity present in the embryo axis of peanut and castor bean or in the endosperm of castor bean.

Absorption studies were also conducted with various seed parts. The initial rate of uptake of chlorpropham by excised soybean cotyledons was faster than that by cotyledons of intact seeds, indicating that the seed coat had a certain degree of impermeability to the herbicide. Desorption experiments indicated that most of the herbicide which was released by the intact seed of soybean was from the seed coat.

The distribution of radioactivity in 16-day old seedlings whose seeds had been incubated in a ^{14}C-chlorpropham solution for 12 hours was determined by autoradiographic techniques. Most of the radioactivity was present in either the cotyledons of soybean and peanut or in the pericarp and endosperm of corn. Very little of the radioactive material was translocated from these organs of greatest initial uptake to other organs of the developing seedling. That which was translocated appeared to be moving apoplastically and no major sites of accumulation were evident.

The distribution of ^{14}C-chlorpropham in various subcellular fractions was determined by sequential centrifugations of soybean-cotyledon homogenates at 1000 g, 20,000 g, and 90,000 g. Almost all the radioactivity was present in the supernatant fluid following the 1000 and 20,000 g centrifugation. After a 90,000 g centrifugation, the activity was distributed between the fatty layer, the supernatant fluid and the "microsomal" pellet. The amounts of radioactivity associated with the fatty layer and "microsomal" pellet decreased with age of the cotyledons and this was accompanied by a corresponding increase in the supernatant fluid.

Rieder et al. (1970) also studied the uptake of chlorpropham by soybean seeds. They found a direct relationship between uptake and concentration. Increasing temperature between 10° and 30°C increased uptake. Uptake rates were similar in living and dead seeds. Absorption could not be associated with absorption of water except during the first few hours when the seeds were rapidly imbibing water. It was concluded that the uptake of chlorpropham was largely a physical process that required seed hydration, but chlorpropham uptake continued after water uptake ceased. However, Helfgott (1969) showed that chlorpropham vapors could be absorbed by dry seeds. Therefore, the requirement of seed hydration, as suggested by Rieder et al. (1970) for chlorpropham is subject to question.

Foy (1961) investigated the absorption and translocation of ^{14}C-barban following foliar treatment to barley, oat, and wild oat. He found that the ^{14}C-label was absorbed and translocated in small amounts when applied to upper surface of the first leaf, the greatest movement apparently being in the transpiration stream toward the tip of the leaf. Much larger amounts were absorbed and translocated when barban was applied to the axil of the first leaf. The total amount of the ^{14}C-label absorbed and translocated by either route was proportional to the period of treatment; 3.5 hours, 24 hours, or 1 week. There were no detectable differences in absorption and translocation of the ^{14}C-label among the three species. Apparently differential absorption and translocation of these species does not account for their differential selectivity to barban.

Phenmedipham absorption and translocation has been studied by Kassenbeer (1969) and Koch et al. (1969). Utilizing [14]C- and ^{3}H-labeled phenmedipham, Kassenbeer (1969) found that most of the herbicide entered the leaf within the first 4 hours in beet, *Sinapis arvensis, Alopecurus myosuroides*, hairy galinsoga (*Galinsoga ciliata*), catchweed bedstraw (*Galium aparine*), and common chickweed (*Stellaria media*). During this time, *A. myosuroides* absorbed the least, about 20% of that applied, where *S. arvensis* absorbed about 50%. After 6 days, *A. myosuroids*, catchweed bedstraw, and common chickweed had absorbed about 30%; beet about 50%, and hairy galinsoga about 70%. Translocation was evident within 8 hours and movement was mainly in the direction of the transpiration current. Translocation was particularly rapid in *S. arvensis*. In the Koch et al. (1969) investigation, foliar absorption of phenmedipham by beet and *S. arvensis* was quite rapid during the first 2 hours and continued slowly thereafter. The rate of absorption increased with increasing light intensity, 2 to 90 lux, and increasing temperature, 10° to 40°C.

Molecular Fate

The most obvious fate of propham and chlorpropham relates to their volatility. These compounds were early recognized to have an appreciable vapor pressure, and the fact that the longevity of activity of pelleted compounds against dodder depends upon molecular structure would seem to indicate that a large portion of an applied dosage may be lost to the atmosphere.

Hodgson (1967) studied the metabolism of chlorpropham in seedlings of maize, cucumber, pea, and soybean. Treating with molecules labeled with ^{14}C either uniformly in the ring or at the number 2 carbon on the isopropyl group, he autoradiographed the plants and analyzed the extracts. The relative distribution of the ^{14}C depended upon the labelling position; evidently the chlorpropham molecule is cleaved in soybean roots but not in maize seedlings. Further tests on soybean suggested that the molecule is split into a water-soluble aniline-containing moiety and an isopropyl moiety both of which may be conjugated or metabolized. The author considered that the chloroform extracts of treated plants contained unaltered chlorpropham; the water extracts contained two major metabolites which were not identified.

Eshel and Warren (1967) studying the postemergence action of chlorpropham found that redroot pigweed and large crabgrass (*Digitaria sanguinalis*) responded rapidly; injury appeared in 1 or 2 days. On pale smartweed inhibition was slow showing only after 2 weeks. Studies on metabolism of chlorpropham showed that the fast action on redroot pigweed was accompanied by reduction in photosynthesis, decrease in chlorophyll content, and increase in respiration. Such symptoms occurred in pale smartweed only after about 2 weeks.

Prendeville et al. (1968) used redroot pigweed, pale smartweed, two susceptible species of differing response time, and parsnip, a species tolerant to chlorpropham in studies on the metabolism of this herbicide. Water-soluble metabolites of differing Rf values were isolated from extracts of pigweed and smartweed. Two metabolites were found in pigweed; one in smartweed; three metabolites were obtained from parsnip; however when higher application rates were used on a greater number of plants only one metabolite was found; less conversion of chlorpropham to water soluble metabolites occurs in parsnip than in pigweed and smartweed. Use of ring- and chain-labeled tracers yielded identical metabolites in the work of Prendeville et al. These writers concluded that since the metabolites isolated following either ring- or chain-labeled treatments were identical, the metabolites did not result from cleavage of the chain from the ring but were more likely conjugates with natural plant components. Different metabolites were formed by all three plant species. Very little $^{14}CO_2$ was liberated from the treated plants indicating little tendency for breakdown of chlorpropham within plants.

Studies on the metabolism of swep have been primarily residue-type investigations on rice with harvests made 75 to 149 days after treatment. Chin et al. (1964) reported that in such studies, a metabolite of swep was detected in straw, hulls, and in the bran layer of rice but not in the polished rice grain or bran oil. This metabolite was identified as a swep-lignin complex which was quite stable and not extractable from plant tissues by common organic solvents. With HCl-dioxane extraction, a portion of this complex is hydrolyzed to swep and soluble products of lignin. They state that the unusual stability of the swep-lignin complex and its relative molecular size suggest that the swep molecules are deeply trapped within the lignin polymer, and the molecular trapping between swep and lignin may occur within the living cells of rice through some special enzymatic anabolic activity. Similar results were obtained with carrots, oats, wheat, maize, and barnyardgrass.

Barban appears to be degraded quite rapidly in most plant species (Riden and Hopkins, 1962). They found that barban was metabolized into a 3-chloro-analine-containing substance (I). The amount of I formed reached a peak on the second or third day and then slowly declined, approaching a zero value well in advance of crop maturity. As much as 60% of the applied barban was converted to I. The rate of I formation initially increased with increasing temperature (70°F night to 80°F day vs. 40°F night to 60°F day) but the total amount of I was about the same after 2 weeks.

Jacobson and Anderson (1972) used two wild oat biotypes and two barley varieties to study the differential responses to barban. Root application retained the differential responses but to a lesser extent than foliar application; wild oat maintained a greater differential response than barley. Differential response to foliar applications was not caused by differential uptake but may result from reduced ability to degrade barban by the susceptible plants. A non-phytotoxic compound-X is found in all plants treated with barban; buildup of this may reduce the metabolism of barban resulting in a greater amount of free barban in the treated susceptible plants 12 to 24 hr after treatment; this free barban may account for the differential response to foliar application.

Phenmedipham has been shown to be degraded in sugar beets (Kossmann, 1969). Although no unchanged phenmedipham was detected at harvest, traces (<0.1 ppm) of 3-methylaniline were detected after hydrolysis of the tissue.

Dichlormate appears to be rapidly modified in plants and only trace amounts can be detected within 7 days of treatment (Herrett et al., 1968). Dichlormate was converted to a water-soluble substance containing the intact molecule in sensitive plants; whereas tolerant plants degrade dichlormate or the water-soluble substance so rapidly that any trace of dichlormate is difficult to detect 48 hours after treatment.

Herrett (1969) writing on the methyl and phenyl carbamates in Kearney and Kaufman's (1969) book discusses various possible pathways of biotransformation that may take place in soils and plants. The following organization is used:

A. Hydrolysis
 1. Ester hydrolysis
 2. Amide hydrolysis
B. Hydroxylation
 1. *N*-alkyl hydroxylation
 2. Aryl hydroxylation
 3. *N*-hydroxylation
C. *N*-dealkylation
D. Sulfur oxidation
E. Conjugate formation

Regarding degradation in plants Herrett suggests that hydrolysis may be a major degradative pathway for chlorpropham. However, the small amounts of hydrolytic products found in analyzing plants treated with swep and barban indicate other metabolic transformations. In the case of dichlormate the biotransformation in sensitive plants is rapid and involves a water soluble product containing the intact carbamate. Tolerant plants break down the carbamate so rapidly that it is hard to find after 48 hours.

In summarizing his discussion on pathways of metabolic degradation Herrett presents the following scheme covering three commonly used herbicides (Figure 13-1).

The hypothesis involved in this scheme is developed on the basis of competing reactions. The hydrolytic pathway C and the biotransformation A to a hydrophilic compound compete for the parent compound. In sensitive plants the biotransformation is favored; in the tolerant plant the hydrolytic pathway is preferred and little or none of the hydrophilic product is found. When an inhibitor is added it blocks the hydrolytic pathway and favors the formation of the water-soluble product. With such a mechanism involved, species differences, sensitivity to plant age, and herbicide dosage as well as environmental factors such as temperature and humidity, can be appreciated. Availability of degradative enzymes such as differences in hydroxylating phenylalanine and in the acyl hydrolase enzyme between resistant rice and sensitive barnyard-

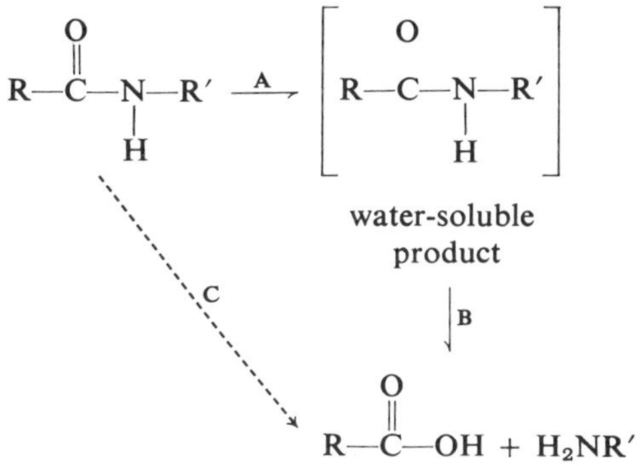

FIGURE 13-1. Proposed pathway for carbamate herbicide degradation in higher plants (Herrett, 1969). Reprinted from P. C. Kearney and D. D. Kaufman, eds, *Degradation of Herbicides*, p. 140, by courtesy of Marcel Dekker, Inc.

grass may be altered. Thus in studying these potential differences in metabolism it is essential that both susceptible and resistant species be used. The carbamates are effectively broken down in both plants and soils, a factor that is in their favor in their use in agriculture.

Biochemical Responses

As with structural responses, biochemical responses to the carbamate herbicides are varied. They have been shown to inhibit oxidative phosphorylation, RNA synthesis, protein synthesis, and the Hill reaction of photosynthesis, as well as reduce the ATP content of tissue sections.

Perhaps protein synthesis and the associated nucleic acid synthesis, as well as the energy requirements for these reactions in the form of ATP are primarily involved in the action of those carbamate herbicides which interfere with cell division. Mann et al. (1965a) reported that propham, chlorpropham, barban, and swep inhibited radioactive methionine incorporation into a polymer fraction of etiolated seedling sections. The polymer fraction was mainly protein but also contained some lignin. Several plant species were used and degree of inhibition with chlorpropham was related to susceptibility of the species to this herbicide. They (Mann et al., 1965b) also showed that chlorpropham markedly inhibited the incorporation of ^{14}C-leucine into protein using barley coleoptiles and sesbania hemp (*Sesbania exaltata*) hypocotyls. Mann et al. (1967) found that chlorpropham and barban blocked the gibberellin-induced synthesis of amylase in barley half-seeds and the decapitation-induced rise in rate of protein synthesis in sesbania hemp hypocotyls. In contrast, these compounds mimicked the action of phytokinins. They preserved chlorophyll levels in excised barley leaves and maintained the synthesis of tyramine methylpherase in roots of barley seedlings.

Moreland et al. (1969) confirmed that chlorpropham markedly inhibits ^{14}C-leucine incorporation into protein. However, in contrast to the Mann et al. (1967) findings, chlorpropham did not significantly inhibit gibberellin-induced alpha-amylase synthesis in barley half-seeds. Moreland et al. (1969) also showed that chlorpropham markedly inhibited RNA synthesis measured by either ^{14}C-6-orotic acid incorproation or ^{14}C-8-ATP incorporation.

Gruenhagen and Moreland (1971) studied the effect of chlorpropham on the level of ATP in excised soybean hypocotyls. They found that chlorpropham reduced the ATP level to 12% of the control. When these results were compared to previously determined effects of chlorpropham on protein synthesis, ATP synthesis, and oxidative phosphorylation, they suggested that chlorpropham inhibits RNA and protein synthesis by interference with the production of energy (ATP) required to drive biosynthetic reactions.

Lotliker et al. (1968) measured the effect of propham and chlorpropham on oxidative phosphorylation of mitochondria isolated from cabbage. They found that propham at 5×10^{-4}M inhibited phosphorylation by about 50% without comparable effect on O_2 uptake. Phosphorylation was inhibited almost completely at 1×10^{-3}M while O_2 uptake was reduced by about 35%. Using chlorpropham, they reported that at either 1×10^{-3}M or 1×10^{-2}M oxidative phosphorylation was severely inhibited with O_2 uptake inhibition being slightly greater than Pi esterification.

Briquet and Wiaux (1967) reported that propham caused a marked stimulation of RNA synthesis in pea roots, a somewhat resistant species, while chlorpropham showed an inhibiting action; they both completely inhibited RNA synthesis in oat roots, a susceptible species.

In a series of experiments, 2-day old etiolated seedlings of wild oat were sprayed with barban at 25, 250, and 1000 ppm; oxidative phosphorylation (Ladonin, 1966), nucleotide metabolism (Ladonin, 1967a), and metabolism of RNA, protein, and nucleoproteins (Ladonin and Svittser, 1967) were measured. After similar treatment, RNA and protein metabolism in wheat seedlings was investigated (Ladonin, 1967b). Barban was reported to inhibit O_2 uptake and Pi esterification by mitochondria isolated from treated wild oat plants, after initially showing some stimulation at the lower two concentrations. Similar results were obtained when barban was added to mitochondria isolated from plants which had not received the herbicide treatment. In these *in vitro* studies there was also evidence of uncoupling of the reaction. Chromatographic analysis showed that barban caused a sharp increase in ATP consumption. Ladonin (1967a) determined that barban increased the nucleotide content of wild oat shoots and this was considered to be associated with a disruption of RNA and protein synthesis brought about by a deficit of ATP. He suggested that barban may increase ATPase activity. However in another study (Ladonin and Svittser, 1967), it was reported that barban caused an increase in the DNA, protein, and nucleotide contents of 36 to 51, 40 to 50, and 10% respectively in whole plants, and 69 to 162, 200 to 300, and 32 to 33% respectively in the area of stem swelling. They suggested that barban may disrupt the synthesis of mRNA causing abnormal cell division.

Although Hill reaction inhibition has been noted for certain alkyl-*N*-phenylcarbamates (Moreland and Hill, 1959; Wessels and van der Veen, 1956), the very nature of their phytotoxicity would seem to preclude this as a major factor in their use as herbicides. Most of them are recommended as preplant or preemergence materials, and those that kill from foliar application appear to inhibit meristematic activity in some way. Shaw and Swanson (1953) reported that *N*-phenyl carbamates having methyl or methoxy substitutions on the phenyl ring produce chlorosis. Among the carbamate herbicides described above, dichlormate inhibits chlorophyll formation in treated seedlings. In this molecule the two chlorines are on the phenyl ring.

Using the Hill reaction in a screening program, Moreland and Hill (1959) found isopropyl *N*-(3,4-dichlorophenyl) carbamate to be the most active of many compounds tested; Shaw and Swanson (1953) had found this compound to be highly phytotoxic.

Testing a series of alkyl esters of *N*-(3-chlorophenyl) carbamic acid Moreland and Hill (1959) found the *sec*-butyl most active followed by the *n*-butyl ester, the *n*-propyl, isopropyl, and amyl esters in decreasing order. Replacement of the imino hydrogen of ethyl phenylcarbamate by an ethyl, a phenyl, or a benzyl radical resulted in loss of inhibitory activity.

Proposing that molecules of ethyl *N*-phenyl carbamate may form association polymers by bonding between imino hydrogens and carbonyl oxygens, Moreland and Hill point out that association should increase the concentration and that marked inhibitory action is obtained as the reaction mixture approaches saturation with the carbamate. Replacement of the carbonyl oxygen with sulfur greatly reduces the inhibitory activity; sulfur may not form hydrogen bonds. Action of the carbamate molecule may also involve hydrogen bonding between carbonyl oxygen of the carbamate molecules and imino hydrogens of the peptide nitrogens on proteins at reactive sites. The authors have diagrammed these possible bonding reactions, Figure 13-2. Moreland and Hill emphasized the important role played by the imino hydrogen in this sort of reaction. In inhibiting the Hill reaction it may take part in hydrogen bond formation with some electronegative constituent located at or near the reactive center of the chloroplast. In support of this is evidence that compounds having the imino hydrogen replaced do not inhibit the photochemical reaction, and derivatives with chlorine substituted at an ortho position on the benzene lack inhibitory activity. An ortho chlorine may be able to form a hydrogen bond intramolecularly with the imino hydrogen to form a chelate ring; through steric hindrance it could prevent an electronegative group from approaching close enough to the imino hydrogen to form a hydrogen bond. Chlorines in the meta or para positions of the ring are too far removed from the imino hydrogen to exert such influence.

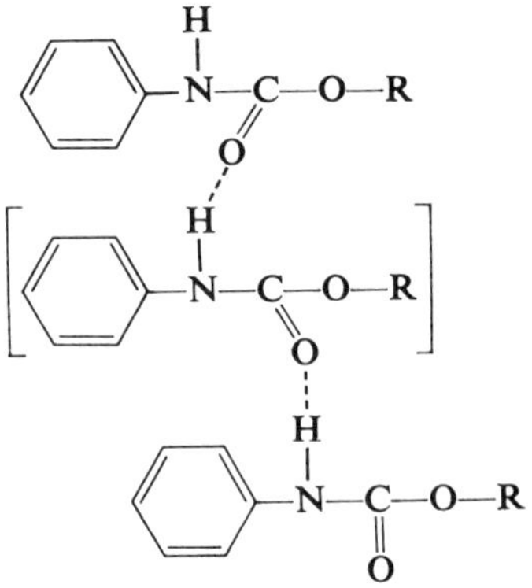

FIGURE 13-2.   Possible bonding of carbamate molecules (Moreland and Hill, 1959).

Mode of Action

While the above discussion may be relevant to the activity of carbamates as mitotic poisons, other more subtle factors must enter in the mode of action of some of the more complex carbamate herbicide molecules. For example, although barban may inhibit meristematic activity in the axillary buds in wild oat, the distinct selectivity seems more closely related to other aspects of its molecular configuration. Hopkins (1959) found that placing the chlorine on the phenyl ring in the 2 or 4 position altered selectivity against oats. And reduction of the triple bond to a double or single bond has a similar result. Replacement of the alkyl chlorine with other substituents eliminates selective toxicity; substitution of the ring has a similar result.

The highly selective nature of phenmedipham is another example where a basic phytotoxicity is channeled in such a way that sugar beets are tolerant whereas most common weeds are susceptible. Swep, with its relatively simple structure, is another case where extremely high selectivity, as related to the unique combination and structural positioning of substituents, is critical. Karbutilate, on the other hand, so lacks selectivity that it is recommended for general weed control through its combined contact and soil activity. Dichlormate produces chlorosis in weed seedlings and hence must act in some way on chlorophyll formation and/or chloroplast development.

When so diverse a group of compounds with respect both to chemical structure and phytotoxic selectivity is under consideration it is very difficult to generalize on mode of action. Some carbamate herbicides cause chlorosis while others do not. Some are highly selective between plant species of similar botanical characteristics, others have a general herbicidal action. Some carbamates are quite volatile while others have virtually no volatility under conditions of use in the field. In spite of these limitations, an attempt to summarize certain aspects is presented below.

Although most of the carbamate herbicides are applied to the soil and enter the plant either through the roots and/or the emerging shoot, barban and phenmedipham are applied to the foliage. It has been shown that chlorpropham enters the emerging shoot more readily than the roots, however the most favorable site of entry of the other soil applied carbamates is not known. Although chlorpropham enters the cotyledons of seeds rapidly, little transport from the cotyledons occurs. Chlorpropham has also been shown to enter certain plants in vapor form. Barban is absorbed by leaf surfaces, but is taken up more rapidly when applied to the axil of the first leaf of grasses. Phenmedipham is rapidly absorbed by leaves. In general differential rates of absorption and translocation have not been found to be responsible for the selectivity between susceptible and tolerant species with the carbamate herbicides. The translocation studies suggest that the carbamate herbicides are almost exclusively distributed via the apoplastic system.

Most of the carbamate herbicides are rapidly degraded in higher plants to an acid and an aniline by hydrolysis at the —C—N— bond. These may be further metabolized; however the aniline moiety may also be conjugated with some endogenous substance(s). The intact molecule of swep appears to form a complex with lignin. The basis of selectivity between susceptible and tolerant species has frequently been attributed to differential rates of molecular degradation with the carbamate herbicides.

Certain carbamate herbicides have been shown to inhibit oxidative phosphorylation, RNA synthesis and protein synthesis, as well as to reduce the ATP content of tissue sections. These effects may well explain the action of those carbamate herbicides which inhibit cell division but may not be applicable to the others. Certain carbamate herbicides have also been shown to inhibit the Hill reaction of photosynthesis.

REFERENCES

Ashton, F. M. and S. Helfgott. 1966. Uptake and distribution of CIPC in seeds during imbition. *Proc. 18th Calif. Weed Conf.*, p. 8.

Baldwin, R. E., V. H. Freed, and S. C. Fang. 1954. Herbicide action: absorption and translocation of carbon-14 applied as O-isopropyl *N*-phenyl carbamate in *Avena* and *Zea. J. Agr. and Food Chem.* **2**:428–430.

Barrentine, J. L. and G. F. Warren. 1970. Isoparaffinic oil as a carrier for chlorpropham and terbacil. *Weed Sci.* **18**:365–372.

Bartels, P. G. and E. J. Pegelow. 1967. An ultrastructural and ultracentrifugal study of chloroplast development in wheat seedlings treated with 3,4-dichlorobenzyl methylcarbamate. *Plant Physiol. Abstr.* **42**:S–28.

Bartels, P. G. and E. J. Pegelow. 1968. The action of Sirmate (3,4-dichlorobenzyl methylcarbamate) in chloroplast ribosomes of *Triticum vulgare* L. seedlings. *J. Cell Biol.* **37**:1–6.

Briquet, M. V. and A. L. Wiaux. 1967. Herbicides and RNA synthesis of *Pisum* and *Avena* roots. *Meded. Rijksfac. LandbWet. Gent.* **32**:1040–1049.

Canvin, D. T. and G. Friesen. 1959. Cytological effects of CDAA and IPC on germinating barley and peas. *Weeds* **7**:153–156.

Chin, W. T., R. P. Stanovick, T. E. Cullen, and G. C. Holsing. 1964. Metabolism of swep by rice. *Weeds* **12**:201–205.

Danielson, L. L. 1959. Mode and rate of release of isopropyl *N*-(3-chlorophenyl) carbamate from several granular carriers. *Weeds* **7**:418–426.

Dawson, J. H. 1969. Longevity of dodder control by soil-applied herbicides in the greenhouse. *Weed Sci.* **17**:295–298.

Ennis, W. B. Jr. 1948. Some cytological effects of *O*-isopropyl *N*-phenyl carbamate upon *Avena*. *Am. J. Bot.* **35**:15–21.

Ennis, W. B. Jr. 1949. Histological and cytological responses of certain plants to some aryl carbamic esters. *Am. J. Bot.* **36**:823.

Eshel, Y. and G. F. Warren. 1967. Postemergence action of CIPC. *Weeds* **15**:237–241.

Foy, C. L. 1961. Uptake of radioactive 4-chloro-butynyl *N*-(3-chlorophenyl) carbamate (barban) and translocation of ^{14}C in *Hordeum vulgare* and *Avena* spp. Research Progress Report, *Western Weed Control Conf.* pp. 96–97.

Freed, V. H. 1951. Some factors influencing the herbicidal efficacy of isopropyl *N*-phenyl carbamate. *Weeds* **1**:48–60.

Gruenhagen, R. D. and D. E. Moreland. 1971. Effects of herbicides on ATP levels in excised soybean hypocotyls. *Weed Sci.* **19**:319–323.

Helfgott, S. and F. M. Ashton. 1966. Absorption of CIPC by seeds. *WSSA Abstr.* p. 47.

Helfgott, S. 1969. Absorption, translocation, and metabolism of chlorpropham (isopropyl-*m*-carbanilate) by germinating seeds. Ph.D. Dissertation. University of Calif. Davis, 207 pp.

Hepler, P. K. and W. T. Jackson. 1969. Isopropyl *N*-phenylcarbamate affects spindle microtubule orientation in dividing endosperm cells of *Haemanthus Katherinae* Baker. *J. Cell Sci.* **5**:727–743.

Herrett, R. A. 1969. Methyl- and phenylcarbamates. In: *Degradation of Herbicides*. pp. 113–145. P. C. Kearney and D. D. Kaufman Ed. Marcel Dekker, Inc. New York.

Herrett, R. A., J. A. Kramer, Jr., and W. P. Bagley. 1968. A biochemical basis for the selective herbicidal action of 3,4-dichlorobenzyl methylcarbamate. *Abstr. 155th Meeting Am. Chem. Soc.*, p. A023.

Hodgson, R. H. 1967. Absorption, translocation of CIPC in plants. *Abstr.* USA 65–66.

Hopkins, T. R. 1959. *A Historical Development of Carbyne.* Spencer Chemical Co. Merriam, Kansas.

Ivens, G. W. and G. E. Blackman. 1949. The effects of phenyl carbamates on the growth of higher plants. *Symposia Soc. Exptl. Biol.* 3:266–282.

Jacobsohn, R. and R. N. Andersen. 1972. Intraspecific differential response of wild oat and barley to barban. *Weed Sci.* 20:74–80.

Jordan, L. S., J. M. Lyons, W. H. Isom, and B. E. Day. 1968. Factors affecting performance of preemergence herbicides. *Weed Sci.* 16:457–462.

Kassenbeer, H. 1969. The absorption and decomposition of phenmedipham marked ^{14}C and ^{3}H in young beets (*Beta vulgaris*) and weeds. *Schering AG, Conf.*, Berlin, pp. 5–6.

Kearney, P. C. and D. D. Kaufman. 1969. *Degradation of Herbicides.* Marcel Dekker Inc. New York. 394 pp.

Knake, E. L. and L. M. Wax. 1968. The importance of the shoot of giant foxtail for uptake of preemergence herbicides. *Weed Sci.* 16:393–395.

Koch, W., J. Majumdar, H. Fykse, and B. Rademacher. 1969. Phytotoxic action of phenmedipham in dependence of certain factors. *Schering AG, Conf.*, Berlin, pp. 12–14.

Kossmann, K. 1969. Phenmedipham residues in plants and soil. *Schering AG, Conf.*, Berlin, pp. 16–20.

Ladonin, V. F. 1966. The effect of Carbyne (barban) on oxidative phosphorylation of mitochondria from etiolated wild oat seedlings, Vest. sel'.-kohz. *Nauki, Mosk.* 11:137–141.

Ladonin, V. F. 1967a. The effect of Carbyne (barban) on nucleotide metabolism in seedlings of *Avena fatua. Agrokhimiya* 2:85–95.

Ladonin, V. F. 1967b. The effect of Carbyne on the metabolism of ribonucleic acid (RNA) and protein in spring wheat seedlings. *Khimiya sel'. Khoz.* 5:36–40.

Ladonin, V. F. and K. M. Svittser. 1967. Influence of Carbyne on the metabolism of RNA, protein, and nucleoproteins in wild oat seedlings. *Soviet Plant Physiol.* 14:853–860.

Lotlikar, P. D., L. F. Remmert, and V. H. Freed. 1968. Effects of 2,4-D and other herbicides on oxidative phosphorylation in mitochondria from cabbage. *Weed Sci.* 16:161–165.

Mann, J. D., L. S. Jordan, and B. E. Day. 1965a. The effects of carbamate herbicides on polymer synthesis. *Weeds* 13:63–66.

Mann, J. D., L. S. Jordan, and B. E. Day. 1965b. A survey of herbicides for their effect upon protein synthesis. *Plant Physiol.* 40:840–843.

Mann, J. D., E. Cota-Robles, K. H. Yung, M. Pu, and H. Haid. 1967. Phenyl-urethane herbicides: Inhibitors of changes in metabolic state. *Biochim. Biophys. Acta* **138**:133–139.

Moreland, D. E., S. S. Malhotra, R. D. Gruenhagen, and E. H. Shokraii. 1969. Effects of herbicides on RNA and protein synthesis. *Weed Sci.* **17**:556–563.

Moreland, D. E. and K. L. Hill. 1959. The action of alkyl-*N*-phenyl carbamates on photolytic activity of isolated chloroplasts. *J. Agr. and Food Chem.* **7**:832–837.

Prendeville, G. N., Y. Eshel, C. S. James, G. F. Warren, and M. M. Schreiber. 1968. Movement and metabolism of CIPC in resistant and susceptible species. *Weed Sci.* **16**:432–435.

Riden, J. R. and T. R. Hopkins. 1962. Formation of a water-soluble, 3-chloro-aniline-containing substance in barban-treated plants. *J. Agr. and Food. Chem.* **10**:455–458.

Rieder, G., K. P. Buchholtz, and C. A. Kust. 1970. Uptake of herbicides by soybean seed. *Weed Sci.* **18**:101–105.

Scott, M. A. and E. B. Struckmeyer. 1955. Morphology and root anatomy of squash and cucumber seedlings treated with isopropyl *N*-(3-chlorophenyl)-carbamate. *Bot. Gaz.* **117**:37–45.

Shaw, W. C. and C. R. Swanson. 1953. The relation of structural configuration to the herbicidal properties of several carbamates and other chemicals. *Weeds* **2**:43–65.

Slater, C. H., J. H. Dawson, W. R. Furtick, and A. P. Appleby. 1969. Effects of chlorpropham vapors on dodder seedlings. *Weed Sci.* **17**:238–241.

Storey, W. B., L. S. Jordan, and J. D. Mann. 1968. Carbamate herbicides—new tools for cytological studies. *Calif. Agr.* **22**:12–13.

Templeman, W. G. and W. A. Sexton. 1945. Effect of some aryl carbamic esters and related compounds upon cereals and other plant species. *Nature* **156**:630.

Wessels, J. S. C. and R. van der Veen. 1956. The action of some derivatives of phenyl urethan and 3-phenyl-1,1-dimethylurea on the Hill reaction. *Biochim. et Biophys. Acta* **19**:548–549.

CHAPTER 14

Dinitroanilines

Representatives of Eli Lilly and Company reported on the herbicidal properties of the 2,6-dinitroanilines in 1960 (Alder et al., 1960). Other companies have, or are now developing herbicides from this class of compounds. In general, they are yellow-orange, have a low water solubility, and are somewhat volatile. They are usually used for selective weed control as a preplant soil-incorporation treatment prior to weed seed germination. Their greatest use has been in cotton, but some of them are currently registered for use in several agronomic, vegetable, and tree fruit crops. Table 14-1 gives the chemical structure of these herbicides.

TABLE 14-1. **Common Name, Chemical Name, and Chemical Structure of the Dinitroaniline Herbicides**

$$R_2\text{---}N\text{---}R_3$$

$$O_2N \qquad NO_2$$

$$R_1$$

Common name	Chemical name	R_1	R_2	R_3
benefin	N-butyl-N-ethyl-α,α,α-trifluoro-2,6-dinitro-p-toludine	$-CF_3$	$-C_2H_5$	$-C_4H_9$
isopropalin	2,6-dinitro-N,N-dipropylcumidine	$-iso\text{-}C_3H_7$	$-C_3H_7$	$-C_3H_7$
nitralin	4-(methylsulfonyl)-2,6-dinitro-N,N-dipropylaniline	$\overset{\displaystyle O}{\underset{\displaystyle O}{-\overset{\uparrow}{\underset{\downarrow}{S}}-CH_3}}$	$-C_3H_7$	$-C_3H_7$
trifluralin	α,α,α-trifluro-2,6-dinitro-N,N-dipropyl-p-toluidine	$-CF_3$	$-C_3H_7$	$-C_3H_7$

Growth and Plant Structure

The dinitroaniline herbicides inhibit growth of the entire plant. However, this is apparently largely brought about by initially limiting root growth, especially the development of lateral or secondary roots. The roots which do develop, often only the primary roots, are somewhat thickened, stubby, and are devoid or have only a limited number of secondary roots.

Many workers have shown that trifluralin inhibits the growth of roots of many species; cotton (Fishher, 1966; Bayer et al., 1967; Normand et al., 1968; Hacskaylo and Amato, 1968; Arle, 1968; Oliver and Frans, 1968; Lund et al., 1970); soybeans (Talbert, 1965; Kirby et al., 1968; Oliver and Frans, 1968; Lund et al., 1970); maize (Hackaylo and Amato, 1968; Schultz et al., 1968); oats (Feeny, 1966); onions, safflower, and barnyardgrass (*Echinochloa crusgalli*) (Bayer et al., 1967). Nitralin has also been reported to inhibit the root growth of many species including wheat (Schieferstein and Hughes, 1966); cotton (Normand et al., 1968; Lund et al., 1970), and maize (Gentner and Burk, 1968; Lund et al., 1970). Wheat root growth, which is inherently quite sensitive, was completely inhibited by 6×10^{-8}M nitralin in culture solution (Schieferstein and Hughes, 1966).

Several of these investigations report that the inhibition of root growth is accompanied by an increase in diameter or swelling of the root near the root tip or in the meristematic region, as well as an inhibition of lateral or secondary root development.

Talbert (1965) reported that the root growth inhibition in soybean was associated with cessation of cell division in the meristematic tissue. He found that 2 hours after placing soybean plants in a nutrient culture solution containing trifluralin (0.5 to 5.0 ppm) the proportion of cell nuclei in prophase stage of cell division increased relative to control plants, and after 8 hours there was a further increase in prophase figures but practically no other stages of cell division detectable. He suggested that trifluralin interfered with the normal function of the spindle. After 24 hours, numerous polynucleate cells were present. He concluded that trifluralin acts as a mitotic poison. At the lower concentrations, growth and cell division resumed after 24 hours, apparently the herbicide was removed by volatilization. However, it further shows that the inhibition is reversible under certain conditions.

Bayer et al. (1967) conducted a comprehensive investigation on the morphological and histological effects of trifluralin on root development of cotton, safflower, onion, and barnyardgrass. They concluded that the most obvious external effect of trifluralin in the primary root was an increase in the radial expansion near the root tip. Trifluralin disrupted the mitotic process (Figure 14-1); however, no one type of mitotic figure prevailed. This is in contrast to the findings of Talbert reported above, in which he found that the prophase

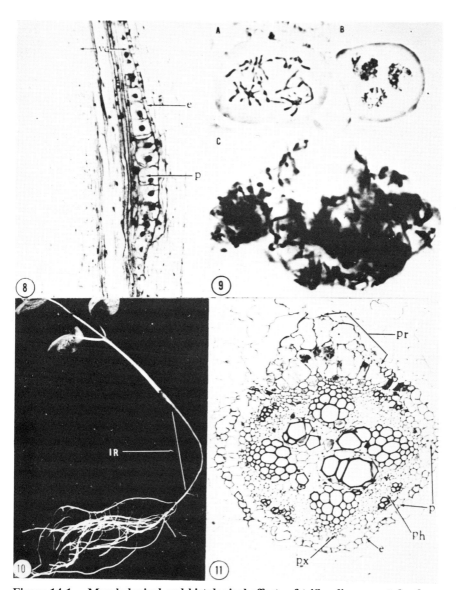

Figure 14-1. Morphological and histological effects of trifluralin on root development. Plate 8, longitudinal section of cotton root after 72 hr treatment with a 10^{-4} M
solution; enlarged pericyclic cells. Plate 9, root tip squashes of onion roots after 36 hr
treatment with 10^{-4} M solution; plate 9A, multipolar anaphase cell; plate 9B, three
micronuclei; plate 9C, numerous chromosomes and polyploid condition. Plate 10,
three-week-old cotton seedling grown in soil which had a $2\frac{1}{2}$ inch layer of soil containing
a 2 lb/A treatment; absense of lateral roots in treated zone. Plate 11, cross section of
primary root of cotton root from area where herbicide was incorporated into the soil;
enlarged pericyclic cells opposite the protoxylem indicate that lateral root formation
was initiated but stopped due to the inhibitory effect of trifluralin on cell division in the
pericycle. (e) endodermis, (ir) incorporation region, (p) pericycle, (ph) phloem, (pr)
primordiomorph, (px) protoxylem pole, (vc) vascular cylinder. (Bayer et al., 1967.)

stage was predominant. Mitotic activity was not affected in all cells, some appeared to be undergoing normal mitosis. Rates of trifluralin which inhibited lateral root emergence without interfering with primary root elongation affected the pericycle and portions of the endodermis of the primary root. The pericyclic cells were much enlarged in regions opposite the protoxylem and had undergone some of the initial phases of lateral root formation (Figure 14-1). Similar findings have been reported by Hacskaylo and Amato (1968) for maize and cotton, and by Schultz et al. (1968) for maize.

Schieferstein and Hughes (1966) reported that nitralin inhibits cell division and causes swelling of cells in the meristematic region of the root. Gentner and Burk (1968) found that nitralin promoted the development of digitate to globose swelling in the region of active cell division of maize roots. Cytological examination of the affected area showed that the effects of the herbicide were prevention of cell wall formation, enlargement of cells, and extensive replication of nuclei. They suggested that nitralin either prevents or limits spindle formation.

Normand et al. (1968) cultured excised cotton roots on agar-solidified White's media containing 1, 3, or 5 ppm trifluralin or nitralin. They reported that both herbicides inhibited root growth; however nitralin was somewhat more inhibitory than trifluralin, especially at the lowest concentrations. These herbicides also inhibited mitosis at all concentrations; the number of mitotic figures was reduced 95% with nitralin and 50% with trifluralin at the lowest concentration. Cross sections of roots treated with either herbicide showed more xylem differentiation than in the control. This is frequently observed with compounds that inhibit root growth; although the root elongation is reduced, differentiation continues resulting in greater differentiation at a given distance from the root tip.

Arle (1968) found an interesting interaction between certain systemic insecticides and trifluralin on the growth of cotton roots. The insecticides, phorate (*o,o*-diethyl S-[(ethylthio)-methyl]-phosphorodithioate) or disulfaton (*o,o*-diethyl S-[(ethylthio)-ethyl]-phosphorodithioate), counteracted the inhibition of shoot growth by trifluralin when incorporated into the soil with the herbicide. This result apparently was a secondary effect and the primary effect was a counteraction of the inhibiting effect of trifluralin on secondary root development. Phorate was more effective than disulfaton in overcoming the inhibitory effect of trifluralin on secondary roots.

Schultz et al. (1968) reported that trifluralin inhibited the elongation of *Avena* coleoptile sections. Kirby et al. (1968) studied the effect of trifluralin on soybean stem tissue as associated with an increased tendency of the crop to lodge with the herbicide treatment. They found that for the first 21 days after planting, a marked decrease in xylem differentiation occurred with herbicide treatment. After 63 days, or at about the time lodging began to occur in the field, the herbicide treated plants showed an increase in the area of xylem. However, the pattern of differentiation was erratic and there were sites on the periphery of the vascular cylinder with essentially no xylem. They concluded the lack of uniformity of xylem differentiation was a response to the trifluralin treatment and the susceptibility of the plant to lodging was influenced by the pattern of xylem distribution.

In 7-week-old soybean plants, Kust and Struckmeyer (1971) found trifluralin to alter cell arrangement of leaves and internodes; walls of pericyclic fibers of stems were abnormally thickened. Starch accumulated in nodule and xylem parenchyma of roots, and bi- and trinucleate cells characterized root tips and lateral roots. Trifluralin reduced nodulation of soybean and seemed to inhibit utilization of cotyledonary reserves and redistribution of organic and mineral constituents of unifoliate leaves. Hassawy and Hamilton (1971) found that when IAA and kinetin were both applied with 5 ppm trifluralin to cotton seedlings lateral roots developed.

Germination and tube growth of *Orobanche ramosa* seeds in stimulation fluid is inhibited by trifluralin (Saghir and Abu-Shakra, 1971). All concentrations from 10 ppm to 100 ppm reduced tube length; there was no germination at 1000 and 3000 ppm of trifluralin.

Absorption and Translocation

The absorption and translocation of the dinitroanilines has not been adequately investigated to come to any concrete conclusions. Obviously they must be absorbed by the plant to exhibit their phytotoxic effects. Although translocation to shoots following root application and to roots following shoot application have been reported, they are not usually considered to be readily translocated.

Parker (1966) demonstrated that trifluralin was absorbed by both the root and emerging shoot of germinating sorghum seedlings. When applied to the roots, 0.065 ppm were required to inhibit root growth 50%, whereas shoot growth was inhibited 50% by 2.7 ppm applied to the shoot. Whether these results mean that roots absorb more trifluralin than shoots or whether the roots are merely more sensitive is not obvious. When Knake et al. (1967) germinated green foxtail (*Setaria viridis*) seeds in a system which exposed either the roots or shoots or both to 1 ppm trifluralin as they elongated, they found that shoot exposure inhibited shoot growth completely whereas root exposure had essentially no effect on shoot growth. These results confirm that trifluralin is absorbed by emerging shoots and furthermore suggest that the translocation of triflurlin from the roots to the shoots is not of sufficient quantity to be of physiological consequence. Negi and Funderburk (1968) showed that when only shoot or root of maize was treated, there was very little or no effect on the other organ. Swann and Behrens (1969) reported that foxtail millet (*Setaria italica*) and proso millet (*Panicum miliaceum*) showed more extensive injury from shoot uptake of trifluralin compared to root uptake. Negi and Funderburk (1968) suggested that trifluralin vapors from soil may be absorbed by shoots of certain species.

In addition to the evidence presented above, based on growth response, which indicates relatively little absorption and translocation of trifluralin from root to shoot and vice versa, Probst et al. (1967) cite analytical residue data to support this point of view. They have not detected trifluralin or its degradation products in leaves, seeds, or fruit of a wide variety of tolerant crops. Root crops, such as onion and garlic, contain the trifluralin residue only in the outer shell. However, carrot roots incorporate trifluralin.

In microradiographic studies on 2-week-old cotton and soybean seedlings, Strang and Rogers (1971) noted that ^{14}C-trifluralin was retained primarily on the root surfaces. Entrance into roots was greatly facilitated by breaks in the epidermis such as might occur from disease or mechanical damage. Little movement out of soybean roots was observed; there was limited translocation into leaves of cotton. Radioactivity accumulated in the protoxylem of the cotton stem where many elements seemed to be plugged.

Ketchersid et al. (1969) studied the absorption and translocation of trifluralin following its application to various parts of peanut plants. Benefin and nitralin were also used in some of their research. They blended the various plant parts with acetone, filtered, added hexane, washed the organic-solvent extracts with water, discarded the water, concentrated the organic phase, and analyzed for the dinitroaniline by means of gas chromatography. Plants of three different ages were placed into sand containing 1 ppm trifluralin. They were harvested 2 weeks later. Absorption of trifluralin was greater when seedlings were germinated in untreated sand and transplanted into treated sand than when germinated in treated sand. This probably resulted because seedlings germinated in treated sand developed almost no lateral roots. Although herbicide uptake was greater in the transplanted seedlings, translocation was apparently reduced with increasing age. At all three ages trifluralin was detectable in all plant parts (leaves, epicotyl, cotyledons, hypocotyl, roots) showing extensive translocation. When trifluralin was applied to various plant parts it was absorbed and translocated to the untreated organs, Table 14-2. Although trifluralin was found in all organs of the plant, it was found in the highest concentration in the cotyledons. These results suggest that trifluralin may actually circulate in the plant's translocation systems (apoplastic and symplastic) and accumulate in areas of high lipid content, i.e., cotyledons.

The predominant evidence suggests that the dinitroaniline herbicides are readily absorbed by roots and shoots but translocated poorly. However, the research of Kerchersid et al. (1969) indicates extensive translocation in peanuts.

TABLE 14-2.   Trifluralin Concentration in Peanut Seedlings 2 weeks after Treatment with 50 μg in 20 μl of water (Ketchersid et al., 1969)

Treated area	Concentration in ppm			
	Cotyledon	Terminal bud	Hypocotyl	Root
Cotyledon	10.95	2.55	1.01	1.12
Terminal bud	16.32	3.52	1.16	0.69
Hypocotyl	9.94	2.34	4.42	7.75
Root	10.34	1.00	4.74	30.73
Check	0.00	0.00	0.00	0.00

Molecular Fate

As in the case with the absorption and translocation of the dinitroaniline, the molecular fate of these compounds in higher plants is not well understood. Probst and Tepe (1969), in their recent review, indicated that the present evidence is inadequate to claim that trifluralin is actively metabolized by plants. However, the work of Biswas and Hamilton (1969) indicates that trifluralin was extensively degraded by intact peanut and sweet potatoes in 72 hours.

Probst et al. (1967) grew soybean and cotton plants in soil containing ^{14}C-trifluralin labeled in the n-isopropyl or trifluoromethyl group. The radioactivity was distributed in lipids, glucosides, hydrolysis products, proteins, and cellular fractions. Small amounts of radioactive carbon dioxide were liberated from the soil-plant system. They concluded that the universal distribution of the radioactivity without definite identification of trifluralin or recognizable metabolites suggests nondescript incorporation of the radioactivity. However, it is not clear from their presentation whether these molecular changes in trifluralin occurred in the plants or whether they could have taken place in the soil and then been absorbed by the plants. Funderburk et al. (1967) treated cotton, corn, and soybean plants with ^{14}C-labeled trifluralin by foliar applications and in nutrient solutions to the roots. Gas chromatography and autoradiographs of thin-layer chromatograms revealed only trifluralin; no degradation products were detected.

Since carrots are quite resistant to trifluralin and a residue can be detected on the surface of their roots following a soil incorporation application of the herbicide, the metabolism of trifluralin in this species has been rather extensively studied by Golab et al. (1966), Golab et al. (1967), and Probst et al. (1967). Carrots were planted in greenhouse soil into which trifluromethyl labeled ^{14}C-trifluralin had been incorporated into the surface layer at a rate of 0.75 lb/A in a 2-inch layer (1.38 μg per gram of soil). After 110 days the plants were harvested, the root systems washed with water, the tops separated from the roots, and frozen for assay by thin-layer or gas chromatography.

The total radioactivity in the carrot root was 0.65 ppm, expressed as trifluralin. This value for the tops was 0.25 ppm. Two thirds of the radioactivity in the root was in the surface layer. In general the amount of trifluralin progressively decreased from the surface to the center of the root; however, a somewhat higher amount was found in the layer containing phloem-xylem junction. Chromatographic results showed that unaltered trifluralin was the major source of radioactivity (89%) and α,α,α-trifluoro-2,6-dinitro-N-(n-propyl)-p-toluidine was the major degradation product (4.7%) in the roots. Small amounts of α,α,α-trifluoro-5-nitro-N^4-(n-propyl)-toluene-3,4-diamine and 4-(di-n-propylamino)-3,5-dinitrobenzoic acid were also indicated by thin layer chromatography, but identification was not verified by other methods. Although the degradation of trifluralin to some other products may have occurred in the soil and these may have been absorbed by the plant, the 4-(di-n-propylamino)-3,5-dinitrobenzoic acid appears to be exclusively a higher plant product since it has not been found in soil degradation studies. In the leaves, only about 40% of the radioactivity appeared to be trifluralin (Figure 14-2).

The molecular fate of trifluralin in peanut and sweet potato plants grown in culture solutions containing 7 ppm of ^{14}C-trifluralin labeled in either the trifluoromethyl or N,N-di-n-propyl moiety was studied by Biswas and Hamil-

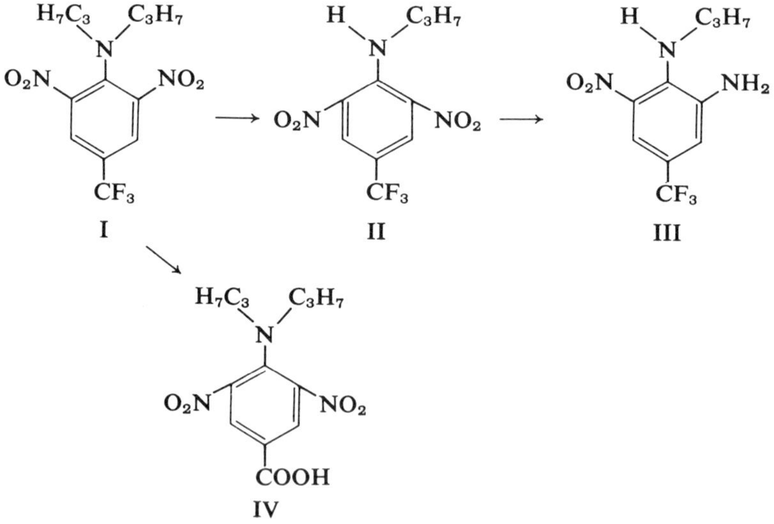

FIGURE 14-2. Proposed reaction sequence of trifluralin degradation in carrot root (Probst and Tepe, 1969): (I) α,α,α-trifluoro-2,6-dinitro-N,N-dipropyl-p-toluidine; (II) α,α,α-trifluoro-2,6-dinitro-N-(n-propyl)-p-toluidine; (III) α,α,α-trifluoro-5-nitro-N-propyltoluene-3,4-diamine; (IV) 4-(dipropylamino)-3,4-dinitrobenzoic acid. Reprinted from P. C. Kearney and D. D. Kaufman, eds, *Degradation of Herbicides*, p. 267, by courtesy of Marcel Dekker, Inc.

ton (1967, 1969) and Hamilton and Biswas (1967). The plants were harvested 72 hours after treatment with the herbicide. The entire plant was then homogenized in hot 80% ethanol and the ethanol fraction subjected to ligroin (petroleum ether) extraction. The ligroin extract was concentrated and subjected to thin-layer chromatography. In their experiments with ^{14}C-trifluoromethyl-labeled trifluralin, less than 1% of the radioactivity was trifluralin in the peanut plant, whereas this value was about 17% in the sweet potato plant. Most of the radioactivity was in the chromatographic fraction which remained at the point of application of the extract. α,α,α-trifluoro-2,6-dinitro-N-(n-propyl)-p-toluidine was a degradation product in peanuts but was not detected in sweet potato. Two other unidentified degradation products were found in peanut and three unidentified compounds were found in sweet potato. In the N,N-di-n-propyl-^{14}C-trifluralin studies, about 1% of the radioactivity was present as trifluralin, whereas this value was about 30% in sweet potato. Degradation of ^{14}C-trifluralin labeled in both positions was also detected using crude leaf extracts of both species.

Probst and Tepe (1969) cited some of the unpublished work of Herberg et al. (1967) concerning the degradation of benefin by higher plants. Peanuts and alfalfa plants were grown in soil treated with benefin, harvested, extracted, and subjected to chromatographic analysis. More than 93% of the total radioactivity was nondescript, not associated with recognizable compounds, in both species. In peanuts, benefin was the major identifiable compound and amounted to 3% of the total radioactivity; in alfalfa this value was 1.5%. Although several degradation products were detected, none were present which were not found in the soil. They take the point of view that little if any degradation of benefin occurs in plants—the degradation occurs in the soil and the plants absorb both benefin and the degradation products from the soil.

Biochemical Responses

Dukes and Biswas (1967) reported that the carbohydrate content of sweet potatoes and peanuts grown for 72 hours in trifluralin solutions, increased over that of the controls at dosages of 5 and 10 ppm but decreased at concentrations of 20, 50, and 100 ppm. The sugar content was lowered and the nitrogen content increased in the aberrant sugarbeet hypocotyledonary neck and root tissues following applications of 0.75 lb/A of trifluralin (Schweizer, 1970). Diem et al. (1968) found that trifluralin caused an increase in the nitrogen content of shoots and roots of grain sorghum and maize. Johnson and Jellum (1969) reported that 0.75 lb/A of trifluralin had no effect on the protein, oil, or fatty acid composition of soybean seeds. Similarly, Penner and Meggitt (1970) found no effect of trifluralin at 0.75 and 1.0 lb/A on the total oil content of soybean seeds but a small but significant effect on oil quality at 1 lb/A of trifluralin. Stearic acid content was reduced whereas there was an increase in linoleic acid content.

The effect of the dinitroanilines on nucleic acids may be the key to their mechanism of action. However, the reported studies have not been consistent in their findings and the results appear to be dependent on experimental material, concentration of herbicide used, and time of exposure before harvest. Dukes and Biswas (1967) observed that trifluralin at 5 and 10 ppm in culture solutions increased the nucleic acid content of intact peanut and soybean plants after 72 hours but gradually nucleic acid decreased with increasing concentrations of 20, 50, and 100 ppm. Diem et al. (1968) reported that the RNA content was slightly increased in trifluralin treated shoots and roots of intact sorghum 48, 72, and 96 hours after treatment. The DNA content of treated shoots was higher at 48 and 72 hours but lower at 96 hours and treated roots contained less DNA at 48 hours than at 72 and 96 hours. Schultz et al. (1968) found that the RNA and DNA content of root tips from intact maize seedlings germinated in 5 ppm of trifluralin for 3 days was decreased about 18 and 31%, respectively; however they remained similar to the controls in the shoots. When they studied the effect of trifluralin on the synthesis of the nucleic acids in intact maize plants by supplying ^{32}P via the roots 8 hours before harvest, they found that 5 ppm of trifluralin inhibited ^{32}P incorporation into sRNA, rRNA, and DNA in roots at 48, 72, and 96 hours after the herbicide treatment and at 48 hours after the herbicide treatment in the shoots (Table 14-3). However, 72 and 96 hours after the herbicide treatment the syntheses of these nucleic acids in the shoots were markedly stimulated. This increase was mainly in sRNA and DNA although rRNA also increased somewhat. The roots from these studies showed the characteristic radial enlargement of the cortical cells and multinucleate cells in the meristematic regions. Although the above reports indicate that trifluralin influences the nucleic acid content and synthesis in intact plants, Moreland et al. (1969) were unable to demonstrate any effect of trifluralin at 2×10^{-4}M on RNA synthesis in maize mesocotyl sections (^{14}C-6-orotic acid incorporation) or soybean hypocotyl sections (^{14}C-8-ATP incorporation).

If the dinitroanilines do have a pronounced effect on nucleic acid metabolism, one would expect that they would also influence protein synthesis. However, here again there are some discrepancies between intact plants and tissue sections. Mann et al. (1965) observed little effect of trifluralin on ^{14}C-leucine incorporation into barley coleoptile sections or hemp sesbania (*Sesbanea exaltata*) hypocotyl sections. Likewise, Moreland et al. (1969) found little effect on ^{14}C-leucine incorporation into protein of soybean hypocotyl sections. However, trifluralin has been shown to inhibit the development of the activity of several enzymes in germinating seedlings. Ashton et al. (1968) demonstrated that 1.5×10^{-6}M trifluralin inhibited the development of

TABLE 14-3. Effect of Trifluralin on ^{33}P
Incorporation into Nucleic Acids in Maize, %
of Control (Adapted from Schultz et al., 1968)

Time (hr)	sRNA[a]	rRNA[a]	DNA[a]
	ROOT		
48	74	75	62
72	66	55	72
92	62	61	84
	SHOOT		
48	53	54	41
72	140	111	215
96	172	135	204

[a] Percent of control.

proteolytic activity 41% in the cotyledons of 3-day-old squash seedlings. The development of phytase activity during germination has also been shown to be inhibited by the presence of 10^{-5}M trifluralin in barley seedlings, squash cotyledons, and maize embryos by 12 to 24% (Penner, 1970). Moreland et al. (1969) showed that trifluralin inhibited the development of gibberellic acid induced amylase activity in distal halves of barley seeds by 32%. The development of dipeptidase activity in squash cotyledons was inhibited 16% at 1.5×10^{-6}M trifluralin 4 days after treatment of the seeds (Tsay and Ashton, 1971). Since all of these enzymes appear to be synthesized *de novo* and are induced by the hormones gibberellic acid or cytokinins, depending on the species, it is possible that trifluralin interferes with the action, development, or transport of hormones rather than protein synthesis *per se*.

Respiration of tissues does not seem to be affected at concentrations of trifluralin which inhibit growth (Feeny, 1966; Talbert, 1967; Negi et al., 1967, 1968; Hendrix and Muench, 1969). However, Negi et al. (1968) showed that oxygen uptake and phosphorous esterification were inhibited by 10^{-4}M trifluralin in mitochondria isolated from several species. An uncoupling of oxidative phosphorylation was also demonstrated at 10^{-4}M trifluralin, however, lower concentrations had little if any effect. Mitochondria responded to nitralin in a similar manner. Gruenhagen and Moreland (1971) showed that trifluralin did not alter the ATP content of excised soybean hypocotyls.

Mode of Action

The dinitroaniline herbicides inhibit both root and shoot growth of many species when absorbed by the roots. The inhibition of root growth is a direct effect and the most observable symptom is the inhibition of lateral root formation. The inhibition of shoot growth following root absorption is probably a secondary effect caused by limited root growth. However, when the herbicide is absorbed by the shoot it inhibits shoot growth directly. They interfere with cell division.

These herbicides are readily absorbed by both roots and shoots but appear to be only slightly translocated to other plant parts in most species. However, one report indicates that trifluralin is translocated extensively in peanuts.

Research from the Eli Lilly and Company laboratories indicate that the dinitroaniline herbicides are degraded very slowly, if at all, by most higher plants. When ^{14}C-trifluralin or ^{14}C-benefin was incorporated into soil and seeds or seedlings planted in the soil and the mature plant harvested, they found radioactivity in lipids, glycosides, protein, cellular fractions as well as in hydrolysis products of the herbicides. This broad distribution of radioactivity indicates extensive degradation; however, they suggest that much of this degradation occurred in the soil and the degradation products were subsequently absorbed by the plant. Although, from a practical residue point of view this type of experiment is excellent, it does not allow us to determine what degradation occurs in plants only. In carrots, one degradation product was identified which has not been found in soil studies. One study of short duration (72 hours) in the absence of soil has reported extensive degradation of trifluralin in peanuts and sweet potato plants. Additional research is required before definite conclusions can be made on the rate and extent of the degradation of the dinitroaniline herbicides in higher plants.

Although trifluralin has been reported to induce several biochemical responses in higher plants, including changes in carbohydrate, lipid, and nitrogen content, apparently the most relevant effect as related to the phytotoxic symptoms is the change in nucleic acids. Uncoupling of oxidative phosphorylation and the inhibition of the development of several hormone-induced enzymes has also been observed.

REFERENCES

Alder, E. F., W. L. Wright, and G. F. Soper. 1960. Control of seedling grasses in turf with diphenylacetonitrile and a substituted dinitroaniline. *Proc. 17th North Central Weed Control Conf.* p. 24.

Arle, H. F. 1968. Trifluralin-systemic insecticide interactions on seedling cotton. *Weed Sci.* **16**:430–432.

Ashton, F. M., D. Penner, and S. Hoffman. 1968. Effect of several herbicides on proteolytic activity of squash seedlings. *Weed Sci.* **16**:169–171.

Bayer, D. E., C. L. Foy, T. E. Mallory, and E. G. Cutter. 1967. Morphological and histological effects of trifluralin on root development. *Amer. J. Bot.* **54**:945–952.

Biswas, P. K. and W. Hamilton. 1967. Degradation of trifluralin in sweet potato and peanut plants. *Abstr. 6th Int. Congr. Pl. Prot.* Vienna, pp. 388–389.

Biswas, P. K. and W. Hamilton. 1969. Metabolism of trifluralin in peanuts and sweet potatoes. *Weed Sci.* **17**:206–211.

Diem, J. R., H. H. Funderburk, and N. S. Negi. 1968. Effect of trifluralin on the nitrogen metabolism of grain sorghum and corn. *Proc. 21st Southern Weed Conf.* p. 342.

Dukes, I. E. and P. K. Biswas. 1967. The effects of trifluralin on the nucleic acids and carbohydrates content of sweet potatoes and peanuts. *WSSA Abstr.* p. 59.

Feeny, R. W. 1966. Effect of trifluralin on the growth of oat seedlings and respiration of excised oat roots. *Proc. 20th Northeast Weed Control Conf.* pp. 595–603.

Fischer, B. B. 1966. The effect of trifluralin on the root development of seedling cotton. *Aust. J. Exp. Agric. Anim. Husb.* **6**:214–218.

Funderburk, H. H. Jr., D. P. Schultz, N. S. Negi, R. Rodriguez-Kabana, and E. A. Curl. 1967. Metabolism of trifluralin by soil microorganisms and higher plants. *Proc. 20th Southern Weed Conf.* p. 389.

Gentner, W. A. and L. G. Burk. 1968. Gross morphological and cytological effects of nitralin on corn roots. *Weed Sci.* **16**:259–260.

Golab, T., R. Herberg, S. Parka, and J. Tepe. 1966. Metabolism of trifluralin in carrots. *WSSA Abstr.* p. 40.

Golab, T., R. J. Herberg, S. J. Parka, and J. B. Tepe. 1967. Metabolism of carbon-14 trifluralin in carrots. *J. Agric. and Food Chem.* **15**:638–641.

Gruenhagen, R. D. and D. E. Moreland. 1971. Effects of herbicides on ATP levels in excised soybean hypocotyls. *Weed Sci.* **19**:319–323.

Hacskaylo, J. and V. A. Amato. 1968. Effect of trifluralin on roots of corn and cotton. *Weed Sci.* **16**:513–515.

Hamilton, W. and P. K. Biswas. 1967. The metabolism of trifluralin by peanuts and sweet potato. *WSSA Abstr.* p. 62.

Hassawy, G. S. and K. C. Hamilton. 1971. Effects of IAA, kinetin, and trifluralin on cotton seedlings. *Weed Sci.* **19**:265–268.

Hendrix, D. L. and S. R. Muench. 1969. The effects of trifluralin on barley. *Plant Physiol. Abstr.* **44**:S-26.

Herberg, R. J., J. V. Gramlich, and T. Golab. 1967. (unpublished.)

Johnson, B. J. and M. D. Jellum. 1969. Effects of pesticides on chemical composition of soybean seed (*Glycine max* (L.) Merill). *Agron. J.* **61**:185–187.

Kerchersid, M. L., T. E. Bosell, and M. G. Merkle. 1969. Uptake and translocation of substituted aniline herbicides in peanut seedlings. *Agron. J.* **61**:185–187.

Kirby, C. J., L. C. Standifer, and W. C. Normand. 1968. Changes in soybean stem tissue apparently induced by trifluralin. *Proc. 21st Southern Weed Conf.* p. 343.

Knake, E. L., A. P. Appleby, and W. R. Furtick. 1967. Soil incorporation and site of uptake of preemergence herbicides. *Weeds* 15:228–232.

Kust, C. A. and B. E. Struckmeyer. 1971. Effects of trifluralin on growth, nodulation, and anatomy of soybeans. *Weed Sci.* 19:147–152.

Lund, Z. F., R. W. Pearson, and G. A. Buchanan. 1970. An implanted soil mass technique to study herbicide effects on root growth. *Weed Sci.* 18:279–281.

Mann, J. D., L. S. Jordan, and B. E. Day. 1965. A survey of herbicides for their effect on protein synthesis. *Plant Physiol.* 40:840–843.

Moreland, D. E., S. S. Malhotra, R. D. Gruenhagen, and E. H. Shokraii. 1969. Effects of herbicides on RNA and protein synthesis. *Weed Sci.* 17:556–563.

Negi, N. S. and H. H. Funderburk, Jr. 1968. Effect of solutions and vapors of trifluralin on growth of roots and shoots. *WSSA Abstr.* pp. 37–38.

Negi, N. S., Funderburk, H. H., D. P. Schultz, and D. E. Davis. 1967. Effect of trifluralin on oxygen uptake and oxidative phosphorylation in isolated mitochondria. *WSSA Abstr.* p. 58.

Negi, N. S., H. H. Funderburk, D. P. Schultz, and D. E. Davis. 1968. Effect of trifluralin and nitralin on mitochondrial activities. *Weed Sci.* 16:83–85.

Normand, W. C., T. Y. Rizk, and C. H. Thomas. 1968. Some responses of excised cotton roots to treatment with trifluralin or nitralin. *Proc. 21st Southern Weed Conf.* p. 344.

Oliver, L. R. and R. E. Frans. 1968. Inhibition of cotton and soybean roots from incorporated trifluralin and persistence in soil. *Weed Sci.* 16:197–203.

Parker, C. 1966. The importance of shoot entry in the action of herbicides applied to the soil. *Weeds* 14:117–121.

Penner, D. 1970. Herbicide and inorganic phosphate influence on phytase in seedlings. *Weed Sci.* 18:360–364.

Penner, D. and W. F. Meggitt. 1970. Herbicide effects in soybean (*Glycine max* (L.) Merrill) seed lipids. *Crop Sci.* 10:553–554.

Probst, G. W. and J. B. Tepe. 1969. Trifluralin and related compounds, pp. 255–282. In P. C. Kearney and P. D. Kaufman, *Degradation of Herbicides.* Marcel Dekker, Inc., New York.

Probst, G. W., T. Golab, R. J. Herberg, F. J. Halzer, S. J. Parka, C. van der Schans, and J. B. Tepe. 1967. Fate of trifluralin in soils and plants. *J. Agr. and Food Chem.* 15:592–599.

Saghir, A. R. and S. Abu-Shakra. 1971. Effect of diphenamid and trifluralin on the germination of *Orobanche* seeds *in vitro. Weed Res.* 11:74–76.

Schieferstein, R. H. and W. J. Hughes. 1966. SD11831—a new herbicide from Shell. *Proc. 8th Br. Weed Cont. Conf.* 377–381.

Schultz, D. P., H. H. Funderburk, and N. S. Negi. 1968. Effect of trifluralin on growth, morphology, and nucleic acid synthesis. *Plant Physiol.*, 43:265–273.

Schweizer, E. E. 1970. Aberrations in sugarbeet roots as induced by trifluralin. *Weed Sci.* **18**:131–134.

Strang, R. H. and R. L. Rogers. 1971. A microradioautographic study of ^{14}C-trifluralin absorption. *Weed Sci.* **19**:363–369.

Swann, C. W. and R. Behrens. 1969. Phytotoxicity and loss of trifluralin vapors from soil. *WSSA Abstr.* No. 222.

Talbert, R. E. 1965. Effects of trifluralin on soybean root development. *Proc. 18th Southern Weed Control Conf.* p. 652.

Talbert, R. E. 1967. Studies on the selective action of trifluralin between soybeans and sorghum. *WSSA Abstr.* pp. 50–51.

Tsay, R. and F. M. Ashton. 1971. Effect of several herbicides on dipeptidase activity of squash cotyledons. *Weed Sci.* **19**:682–684.

CHAPTER 15

Nitriles

General Properties

The nitrile herbicides are compounds of rather recent development. Dichlobenil was first described as phytotoxic in 1958; ioxynil was discovered in 1959 (Wain, 1964) and patented as a herbicide in 1960; discovery of the phytotoxicity of bromoxynil followed (Hart et al., 1964) and U.S. Patent 3,397,054 covers use of bromoxynil and ioxynil as herbicides. All three occur as solids; the latter two have hydroxyl groups on the benzene ring and hence are weak acids. Bromoxynil and ioxynil are available as sodium salts and esters of octanoic acid; bromoxynil is also available as the potassium salt, ioxynil as an oil-soluble amine salt.

Screening studies on eleven weed species and four crops proved that ioxynil kills a number of weeds that are relatively tolerant of 2,4-D. Comparative tests on herbicidal activity of some 24 halogenated derivatives of ioxynil revealed none that equaled this compound (Cooke et al., 1965).

Dichlobenil is available in the form of a wettable powder and in granules. All three of these nitriles are only slightly soluble in water. Salts of bromoxynil and ioxynil are applied as foliage sprays in water to kill broadleaf weeds in wheat, barley, and rye. Bromoxynil may also be used in oats, sorghum, and fall-seeded legumes. Dichlobenil may be applied as a suspension in water or as granules over established crops as a ground or aerial treatment. Action is principally through the soil. Dichlobenil is volatile and is quickly lost from the warm soil surface. On the surface of warm dry soil evaporation is rapid and incorporation is recommended. Parochetti et al. (1971) found that dichlobenil vapor escaped from a Lakeland sand equally at field capacity and saturated; approximately 18% of an application was lost in 3 hours at 40°C; losses from air dry soil were 4% or less. Application should be postplanting or postemergence to the crop and preemergence to the weeds. Dichlobenil is also used in water for aquatic weed control (Comes and Morrow, 1971). Its unique properties as an inhibitor of germination and as a soil-borne herbicide have been elaborated by Koopman and Daams (1965).

Dichlobenil

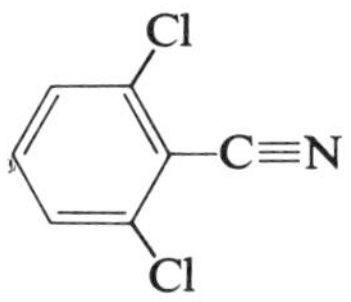

Dichlobenil, 2,6-dichlorobenzonitrile, in pure form is a white crystalline solid with an aromatic odor. It is a powerful inhibitor of germination of seeds and of actively dividing cells of meristems; it acts primarily on buds, growing points, and root tips. It controls both mono- and dicotyledonous weeds by preventing the establishment of seedlings. Growth of sprouts of perennial weeds is halted but the root systems survive and eventually produce new shoots.

Dichlobenil has proved useful for controlling annual weeds in orchards, vineyards, and cane berry crops. It fits well the non-tillage programs being established in these crops. It has also proved effective in cranberries, fruit-tree, and forest-tree nurseries, ornamental plantings of trees and vines and shelterbelt trees; it inhibits sprouting of potatoes and is used in aquatic weed control. Dichlobenil inhibits bud growth of nutsedge tubers; temperature and moisture are important; it has no effect on dormant tubers (Hardcastle and Wilkinson, 1968).

Recently dichlobenil has been found to inhibit sprouting of field bindweed in vineyards when applied to the bare soil as a subsurface layer with a spray blade. If this treatment is made in early spring the bindweed roots and rhizomes cut off by the spray blade become swollen and develop callus tissue; they fail to produce viable shoots and so the weeds are controlled. This treatment is most effective if made in spring or early summer; the inhibited weeds never become competitive during the ensuing growth season. The soil should be prepared to good tilth to insure even, free flow over the spray blade; depth of treatment should be 1.5 to 3.0 inches below the surface of the soil.

Ioxynil

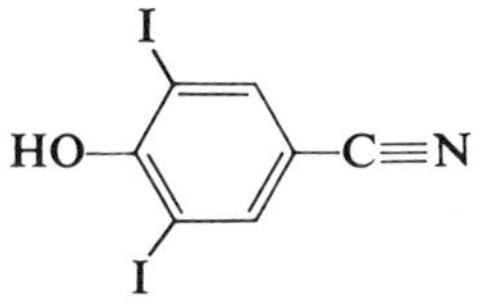

Ioxynil, 4-hydroxy-3,5-diiodobenzonitrile, occurs as a light buff colored odorless powder with a water solubility of 130 ppm. It is sold as the sodium salt, also as an oil soluble amine salt formulation, emulsifiable in water, and as an emulsifiable octanoic acid ester. It is used largely to control broadleaf weeds which do not respond to 2,4-D; some of these are Douglas fiddleneck (*Amsinckia douglasiana*), Tartary buckwheat (*Fagopyrum tataricum*), hempnettle (*Galeopsis tetrahit*), bedstraw (*Galium aparine*), Kochia (*Kochia scoparia*), pineapple weed (*Matricaria matracariodes*), and prostrate knotweed (*Polygonum aviculare*). Many of these weeds are controlled by $\frac{1}{2}$ lb/per acre or less of ioxynil. Cereal crops may be temporarily injured but they usually recover with no loss in yield. Surfactants have not proved beneficial.

Volume of spray solution may be a critical factor; volumes below 10 gal per acre (gpa) reduce the effectiveness; 20 gpa are better. Stage of growth of the weeds may be even more critical; best control in many locations has been obtained on young weeds in the 3 to 4 leaf stage. Beyond this stage higher rates of chemical are required and weed control becomes uncertain.

The first symptom observable within a few hours is collapse of the leaf cells. Chlorosis, which may appear later around necrotic areas of the leaf, is a secondary effect which may enhance phytotoxicity with time.

Bromoxynil

Bromoxynil, 3,5-dibromo-4-hydroxybenzonitrile, occurs as a buff colored powder of very low water solubility. It is produced as the sodium and potassium salts and an ester of octanoic acid. It is used like ioxynil to kill broadleaf weeds that tolerate the phenoxy herbicides. Weeds controlled include blue mustard (*Chorispora tenella*), corn gromwell (*Lithospermum arvense*), cow cockle (*Sapenaria vaccaria*), coast fiddleneck (*Amsinckia intermedia*), field pennycress (*Thlaspi arvense*), green smartweed (*Polygonum scabrum*), common lambsquarters (*Chenopodinum album*), London rocket (*Sisymbrium irio*), shepherdspurse (*Capsella bursapastoris*), silverleaf nightshade (*Solanum elaeagnifolium*), Tartary buckwheat, tarweed (*Hemizonia* spp.), tumble mustard (*Sisymbrium altissimum*), wild buckwheat (*Polygonum convolvulus*), and wild mustard (*Brassica kaber*).

Bromoxynil should be applied to the weeds in early growth stages, preferably having 3 to 4 leaves; if applied later a heavier dosage is needed; it should not be used in cereal crops that have reached the boot stage. Where broadspectrum weed control is required, bromoxynil may be combined with phenoxy-type herbicides; sufficient water as carrier should be used to give good coverage, usually 10 to 20 gpa.

Growth and Plant Structure

Bromoxynil and ioxynil are postemergence contact herbicides used to control broadleaf weeds in cereal and grass crops. Injury shows as a burning of the foliage and overdose may result in some burning of the foliage of the crop; however, recovery is rapid and crop loss is seldom experienced. Treatment has to be made during the early growth stages of the crop, and weeds become more resistant to injury as they mature. Cereals should not be treated in the boot stage but injury at earlier stages does not result in abnormal or stunted growth.

Injury appears as blistered or necrotic spots on leaves within 24 hours and with the passing of time there is extensive destruction of all leaf tissue and the plants die. These symptoms appear on susceptible broadleaf plants; cereal and grass plants show no injury at recommended dosages. Broadleaf plants having waxy cuticles show an intermediate susceptibility.

Chlorosis surrounding treated spots, and chlorosis of untreated leaves indicate some translocation of the chemical but autoradiographic studies indicate only limited phloem transport. Corn spurry (*Spergula arvensis*) has reacted to ioxynil by showing slight epinasty. Although barley may retain only about one-fifth the amount of chemical as charlock (*Sinapis arvensis*), it has been shown that differential spray retention will not explain the selectivity of these nitriles between common broadleaved plants and graminaceous species. Use of wetting agents reduces such selectivity but does not eliminate it (Carpenter et al., 1964).

Dichlobenil is a preemergence herbicide applied via the soil and its principal site of action is the meristems of embryos as these start activity in germination. Unlike the situation with the triazine and substituted urea herbicides, dichlobenil treated seedlings fail to emerge; any that receive only sublethal doses do emerge, and though growth is retarded the seedlings appear normal.

Milborrow (1964) has pointed out the obvious similarities between symptoms of boron deficiency and the effects of dichlobenil on plants. Inhibition of growth followed by blackening and death of apical meristems were found in a number of crop plants treated with dichlobenil; in sugar beet seedlings there was a large amount of dark brown material in the apical meristem, phloem, and cortical tissues; the brown material Milborrow considered as possibly melanin. Another effect of dichlobenil, seen when sublethal amounts are used, is a darker-than-normal appearance of leaves, resulting from a higher chlorophyll content. Dichlobenil was also found to inhibit active transport, to cause a cessation of growth of roots with attendant swelling and discoloration, and a rapid cessation of cell divisions in root tips; the number of cells in the early stages of division fell to 25% of normal in 2 hours. Nuclei of affected cells were granular, they stained less readily; and chromosomes were mottled.

Pate (1966) found in bean and alligator weed (*Alternanthera philoxeroides*) that dichlobenil caused destruction of phloem, cambium, and associated parenchyma tissues, above and within nodes.

Absorption and Translocation

Work on spray retention and plant morphology with respect to phytotoxicity of ioxynil by Davies et al. (1967) proved that although spray retention by leaves was a major factor in determining selectivity, it was not the only basis for selectivity. Addition of a surfactant increased retention on all species; it eliminated selectivity between mustard and peas; it increased retention so that a 26-fold retention difference between barley and mustard with no surfactant was reduced to a 10-fold difference; selectivity however still existed.

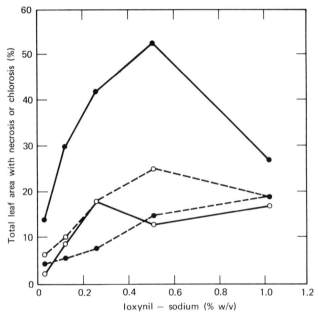

FIGURE 15-1. Effect of herbicidal concentration on necrosis (– – –) and chlorosis (————); ●, *Sinapis;* ○, *Stellaria*. (Carpenter et al., 1964.)

Savory (1968) found that solar radiation was negatively correlated with herbicidal activity of ioxynil-Na; high post-spraying activity was positively correlated with increase in activity. Field results show that the activity of ester formulations is not greatly affected by environmental conditions.

Translocation of ioxynil was studied by Carpenter et al. (1964). Using plants of charlock (*Sinapis arvensis*) and common chickweed (*Stellaria media*) for testing, they applied droplets to all leaves and four days later they measured the necrotic lesions. They also estimated the extent of the chlorotic lesions which appeared on untreated leaves. Figure 15-1 shows the effect of ioxynil concentration on necrosis and chlorosis on these two weed species. Since the chlorosis results from translocation to untreated leaves the upper curve in Figure 15-1 is a measure of the translocation that took place in wild mustard.

Carpenter et al. (1964) also tested the movement of ioxynil in pale smartweed (*Polygonum lapathifolium*) by enclosing the plants at the 5-leaf stage in polythene tubes with the fourth leaf protruding. The exposed leaves were sprayed with sublethal doses of ioxynil and then the polythene covering was removed. Three weeks later they noted the chlorotic leaves that developed; there was some chlorosis in leaf three, making up 24% of the total leaf area; all leaves in the leaf 4 position were chlorotic; 16 leaves out of 24, in the number 5 position, were chlorotic constituting 70% of the total leaf area.

Using leaf disks Davies et al. (1968a) found mustard to absorb ^{14}C-ioxynil rapidly during the first 4 hours, more slowly thereafter. Figure 15-2 shows the results of one of their experiments. There was no loss of activity from leaf segments over 4 days following removal of the labeled solution; evidently there was little metabolism of ioxynil during this time. All solutions used in

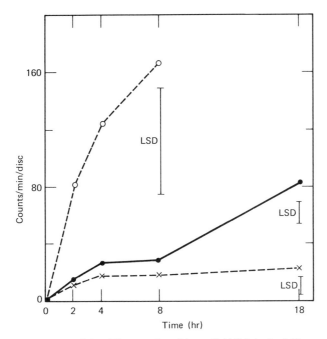

FIGURE 15-2.   The uptake of ioxynil-^{14}C into leaf discs or segments of mustard (○), pea (×), and barley (●) from a 371 ppm (0.001 M) solution of ioxynil in 0.1% Tween 20. (Davies et al., 1968a.)

this experiment contained 0.1% Tween 20; uptake was drastically reduced when the Tween 20 was not used.

By autoradiography Davies et al. (1968a) proved that translocation of ^{14}C-ioxynil was limited. Movement down to the base of the petiole of the treated leaf could be detected in 5 hours; there was some apoplastic movement to leaf tips. With more exposure time, the label could be followed up the stem with accumulation in young leaves at the apex; if mature leaves occurred between the treated leaf and the young importing leaves these were bypassed. This indicates symplastic movement in sieve tubes. When treatment was to the roots via the culture medium, there was heavy labeling of roots but only slight apoplastic movement to the foliage.

Absorption through incisions on stems resulted in strong upward movement, particularly marked over the veins. Results of a comparison of leaf-tissue removal vs. ioxynil treatments on mustard, peas, and barley are shown in Figure 15-3. It is apparent that ioxynil does something more than simply provide localized contact activity. Injury in the case of mustard and pea is almost twice as high from ioxynil as from removal by cutting; the difference in the case of barley amounts to only about 25% of the ioxynil value.

Since ioxynil is more active than bromoxynil in the less susceptible species of weeds while the two compounds are equally active on the very susceptible ones, one might presume that uptake and transport is somewhat inhibited in the case of bromoxynil. Minimum effective doses on some eight weed species were equivalent except for common chickweed where twice as much bromoxynil as ioxynil was required. There is little published research on the comparative mobility of these two compounds.

Since dichlobenil is strictly a preemergence herbicide that kills weed seedlings before or soon after they emerge, it seems evident that translocation can have little significance with respect to the practical use of this material. Massini (1961) found that dichlobenil is absorbed almost uniformly by all aerial parts of bean plants exposed to a saturated atmosphere of the chemical. Uptake was constant for 3 days, then decreased; in this time each plant had accumulated about 350 μg of dichlobenil per gram dry weight.

When dichlobenil was applied to cut stems, uptake occurred but movement along the xylem was slow; Massini compared it with a paper chromatograph of a compound with a low Rf value. Since dichlobenil adsorbs strongly to lignin this result is to be expected; basic dyes give this same response. Uptake by roots also gave restricted apoplastic movement. In all of his work Massini found no evidence for phloem transport of dichlobenil; any movement independent of the transpiration stream may have occurred by vapor diffusion through the intercellular space system.

Pate and Funderburk (1965) studied the absorption, translocation, and metabolism of ^{14}C-labeled dichlobenil in bean and alligator weed. This compound moved apoplastically with ease; symplastic movement did not occur.

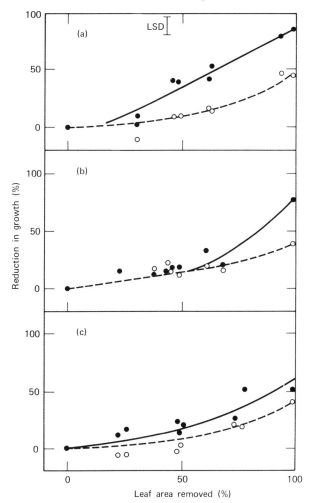

FIGURE 15-3 The reduction of growth in dry weight
following removal by cutting (○) or ioxynil treatment
(●) of different proportions of the leaf area of 16-day-
old plants of (a) mustard, (b) pea, and (c) barley
assessed 19 days after treatment. (Davies et al., 1968a.)

In studies of metabolism in the two plants named above and in two fungus
species they found, in extracts of treated plants and fungi, a ^{14}C-labeled
metabolite having an Rf of 0.25; this they considered to be a 2,6-dichloro-
benzoic acid.

Although dichlobenil does not move symplastically, its presence in plants
seriously affects the movement of normal assimilates. Glenn et al. (1971) ran

a series of experiments, exposing roots of cotton seedlings to dichlobenil in the culture solution and measuring the movement of ^{14}C-labeled assimilates from leaves exposed to $^{14}CO_2$. When the culture medium contained 20 ppm of the dichlobenil the results presented in Table 15-1 were obtained. Transport to apical buds, stems, and roots was greatly reduced 1 day after exposure.

TABLE 15-1.   % Distribution of Labeled Assimilates in Cotton after 0 and 24 hr Transport Periods. Based on s.a. (c/m/mg) (Glenn et al., 1971)

Days following 24 hr root treatment with 20 ppm dichlobenil solution	Leaves		Apical buds		Stems		Roots	
	0 hr	24 hr	0 hr	24 hr	0 hr	24 hr	0 hr	24 hr
Control	96.0	32.7	0.3	18.7	3.4	20.1	0.3	28.5
0 day	97.0	40.5	0.2	19.0	2.3	16.3	0.5	24.2
1 day	98.2	83.9	1.4	1.5	0.3	4.9	0.1	9.7
7 days	94.1	83.6	4.0	11.1	1.6	4.7	0.2	0.6

Molecular Fate

The most pertinent fact about the fate of dichlobenil is that it is quite volatile; under high temperature conditions any dichlobenil that falls on foliage or soil may be lost in the form of vapor; persistence of dichlobenil under tropical and temperate conditions is short (Barnsley and Rosher, 1961).

Price and Putnam (1969) studied the loss of dichlobenil vapor from maize seedlings grown with the herbicide dissolved in the culture solution. When the concentration of the herbicide was maintained at 0.3 ppm the maximum quantity was obtained in the maize roots after 12 hours; in shoots the quantity increased up to 72 hours. Vapor of ^{14}C-dichlobenil was emitted from pre-loaded maize plants the same in light and dark; efflux was temperature dependent. The quantity emitted in 24 hours was 70 to 80% of the amount absorbed. Loss from roots was sevenfold that from shoots over a 24-hour test period.

Incorporation into the soil lengthens the period of herbicidal activity of dichlobenil. When the chemical is rapidly incorporated it may remain active in the soil for several months; it is chemically stable and not subject to appreciable loss by leaching, hence quite persistent but subject to slow microbiological detoxication. Dichlobenil, when incorporated, tends toward absorption on lignin or humic substances and therefore organic soils remove it from solution; in a highly organic soil dichlobenil was not leached below the 4 inch depth despite extensive leaching by rainfall (Swanson, 1969).

Massini (1961) presents evidence for the metabolism of dichlobenil in plants. Steam distillation of freeze dried plants extracted only half of the radioactivity applied as nitrile-^{14}C-dichlobenil; extraction of another sample with 2N HCl in ethanol for 10 minutes yielded 90% of the activity; chromatography of this extract gave two spots, one with an Rf of 0.50 another with an Rf of 0.10; that having the Rf of 0.50 corresponded to dichlobenil; that of Rf 0.10 was an unknown. In later work on the metabolism of dichlobenil, 2,4-dichlorobenzoic acid and methyl-2,6-dichlorobenzoate have been identified as metabolites. Swanson (1969) considers that 2,6-dichlorobenzoic acid may be accepted as one metabolite in plants.

Pate and Funderburk (1965) and Pate (1966) found 2,6-dichlorobenzoic acid as one metabolite of dichlobenil but Verloop and Nimmo (1969) question the importance of this compound since it would be a reasonable product of conversion of dichlobenil to 3-hydroxy-2,6-dichlorobenzonitrile which in turn by hydrolysis would go through 3-hydroxy-2,6-dichlorobenzamide to 2,6-dichlorobenzoic acid. This pathway Verloop and Nimmo consider to be of secondary importance.

By detailed studies Verloop and Nimmo (1969) found that hydroxylation followed by conjugation forms the principal metabolic conversion of dichlobenil in bean seedlings. From the conjugates the phenols can be liberated by acid hydrolysis. After five days uptake of a 12 ppm dichlobenil solution by bean seedlings, 20% of the phenols formed are present as such, 60% are in the form of directly extractable glucosides, and 20% are bound to insoluble plant constituents; after longer times more of the insoluble conjugates are formed. The ratio between 3- and 4-hydroxylation in bean plants is about 4 to 1 so that 3-hydroxylation predominates.

Figure 15-4 gives the scheme of dichlobenil metabolism worked out by Verloop and Nimmo (1969) for bean seedlings. Studying the phytotoxicity of dichlobenil and its metabolites Verloop and Nimmo found the 3- and 4-hydroxy derivatives to be about as toxic as the parent compound; 2,6-dichlorobenzamide had little effect except to inhibit germination; 2,6-dichlorobenzoic acid showed growth-regulating activity in bean stem tests and oak root uptake studies.

From experiments with isolated root systems Verloop and Nimmo (1969) determined that hydroxylation in these is very slow; in leaves it is rapid. In roots, therefore, the principal process is a three-fold accumulation of dichlobenil followed by translocation to leaves; here a large portion of the compound is lost by volatilization; 90% of the dichlobenil absorbed in five days is lost in this way. Of the compound left, around 95% is converted to 3-hydroxy- or 4-hydroxy-dichlobenil, which go on to form the conjugates found in quantity in the bean seedlings.

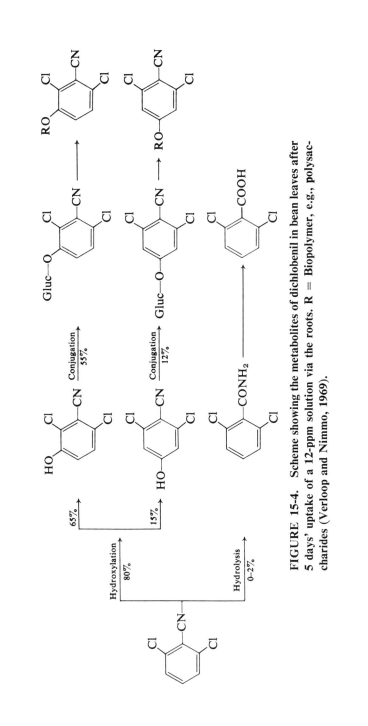

FIGURE 15-4. Scheme showing the metabolites of dichlobenil in bean leaves after 5 days' uptake of a 12-ppm solution via the roots. R = Biopolymer, e.g., polysaccharides (Verloop and Nimmo, 1969).

In further studies Verloop and Nimmo (1970) treated wheat and rice seedlings with ^{14}C-labeled dichlobenil; they found ready uptake by roots and translocation to shoots; rice accumulated to a higher level than wheat. In wheat, dichlobenil was hydroxylated to both 3- and 4-hydroxy-2,6-dichlorobenzonitrile.

These conversion products were present mainly as conjugates. The insoluble fraction, made up apparently of phenols conjugated with plant polymers, increased with time. The conversion rate was lower in rice; the dichlobenil content was correspondingly higher; there was no evidence of hydrolysis of dichlobenil.

Wain (1964b) has suggested that breakdown of ioxynil may lead to benzoic acid and iodide ions; iodide has been identified in bean plants treated with ioxynil (Swanson, 1969). When the commercial formulation as the salt is applied to an acid soil, most of the ioxynil was removed, possibly as the precipitated free phenol. Whereas in sterile soil no breakdown was found, in non-sterile soil nearly complete degradation occurred in about 3 weeks; the compound may break down to the amide and then to the corresponding acid; possibly loss of iodine from the molecule also occurs (Swanson, 1969). Presumably bromoxynil would undergo similar reactions.

Biochemical Responses

Devlin and Demoranville (1968) found that dichlobenil increased the anthocyanin content of cranberries.

Bromoxynil and ioxynil are primarily contact herbicides used to control certain hard-to-kill weed species in cereal crops; outstanding examples are bedstraw in Britain and continental Europe and fiddleneck (*Amsinckia* spp.) in the western USA. Since these compounds kill by contact, it is pertinent to study the responses to their presence in plant cells, particularly those concerned with growth. Various plant cell fractions, isolated by centrifugation have been used in such studies.

Paton and Smith (1965) employed isolated whole chloroplasts and fragmented chloroplasts or grana to study CO_2 fixation, ATP formation, and NADP reduction with and without ioxynil in the culture medium. They reported their results as percentage of control for each effect. Figure 15-5 shows their results. It is evident that electron transport and non-cyclic phosphorylation are inhibited. No oxygen was evolved during the reaction indicating that the oxygen evolving mechanism was probably inhibited. When phenazine methosulfate, a catalyst of cyclic phosphorylation was used, inhibition of ATP formation was only reduced by 10%. Since, in the electron carrier system in chloroplasts, phenazine methosulfate was unable to overcome ATP inhibition, it seems that there must be a sensitive phosphorylation site

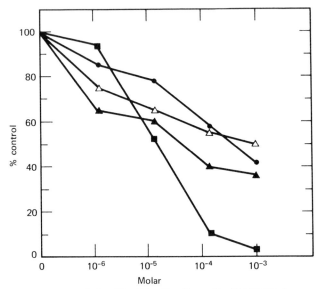

FIGURE 15-5.   ■, CO_2 fixation; ●, NADPH formation; ▲, ATP formation; △, ATP formation in the presence of phenazine methosulphate; all measurements are taken in relation to a control = 100%. Details of the reaction mixtures are indicated in the text. (Paton and Smith, 1965.)

between the plant cytochromes and chlorophyll-a; thus CO_2 fixation was inhibited as shown in Figure 15-5.

Wain (1964b) reported before the British Weed Control Conference that ioxynil inhibits the Hill reaction, that it uncouples oxidative phosphorylation, that its toxic action is expressed more vigorously in the light than in darkness, that dilute solutions undergo decomposition in sunlight, and that it combines with benzene when exposed to ultraviolet light. Also, in ultraviolet light it combines with benzene to produce 3,5-diphenyl-4-hydroxybenzonitrile. This reaction may involve formation of free radicals. In Wain's laboratory it was established that the sodium salt of ioxynil applied through roots to bean plants translocates to the tops and is broken down in the process; the corresponding benzoic acid could be detected in the stems by chromatography; iodide was also present. Fission of the ring might then lead to liberation of simpler, highly toxic compounds; Wain suggests the possible production of iodoacetic acid, a potent enzyme poison. Dichlobenil inhibits germination of seeds and it slows down or stops meristematic growth. That material which gets into developed plants is metabolized, rapidly in leaves, more slowly in roots.

Penner and Ashton (1966, 1967) found that the removal of the axis tissue from the embryos of squash prior to germination resulted in reduced production of proteolytic activity in the cotyledons of 2-day-old seedlings. The addition of benzyladenine to the culture medium prevented this inhibition. Use of puromycin and actinomycin D suggested *de novo* synthesis of the proteolytic enzyme and the mediation of this synthesis by RNA, presumably mRNA. Perhaps kinins control the synthesis of proteolytic activity in cotyledons by their influence on mRNA. In 1968 they (Penner and Ashton, 1968) studied the influence of dichlobenil and two other herbicides on kinin control of proteolytic activity in the cotyledons of squash. None of the herbicides affected the activity of proteolytic enzymes. The presence of dichlobenil in the culture solution during germination reduced the development of the activity in both excised cotyledons and the cotyledons of intact embryos; addition of benzyladenine partly reversed this inhibition. Interference with proteinase synthesis could result from inhibition of the development of DNA, tRNA, ATP, GTP, or ribosomes. Since Foy and Penner (1965) found that dichlobenil uncouples oxidative phosphorylation in mitochondria it seems possible that part of the effect on the development of proteolytic activity in squash cotyledons results from its effect on ATP formation.

Ioxynil and bromoxynil are contact herbicides. Biochemical responses to these compounds should presumably lead to fairly rapid death of the foliage to which they are applied.

Kerr and Wain (1964), using respiring isolated pea shoot particles found that ioxynil and several of its analogues will depress the uptake of inorganic phosphate, and, to a lesser extent, of oxygen; the P/O ratio was eventually reduced to zero at 4×10^{-5}M concentration. The effect of ioxynil far exceeded that of DNP and was comparable to that of dinitro-*ortho*-cresol.

Foy and Penner (1965) studied the effect of several herbicides, including ioxynil upon the oxidation of tricarboxylic acid substrates by isolated cucumber mitochondria. Table 15-2 gives their results. In contrast to ioxynil, dichlobenil increased oxygen consumption to around 150% of the controls at 1.45×10^{-4} and 0.73×10^{-4}M concentrations.

Smith, Paton, and Robertson (1966) in continued studies on the effects of herbicides on electron transport in plant cells, have summarized results of experiments on photosynthesis by the relations shown in Figure 15-6. This figure shows that the light phase of photosynthesis causes the formation of three important products: NADP, ATP, and molecular oxygen. Having shown already that ioxynil inhibits the uptake of CO_2 (Figure 15-5), and that it also inhibits the electron transfer of the oxidant dichlorophenol-indophenol (DCPIP) by 50% at a concentration of 8×10^{-7}M (Paton and Smith, 1965), they state that ioxynil appears to be a more efficient inhibitor of the Hill reaction than the common herbicides 2,4,5-T, 3,4,5-TBA, and CIPC, and that

TABLE 15-2.   Effect of Ioxynil on Tricarboxylic Acid Cycle Substrate Oxidation by Cucumber Mitochondria (Foy and Penner, 1965)

Concentration of ioxynil (M)	O_2 consumed % of control substrate			
	succinate		α-ketoglutarate	
	30 min	60 min	30 min	60 min
0.75×10^{-3}	0	0	0	0
0.75×10^{-4}	17	16	6	8
0.75×10^{-5}	69	53	8	9
0.75×10^{-6}	92	83	73	84

it is comparable with diuron in this respect. Since electron transport is an essential feature of photosynthesis they conclude that this type of blockage will lead to cessation of ATP synthesis, to drastic changes in the physiology of the plant, and eventually to death. This inhibitory effect may be partially reversed by cysteine and almost completely reversed by glutathione. They consider the actual site of action may be near plastoquinone, an essential component of the electron transport system in higher plant photosynthesis.

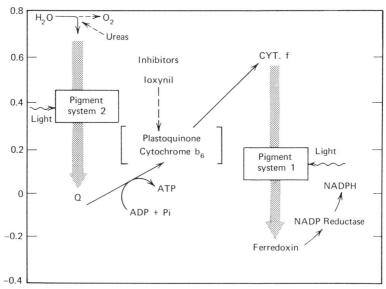

FIGURE 15-6.   The pathway for electron flow in higher plant photosynthesis and sites of action of some herbicides, (Smith, Paton, and Robertson, 1966.)

The pathway for the oxidation of cellular constituents and some possible sites of herbicidal phytotoxicity are shown in Figure 15-6 from Smith, Paton, and Robertson (1966). They have found that ioxynil at very low concentrations (ca. 10^{-8}M) can uncouple oxidative phosphorylation in plant tissue. Further they suggest that the probable site of inhibition in the respiratory chain occurs in the region of ubiquinone. Thus ioxynil seems to inhibit electron flow near the quinone component of electron transport; Robertson and Smith (unpublished) found this to be true in *Vicia faba* plants (Figure 15-7).

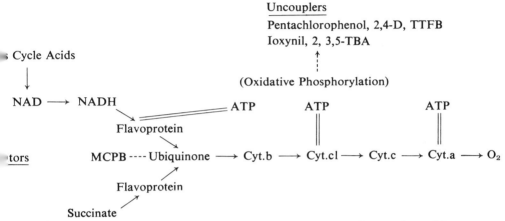

FIGURE 15-7. The pathway for the oxidation of cellular products and possible sites of action of some herbicides (Smith, Paton, and Robertson, 1966).

Smith et al. (1966) derive from these studies the concept that a single lone site of primary action for a herbicide is not acceptable and the selective properties of a molecule cannot relate only to its ability to function as a metabolic toxin. Selectivity of a herbicide, they conclude, cannot be connected entirely to the inhibition of a biochemical reaction; it must be considered with the biophysical factors of uptake into cells, migration across membranes and tissues, and translocation.

Paton and Smith (1967) studied the effect of ioxynil on the component parts of the respiratory electron transport chain by use of mitochondria isolated from roots of *Vicia faba*. This herbicide was found to be capable of stimulating oxygen evolution at low concentrations (5×10^{-8} to 5×10^{-7}M) and to inhibit evolution at higher values (5×10^{-6} to 5×10^{-3}). At an ioxynil concentration of 5×10^{-7}M the P/O ratio is decreased by about 50%. Paton and Smith conclude that the ability to uncouple oxidative phosphorylation at low concentrations and to block electron flow at higher values may contribute to the herbicidal activity of ioxynil.

On studies on the inhibition of photosynthetic electron transport by ioxynil, Friend and Olsson (1967) used chloroplasts from broad bean and sugar beet. They found that ioxynil inhibited the photoreduction of endogenous plasto-quinone in the absence of added cofactors, and in the case of sugar beet chloroplasts they found a photooxidation of plastoquinone in the presence of ioxynil. They considered that ioxynil inhibits plastoquinone photoreduction at the same site as o-phenanthroline and diuron.

Since both o-phenanthroline and diuron are considered to inhibit photo-synthetic electron transport at the site of oxygen evolution and this inhibition can be reversed by ascorbate and indophenol, Friend and Olsson studied the effects of ioxynil on the Hill reaction with 2,6-dichlorophenol-indophenol as electron acceptor, on the Hill reaction using NADP as electron acceptor, and the photoreduction of NADP using ascorbate and indophenol as electron donors. They found that whereas ioxynil inhibited both Hill reactions, it had little effect on NADP reduction in the presence of ascorbate and indophenol as electron donors to bypass the oxygen-evolving site. Ioxynil in these reac-tions behaved like o-phenanthroline and diuron. The authors concluded that ioxynil inhibits the oxygen evolving step in photosynthetic electron transport.

Davies et al. (1968) found that ioxynil destroys chlorophyll in leaf segments of mustard, pea, and barley; the latter was considerably more tolerant, but lost most of its chlorophyll from 48-hour exposure to ioxynil at 100 ppm. Complete reduction of 2,3,5-triphenyl tetrazolium chloride (TTC) occurred at 200 ppm ioxynil in mustard leaf segments; at 500 ppm in pea, and at 1000 ppm in barley. While analysis of plant extracts for ^{14}C-ioxynil and its deriva-tives usually showed that most of the radioactivity recovered was unaltered, traces of amide were present in pea and mustard leaves after 4 days and about 1% of the radioactivity recovered from barley at 4 days consisted of amide and benzoic acid derivatives.

Davies et al. (1968b) propose that the higher than normal reduction of TTC in mustard and barley at low ioxynil concentrations may indicate an effect on the respiration of the tissues. Since DNP may uncouple ATP synthesis from ADP during oxidative phosphorylation, ioxynil, which Kerr and Wain (1964) found to be a stronger uncoupler, may increase the reduction potential and hence bring about the effect on TTC noted above.

Using mitochondria from white potato tubers, Ferrari and Moreland (1969) studied the effects of bromoxynil, chloroxynil, ioxynil, and DNP on the activity of white potato tuber mitochondria. With succinate as substrate, ioxynil stimulated both ADP-limited and ADP + phosphate-deficient oxygen util-ization. Ioxynil also circumvented oligomycin-inhibited respiration and stimulated slightly ATPase activity. Ioxynil produced responses on respiration similar to those of DNP but at much lower concentrations. Analogues of ioxynil produced similar results in the following order: I > Br > Cl.

Mode of Action

Any discussion of mode of action of the nitrile herbicides must consider the anomaly that dichlobenil is a general preemergence herbicide that can be used selectively on established crop plants because it is not readily leached and acts only in the top soil, whereas bromoxynil and ioxynil are selective contact materials sprayed over the foliage of growing crops and weeds (Carpenter and Heywood, 1963). Thus, since both contain the $C\equiv N$ group, this is not the part of the molecule that is responsible for their unique differences.

Much of the work on metabolism of dichlobenil has been done on well-developed plants (Massini, 1961; Swanson, 1969; Verloop and Nimmo, 1969). Since the primary products of metabolism are as toxic as the parent compound, and since the principal herbicidal action involves inhibition of germination and seedling growth, the work done so far has little relevance to mode of action as exhibited in the field. Probably dichlobenil and its immediate metabolites (Figure 15-4) are toxic in the sense that they injure the cells, particularly of meristems in germinating seeds and young seedlings and thus prevent the survival of the weeds; having little contact action on foliage, certain developed crop plants are unharmed; being retained in the surface layer of soil, trees and vines having deep root systems escape injury.

Dichlobenil has a profound effect on transport of assimilates in plants as shown by Glenn et al. (1971), (see Table 15-1).

Bromoxynil and ioxynil are strong inhibitors of oxidative phosphorylation; they also prevent CO_2 fixation. While these effects probably do not entirely explain the rather rapid contact action of these herbicides, they may be indicators of other, more drastic effects upon plant cells, particularly their membrane systems.

The chlorosis which may appear later than primary necrosis has not been studied in sufficient detail to provide clues on mechanism. It is secondary in the phytotoxicity of ioxynil.

The retention of selectivity between broadleaved weeds and cereals when surfactants are used indicates a truly protoplasmic type of selectivity. While this undoubtedly involves differences in metabolism of these two groups of plants there has been no research to elucidate the possible mechanism.

REFERENCES

Barnsley, G. E. and P. N. Rosher. 1961. The relationship between the herbicidal effect of 2,6-dichlorobenzonitrile and its persistence in the soil. *Weed Res.* 1:149–158.

Carpenter, K. and B. J. Heywood. 1963. Herbicidal action of 3,5-dihalogen-4-hydroxy-benzonitriles. *Nature* 200:28–29.

Carpenter, K., H. J. Cottrell, W. H. de Silva, B. J. Heywood, W. Gleeds, K. F. Rivett, and M. L. Soundy. 1964. Chemical and biological properties of two new herbicides—ioxynil and bromoxynil. *Weed Res.* **4**:175–195.

Comes, R. D. and L. A. Morrow. 1971. Control of waterlilies with dichlobenil. *Weed Sci.* **19**:402–405.

Cooke, A. R., R. D. Hart, and N. E. Achuff. 1965. Biological activity of various halogenated derivatives of ioxynil. *Proc. 19th Northeast Weed Control Conf.* pp. 321–323.

Davies, P. J., D. S. H. Drennan, J. D. Fryer, and K. Holly. 1967. The basis of differential phytotoxicity of 4-hydroxy-3,5-di-iodobenzonitrile. I. Influence of spray retention and plant morphology. *Weed Res.* **7**:220–233.

Davies, P. J., D. S. H. Drennan, J. D. Fryer, and K. Holly. 1968a. The basis of the differential phytotoxicity of 4-hydroxy-3,5-di-iodobenzonitrile. II. Uptake and translocation. *Weed Res.* **8**:233–240.

Davies, P. J., D. S. H. Drennan, J. D. Fryer, and K. Holly. 1968b. The basis of the differential phytotoxicity of 4-hydroxy-3,5-di-iodobenzonitrile. III. Selectivity factors within plant tissues. *Weed Res.* **8**:241–252.

Devlin, R. M. and I. E. Demoranville. 1968. Influence of dichlobenil on yield, size, and pigmentation of cranberries. *Weed Sci.* **16**:38–39.

Ferrari, T. E. and D. E. Moreland. 1969. Effects of 3,5-dihalogenated-4-hydroxy benzonitriles on the activity of mitochondria from white potato tubers. *Plant Physiol.* **44**:429–434.

Foy, C. L. and D. Penner. 1965. Effect of inhibitors and herbicides on tricarboxylic acid cycle substrate oxidation by isolated cucumber mitochondria. *Weeds* **13**:226–231.

Friend, J. and R. Olsson. 1967. Inhibition of photosynthetic electron transport by ioxynil. *Nature* **214**:942–943.

Glenn, R. K., O. A. Leonard, and D. E. Bayer. 1971. Some effects of dichlobenil on transport of assimilates in cotton. *Res. Prog. Report, Western Soc. Weed Sci.* pp. 150–152.

Hardcastle, W. S. and R. E. Wilkinson. 1968. Response of purple and yellow nutsedge to dichlobenil. *Weed Sci.* **16**:339–340.

Hart, R. D., J. R. Bishop, and A. R. Cooke. 1964. Discovery of ioxynil and its development in the United States. *Proc. 7th Brit. Weed Control Conf.* pp. 3–9.

Kerr, M. W. and R. L. Wain. 1964. The uncoupling of oxidative phosphorylation in pea shoot mitochondria by 3,5-diiodo-4-hydroxy benzonitrile (ioxynil) and related compounds. *Ann. Appl. Biol.* **54**:441–446.

Koopman, H. and J. Daams. 1965. Reaction between structure and herbicidal activity of substituted benzonitriles. *Weed Res.* **5**:319–326.

Milborrow, B. V. 1964. 2,6-dichlorobenzonitrile and boron deficiency. *J. Exptl. Bot.* **15**:515–524.

Parochetti, J. V., E. R. Hein, and S. R. Colby. 1971. Volatility of dichlobenil. *Weed Sci.* **19**:28–31.

Pate, D. A. 1966. Degradation of ^{14}C-dichlobenil in bean, alligator weed, and certain microorganisms. Dissertation in partial fulfillment of the requirements for the Ph.D. degree. Auburn University, Auburn, Ala.

Pate, D. A. and H. Funderburk. 1965. Absorption, translocation, and metabolism of ^{14}C-labeled dichlobenil. *IAEA Symposium.* Use of isotopes in weed research. Vienna, 1965.

Paton, D. and J. E. Smith. 1965. The effect of 4-hydroxy-3,5-di-iodobenzonitrile on CO_2 fixation, ATP formation and NADP reduction in chloroplasts of *Vicia faba* L. *Weed Res.* **5**:75–77.

Paton, D. and J. E. Smith. 1967. The effect of 4-hydroxy-3,5-di-iodobenzonitrile (ioxynil) on respiratory electron transport in *Vicia faba* L. *Can. J. Biochem.* **45**: 1891–1899.

Penner, D. and F. M. Ashton. 1966. Proteolytic enzyme control in squash cotyledons. *Nature* **212**:935–936.

Penner, D. and F. M. Ashton. 1967. Hormonal control of proteinase activity in squash cotyledons. *Plant Physiol.* **42**:791–796.

Penner, D. and F. M. Ashton. 1968. Influence of dichlobenil, endothall, and bromoxynil on kinin control of proteolytic activity. *Weed Sci.* **16**:323–326.

Price, H. C. and A. R. Putnam. 1969. Efflux of dichlobenil from the shoots and roots of corn. *J. Agr. and Food Chem.* **17**:135–137.

Savory, B. M. 1968. Some investigations into the effects of environment on the activity of the hydroxy benzonitriles. *Proc. 9th Brit. Weed Control Conf.* **1**:102–107.

Smith, J. E., D. Paton, and M. M. Robertson. 1966. Herbicides and electron transport. *Proc. 8th British Weed Control Conf.* **1**:279–282.

Swanson, C. R. 1969. The benzoic acid herbicides. In: P. C. Kearney and D. D. Kaufman, *Degradation of Herbicides.* Marcel Dekker Inc., New York, N.Y. 394 pp.

Verloop, A. and W. B. Nimmo. 1969. Absorption, translocation, and metabolism of dichlobenil in bean seedlings. *Weed Res.* **9**:357–370.

Verloop, A. and W. B. Nimmo. 1970. Transport and metabolism of dichlobenil in wheat and rice seedlings. *Weed Res.* **10**:65–70.

Wain, R. L. 1964a. Ioxynil—a new selective herbicide. *Proc. 7th Brit. Weed Control Conf.* **1**:1–2.

Wain, R. L. 1964b. Ioxynil—some considerations on its mode of action. *Proc. 7th Brit. Weed Control Conf.* **1**:306–311.

CHAPTER 16

Phenols

General Properties

The substituted phenolic herbicides are a group of contact toxicants some of which have been in use for over thirty years. Sodium dinitro-*ortho*-cresylate was introduced into the United States from France in 1938 and soon became the major selective contact material for use against broadleaf annual weeds in cereal crops, flax and peas (Westgate and Raynor, 1940). Pentachlorophenol was the next phenolic compound to enter the herbicide field. It was introduced as an activator for sodium chlorate (Hance, 1940) but soon became recognized as an effective herbicide in its own right (Crafts and Reiber, 1945). Like sodium dinitro-cresylate, sodium pentachlorophenate may be activated by including an acid salt such as ammonium sulfate, aluminum sulfate, or sodium bisulfate in the spray solution.

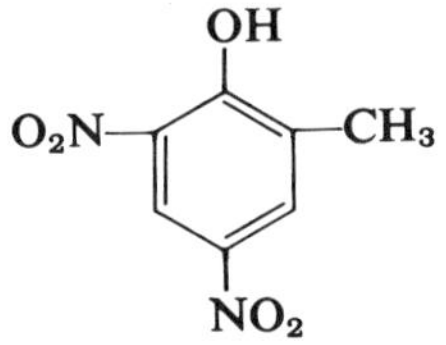

DNOC
4,6-dinitro-*o*-cresol

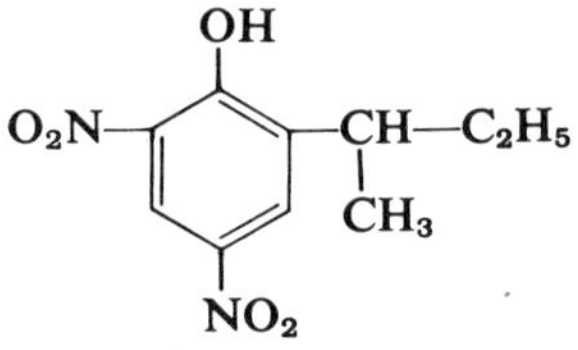

dinoseb
2-*sec*-butyl-4,6-dinitrophenol

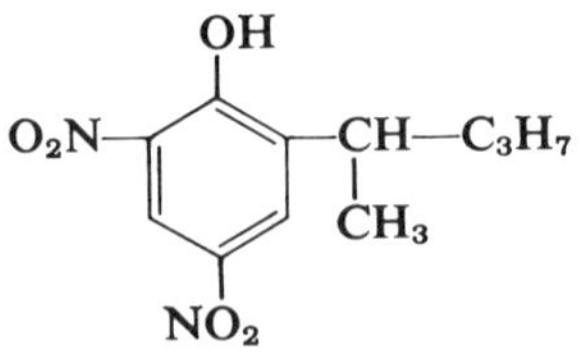

dinosam
2-(1-methylbutyl)-4,6-dinitrophenol

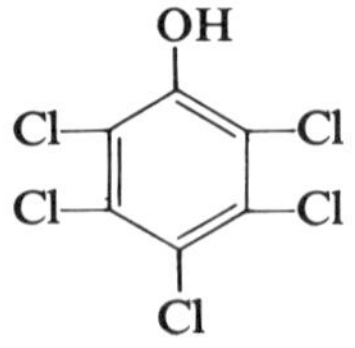

PCP
pentachlorophenol

FIGURE 16-1. Common name, chemical name, and chemical structure of the phenol herbicides.

Dinoseb and dinosam were described as potential herbicides in 1945 (Crafts, 1945). As with PCP and DNOC, the parent molecules of all of these dinitro compounds are oil soluble, their sodium, potassium, ammonium, and amine salts are water soluble. Since the alkali salts require activation by the addition of an acid salt to the spray solution, it soon became common practice to supply the dinitrophenols in the form of the ammonium or amine salts as selective herbicides. The parent compounds in oil solution with added surfactant were put on the market as general-contact herbicides. The surfactant primarily acted as an emulsifier. While a great many new compounds have been found to be phytotoxic in the intervening years the phenolic herbicides are still widely used.

When the use of 2,4-D as a preemergence herbicide was discovered (Anderson and Wolf, 1947; Anderson and Ahlgren, 1947) it was apparent that this type of herbicide held great promise. Soon the Dow Chemical Company formulated an amine salt of dinitro-*sec*-butylphenol for preemergence application through the soil; they named this formulation Premerge. It was aimed primarily as a preemergence toxicant for use in cotton, in which 2,4-D could not be used because of its extreme toxicity. Since that time Premerge has been adopted for use in a number of our major crops, including peanuts, beans, potatoes, maize, peas, pumpkins, squash, strawberries, and forage crops.

DNOC is the basic compound which, as the sodium salt, became the first organic herbicide to find wide acceptance in Europe and the United States. The parent compound, in its pure form, is an explosive and a yellow dye. This compound is only slightly soluble in water (250 ppm). Its sodium salt is quite soluble.

The commercial product Sinox was sold as a dense suspension of fine crystals made up of 30% sodium dinitro-*o*-cresylate and 70% water. It is non-corrosive and relatively non-poisonous; dust masks are recommended for applicators.

As commonly used to selectively kill broadleaf weeds in cereal crops, peas and flax, this herbicide is diluted with water to field strength and activated by adding ammonium sulfate or aluminum sulfate at a rate of 3 lb per gal of Sinox. Volume rates of application range from 8 gal for airplane to 100 gal or more per acre when applied by ground rig. Sinox dosages vary between $\frac{1}{2}$ and $1\frac{1}{2}$ gal per acre, depending upon the size, succulence, and species of weeds.

Dinoseb is a dark brown solid at 32°C or below; a dark orange liquid at temperatures above 32°C. It is readily soluble in oil and is formulated as an oil solution with added surfactant so that it emulsifies readily in water; this is the form in which it is used as a general contact herbicide.

The ammonium salt of dinoseb is water soluble; it is used as a selective herbicide to control broadleaf weeds in cereal crops, peas, flax seedling, onions, and alfalfa. Its selectivity is due to differential wetting; if a surfactant is added to the spray solution, phytotoxicity is increased but selectivity is largely lost.

Premerge is a water-soluble alkanolamine (ethanol and isopropylanol series) salt of dinoseb. It is used as a preemergence herbicide against annual broadleaf and grassy weeds in peanuts, beans, potatoes, peas, maize, sorghum, crucifer crops, and strawberries. It is not translocated in plants. It has fungicidal and insecticidal properties and is rapidly decomposed in soils; its average persistence in soils is 2 to 4 weeks.

Dinosam is very much like dinoseb in herbicidal properties; it may be a bit less phytotoxic (Crafts, 1945). However dinosam is more oil-like than dinoseb, so much so that it does not crystallize but remains liquid at ordinary temperatures. For this reason it has been combined with dinoseb in some general contact herbicidal formulations to prevent crystallization of the formulation under cold conditions of storage.

Dinosam resembles dinoseb in phytotoxicity and can be used for the same weeds and under the same conditions.

PCP is a gray flaky solid with needle-like crystals mixed in. It has a pungent odor and causes acute sneezing when the dust is inhaled. It may be absorbed through the skin and may cause dermatitis when not handled carefully. In oil solution it is commonly used as a wood preservative. It is not translocated in plants and produces direct necrosis of plant cells.

PCP may be used in many places where DNOC is recommended but its disagreeable properties have limited its use. It is cheap and effective as both a selective and a general herbicide. Formulations of PCP in a fairly aromatic oil have been placed on the market; these usually contain a surfactant to permit emulsification in water. At one time such formulations were extensively used to control weeds in sugar cane in Hawaii, Puerto Rico, Cuba, and the Southern U.S.A. Sodium pentachlorophenate has found extensive use to control rice weeds in Japan.

Yoshida and Takahashi (1967) found that PCP lowers transpiration of water-culture grown rice plants. Uptake of ammonium, potassium, silicon, and especially phosphate ions are curtailed in rice and barnyardgrass plants. The authors consider that PCP injures plants by inhibiting oxidative phosphorylation.

Matsunaka (1970) includes PCP among recommended herbicides used in transplanted rice in Japan; 6.25 to 7.5 kg/ha applied 3 days after transplanting are to be followed in 20 to 30 days by 2,4-D or MCPA. PCP controls weedy grasses soon after application; the phenoxy herbicides follow in 3 to 4 weeks to kill broadleaf weeds.

Dawson (1971) used dinoseb, both as the amine salt and as the parent phenol to control dodder in alfalfa. At 12 lb/A properly timed treatments were effective. Vapors rising from soil treated with either form of dinoseb killed dodder seedlings, even in soil beyond the bounds of treatment. Dinoseb has one advantage over chlorpropham in dodder control; it will kill many weeds that are resistant to the latter chemical.

The dinitrophenol herbicides are often used in combination with other herbicides, such as naptalam and propachlor.

Growth and Plant Structure

There are two general reactions of plant organs to the phenolic herbicides; if sufficient dosage is applied the bulk of the plant is rapidly killed; the green tissues turn brown and dessicate. Where the dosage is sublethal because of too dilute a spray solution, or application of a normal strength solution during cold and/or dry weather, the plants simply turn a dull gray color and stop growing. Such plants cannot compete with crop plants for light, water, and nutrients and may constitute no threat to the crop. If the spray solution is too dilute, or if growing conditions improve before the weeds are dead, weeds may recover by growth of axillary buds; such weeds may again compete with the crop and reduce yields.

The phenolic herbicides have no growth regulatory activity; they do uncouple oxidative phosphorylation, and the slow stunted growth described above may result from the effects of such action; the weeds simply cannot produce the energy required for growth in competition with the crop.

Absorption and Translocation

Much of the early work on the mechanism of absorption of organic herbicides was done with the phenolic compounds. Working under the assumption that the cuticle of plants is lipoid in nature and carries a residual negative charge in the presence of water, Crafts and Reiber (1945) studied the effect of pH of the spray solution on herbicidal effectiveness. They found that the activation of the salts of phenolic herbicides resulted from the shift in pH toward the acid side. This shift did not have to go all the way to the isoelectric value; if the solution was buffered so that even a relatively small proportion of the phenolic molecules were in the form of the parent phenols rather than anions, penetration of the herbicide was enhanced and herbicidal action was strengthened. More recent work (Crafts, 1948) has proved that this same principle applies for 2,4-D and many other anionic phytotoxic compounds.

As with many other herbicides, climatic factors of temperature, sunlight, and humidity play roles in the absorption of the phenolic herbicides; warm temperatures, direct sunlight, and high humidity all promote rapid and complete killing of weeds. Also, as with other herbicides, small weeds in general are more susceptible than large ones and while the parent phenol compounds in oil are highly phytotoxic, old, tough annual grasses are difficult to eliminate; perennials of course survive and usually resprout to reinfest the treated area.

None of the phenolic herbicides are translocated symplastically to any appreciable extent; their rapid contact action destroys the organs responsible for both apoplastic and symplastic transport. The soluble salts in solution will move into cut stems and distribute apoplastically; when used as pre-emergence treatments the phenolics apparently destroy the roots of young weeds before they develop appreciable foliage. Bruinsma (1967) found that roots of young rye plants take up DNOC readily and transport it into the xylem where it is transported throughout the plant. Much of the DNOC taken up is adsorbed to walls of tracheal elements; none is metabolized.

Molecular Fate

Little work has been done on the pathway of breakdown of the phenolic herbicides in plants. Early work on mammalian toxicity of sodium dinitro-*o*-cresylate proved that the minimum lethal dose as administered by stomach tube to sheep was non-toxic when sprayed on grass and fed. This indicates that the toxic properties of the compound are altered as a result of its reaction with plant material. The very rapid destruction of foliage when these compounds are sprayed on plants and their complete lack of translocation proves that they react with the tissues with which they come into immediate contact. And the nature of the killing process suggests that they destroy the membranes of the cells that they contact. Since these phenolic compounds resemble the phenols of aromatic oils, it may be that they solubilize these membranes and that they do this at appreciably lower dosages than are required of oils. Davis and Funderburk (1958) found that the vapor of dinoseb will kill plants just as Currier (1951) and Currier and Peoples (1954) found for benzene and substituted benzenes. Crafts (1945) determined that toxicity to plants increases through the series benzenes, phenols, substituted phenols, and that dinitrophenols are more toxic than nitro-, chloro-, or nitrochlorophenols with dinitro-*o*-*sec*-butylphenol being the most toxic of the *o*-aliphatic substituted compounds. Thus dinoseb plays a unique role as a herbicide. Whether this unique role results from penetration and active uptake by living cells or upon some characteristic biochemical and/or biophysical effect on protoplasm is not known. It seems hardly possible that the very rapid acute toxic action results from uncoupling of oxidative phosphorylation; oxygen consumption without energy release may well explain the slow deterioration of plants that do not receive the normal spray load.

Since dinoseb seems to have a unique position among the phenolic herbicides, which depends on both the *ortho*-substituted aliphatic chain and the *ortho*- and *para*-substituted dinitro groups it is possible that the reaction with plant cells mentioned above results in removal of the nitro groups and their conversion to ammonium or amine ions. Oxidation to nitrate should produce nitrate poisoning in animals; this does not occur.

Biochemical Responses

It is widely recognized that the phenolic herbicides are uncoupling agents. As van Overbeek (1964) explains, ATP, a high energy phosphate compound generated inside mitochondria, is required for biological activities that require energy. The mitochondria generate ATP by directing a stream of electrons from stored food, sugars for example, to oxygen from the air. This flow of electrons is coupled to the generation of ATP; thus stored energy from foods is converted to readily usable chemical energy in the form of ATP. Uncouplers unmesh this sequence of reactions with the result that respiration proceeds without ATP generation.

Kandler (1958) found that 2,4-dinitrophenol (DNP) concentrations which inhibit oxidative phosphorylation in Chlorella up to 50% and increase endogenous respiration maximally do not inhibit photosynthesis. Only concentrations high enough to inhibit respiration also inhibit photosynthesis.

Gaur and Beevers (1959) studied the respiratory and associated responses of carrot tissue to four nitrophenols, four chlorophenols, three bromophenols, phenol, methylene blue, and sodium azide. The response of each compound was a respiratory stimulation which reached a maximum and then declined sharply as the concentration was raised over a narrow range. The marked changes in respiration with increasing phenol concentration were paralleled by progressive inhibition of glucose absorption. The concentrations which produced maximum respiration drastically reduced glucose uptake. Since the various compounds tested induced stimulated O_2 uptake by castor bean mitochondria in a manner similar to that of DNP, it was assumed that they were acting with varying degrees of efficiency as uncoupling agents. DNP was the most active compound followed by 2,4-dichlorophenol, *p*-nitrophenol, the bromophenols, *m*- and *p*-chlorophenol, *m*- and *o*-nitrophenol, *o*-chlorophenol, and phenol itself. Methylene blue induced changes similar to those invoked by the substituted phenols; azide induced only respiration inhibition and lowering of glucose accumulation; $^{14}CO_2$ evolution paralleled inhibition of ^{14}C-glucose uptake.

Wessels (1959) demonstrated that DNP is able to catalyze the generation of ATP by chloroplasts in the light. In his tests DNP was even more active than FMN in catalyzing ATP synthesis in the light. This synthesis peaked at 0.6μM of DNP; either more or less in the reaction mixture reduced ATP production. Wessels, recognizing that many organic compounds serve as cofactors of photosynthetic phosphorylation, proposed that DNP may act as an intermediate electron carrier across some gap in electron transport in isolated chloroplasts. He also poses the possibility that DNP may inhibit some reaction which, in the absence of FMN or vitamin K_3, competes with photophosphorylation. Finally, he mentions the possibility that DNP catalysis of ATP synthesis may bear some relation to its uncoupling action in oxidative phosphorylation; however, other uncoupling agents such as PCP are unable to catalyze photosynthetic phosphorylation. Also it should be recognized that when the concentration of DNP was increased by a factor of 5, ATP generation was cut to 25% of that induced by the $0.6\mu M$ level.

Weinbach (1956) studying the molluscicidal activity of PCP, found that it inhibits oxidative phosphorylation completely; anaerobic (glycolic) phosphorylation is much less sensitive. The same low concentrations that dissociate oxidative phosphorylation *in vitro* increase respiration and glycolysis in snails. Weinbach found PCP much more effective than DNP in this respect. PCP also inhibits mitochondrial adenosine triphosphatase, further indicating a role in the disturbance of phosphate metabolism.

Mitchell (1961) concluded that uncouplers act on the membranes of organelles in which phosphorylation (ADP + Pi $\rightarrow$ ATP) takes place. These membranes are able to keep H^+ ions and OH^- ions separated, a necessary part of the phosphorylation mechanism. Uncoupling agents penetrate the membrane, destroying its selective impermeability. Thus respiration may continue without the attendant provision of energy for the energy-requiring processes; the plant deteriorates and dies from lack of energy from its foods.

Since variability in susceptibility to phenolic herbicides might reflect differences in ATP reserves it seems logical to examine the evidence for such variability. Davis and Funderburk (1958) studied the variability in response of some eight crop species to preemergence application of dinoseb; their susceptibility was as follows: cotton = oats = soybeans > grain sorghum > English peas = maize = snap beans > peanuts. The types of injury induced by dinoseb varied; cotton, soybeans, snap beans, and peanuts all were killed at the soil level; the plants fell over and the tops died from failure of the vascular systems. When maize, oats, and sorghum received sublethal doses the outer leaves died but the inner leaves survived and the plants recovered. At very high dosage rates all plants died.

Davis and Funderburk concluded that resistance to dinoseb is correlated with seed size and seedling size. The larger the diameter of a seedling the greater the ratio of internal tissues to exposed surface. Since shade-grown plants were more susceptible than sun-grown plants they proposed that cuticle development could also be involved.

Meggitt et al. (1956) using soybean plants studied the effects of temperature, light, and formulation on susceptibility to injury by dinoseb. They found that the activity of dinoseb increased as temperature after treatment increased from 60° to 96°F. Growing temperatures of 70° and 80°F made plants more susceptible than either 60° or 90°F. Light conditions prior to treatment had little effect in their tests; light following treatment reduced the apparent activity of dinoseb; plants grown under low light were injured more than those grown under higher light. The ammonium salt of dinoseb was somewhat more phytotoxic than amine salts; the triethanolamine salt was more toxic than the alkanolamine.

Using bean leaves, Ross and Salisbury (1960) found that light reduced the action of DNP and they suggested that this protective action might result from a restoration of ATP by photosynthetic phosphorylation. When they treated leaves with DNP they found a decrease in ATP which was greater in the dark than in the light. Colby et al. (1965) found that simetone, a potent inhibitor of photosynthesis, has a strong inhibiting action on cucumber seedlings in the light whereas DNP inhibits growth much more in the dark. When they treated these plants with a solution containing 400 ppm of simetone plus 250 ppm of DNP they obtained strong inhibition in both light and dark. The effectiveness of this combination in the light provides evidence that photosynthetic phosphorylation, by producing ATP in the light, is responsible for the protective effect of light against DNP.

Wojtaszek (1966) studied the relationship between susceptibility of certain plant species to dinoseb and their capacity to accumulate externally applied ^{32}P and to form ATP in leaf discs. To determine the relative susceptibility of species to dinoseb they sprayed 24 weed and crop species with dosages equivalent to $\frac{1}{16}, \frac{1}{8}$, and $\frac{1}{4}$ lb of dinoseb per acre in 100 gal. They found about a 50-fold difference in the amount of dinoseb required to produce 50% injury on resistant and susceptible plants.

Having found that in leaf disks of tomato, the greater the amount of ^{32}P accumulated the higher the amount of ATP generated, Wotjaszek selected 13 species ranging from the most to the least responsive to dinoseb and studied ^{32}P accumulation by leaf disks in the light and in the dark. Much greater amounts of ^{32}P were accumulated in leaf disks highly resistant to dinoseb than in those that are highly susceptible. In the light the correlation of these phenomena is 0.91; in the dark, 0.70. Using crabgrass as a highly resistant and tomato as a highly susceptible species, Wojtaszek showed that under light or dark some part of the accumulated ^{32}P was incorporated into ATP but both the total ^{32}P accumulated and the amount of ATP generated was much greater in crabgrass than in tomato.

In order to rule out the possibility that the differences in susceptibility found in his studies were the result of differential penetration, Wojtaszek ran a test in which the chemical was applied to soil for uptake through roots;

the order of susceptibility was about the same and Wojtaszek concluded that he was dealing primarily with internal differences rather than differential penetration. He postulated that susceptibility to dinoseb depends upon the level of ATP in the tissues. The level of ATP may depend in turn, either on its formation or storage in the plant. In the dark the amount of ATP decreases for two reasons: (1) it is used in normal metabolism, and (2) synthesis is lower in the dark than in the light because ATP is produced only by oxidative phosphorylation.

Mode of Action

Apparently there are two possible ways in which the phenolic herbicides may kill plants: (1) at high dosage rates by destroying membranes of treated tissues with resultant leakage of sap and dessication of the plant and (2) by preventing the formation of ATP from inorganic phosphorus by oxidative phosphorylation. Wojtaszek (1966) was dealing primarily with the second mechanism because he was using relatively low doses in order to find the LD_{50} values for comparative purposes. In his leaf disk experiments he used dinoseb at 2.5 ppm in the bathing solutions and he showed greatly reduced ATP production in both the resistant crabgrass and in susceptible tomato.

In the field when either the selective sprays of salts of the phenolic herbicides are used, or the parent compounds in oil emulsion, young tender leaves of susceptible weeds show flaccidity and start drying within minutes after application. This happens most visibly in warm, moist weather on leaves in the sunlight. It certainly must result from loss of membrane integrity.

On cold, dark days the effects of phenolic herbicide sprays are much less dramatic; the weeds may take several days to start breaking down and unless dosage is excessive they may live for weeks; however, growth slows and stops, tillers or branches fail to grow and the plants lose vigor and have little competitive effect. This sort of response may very well reflect the internal effect of the herbicide on ATP formation.

REFERENCES

Anderson, J. C. and G. Ahlgren. 1947. Growing corn without cultivating. *Down to Earth* 3:16.

Anderson, J. C. and D. E. Wolf. 1947. Pre-emergence control of weeds in corn with 2,4-D. *Amer. Soc. Agron. J.* 39:341–342.

Bruinsma, J. 1967. Uptake and translocation of 4,6-dinitro-*o*-cresol (DNOC) in young plants of winter rye (*Secale Cereale* L.) *Acta Bot. Neerl.* 16:73–85.

Colby, S. R., T. Wojtaszek, and G. F. Warren. 1965. Synergistic and antagonistic combinations for broadening herbicidal activity. *Weeds* 13:87–91.

Crafts, A. S. 1945. A new herbicide, 2,4-dinitro, secondary butyl phenol. *Science* **101**:417–418.

Crafts, A. S. 1948. A theory of herbicidal action. *Science* **108**:85–86.

Crafts, A. S., and H. G. Reiber. 1945. Studies on the activation of herbicides. *Hilgardia*, **16**:487–510.

Currier, H. B. 1951. Herbicidal properties of benzene and certain methyl derivatives. *Hilgardia* **20**:383–406.

Currier, H. B. and S. A. Peoples. 1954. Phytotoxicity of hydrocarbons. *Hilgardia* **23**:155–173.

Davis, D. E. and H. H. Funderburk, Jr. 1958. Variability in susceptibility to injury by DNBP. *Weeds* **6**:454–460.

Dawson, J. H. 1971. Dodder control in alfalfa with dinoseb and D(-) (3-chloro-phenylcarbamoyloxy)-2*N*-isopropylpropionamide. *Weed Sci.* **19**:551–554.

Gaur, B. K. and H. Beevers. 1959. Respiratory and associated responses of carrot discs to substituted phenols. *Plant Physiol.* **34**:427–432.

Hance, F. E. 1940. The factor of synergism in chemical weed control. *Hawaii Planters Record* **44**:253–272.

Kandler, O. 1958. The effect of 2,4-dinitrophenol on respiration, oxidative assimilation and photosynthesis in Chlorella. *Physiol. Plant* **11**:675–684.

Matsunaka, S. 1970. Weed control in rice. *Technical Papers of the FAO International Conf. on Weed Control.* pp. 7–23.

Meggitt, W. F., R. J. Aldrich, and W. C. Shaw. 1956. Factors affecting the herbicidal action of aqueous sprays of salts of 4,6-dinitro-ortho-secondary butyl phenol. *Weeds* **4**:131–138.

Mitchell, P. 1961. Coupling of phosphorylation to electron and hydrogen transfer by a chemi-osmotic type of mechanism. *Nature* **191**:144–148.

Overbeek, J. van. 1964. Survey of mechanisms of herbicide action. pp. 387–400. In: *The Physiology and Biochemistry of Herbicides.* Ed. L. J. Audus. Academic Press, London.

Ross, M. A. and F. B. Salisbury. 1960. Effects of herbicides on estimated adenosine triphosphate levels in whole bean leaves, *USSA Abstr.* p. 43.

Weinbach, E. C. 1956. Biochemical basis for the toxicity of pentachlorophenol. *Science* **124**:940.

Wessels, J. S. C. 1959. Dinitrophenol as a calalyst of photosynthetic phosphorylation. *Biochim. Biophys. Acta* **36**:264–265.

Westgate, W. A. and R. N. Raynor. 1940. A new selective spray for the control of certain weeds. *Calif. Agr. Expt. Sta. Bul.* **634**:1–36.

Wojtaszek, T. 1966. Relationship between susceptibility of plants to DNBP and their capacity for ATP generation. *Weeds* **14**:125–129.

Yoshida, T. and J. Takahashi. 1967. Effect of pentachlorophenol on uptake of mineral salts by paddy rice (Oryza *sativa*) and barnyardgrass (*Echinochloa crusgalli* var. *oryzicola*). *J. Sci. Soil Manure.* **38**:372–344.

CHAPTER 17

Phenoxys

General Properties

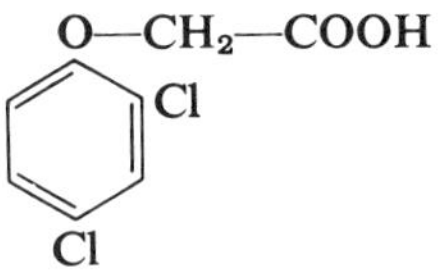

O—CH₂—COOH
Cl

Cl

2,4-D

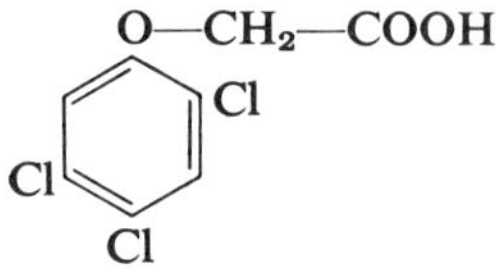

O—CH₂—COOH
Cl

Cl

Cl

2,4,5-T

O—CH₂—COOH
CH₃

Cl

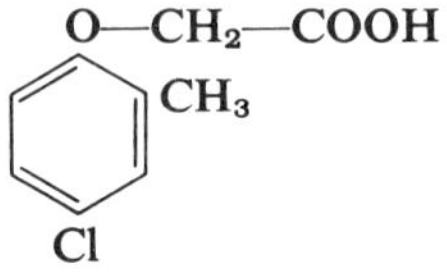

MCPA

The chlorinated phenoxy acid compounds comprise a whole family of phytotoxic substances that are used as herbicides in the form of the parent acids, as salts and as esters. The type compound of this family is 2,4-D [(2,4-dichlorophenoxy)acetic acid]; two other important compounds are 2,4,5-T [(2,4,5-trichlorophenoxy)acetic acid] and MCPA [(4-chloro-o-tolyl) oxy]acetic acid, the molecular structures of which are shown above.

The chlorophenoxy compounds along with certain benzoic and picolinic acid derivatives are growth regulators with hormone-like activity. This means that, at relatively low dosage, they bring about growth responses in regions distant from the point of application. Practically, this property of translocation is responsible for the fact that they can be used by the low volume method and that they can be used to kill root systems of perennial weeds by foliar application. Used at somewhat higher dosage, they are occasionally employed as preemergence herbicides by application to the soil and post-emergence to flooded rice by application through the water (De Datta et al., 1971).

Added to these virtues is the selectivity shown between broadleaf and grass species, enabling 2,4-D, for example, to be used in the control of many broadleaf weeds in cereal and grass crops.

When 2,4-D was first introduced in 1944, chemists immediately started work on the problem of formulation. Of nominal importance in use against annual weeds, formulation proved to be paramount to success when the chlorophenoxy compounds were tested against perennials. It soon became apparent that these compounds display optimum dosage responses; below

266

this optimum dosage provides too little toxicant to be effective at a distance from the treated foliage, too much produces excessive contact injury to the foliage resulting in little translocation.

To indicate the nature of the formulation problem let us designate the

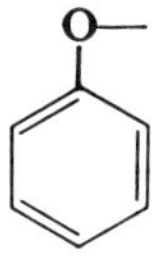

phenoxy group as R. Then examples of the various salts and esters are as follows:

Salts

Sodium salt, R—CH_2COONa
Potassium salt, R—CH_2COOK
Ammonium salt, R—CH_2COONH_4
Triethylamine salt, R—$CH_2COONH(C_2H_5)_3$
Triethanolamine salt, R—$CH_2COONH(C_2H_4OH)_3$

Alkyl Esters

Methyl ester, R— CH_2COOCH_3

Isopropyl ester, R—$CH_2COOC \overset{\displaystyle H \quad CH_3}{\underset{\displaystyle CH_3}{\diagup}}$

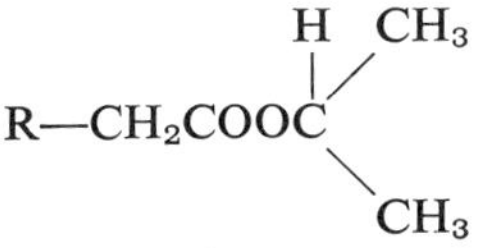

Butyl ester, R—$CH_2COOC_4H_9$
Octyl ester, R—$CH_2COOC_8H_{17}$

Heavy or Low Volatile Esters

Butoxyethanol ester, R—$CH_2COOC_2H_4$—O—C_4H_9
Propyleneglycolbutylether ester, R—$CH_2COOCH_2CH(OH)CH_2O$—C_4H_9

Tetrahydrofurfuryl ester, R—CH_2COO—$C \overset{\displaystyle H \quad O}{\diagdown \diagup} CH_2$
$\quad\quad\quad\quad\quad\quad\quad\quad\quad\quad\quad\quad C—C$
$\quad\quad\quad\quad\quad\quad\quad\quad\quad\quad\quad\quad H_2 \quad H_2$

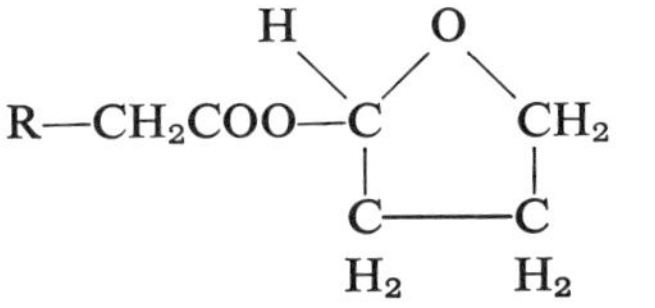

The above examples represent only a few of the common compounds that have been tested. Great numbers of alkyl and alkanol amine salts have been studied; the same applies to the esters. And it should be apparent that R in the above examples could be the 2,4-D, the MCPA, the 2,4,5-T, the 3,4-D or the 4-chloro radical all of which have weed killing properties. And in place of the phenoxyacetic grouping the phenoxypropionic, the phenoxybutyric or higher analogues may be substituted. The propionic acid analogues have proven to be highly toxic against woody species and certain aquatic weeds. The butyric acid series control weeds that have an effective beta-oxidation system capable of degrading the butyric acids to acetic acids; they are safe to use on a number of leguminous crops that have a less active beta-oxidation system.

In the intervening years since 1944, a very great number of compounds involving the phenoxy grouping have been tested; these involve amino acid derivatives, variously substituted phenoxy alkyl carboxylic acids including α-phenoxypropionic, β-phenoxypropionic, β-phenoxybutyric, and α-phenoxybutyric acids. The butyric acid analogues have been explored and the even-numbered members through the octanoic have growth regulating properties; only the straight-chain members are active; the branched-chain compounds are not degraded to acetic acid (Wain, 1955).

Growth and Plant Structure

The chlorophenoxy herbicides have profound effects upon the growth and structure of plants. Epinastic bending may follow foliar application within minutes, growth may cease within hours and over days of exposure formation of tumors, secondary roots, and fasciated structures is pronounced. As described by Hanson and Slife (1961), when susceptible seedlings are sprayed with 2,4-D the normal growth pattern changes rapidly; meristematic cells cease dividing; elongating cells stop length growth but continue radial expansion. In mature plant parts parenchyma cells swell and soon begin to divide, producing callus tissue and expanding root primordia. Root elongation stops and root tips swell. Young leaves stop expanding and develop excessive vascular tissue, very compact mesophyll low in chlorophyll, and often fasciation. Roots lose their ability to absorb water and salts, photosynthesis is inhibited and the phloem becomes plugged. All such disruptions contribute to the death of the plants. At the cellular level, 2,4-D prevents immature cytoplasm from maturing, mature cytoplasm reverts to the immature stage and the number of ribosomes increases. Attending these changes, RNA increases in expanding tissues but decreases in necrotic cells.

As noted by Overbeek (1964), the abnormal quality of the 2,4-D-induced growth results from hormonal imbalance caused by saturating concentrations of 2,4-D; the imbalance could be in the auxin-kinin relation.

Hanson and Slife (1969) report that auxins can alter nucleic acid metabolism of plants. IAA application is known to increase RNA and DNA in tobacco pith callus. Remobilization of phosphorus and nitrogen has been found in 2,4-D treated plants. While the swelling growth of seedling stems induced by 2,4-D involves increase in ribosomal RNA, elongation growth depends on synthesis of a DNA-like RNA, probably messenger RNA (Key and Ingle, 1964). When herbicidal concentrations of 2,4-D are used on stem sections there is a massive production of ribosomal and other RNA; excised tissue is blocked in its RNA catabolism and ribosome nuclease is hindered. This work showing that nucleic acid metabolism is accelerated where growth is stimulated and suppressed where growth is suppressed provides a key to the above observations of Hanson and Slife on seedling growth effects. Resistance of grasses to 2,4-D growth responses may relate to the well-known high ribonuclease levels in these plants.

Schröder (1970) et al. found that 2,4-D at 10^{-3}M increased DNA concentration in *Neurospora crassa* to 140% when the control was rated at 100%. Protein synthesis was also increased but only to a value of about 117%.

Using mitochondria from cabbage, Lotlikar et al. (1968) found that 2,4-D inhibited an oleate-stimulated ATPase activity and ATP-^{32}Pi exchange. They suggest that their results indicate that 2,4-D may inhibit respiration by an effect on a reaction involved in coupling phosphorylation with electron transport.

As stated by Hanson and Slife (1969) a new dimension has been added to the biochemistry of 2,4-D action. Morgan and Hall reported in 1962 that 2,4-D brings about the release of ethylene from cotton plants; ethylene production by treated plants was 26 times that of controls. Sorghum, a 2,4-D resistant plant, did not give this response. Since the side chain of 2,4-D was not converted to ethylene, Morgan and Hall concluded that the release of ethylene by cotton is an indirect effect.

Recent works on the mechanism of auxin action have shown that the initial response of cell elongation exhibited in epinastic bending of petioles and elongation of stem sections is not an ethylene effect; the later cell divisions and cell proliferation are caused by ethylene.

Holm and Abeles (1968) found that ethylene and 2,4-D both inhibited the growth of etiolated soybean seedlings, causing tissue swelling and increases in RNA, DNA, and protein in the subapical hypocotyl tissue. Ethylene production was increased by 2,4-D and some of the response of soybeans to 2,4-D results from increased ethylene production. Abscisic acid inhibits 2,4-D induced tissue swelling and ethylene production. The cotyledons appear to regulate the 2,4-D induced production of ethylene and roots are necessary for the 2,4-D induced tissue swelling to take place. In the soybean seedlings used, the cotyledons constitute the source and the roots the sink for food transport; 2,4-D is dependent upon food transport for translocation from cotyledons to roots. This may explain the results of Holm and Abeles who found that 2,4-D increased the length of excised elongating sections with attached cotyledons while as little as 10^{-6}M of 2,4-D reduced hypocotyl length in intact seedlings.

Abeles (1968), using light-grown maize and soybean seedlings, found that ethylene production was stimulated by 2,4-D treatment in both species when sufactant was used to insure thorough wetting. Ethylene had an inhibitory effect on the growth of both test plants, but a reversal of the ethylene inhibition could not be demonstrated using the competitive inhibitor CO_2. Abeles concluded that ethylene does not account for the herbicidal activity of 2,4-D since it did not kill his test plants. It may play a role in the proliferative growth in mature regions of hypocotyl and roots, particularly when sublethal amounts of 2,4-D are used. Such proliferative growth may in turn bring about phloem plugging and eventually death.

Absorption and Translocation

As explained in Chapter 4 when 2,4-D in the acid, salt, or ester form is applied to the surface of a leaf, the molecules diffuse into the cuticle, move through it into the aqueous apoplast and finally enter the living cells by penetrating the plasmolemma. In the form of the highly dissociated ions of the sodium or potassium salts, the negatively charged anions approach with reluctance the negatively charged cuticle surface; these formulations are not readily absorbed. The salts of weak bases such as ammonium or the amines are less completely dissociated; the few undissociated parent acid molecules dissolve in the cuticle and pass on through; they become more completely dissociated in the process. When 2,4-D is applied in the form of the suspended or emulsifiable acid, the molecules readily dissolve in and pass through the cuticle. The esters also dissolve in the cuticle and there is evidence (Crafts, 1960; Morre and Rogers, 1960; Szabo, 1963) that they are hydrolysed as they pass into the symplast. Thus, whatever the formulation, it seems that it is 2,4-D in the anion form that moves symplastically in the plant. Emulsifiable acid and heavy ester formulations have proved particularly effective for the control of perennial herbaceous and woody species. Surfactants have improved penetration of 2,4-D from any formulation.

Translocation of phloem-mobile herbicide molecules is recognized to take place along with the assimilate stream through the phloem system (Wills and Basler, 1971). For this reason it seems apparent that active movement of foods in the plant is a requirement for thorough distribution of a herbicide. This is particularly true of 2,4-D because the 2,4-D ions are readily absorbed by all living cells. As a result, the concentration of 2,4-D in the assimilate stream is constantly diminishing; if assimilate transport is slow the 2,4-D from a foliar application may all be absorbed in the stems and upper roots; none may reach the absorbing portion of the root systems.

Coble, Slife, and Butler (1970) stress the importance of sink activity in the movement of 2,4-D into the roots of honeyvine milkweed (*Ampelamus albidus*).

Plants having no root growth failed to move 2,4-D into roots; plants with an average of 5 and 20 new roots translocated 2.5 and 7.7% of the applied dose into the roots.

High sink activity of parasitic organs has been noted in autoradiographic studies. The endophytic system of dwarf mistletoe was found to import labeled assimilates at all seasons of the year (Leonard and Hull, 1965). Rust pustules on leaves of cereals are strongly labeled with ^{14}C-assimilates. ^{14}C-labeled photosynthate was found to accumulate in nodules of legumes (Small and Leonard, 1968). Coaldrake (1970) found ^{14}C-labeled 2,4,5-T to accumulate in nodules on roots of brigalow (*Acacia harpophylla*) seedlings to concentrations above that of adjacent root tissues.

When a phloem-mobile material is applied to the upper foliage most of it is translocated into the shoot tips, whereas molecules applied to lower leaves move into the roots. Therefore it is important, in the use of 2,4-D against perennial species, to make a thorough application covering all basal portions of the plant; this is particularly true in spraying or making basal applications to brush or trees.

Brady (1969) found that light intensity affects absorption and translocation of the isooctyl ester of 2,4,5-T. Post oak (*Quercus stellata*) and water oak (*Quercus alba*) showed optimum uptake at 2680 ft-c; longleaf pine (*Pinus palustris*) and American holly (*Ilex opaca*) absorbed less of the herbicide but uptake continued to increase up to 4000 ft-c, the highest intensity employed.

Upchurch, Keaton, and Coble (1969) found that diesel oil is useful as an additive for 2,4,5-T sprays used on turkey oak (*Quercus laevis*), especially where treatment is made in the latter half of the growing season, and particularly where the lower stems can be treated.

Eaton, Elwell, and Santelmann (1970) reporting a 3-year study of aerial brush control treatments at 79 sites found that the stage of growth of blackjack oak (*Quercus marilandica*) and post oak was of primary importance. The plants were most susceptible 6 to 8 weeks after the last killing frost when the leaves had reached full size. High air temperature, low relative humidity and poor spray coverage all reduced the effectiveness of the sprays. Rainfall during the month preceding treatment provided soil moisture for root growth and enhanced the treatments; at the other times rainfall had no effect; dosage of 1.75 lb/A proved optimum.

Schmutz (1971) found maximum absorption and translocation of ^{14}C-labeled 2,4,5-T by creosotebush occurred about 30 days after start of effective summer rains and coincided with maximum killing of creosotebush by sprays.

Molecular Fate

The degradation of the phenoxyalkyl acid derivatives by higher plants and microorganisms has recently been reviewed by Loos (1969). Weintraub et al.

(1952) were among the first to study the breakdown of 2,4-D in plants. Using 2,4-D labeled with ^{14}C in the carboxyl position, in the ethylene carbon, and in the ring, they found the carboxyl carbon to be most easily oxidized to ^{14}CO$_2$; the ethylene carbon was much less subject to oxidation and there was little or no release of ^{14}CO$_2$ from the ring. Weintraub et al. (1952) found a wide array of metabolites from 2,4-D breakdown in the plant.

Fang (1958) found two 2,4-D protein complexes in pea and tomato plants; these he termed "unknown 1" and "unknown 2." Unknown 1 was more prominent in peas, 2 predominated in tomato. Studying the metabolism of IAA in cells Andreae and Good (1955) found that conversion of tryptophan does not stop at IAA but may continue to form a conjugate of IAA with aspartic acid. This would avoid the breakdown of IAA by IAA oxidase. Such a conjugate does not form in the case of 2,4-D (Andreae and Good, 1957). This results in phytotoxicity that involves both resistance to detoxication and freedom from complexing.

Luckwill and Lloyd-Jones (1960a,b) used 2,4-D in studies on selectivity between red and black currants and varieties of apples and strawberries. They found differences to be strongly correlated with ability to oxidize the carboxyl and methylene carbons from the side chain. Red currants oxidized around 50% of the carboxyl carbon and 20% of the ethylene carbon of 2,4-D fed to leaves through cut petioles; red currant is a tolerant plant. Black currants, a susceptible variety, oxidized only 2% of the 2,4-D under comparable circumstances. Red currant is able to metabolize 2,4,5-T, 4-CPA, and MCPA, but not 2-CPA. In both currant varieties 5 to 10% of the 2,4-D absorbed by leaves is converted to water-soluble compounds, and 10 to 30% is bound in a non-extractable form.

In apple varieties, Cox, a resistant variety, decarboxylated 57% of the applied 2,4-D in 92 hours; Bramley's Seedling, a susceptible variety, metabolized only 2%. High rates of 2,4-D decarboxylation were also found in strawberries and lilac. Sixteen other plant species including tolerant and susceptible ones carried out only low rates of decarboxylation. Luckwill and Lloyd-Jones concluded that oxidative breakdown of 2,4-D in tolerant plants may partially explain selectivity, but that other mechanisms are also involved.

Bach and Fellig (1961a) have observed that growth of excised bean stems treated with 2,4-D ceased when the free 2,4-D disappeared. The radioactivity of carboxyl-labeled ^{14}C-2,4-D remained largely in the tissue in an ethanol-soluble form, chromatographically different from 2,4-D; only small amounts of radioactive ^{14}CO$_2$ were found. In another report Bach and Fellig (1961b) state that as much as 60% of the radioactivity of extracts of 2,4-D treated bean stems was in the form of the ethanol-soluble fraction with an Rf value of 0.5. This fraction, as well as Fang and Butts' (1954) "unknown 1" upon hydrolysis yield a compound with the same Rf as 2,4-D. Bach and Fellig doubt that this compound is either a glycoside or a peptide of 2,4-D.

Bach (1961) has isolated a large number of radioactive metabolites from ^{14}C-2,4-D treated bean stems. Half of the original radioactivity was found in ten components of an ether extract; these retained the aromatic ring structure. The ether-insoluble fraction contained six major components. Acid hydrolysis of these yielded two major components of the original ether extract. Ten amino acids were also isolated; these Bach considered to be possible components. Bach questions the cleavage of the phenol-ether linkage and suggests that degradation of the 2,4-D molecule may proceed by a oxidative pathway as proposed by Zenk (1960) for microbial breakdown. This would involve hydroxylation of the ring and oxidation of the hydroxyls to carboxyls with an attendant splitting of the ring. Figure 17-1 illustrates the scheme proposed by Bach.

Zenk (1960) has reported that 2,4-D and other auxins may enter into reactions involving the formation of thioesters of coenzyme A. Bach proposes that if 2,4-D-acetyl-CoA can react with CoA, CO_2, and ATP, a lengthening of the side chain analogous to fatty acid biosynthesis may take place. Depending on the course of this synthesis one might expect to find formation of mono- and dicarboxylic acids with potential sites for hydroxyl or keto linkages. Beta-oxidation might affect such compounds with even numbers of carbon atoms in the side chain; they would not form in plants susceptible to 2,4-DB.

Audus (1961) reasoning from the breakdown of chlorophenoxy acetic acids by soil microorganisms has proposed the schemes shown in Figure 17-2. He sees no reason to suppose that these two pathways should be mutually exclusive; there may be several possible routes of degradation with more than one possibly taking place in the same organism at the same time.

Wilcox, Moreland, and Klingman (1963) found evidence for formation of a ring-labeled metabolite by excised root systems of oats, barley, and maize from phenoxy-n-carboxylic acids having an even number of carbons in the side chain. They tested this series from acetic through hexanoic selecting the compounds possessing an even number of carbons in the side chain. Peanut, soybean, and alfalfa roots did not show this activity; these are all 2,4-D susceptible species. Chromatographic and ultraviolet spectral data indicated that the metabolite was in all cases 4-hydroxyphenoxy acetic acid; no hydroxylated products of longer chain length were found. These observations suggest that β-oxidation occurred first followed by ring hydroxylation. The greater hydroxylation activity of the resistant graminaceous species as contrasted with the susceptible legumes suggests that hydroxylation may constitute a detoxification mechanism.

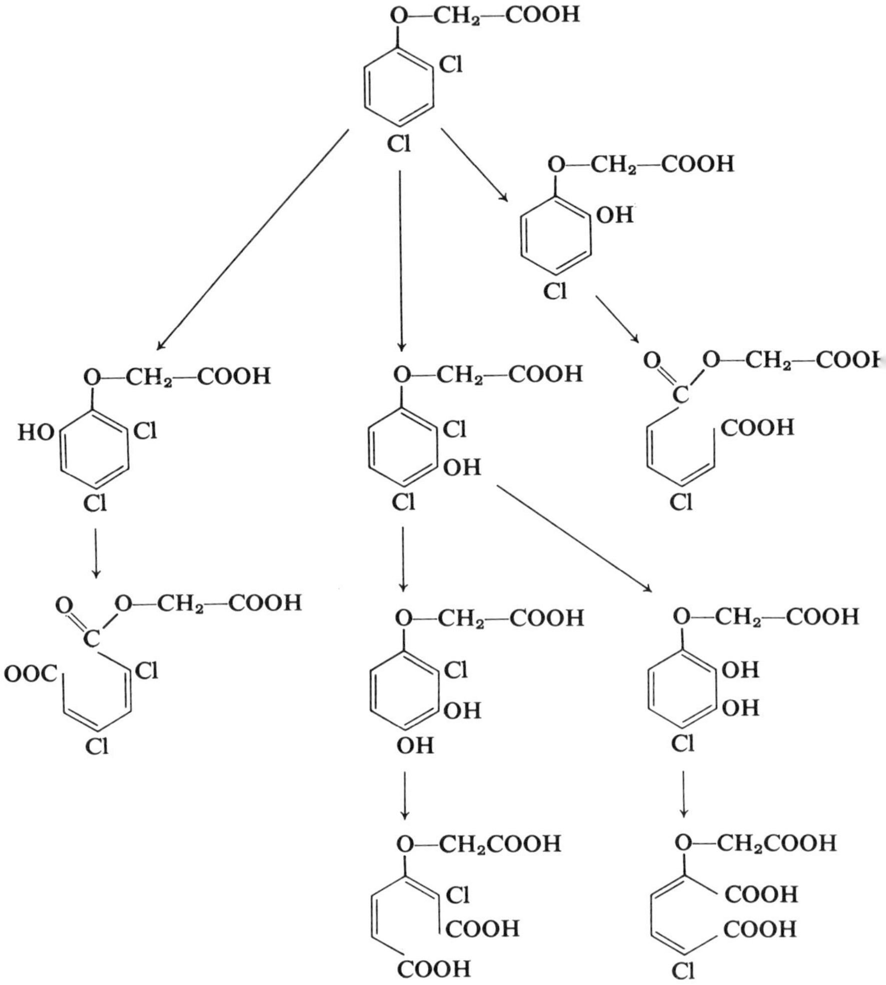

FIGURE 17-1. Proposed degradation pathway of 2,4-D in higher plants (Bach, 1961).

Fleeker and Steen (1971) found hydroxylated products of 2,4-D metabolism in a number of weeds. Table 17-1 shows the amount of 2,4-D absorbed and hydroxylated on the *para* position of the ring by 7 weed species. Table 17-2 shows the percentage of the absorbed 2,4-D metabolized to *para*-hydroxylated compounds. The lack of correlation between the amount of herbicide hydroxylated and herbicide tolerated by the plant suggest that hydroxylation per se does not account for variation in susceptibility to 2,4-D among these weed species.

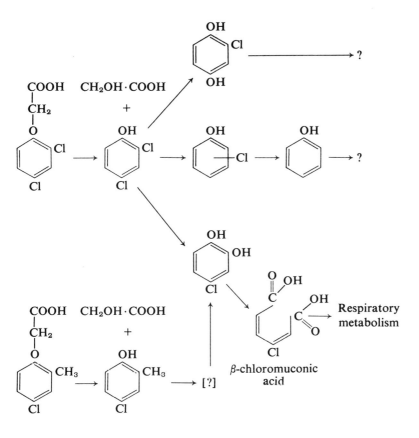

FIGURE 17-2. Proposed degradation pathway of 2,4-D by soil microorganisms (Audus, 1961).

TABLE 17-1. Amount of 2,4-D Absorbed and Hydroxylated on the *para-* Position of the 2,4-D-Ring by Several
Weed Species (Fleeker and Steen, 1971).

	2,4-D absorbed	2,4-D hydroxylated
	(nmole/g dry wt)	(nmoles/g dry wt)
Wild buckwheat	250	8.1[a]
Wild oat	367	9.0[a]
Leafy spurge	63	2.5[b]
Yellow foxtail	234	16.5[a]
Wild mustard	295	trace[c]
Perennial sowthistle	137	trace[c]
Kochia	384	trace[c]

[a] Data includes 4-OH-2,5-D, 4-OH-2,3-D, and 4-OH-2-CPA.
[b] Data includes 4-OH-2,5-D and 4-OH-2-CPA.
[c] Only the 4-OH-2,5-D compound was detected.

TABLE 17-2. The Percentage of Absorbed 2,4-D Metabolized to *para*-Hydroxyl-
ated Compounds by Several Weed Species (Fleeker and Steen, 1971).

	4-OH-2,5-D	4-OH-2,3-D	4-OH-2-CPA
	(%)	(%)	(%)
Wild buckwheat	2.6	0.6	trace[a]
Wild oat	1.8	0.4	0.2
Leafy spurge	3.2	lost	0.6
Yellow foxtail	4.4	1.5	1.1
Wild mustard	trace[a]	0.0	0.0
Perennial sowthistle	trace[a]	0.0	0.0
Kochia	trace[a]	0.0	0.0

[a] The compound was present in the plant, but the ^{14}C content of the carrier diluted
compound was less than two to three times background for a 2 to 5-mg sample.

Biochemical Responses

The biochemical mechanism by which the chlorophenoxy compounds affect
plants has proven to be very elusive. In the first place with differences in
dosage, responses differ; trace amounts of 2,4-D stimulate length growth in
much the same way as IAA. At higher dosage meristems are inhibited, length
growth ceases and lateral expansion produces swollen, tumerous stems and
roots. If these effects are prolonged, phloem plugging and finally xylem
breakdown occur; the plant becomes sick and dies. At even higher dosage

there may be strong contact action; the foliage is killed; translocation stops and only the treated portion of the plant succumbs.

Early studies sought enzymes, disturbance of the function of which might be responsible for the above observations. Wort (1964) in Table IV of his review indicates that 2,4-D may affect the activity of amylase, ascorbic acid oxidase, catalase, cytochrome oxidase, glycolic acid oxidase, IAA oxidase, invertase, pectin methylesterase, peptidase, peroxidase, phosphatase, phosphorylase, polyphenol oxidase, proteinase and proteolytic enzymes. The apparent activity of a number of enzymes could be altered indirectly by the influence of altered amounts of ascorbic acid, or some members of the vitamin B group, and by changes in the amounts of metallic ions, vitamins, and amino acids available for enzyme synthesis. Concentrations of these various factors have all been shown to be changed by application of 2,4-D to plants. Wort concludes: "The concensus is that the activity of enzymes is altered by 2,4-D in an indirect way in the majority of cases. This may be through its effects on the conditions under which the enzymatic reaction progresses, e.g., pH, hydration, etc., on the supply of material for apoenzyme and coenzyme formation or on the supply of energy for endergonic reactions through ATP production. The activity of an enzyme may also be influenced by metabolic products arising from reactions induced by the herbicide." Thus the approach to the mechanism of 2,4-D action via enzymes has not resulted in a satisfying picture.

The demonstration that auxin treatment causes an increase in the RNA and DNA content of tobacco pith suggested that auxin action may be linked with nucleic acid metabolism. Workers using the chlorophenoxy herbicidal compounds soon found that 2,4-D application to seedlings increased the RNA content of the stem. Soybean plants harvested 48 hours after application of $5 \times 10^{-4} M$ 2,4-D contained twice as much RNA in the hypocotyls as control plants, with over half the increase appearing in the microsomal fraction and a fourth in the soluble fraction (Chrispeels and Hanson, 1962). Since RNA and protein synthesis are controlled by DNA, the authors suggested that the primary site of 2,4-D action might well be the nucleus. The cytochemical mechanism of 2,4-D action would then involve renewed nuclear activity and reversion of the tissue to a meristematic state.

Key and Shannon (1964) found that concentrations of 2,4-D and IAA that promoted cell elongation (5 to 25 ppm of 2,4-D) enhanced ADP-8-^{14}C and ATP-8-^{14}C incorporation into RNA of excised soybean hypocotyls. Inhibitory levels (100 to 500 ppm of 2,4-D) decreased ^{14}C-nucleotide incorporation. ADP incorporation into RNA was inhibited by the protein synthesis inhibitor actinomycin D (Key 1963). Key and Shannon (1964) found a net increase in RNA content, primarily ribosomal, in fully elongated cells, as much as 25 to 30% following treatment with 25 ppm of 2,4-D; they concluded that this dosage of 2,4-D brought about a net transfer of ^{14}C-RNA from the nucleus to the ribosomes. Actinomycin D inhibited RNA synthesis and prevented increases in ribosomal RNA.

Using a series of auxin inhibitors Key (1964a) found that the enhancement of cell elongation by 2,4-D required active RNA synthesis and protein synthesis. He later (Key, 1964b) found that auxin-induced elongation appeared to depend upon the synthesis of a fraction of RNA having the general properties of messenger RNA; actinomycin D at low concentration inhibited RNA synthesis but not auxin-induced growth.

Labeled messenger RNA has been isolated from stem sections of pea following one to 2-hour incubations with 10^{-5}M ^{14}C-carboxy-labeled IAA. This labeling was inhibited by 10 ppm of actinomycin D. For both IAA and 2,4-D, the carboxyl label was more effective than the methylene label. The incorporation of the labeled auxin and CO_2 into RNA was stimulated by light, indicating that some of the effect results from decarboxylation and CO_2 recycling (Bendana et al. 1964).

During a 12-hour incubation, 10 ppm of 2,4-D enhanced the metabolic breakdown of RNA in excised maize mesocotyl tissue whereas 800 ppm of 2,4-D almost completely inhibited RNA breakdown after only 4 hours. This enhanced breakdown of RNA at 10 ppm 2,4-D occurred entirely at the expense of microsomal and soluble RNA; there was no change in nuclear or mitochondrial RNA (Key, 1963).

Shannon et al. (1964) using excised maize mesocotyl showed that 2,4-D at up to 50 ppm accelerated growth and RNase activity in parallel; at higher concentration both were inhibited; protein content decreased. In intact tissue normal lateral growth of maize mesocotyl and cucumber hypocotyl was accompanied by an increase in RNase activity (Shannon, 1963). Low levels of 2,4-D promoted growth and RNase activity while herbicidal concentrations inhibited both; development of RNase activity appeared to parallel maturity rather than growth. High levels of 2,4-D were shown to induce synthesis of nucleic acid and protein. Shannon (1963) associated the increases in nucleic acids and proteins with increased cell division, especially in the region of vascular bundles; he considered the action of 2,4-D to be mediated through inhibition of normal maturation, possibly by blocking RNase activity; at the same time the nucleus is activated giving rise to cell division, aberrant growth, and ultimately death.

Basler and Hansen (1964) showed that 10^{-3}M 2,4-D inhibited both RNA synthesis and degradation. Sucrose seemed necessary for increased RNA synthesis to occur in the particulate nucleic acid fraction; sucrose also stimulated loss of nucleic acid in the RNA containing supernatant. The authors concluded that 2,4-D reduced the movement of RNA from cell particulates to the cytoplasmic fraction or that 2,4-D caused preferential incorporation of ^{14}C-orotic acid into DNA of the cell nucleus; this might result from renewed nuclear activity following 2,4-D treatment.

In their excellent review of the mode of action of the phenoxy-acid compounds, Robertson and Kirkwood (1970), following reports on much of the research just discussed, give the following analysis of the effects of 2,4-D on protein and nucleic acid metabolism. Having reviewed the work on RNA increases and protein synthesis they cite Malhotra and Hanson (1966) to the effect that the activity of DNase and RNase bound to membranes was inversely correlated with herbicide sensitivity. Since, (1) Andreae and Good (1957) identified an aspartic acid fraction linked by a peptide bond in peas after incubation with 2,4-D, and (2) Klämbt (1961) found a similar linkage in wheat coleoptile cylinders, and (3) the active auxins IAA and 2,4-D were also reported to promote the enzymatically-catalyzed formation of aspartate complexes in excised pea tissue, (Sudi, 1966) it appears that the formation of such complexes are induced. IAA and 2,4-D are postulated to act as allosteric effectors (Monod, Changeux, and Jacob, 1963); perhaps they act by depressing the gene-regulating synthesis of the coupling enzyme. Work with inbred maize mitochondria suggests that IAA may act as an allosteric effector, possibly exerting its effects on the coenzyme A-acetyl CoA complex. Armstrong (1966) proposed that a specific sRNA fraction must be charged with auxin to derepress RNA synthesis, triggering polypeptide chain initiation. If the sRNA is functionally similar to other types of sRNA, the origin of protein-bound auxin may be understood. Continued synthesis of RNA would take place where the auxin-specific sRNA is saturated, and growth inhibition at high auxin concentrations would result from failure in the specificity of the amino-acyl-RNA synthetase leading to premature chain termination. Such miscoding would prevent meaningful protein synthesis.

Thus the main mechanism of 2,4-D action would involve a complex series of reactions initiated by the derepression of the gene regulating synthesis of the enzyme RNase. If a proper auxin-kinin balance exists, the resultant synthesis of RNA and protein would be accompanied by massive cell proliferation, depending upon 2,4-D dosage. Since the newly synthesized RNA and protein apparently migrate to the stem, proliferation leads to disruption of the transpiration and translocation systems; assimilates accumulate and roots starve. Inhibition of the Hill reaction and oxidative phosphorylation by 2,4-D would seem to be of secondary importance; however, these should not be too seriously discounted because any inhibition of these processes must influence the effectiveness of energy-dependent functions including mineral uptake, assimilate movement, and thus the absorption and translocation of the herbicide itself.

In studies of the effects of 2,4-D on honeyvine milkweed, Coble and Slife (1971) found the commonly observed decrease in starch of roots and the attendant increase in protein. Their treatment consisted of spraying plants in the field with 0.56 kg/ha of the dimethylamine salt of 2,4-D. They also measured α-amylase exudation from root sections using limit-dextrin-agar plates. These results are presented in Table 17-3. Since there was no increase in α-amylase activity despite the four-fold increase in α-amylase leakage as shown by hydrolysis of dextrin-agar the authors interpreted these results as indicating a loss of membrane integrity of the root cells.

TABLE 17-3. Measurements of α-amylase and Nucleotide Leakage from Honeyvine Milkweed Roots at 0, 1, 2, and 3 days after Foliar Treatment with 2,4-D (Coble and Slife, 1971).[a]

Duration after treatment (days)	Hydrolysis of dextrin-agar plates[b] (mm)	α-amylase activity[c] (%)	Nucleotide leakage (μg per g fresh wt per hr[d])
0	1.8	100	15
1	1.9	100	15
2	2.7	99	18
3	7.6	97	25
1sd .05	0.4	NS	1.8

[a] Data presented are averages of one field experiment and two greenhouse experiments of four replications each
[b] Total diameter of unstained area minus diameter of root section
[c] Expressed as % of check
[d] Based on adenosine-5-monophosphate standard curve

TABLE 17-4. Cellulase Activity in Honeyvine Milkweed Roots at 0, 1, 2, and 3 days after Foliar Treatment with 2,4-D (Coble and Slife, 1971).[a]

Duration after treatment (days)	Reduction in viscosity of 1.0% CMC[b] (%)	Reducing power[c] (OD_{540} of CMC solution × 10)
0	2.3	0.00
1	6.3	0.01
2	53.5	0.09
3	65.1	0.25
1sd .05	8.8	0.05

[a] Data presented are averages of three greenhouse experiments of four replications each
[b] Percentages based on differences in viscosity between 1.0% CMC and 1.0% cellobiose solutions, 2 hr after addition of enzyme
[c] OD_{540} is a measure of reducing power using 3,5-dinitrosalicylic acid color reagent

Observing a softening of the root tissues comparable with the ripening of a pome fruit, Coble and Slife (1971) measured cellulase activity. This they found to increase 20-fold in 2 days after treatment, 30-fold in three days. Table 17-4 presents their results shown by reduction in the viscosity of a carboxymethyl cellulose (CMC) solution and by increase in the reducing power of the hydrolyzed CMC solution. Since protein in the treated roots increased by 80% in three days it seems possible that part of this new protein might be the enzyme cellulase. Thus the evidence for starch decrease, protein increase, exudation of α-amylase and nucleotides, and increase in cellulase activity coincides with observations on physical softening and the initiation of decay

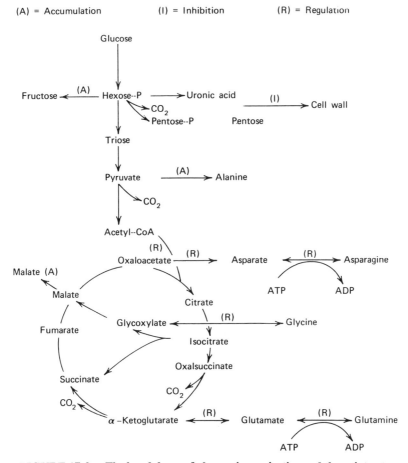

FIGURE 17-3. The breakdown of glucose in respiration and the points at which 2,4-D exerts its influence. (Mostafa and Fang, 1971.)

of the roots of this perennial weed. Taken together these observations indicate that several distinct reactions may be involved in the herbicidal effects of 2,4-D on plants.

In studies on the effects of 2,4-D on metabolism of ^{14}C-glucose in pea and maize tissues, Mostafa and Fang (1971) found that there was a preferential release of carbon-1 as CO_2 in treated plants. In 24 hours total recovery of the radioactivity in the respiratory CO_2 accounted for 50.7% and 35.4% for C-1 and C-6 respectively. The glucuronic acid pathway was stimulated greatly in pea roots, slightly in maize stems; it was inhibited in maize roots. The pentose phosphate pathway was affected in the opposite way. Incorporation of both C-1 and C-6 atoms into alcohol-insoluble residue was also affected by 2,4-D; the degree varied from pea to maize and roots to stems. Analysis of the alcohol-soluble fraction revealed different effects on labelling of some of the amino acids, and an accumulation of malic acid. Figure 17-3 illustrates Mostafa and Fang's conception of the breakdown of glucose in respiration and the points at which 2,4-D exerts its influence.

Mode of Action

Since the discovery of the herbicidal action of 2,4-D in 1944, volumes have been written on the mode of action of the chlorophenoxy compounds (Woodford et al., 1958; Crafts, 1961; Wort, 1961, 1964; Penner and Ashton, 1966).

As discussed by Crafts (1961) herbicides of the chlorophenoxy family affect almost every biological activity of a plant. Wort (1964) described the effects of 2,4-D on uptake, retention, and loss of water and minerals, on vitamin and oil content, on chlorophyll and other pigments, on photosynthesis and carbohydrate content, on respiration, on nitrogen and phosphorus metabolism, and on enzymes and their activities. Probably, most pertinent of all research on mode of action has been that, described in Chapter 6, on nucleotide responses and their possible effects on enzymes.

To recapitulate a bit, the complete foliar action of 2,4-D involves penetration of leaves, stems or roots, absorption into the symplast, migration across parenchymatous tissues to the vascular channels, translocation from sources to sinks of plant foods with the assimilate stream and finally the herbicidal response. Final death of plant tissues may result from contact action, most characteristic of the 2,4-D light ester or amine formulations on foliage. It may result from extreme hormone-like response giving rise to tumerous tissues, excessive production of buds or root initials, softening of root cortex and general degeneration; where concentrations are not sufficient to cause direct death, crushing and plugging of vascular tissues may result in a slower death from lack of nutrients normally supplied via these tissues.

Herbicidal action of the chlorophenoxy compounds through direct application to roots is less well understood. Adsorption of 2,4-D and 2,4,5-T to roots with little or no apoplastic movement to tops is recognized (Crafts 1961). When dosage is increased sufficient to produce strong contact action, these compounds penetrate and move rapidly in the transpiration stream resulting in rapid death (Crafts 1961).

When 2,4-D, 2,4,5-T, 2,4,5-TP, 2,4,5-TB, picloram, dicamba, AMS, and amitrole were applied to tops and roots of white ash (*Fraxinus americana*) and red maple (*Acer rubrum*) seedlings at equimolar concentrations, Perry and Upchurch (1968) found very wide differences in toxicity. For the three trichlorophenoxy compounds root treatment exceeded foliar treatment in effectiveness; results of these tests are shown in Figure 17-4. The AMS treatments showed the opposite effects; foliar application exceeded root treatments in phytotoxicity. For the 2,4-D, dicamba, picloram and amitrole treatments, shoot and root treatments were equally effective. Widely different species susceptibility were observed in these studies.

If, as described in the previous part of this chapter, 2,4-D treatment may

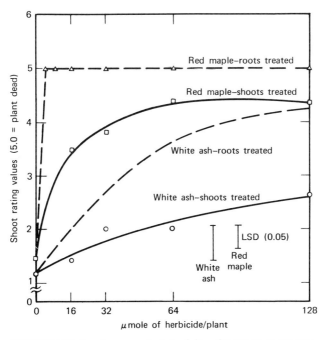

FIGURE 17-4.   Average phytotoxicity of 2,4,5-T, 2,4,5-TP, and 2,4,5-TB to white ash and red maple applied to shoots or roots (curves based on average regression). (Perry and Upchurch, 1968.)

cause an increase in RNA and protein, this may indicate that the primary site of action may be the cell nucleus.

With renewed nuclear activity the reacting tissues may be thought of as reverting to a meristematic condition. Much of the current research on 2,4-D and other growth regulators implicates ethylene in these responses but because of the rapidity of the growth response, ethylene production is probably a secondary rather than a primary response.

Furthermore since the lag time in the response of cell elongation to auxin has been shown to be zero when high temperatures and concentrations are used, induction of protein synthesis or enzyme production may not be a prerequisite to auxin or 2,4-D formative effects (Nissl and Zenk, 1969). Possibly the initial effects are in relaxing the cell walls and increase in osmotic pressure to cause elongation, with protein synthesis and overgrowth following with a lag time of several minutes to an hour or more.

Cardenas et al. (1968) noted three distinct phases in the response of cocklebur to 2,4-D. For the first 2 days after treatment there is a net weight increase due to abnormal growth of the axis; root and shoot growth are drastically curtailed. Between 2 and 7 days, axis growth continues primarily at the expense of leaf tissue which starts to senesce. The last phase between 7 and 10 days leads to collapse, withering, and death. Analyses of nitrogenous constituents showed mobilization to the axis with early large increases in nucleic acid. Both photosynthesis and ion absorption were initially stimulated but they declined sharply after the first 24 hours. Translocation to leaves and roots is reduced in favor of the proliferating axis. These authors considered the death of the plant to result from inhibition of normal apical growth coupled with induction of abnormal axis growth. Failure of new leaf and root growth plus inadequate nutrition because of phloem plugging would seem to account for the final collapse and death of the plant.

If the renewed growth resulting from 2,4-D is maintained for any prolonged period of time, all of the varied structural abnormalities observed following 2,4-D treatment may be explained. And the fact that these are not ordered in the way that normal divisions are may account for the varied growth patterns that have been recorded. Thus the mode of action of the chlorophenoxy compounds must consist of a great number of structural and biochemical reactions revolving around the central theme of prolonged abnormal growth with failure of those changes characteristic of maturity and senescense. In no other way may the great number and diversity of structural and metabolic changes be reconciled.

REFERENCES

Abeles, F. B. 1968. Herbicide-induced ethylene production: role of the gas in sublethal doses of 2,4-D. *Weed Sci.* 16:498–500.

Andreae, W. A. and N. E. Good. 1955. The formation of indolacetylaspartic acid in pea seedlings. *Plant Physiol.* **30**:380–382.

Andreae, W. A. and N. E. Good. 1957. Studies on 3-indoleacetic acid metabolism. IV. Conjugation with aspartic acid and ammonia as processes in the metabolism of carboxylic acids. *Plant Physiol.* **32**:566–572.

Armstrong, D. J. 1966. Hypothesis concerning the mechanism of auxin action. *Proc. Natl. Acad. Sci. U.S.A.* **56**:64–66.

Audus, L. J. 1961. Growth regulators: Metabolism and mode of action. In *Encyclopedia of Plant Physiology*, Vol. **14**. Springer-Verlag, Berlin.

Bach, M. K. 1961. Metabolites of 2,4-dichlorophenoxy acetic acid from bean stems. *Plant Physiol.* **36**:558–565.

Bach, M. K., and J. Fellig. 1961a. Correlation between inactivation of 2,4-dichlorophenoxy acetic acid and cessation of callus growth in bean stem sections. *Plant Physiol.* **36**:89–91.

Bach, M. K. and J. Fellig. 1961b. The uptake and fate of ^{14}C-labeled 2,4-dichlorophenoxyacetic acid in bean stem sections. *Plant Growth Regulation.* Iowa State University Press. Ames, Iowa. pp. 273–289.

Basler, E. and T. L. Hansen. 1964. Effects of 2,4-dichlorophenoxyacetic acid and sucrose on orotic acid uptake in nucleic acids of excised cotton cotyledons. *Bot. Gaz.* **125**:50–55.

Bendana, F., A. W. Galston, R. Kaur-Sawhney, and P. J. Penny. 1964. Recovery of labeled RNA following *in vivo* administration of labeled IAA to green pea stem sections. *Plant Physiol. Abstr.* **39**: *xxxi.*

Brady, H. A. 1969. Light intensity and the absorption and translocation of 2,4,5-T by woody plants. *Weed Sci.* **17**:320–322.

Cardenas, J., F. W. Slife, J. B. Hanson, and H. Butler. 1968. Physiological changes accompanying the death of cocklebur plants treated with 2,4-D. *Weed Sci.* **16**:96–100.

Chrispeels, M. J. and J. B. Hanson. 1962. The increase in ribonucleic acid content of cytoplasmic particles of soybean hypocotyl induced by 2,4-dichlorophenoxyacetic acid. *Weeds* **10**:123–125.

Coaldrake, J. E. 1970. Preferential accumulation of radioactivity in root nodules of a legume after dosage with ^{14}C- 2,4,5-T. *Weed Res.* **10**:293–295.

Coble, H. D. and F. W. Slife. 1971. Root disfunction in honeyvine milkweed caused by 2,4-D. *Weed Sci.* **19**:1–3.

Coble, H. D., F. W. Slife, and H. S. Butler. 1970. Absorption, metabolism, and translocation of 2,4-D by honeyvine milkweed. *Weed Sci.* **18**:653–656.

Crafts, A. S. 1960. Evidence for hydrolysis of esters of 2,4-D during absorption by plants. *Weeds* **8**:19–25.

Crafts, A. S. 1961. *The Chemistry and Mode of Action of Herbicides.* Interscience Publishers, New York. 269 pp.

DeDatta, S. K., R. Q. Lacsina, and D. E. Seaman. 1971. Phenoxy acid herbicides for barnyardgrass control in transplanted rice. *Weed Sci.* **19**:203–206.

Eaton, B. J., H. M. Elwell, and P. W. Santelmann. 1970. Factors influencing commercial aerial applications of 2,4,5-T. *Weed Sci.* **18**:37–41.

Fang, S. C. 1958. Absorption, translocation and metabolism of ^{14}C-2,4-D-1 in pea and tomato plants. *Weeds* **6**:179–186.

Fang, S. C. and J. S. Butts. 1954. Studies in plant metabolism. III. Absorption, translocation and metabolism of radioactive 2,4-D in corn. *Plant Physiol.* **29**:56–60.

Fleeker, J. and R. Steen. 1971. Hydroxylation of 2,4-D in several weed species. *Weed Sci.* **19**:507–510.

Hanson, J. B. and F. W. Slife. 1961. How does 2,4-D kill a plant? *Illinois Res.* **3**:3–4.

Hanson, J. B., and F. W. Slife. 1969. Role of RNA metabolism in the action of auxin-herbicides. In: *Residue Reviews.* **25**:59–67. F. A. Gunther, Ed. Springer-Verlag. New York.

Holm, R. E., and F. B. Abeles. 1968. The role of ethylene in 2,4-D-induced growth inhibition. *Planta* **78**:293–304.

Key, J. L. 1963. 2,4-D–induced changes in ribonucleic acid metabolism in excised corn mesocotyl tissue. *Weeds* **11**:177–181.

Key, J. L. 1964a. Ribonucleic acid and protein synthesis as essential processes for cell elongation. *Plant Physiol.* **39**:365–370.

Key, J. L. 1964b. RNA synthesis and the control of expansive growth of excised soybean hypocotyl. *Plant Physiol. Abstr.* **39**: xxx.

Key, J. L. and J. Ingle. 1964. Requirement for the synthesis of DNA-like RNA for growth of excised plant tissue. *Proc. Nat. Acad. Sci. U.S.* **52**:1382–1388.

Key, J. L. and J. C. Shannon. 1964. Enhancement by auxin of ribonucleic acid synthesis in excised soybean hypocotyl tissue. *Plant Physiol.* **39**:360–364.

Klambt, H. D. 1961. Growth induction and metabolism of growth substances in wheat coleoptile cylinders. 3. Metabolic products of naphthyl-1-acetic acid and 2,4-dichlorophenoxyacetic acid compared with those of indole-3-acetic acid and benzoic acid. *Planta* **57**:339–353.

Leonard, O. A. and R. J. Hull. 1965. Translocation relationships in and between mistletoes and their hosts. *Hilgardia* **37**:115–153.

Loos, M. A. 1969. Phenoxyalkanoic Acids, pp. 1–49. In, Kearney, P. C. and D. D. Kaufman. *Degradation of Herbicides.* Marcel Dekker, Inc., New York.

Lotlikar, P. D., L. F. Remmert, and V. H. Freed. 1968. Effects of 2,4-D and other herbicides on oxidative phosphorylation in mitochondria from cabbage. *Weed Sci.* **16**:161–165.

Luckwill, L. C. and C. P. Lloyd-Jones. 1960a. Metabolism of plant growth regulators. I. 2,4-dichlorophenoxyacetic acid in leaves of red and black currant. *Ann. Appl. Biol.* **48**:613–675.

Luckwill, L. C. and C. P. Lloyd-Jones. 1960b. Metabolism of plant growth regulators. II. Decarboxylation of 2,4-dichlorophenoxyacetic acid in leaves of apple and strawberry. *Ann. Appl. Biol.* **48**:626–636.

Malhotra, S. S. and J. B. Hanson. 1966. Nucleic acid synthesis in seedlings treated with the auxin-herbicide Tordon (4-amino-3,5,6-trichloropicolinic acid). *Plant Physiol. Abstr.* **41**:*vi*.

Monod, J., J. P. Changeux and F. Jacob. 1963. Allosteric proteins and cellular control systems. *J. Molec. Biol.* **6**:306–329.

Morgan, P. W., and W. C. Hall. 1962. Effect of 2,4-dichlorophenoxyacetic acid on the production of ethylene by cotton and grain sorghum. *Physiol. Plantarum.* **15**:420–427.

Morre, D. J. and B. J. Rogers. 1960. The fate of long chain esters of 2,4-D in plants. *Weeds* **8**:136–147.

Mostafa, I. Y. and S. C. Fang. 1971. Effects of 2,4-D on metabolism of ^{14}C-glucose in plant tissues. *Weed Sci.* **19**:248–253.

Nissl, D. and M. H. Zenk. 1969. Evidence against induction of protein synthesis during auxin-induced initial elongation of *Avena* coleoptiles. *Planta* **89**:323–341.

Overbeek, J. van. 1964. Survey of mechanisms of herbicide action, pp. 387–399. In: *The Physiology and Biochemistry of Herbicides.* L. J. Audus, Ed. Academic Press. New York.

Penner, D. and F. M. Ashton. 1966. Biochemical and metabolic changes in plants induced by chlorophenoxy herbicides. *Residue Rev.* **14**:39–113.

Perry, P. W. and R. P. Upchurch. 1968. Growth analysis of red maple and white ash seedlings treated with eight herbicides. *Weed Sci.* **16**:32–37.

Robertson, M. M. and R. C. Kirkwood. 1970. The mode of action of foliage-applied translocated herbicides with particular reference to the phenoxy-acid compounds. II. The mechanism and factors influencing translocation, metabolism and biochemical inhibition. *Weed Res.* **10**:94–120.

Schmutz, E. M. 1971. Absorption, translocation and toxicity of 2,4,5-T in creosotebush. *Weed Sci.* **19**:510–516.

Schröder, I., M. Meyer, and D. Mücke. 1970. Die Wirkunder Herbicide 2,4-D, amitrol, atrazin, chlorpropham und chlorfurenol auf die Nuclein-saure-biosynthese des Ascomyceten *Neurospora crassa. Weed Res.* **10**:172–177.

Shannon, J. C. 1963. The effect of 2,4-dichlorophenoxyacetic acid on the growth and ribonuclease content of seedling tissues. Ph.D. Dissertation. University of Illinois, Urbana.

Shannon, J. C., J. B. Hanson, and C. M. Wilson. 1964. Ribonuclease levels in the mesocotyl tissue of *Zea mays* as a function of 2,4-dichlorophenoxyacetic acid application. *Plant Physiol.* **39**:804–809.

Small, J. G. C. and O. A. Leonard. 1968. Translocation of ^{14}C-labeled photosynthate in nodulated legumes as influenced by nitrate nitrogen. *Amer. J. Bot.* **56**:187–194.

Sudi, J. 1966. Increases in the capacity of pea tissue to form acylaspartic acids specifically induced by auxin. *New Phytol.* **65**:9–21.

Szabo, S. S. 1963. The hydrolysis of 2,4-D esters by bean and corn plants. *Weeds* **11**:292–294.

Upchurch, R. P., J. A. Keaton and H. D. Coble. 1969. Response of turkey oak to 2,4,5-T as a function of final formulation oil content. *Weed Sci.* **17**:505–509.

Wain, R. L. 1955. Herbicidal selectivity through specific action of plants on compounds applied. *J. Agric. and Food Chem.* **3**:128–130.

Weintraub, R. L., J. W. Brown, M. Fields and J. Rohan. 1952. Metabolism of 2,4-dichlorophenoxyacetic acid. I. $^{14}CO_2$ production by bean plants treated with labeled 2,4-dichlorophenoxyacetic acid. *Plant Physiol.* **27**:293–301.

Wilcox, M., D. E. Moreland, and G. C. Klingman. 1963. Aryl hydroxylation of phenoxy aliphatic acids by excised roots. *Physiol. Plant.* **16**:565–571.

Wills, G. D. and E. Basler. 1971. Environmental effects on absorption and translocation of 2,4,5-T in winged elm. *Weed Sci.* **19**:431–434.

Woodford, E. K., K. Holly, and C. C. McCready. 1958. Herbicides. *Ann. Rev. Plant Physiol.* **9**:311–358.

Wort, D. J. 1961. Effects on the composition and metabolism of the entire plant. In *Handbuch der Pflanzen Physiol.* W. Ruhland, Ed. **14**:1110–1136. Springer-Verlag. Berlin.

Wort, D. J. 1964. Effects of herbicides on plant composition and metabolism, pp. 291–330. In *The Physiology and Chemistry of Herbicides.* L. J. Audus, Ed. Academic Press. New York.

Zenk, M. H. 1960. Enzymatische Aktivierung von Auxinen and Ihre Konjugierung mit Glycin. *Zeitschr. f. Naturforschung.* **15**b:436–441.

CHAPTER 18

Thiocarbamates

General Properties

The carbamic acid molecule

$$NH_2-\overset{\overset{\displaystyle O}{\|}}{C}-OH$$

may have one sulfur substituted for one oxygen to form thiocarbamic acid,

$$NH_2-\overset{\overset{\displaystyle O}{\|}}{C}-SH;$$

two sulfur substitutions give dithiocarbamic acid,

$$NH_2-\overset{\overset{\displaystyle S}{\|}}{C}-SH.$$

Derivatives of both of these simple compounds are important among pest-control chemicals. Some dithiocarbamate molecules are toxic to plants and others to insects. The development of the dithiocarbamate herbicides preceded the development of the thiocarbamate herbicides. The two dithiocarbamate compounds which are used as herbicides are CDEC and metham. Their common name, chemical name, and chemical structure are presented in Table 18-1.

CDEC has been used for years to control a variety of weeds in field and vegetable crops. It is applied preemergence and has proved most effective in the more humid regions where rainfall can be depended upon to move it into the soil. Where rainfall is erratic shallow incorporation into soil is recommended. Used in the range of 2 to 6 lb/A CDEC is most commonly applied at 4 lb/A; it is particularly effective against annual grasses but it will also control several pigweeds (*Amaranthus* spp.), henbit (*Lamium amplexicaule*), common chickweed (*Stellaria media*), common purslane (*Portulaca oleracea*), and several other broadleaf species.

TABLE 18-1.   Common Name, Chemical Name, and Chemical Structure of the Dithiocarbamate Herbicides

$$\begin{array}{c} R_1 \\ \diagdown \\ N-C-S-R_3 \\ \diagup \quad \| \\ R_2 \quad S \end{array}$$

		Chemical structure		
Common name	Chemical name	R_1	R_2	R_3
CDEC	2-chloroallyl diethyl-dithiocarbamate	C_2H_5-	C_2H_5-	$\begin{array}{c}Cl\\ \|\\ CH_2=C-CH_2-\end{array}$
Metham	sodium methyl-dithiocarbamate	CH_3-	$H-$	$Na-$

Metham is a water-soluble salt which, upon contact with moist soil, undergoes decomposition to release methyl isothiocyanate, a toxic vapor. It is used as a soil fumigant and under proper conditions of soil moisture and distribution it kills fungi, bacteria, nematodes, insects, weeds, and weed seeds. Metham is most effective in light textured soils having a moisture content near field capacity. It is also effective against growing rhizomes of johnsongrass (*Sorghum halepense*), bermudagrass (*Cynodon dactylon*), and quackgrass (*Agropyron repens*), and against tubers of nutsedge (*Cyperus* spp.), and bulbs of wild onion (*Allium canadense*) and wild garlic (*Allium vineale*).

At proper dosage metham will kill the root systems of deep-rooted perennials such as field bindweed (*Convolvulus arvensis*) and Russian knapweed (*Centaurea repens*). It is used for seed bed fumigation, for spot treatment of perennial weeds, and for field fumigation against nematodes and fungi. It does not require the elaborate sealing process such as polyethylene tarps, commonly used to hold in the more volatile fumigants. It is also applied as a drench in relatively large volumes of water. Moistening to seal the top four inches of soil is sufficient for holding in the vapors. Soil incorporation followed by rolling is also used. Dribbling or spraying into the plow furrow has proved effective. Soil temperature should be between 50° and 75°F.

Metham has proved most useful for fumigating vegetable, tobacco, and ornamental seed beds. It is used in chrysanthemum nurseries, ornamental and nursery propagating beds, areas to be replanted in orchards and vineyards, in home gardens, and hard-to-weed areas such as patios, driveways, walks, tennis courts, etc.

Metham has proved toxic to pigweeds, common chickweed, mallows (*Malva* spp.), goldenrods (*Solidago* spp.), heliotropes (*Heliotropium* spp.), common lambsquarters (*Chenopodium album*), milk thistle (*Silybum marianum*), Russian thistle (*Salsola kali*), fiddlenecks (*Amsinckia* spp.), nightshades (*Solanum* spp.), wild radish (*Raphanus raphanistrum*), barnyardgrass (*Echinochloa crusgalli*), and many shallow-rooted species.

The thiocarbamate herbicides include butylate, cycloate, diallate, EPTC, molinate, pebulate, triallate, and vernolate. Their common name, chemical name, and chemical structure are given in Table 18-2. All of the thiocarbamate herbicides except diallate and triallate were developed by Stauffer Chemical Company. Diallate and triallate were developed by Monsanto.

EPTC was the first thiocarbamate herbicide developed. The volatile nature of this compound resulted in highly variable weed control when it was first used as a preemergence herbicide. The technique of soil incorporation proved to correct this deficiency and provided the first general usage of soil incorporation. This technique has subsequently been used extensively with many herbicides. In arid areas it is also commonly used with non-volatile herbicides to place the toxicant near the germinating weed seeds.

EPTC is a preemergence herbicide which is commonly incorporated mechanically to a depth of 2 to 3 inches. In dry soil EPTC may be incorporated by sprinkler irrigation immediately after application; it has also been applied through the sprinkler system and by metering into furrow irrigation water in treating such crops as deciduous or citrus orchards, alfalfa, mint, clover, etc. EPTC is used against a wide array of weeds including many grasses and a variety of annual broadleaf weeds.

EPTC is rapidly metabolized in plants and hence does not give a residue problem. It is adsorbed by dry soil but may be readily displaced by water. It is rapidly decomposed by microorganisms in the soil and it readily volatilizes from moist soil if not incorporated.

Butylate is a liquid formulated as an emulsifiable concentrate. It should be incorporated into the top 2 to 3 inches of soil immediately after application; it is also available in granular form. It can be used in either form preplant in maize, one of the crops for which it is recommended. Butylate is suggested for use against nutsedge, seedlings of perennial grasses such as quackgrass and johnsongrass, and for control of most annual grasses, as well as common lambsquarters, pigweeds, common purslane, and velvetleaf (*Abulilon theophrasti*).

When applied to the soil butylate is rapidly absorbed by roots and undergoes apoplastic distribution. Its primary site of action is meristems. It is readily metabolized within plants and by soil microorganisms. It has a half-life in soils of from $1\frac{1}{2}$ to 3 weeks and it leaves no harmful residue.

Cycloate is a colorless liquid formulated as an emulsifiable material having

TABLE 18-2. Common Name, Chemical Name, and Chemical Structure of the Thiocarbamate Herbicides

$$R_1-N(R_2)-\overset{\overset{O}{\|}}{C}-S-R_3$$

Common name	Chemical name	Chemical structure R₁	R₂	R₃
Butylate	S-ethyl diisobutylthiocarbamate	$CH_3-CH(CH_3)-CH_2-$	$CH_3-CH(CH_3)-CH_2-$	C_2H_5-
Cycloate	S-ethyl N-ethylthiocyclohexane-carbamate	C_2H_5-	(cyclohexyl)	C_2H_5-
Diallate	S-(2,3-dichloroallyl) diisopropylthiocarbamate	$CH_3-CH(CH_3)-$	$CH_3-CH(CH_3)-$	$Cl-C(H)=C(Cl)-CH_2-$
EPTC	S-ethyl dipropylthiocarbamate	C_3H_7-	C_3H_7-	C_2H_5-
Molinate	S-ethyl hexahydro-1H-azepine-1-carbothioate	(azepine ring, N-)[a]		C_2H_5-
Pebulate	S-propyl butylethylthiocarbamate	C_2H_5-	C_4H_9-	C_3H_7-
Triallate	S-(2,3,3-trichloroallyl) diisopropylthiocarbamate	$CH_3-CH(CH_3)-$	$CH_3-CH(CH_3)-$	$Cl-C(Cl)=C(Cl)-CH_2-$
Vernolate	S-propyl dipropylthiocarbamate	C_3H_7-	C_3H_7-	C_3H_7-

[a] The nitrogen atom in the molinate ring structure is the nitrogen atom of the parent thiocarbamic acid molecule; there is only one nitrogen atom in molinate.

6 lb/gal ai.; applied to the soil it should be quickly incorporated to a depth of 2 to 3 inches. It may also be applied below the soil surface with proper injection equipment. If applied to dry soil it may be incorporated by overhead irrigation.

Cycloate is a selective herbicide effective against nutsedges and annual grasses. It also controls black nightshade (*Solanum nigrum*), henbit, common lambsquarters, common purslane, pigweeds, shepherds purse (*Capsella bursa-pastoris*), burning nettle (*Urtica urens*), and other similar weeds. It is used preplant incorporated on sugar beet beds at 3 to 4 lb/A (Dawson, 1971). It is more selective and less phytotoxic to sugar beets than pebulate, a commonly used herbicide. It is also recommended for controlling annual grasses and certain broadleaf weeds in spinach.

Cycloate resists leaching in clay soils and organic soils. In sugar beets cycloate is metabolized to ethyl-cyclohexylamine, CO_2, and natural plant constituents. In the soil it disappears due to microbial breakdown; under crop growing conditions it has a half-life of 4 to 8 weeks. It is volatile when applied to the surface of moist soil.

Molinate is named as a carbothioate but, except for the substitutions on the nitrogen, it has the features typical of a thiocarbamate. It is a liquid formulated as an emulsifiable concentrate or as a granular material. It has proved most useful as a preplant, soil incorporated herbicide for use on water-seeded rice. It can be applied postemergence to the rice but before the weedy grasses emerge above the water. It is used principally against barnyardgrass, the most serious grassy weed in water-seeded rice.

Molinate is readily metabolized by plants and by soil microorganisms. It is rapidly volatilized from wet soil; it is adsorbed on dry soil and remains inert until the field is flooded. In flooded fields the layer of water, which is normally about 6 inches in thickness, serves to seal in the vapors.

Pebulate is formulated as an emulsifiable concentrate and it should be applied preplant for direct seeded crops and incorporation is essential. It may also be used before transplanting of such crops as tomatoes, or posttransplant. Pebulate is active against nutsedge, barnyardgrass, henbit, common lambsquarter, common purslane, pigweeds, crabgrass (*Digitaria* spp.), foxtails (*Setaria* spp.), wild oat (*Avena fatua*), hairy nightshade (*Solanum sarachoides*), and nettleleaf goosefoot (*Chenopodium murale*).

Pebulate is readily and quickly metabolized by plants to CO_2 and harmless compounds. Pebulate is readily adsorbed onto dry soil and is replaced by water when the soil is moist. It is this loss by volatilization which makes it mandatory to incorporate the thiocarbamate herbicides. Pebulate does not persist over a few weeks in soil and it does not leave a harmful residue. Tillage following use of pebulate should be shallow.

Vernolate is a thiocarbamate with three propyl substitutions making it somewhat simpler in structure than many of the other thio- or dithiocarbamates. It is a liquid sold as an emulsifiable concentrate. It is applied preplant for incorporation to depths of 2 to 3 inches or postemergence in row crops when they are 1 to 2 inches tall. Any subsequent tillage should not penetrate below the depth of incorporation.

Vernolate is used to control crabgrass, barnyardgrass, foxtail grasses, johnsongrass seedlings, nutsedge, goosegrass (*Eleusine indica*), shatter cane (*Sorghum bicolor*), and several broadleaf weeds including pigweeds, common lambsquarters, sicklepod (*Cassia obtusifolia*), carpetweed (*Mollugo verticillata*), and Florida pusley (*Richardia scabra*).

Vernolate is metabolized in plants to CO_2 and naturally occurring plant constituents. It is readily broken down by soil microorganisms and leaves no harmful residue.

Diallate consists of the *cis* and *trans* isomers of diisopropylthiocarbamate. It is applied preplant with soil incorporation. It is particularly phytotoxic against wild oats and is suggested for use in sugar beets, flax, barley, maize, forage legumes, lentils, peas, potatoes, and safflower. Application rates extend from $1\frac{1}{2}$ to 4 lb/A depending upon the crop. It is sold as an emulsifiable concentrate or in a granular formulation. On wild oat, it is absorbed by the emerging coleoptile and it acts as a mitotic poison. It is readily metabolized within plants and by microorganisms in the soil; its half-life is around 30 days in warm moist soil.

Triallate is like diallate except that it has two chlorines on the terminal allyl carbon. It is an oily liquid sold as an emulsifiable concentrate. It is recommended for use at 1 to $1\frac{1}{2}$ lb/A on wild oats in peas, barley, and durum and spring wheats. It should be applied preplant or preemergence and incorporated by cross harrowing or discing.

Triallate is absorbed by the emerging coleoptile as it pushes up through the soil and acts as a mitotic poison. It is metabolized in plants and by microorganisms in the soil. It persists up to 6 weeks in many soils providing time for the crop to compete successfully with any late emerging weeds.

Growth and Plant Structure

The thiocarbamate and dithiocarbamate herbicides are applied to the soil and primarily inhibit the growth of emerging shoots of weeds, especially grasses. Dawson (1963a) studied the effect of EPTC on the development of barnyardgrass seedlings. He reported that most of the elongation responsible for barnyardgrass seedling emergence occurred in the first internode. EPTC caused kinking of the first internode into a zig-zag pattern, Figure 18-1. However, the developing foliar leaves within the coleoptile were the major

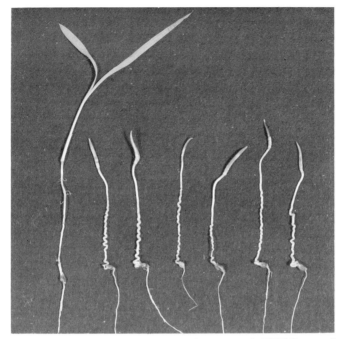

FIGURE 18-1.   Barnyardgrass seedlings grown in EPTC-treated soil showing kinking of the first internode. Untreated plant of the same age on the left. (Dawson, 1963a.)

site of EPTC injury. At low rates, injury involved the first foliar leaf only, which elongated beyond the coleoptile but remained longitudinally rolled. At higher rates, the number of individual leaves affected proceeded centripetally in the bud and the degree of injury to individual leaves increased. As injury to individual leaves increased, the total growth beyond the coleoptile was reduced and this growth became increasingly distorted. At even higher rates, no leaves developed beyond the coleoptile. Evidently the leaf primordia on the apical buds as well as the young leaves were injured. In many seedlings, the expanding foliar leaves within the coleoptile were not able to emerge from the coleoptile in a normal manner, but ruptured the side of the coleoptile. Figure 18-2 shows the several abnormalities of developing barnyardgrass seedlings induced by EPTC. It was also shown that EPTC vapors at concentrations sufficiently high to almost completely inhibit leaf growth had little if any effect on the development of the coronal and first internode adventitious roots, Figure 18-3.

Microscopic studies by Dawson (1963a) on the effect of EPTC on barnyardgrass seedling development showed changes in the mesophyll of the developing

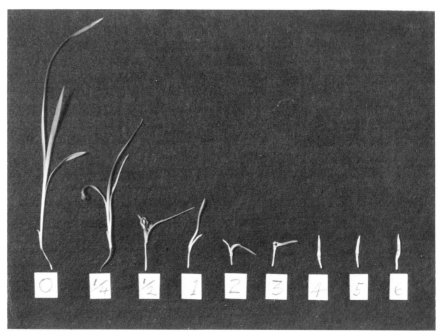

FIGURE 18-2. Representative 4-week-old barnyardgrass seedlings grown in fine sandy loam containing the indicated parts per million of EPTC. (Dawson, 1963a.)

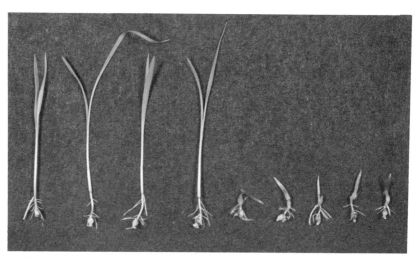

FIGURE 18-3. Barnyardgrass seedlings grown from seeds on the soil surface under sealed plastic bags. Those on the right were exposed to EPTC vapors. Coronal and first internode adventilious roots of treated and untreated plants are similar. (Dawson, 1963a.)

leaves within the coleoptile. The chloroplasts were more concentrated near the cell walls and they had a much greater affinity for various stains in the EPTC-treated plants. Intercellular spaces were also reduced and linear rows of cells were buckled and kinked, Figure 18-4. There was swelling about the coleoptilar node although no cell proliferation was involved. Instead, there was an outward folding of the coleoptile all around its base just above its point of attachment to the stem. Since there were no cellular abnormalities in the first internode or the coleoptiles, Dawson concluded that the injury was concentrated in the mesophyll of the foliar leaves within the coleoptile. Dawson (1963a, 1963b) also reported that EPTC did not inhibit primary root growth or seed germination of barnyardgrass.

Broadleaf weeds treated with EPTC may develop cupped leaves with necrotic tissue around the edges. This results from the apoplastic translocation of these herbicides; phytotoxic amounts at the hydathodes, at the tip and at bundle ends around the leaf margin may result in necrotic spots and somewhat deformed leaves. At higher rates, crops may also develop these phytotoxic symptoms but they usually grow out of this condition.

Gentner (1966) found that EPTC inhibited the deposition of external foliage wax on cabbage leaves; inhibition was correlated directly with rate of application. Reduction of foliage wax resulted in increased retention of

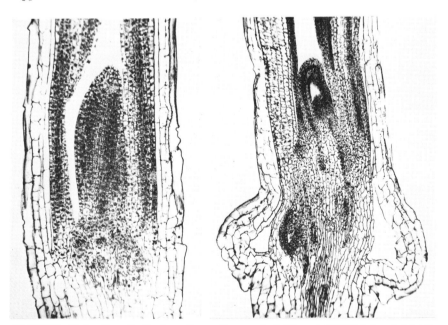

FIGURE 18-4. Longitudinal sections of normal (left) and EPTC-injured (right) barnyardgrass seedlings in region of coleoptilar nodes. (Dawson, 1963a.)

sprays; the spray droplets showed decreased contact angles. EPTC increased transpiration in direct proportion to its rate of application. Wilkinson and Hardcastle (1969) found in sicklepod that petiole cuticle thickness decreased 35% as EPTC concentration increased from 0 to 4.48 kg/ha.

Diallate, if sublethal in dosage, may allow limited emergence of the first leaf of wild oat from the coleoptile, but its shape is somewhat distorted and dark green and glossy in appearance (Parker, 1963) and brittle (Banting, 1967).

Since a number of carbamates have been shown to be mitotic poisons in plants it seems logical to test the thio- and dithiocarbamates for this property. Banting (1967) studied the effect of diallate and triallate vapor on cell division in wild oat (the target species) and spring wheat. Mitotic abnormalities found included short-thick chromosomes, dumbell-shaped nuclei, clumps of chromosome material, doubling of the chromosome number, bridges between chromosomes, and micronuclei. Diallate was twice as effective in producing abnormalities as triallate in the stem apex and leaf meristem cells of wild oat. The number of cell divisions decreased as the concentration of either chemical was increased. Shoot tissues of both wild oat and wheat were more sensitive than root tissues (Figure 18-5). Shoot growth inhibition occurred in both wild oat and wheat at concentrations that did not affect mitosis. Thus mitotic damage from either chemical appeared to be secondary to cell elongation and expansion as a mechanism of action. Since root tips showed little or no effect from exposure to diallate or triallate vapor, and since shoot growth of wild oat was reduced to a greater extent than root growth, Banting considered that the major effect of these herbicides is on cell elongation or expansion.

In some plant species root protrusion from the germinating seed is known to precede mitosis in the root tip. If the same situation holds in the shoot tip, and the first effect of diallate and triallate is on the process of cell elongation, then some portion of the overall phytotoxicity must also include mitosis since mitotic disturbance follows just behind cell elongation.

Absorption and Translocation

The thio- and dithiocarbamates may be divided into two groups on the basis of the presence or absence of the allyl group. CDEC, diallate, and triallate, have this group in their molecules, the remaining compounds do not. CDEC is used principally to control grassy weeds in broadleaf crops; sweet and field maize are exceptions. CDEC is not absorbed by the foliage of plants but it is readily absorbed by roots and moved apoplastically to the foliage. It kills weeds in the early seedling stages.

Diallate absorption and translocation have been studied (Parker, 1963; Appleby et al., 1965; Nalewaja, 1968). Appleby et al. (1965) found that

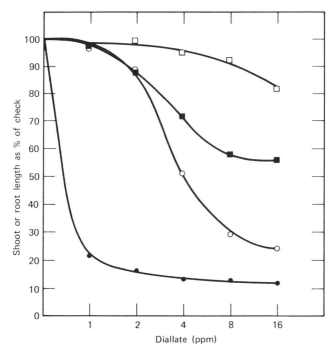

FIGURE 18-5.   Root (□) and shoot length (■) of wheat, and root (○) and shoot length (●) of wild oat as percent of check after 5 days of exposure to vapor from various concentrations of diallate. (Banting, 1970.)

diallate exerts its primary phytotoxic effect when absorbed through the coleoptiles of emerging grass seedlings. Parker (1963) found that absorption of both diallate and triallate is principally through the coleoptile and by making droplet applications on *Avena* seedlings he proved that a region 10 to 15 mm above the coleoptile node gave the maximal retardation of growth; this proved true for wild oat and for commercial varieties of barley and wheat. Parker found that both depth of seeding and depth of incorporation of the herbicide are important in selectivity between wild oat, the principal target species for di- and triallate, and the cereal crops barley and wheat.

Nalewaja (1968), using ^{14}C-labeled diallate, found very little translocation when this tracer was applied to root tips of barley and wild oat. When the ^{14}C-diallate was applied to the coleoptile tips of barley, wheat, and wild oat seedlings, there was movement of ^{14}C through the coleoptiles into the roots. Treated midway on the shoots of the seedlings of these plants, there was acropetal movement to the coleoptile tips and basipetally to root tips.

Treatment at the base of the mesocotyl of wild oat resulted in similar distribution. Exposure of roots of seedlings of flax, wild oat, wheat, and barley resulted in movement throughout the roots and foliage of these seedlings.

Banting (1967) found that diallate vapor in the soil is effective in controlling wild oat seedlings. Table 18-3 shows some of his results. Banting also found that the soil moisture content at somewhat less than the wilting point would activate diallate; at 1 lb/A 15% moisture sufficed in Regina heavy clay having a wilting point at 19.6%. This indicates that diallate may utilize water bound to soil colloids for activation, something the plant cannot do. These findings explain some of the conflicting observations on the variable effects of depth of planting, depth of incorporation, and stage of growth on wild oat control by di- and triallate.

TABLE 18-3.   Control of *Avena fatua* by Di-allate Vapor in Soil at Two Moisture Levels (Banting, 1967)

Oz/ac	Soil moisture in top 5 cm (%)	Germination (%)	Leaf stage	Shoot height [a] (cm)	% control
0	30	77	1.2	17.4	0
24	30	83	1.0	3.0	100
0	40	77	1.2	15.6	0
24	40	70	1.0	3.0	100

[a] Measured from soil surface to tip of first leaf.

Considerable research has been done on the uptake and distribution of EPTC in plants. Using ^{35}S-EPTC, Yamaguchi (1961) found EPTC to be very mobile in plants. Applied as a vapor to leaves this tracer was found to be absorbed and translocated symplastically in lima and red kidney beans, maize, sugar beet, field bindweed, and jimsonweed (*Datura stramonium*). Applied to roots through the culture medium, ^{35}S-EPTC was readily absorbed and distributed apoplastically throughout the foliage (Figures 18-6 and 18-7).

Oliver et al. (1968) studied species differences in the site of root uptake of ^{14}C-EPTC. They found barley to be most tolerant to this herbicide by absorption via root; wheat, oats, sorghum, and giant foxtail (*Setaria faberi*) increased in sensitivity in the order named. The roots were the major site of uptake by barley, but injury to the other species from root exposure was equal to or slightly less than that from shoot exposure. The authors concluded that differences in tolerance can be associated with sites of uptake.

Prendeville et al. (1968) determined the effects of EPTC placed in different shoot zones below the soil surface after emergence of barley, wheat, oats, and

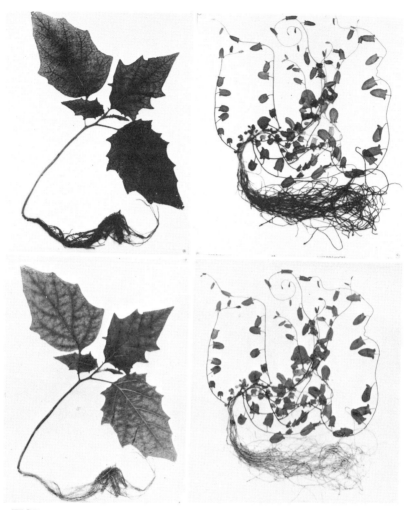

FIGURE 18-6. Plants (below) and autoradiographs (above) of jimsonweed and field bindweed showing the uptake and distribution of ^{35}S-labeled
EPTC from culture solutions by roots. In each case, 20 ml of culture solution
containing 2.5 μc of ^{35}S were used to bathe the roots. The plants were allowed
to absorb this solution until all liquid was gone, then more straight culture
solution was added. The jimsonweed absorbed the original 20 ml of culture
solution in 15 days; the field bindweed absorbed the 20 ml in 3 days; total
treatment time was 4 days. (Yamaguchi, 1961.)

FIGURE 18-7. Plants (below) and autoradiographs (above) of jimsonweed and field bindweed showing the movement of ³⁵S-labeled EPTC from leaf treatment. In the case of the jimsonweed 10 µl containing 2.5 µc of the ³⁵S-labeled EPTC was placed in a small dish, and this was set on the leaf to be treated and covered with a petri dish which was sealed to the leaf with lanolin. Thus the EPTC was applied in the vapor form; treatment time was 4 days. The field bindweed plant received 40 µl containing 3µc of ³⁵S-labeled EPTC applied as droplets on three leaves. Treatment time was 4 days. In both types of leaf treatment there was loss of the labeled EPTC to the atmosphere. (Yamaguchi, 1961.)

sorghum. Wheat, barley, and oats were severely injured when treated at the coleoptilar node; exposure of the remaining shoot did not affect growth. Sorghum was severely injured regardless of the shoot zone exposed. Uptake of ^{14}C-EPTC from the soil by sorghum shoots was double that of wheat. The writers concluded that growth responses of species to EPTC applied to shoots is dependent upon the stage of plant development at which the treatment takes place. This means that for successful control of weeds using EPTC, attention must be given to dosage, depth of incorporation, placement of seeds, and weed species involved.

Using a charcoal barrier method to prevent movement of EPTC vapor in soil, Gray and Weierich (1969) studied the relative importance of root, shoot, and seed exposure to the phytocidal activity of EPTC. Exposing the roots of barley, oats, barnyardgrass, Italian ryegrass (*Lolium multiflorum*), wheat, rice, cotton, and yellow nutsedge (*Cyperus rotundus*) caused more injury than shoot exposure. Shoot exposure was somewhat more effective than root exposure to johnsongrass, sorghum, and peas. Seed exposure brought about variable results. Concentrations of EPTC as low as 1 ppm applied to roots in culture solution inhibited shoot growth of oats, barley, sorghum, and maize.

Because all of the thio- and dithiocarbamates are normally used by soil application for absorption through the roots, coleoptiles, or lower stems, little research has been done upon the effects of foliar application. However, from the notes on these herbicides (metham, vernolate, pebulate, butylate, molinate, and cycloate) that appear in the WSSA Herbicide Handbook (1970) it seems that all are readily absorbed by leaves and moved symplastically. They are also absorbed by roots, coleoptiles, or lower stems and translocated apoplastically.

Molecular Fate

It is recognized that, being volatile, thio- and dithiocarbamates are lost from soils relatively soon after application, that is, within a few days to several weeks. Working with ^{14}C-labeled compounds, it has also been found that the sulfur-containing carbamates are broken down in plants with the evolution of ^{14}CO$_2$. Since they are readily translocated in plants they must be thoroughly distributed and metabolic breakdown may occur in all plant parts.

Metabolism of thiocarbamates in plants has been reviewed by Fang (1969) in Kearney and Kaufman's book (1969). EPTC was found to be degraded by plant seedlings; more in seedlings of resistant than in susceptible plants; ^{35}S was found to be incorporated into sulfur-containing amino acids and other organic plant constituents.

Nalewaja et al. (1964) found that ^{14}C-EPTC was broken down to CO$_2$. ^{14}CO$_2$ may then be incorporated into a number of naturally occurring plant constituents via photosynthesis and/or dark CO_2 fixation; ^{14}C was found in fructose, glucose, and several amino acids. Temperature, concentration, plant species, and many other factors determine the rates and extent of metabolic breakdown.

Butylate is rapidly metabolized in plants to CO_2, diisobutylamine, fatty acids, conjugates of amines and fatty acids, and naturally occurring plant constituents. Cycloate breaks down to ethyl cyclohexylamine and CO_2. The CO_2 may then be incorporated into amino acids, sugars, and other naturally occurring plant constituents via photosynthesis and/or dark CO_2 fixation. Molinate, pebulate, and vernolate apparently undergo the same metabolic transformations and hence experience the same fate.

Bourke and Fang (1968) studied the metabolism of ^{14}C-vernolate in soybean seedlings. Being predominantly a grass killer, one might expect to find this compound to be readily broken down in this broadleaf plant and this agrees with the findings of Bourke and Fang. Having found that vernolate is readily absorbed and translocated by both peanut and soybean seedlings and that the concentration of unchanged chemical drops rapidly after reaching a maximum with no increases in other metabolic products. Bourke and Fang determined the rate and extent of breakdown in soybean seedlings at various stages of germination. They found that degradation of ^{14}C-vernolate to $^{14}CO_2$ was indeed dependent upon the age of the treated seedlings and that pretreatment with cold vernolate resulted in a reduction of catabolic oxidation of ^{14}C-vernolate to $^{14}CO_2$. There was a parallel reduction in cellular incorporation and an increase in ethanol soluble metabolites; they found two major and two minor metabolites, the relative abundance of which again depended upon the age of the seedlings. One major metabolite (their #3) which was of greatest abundance after 48 hours exposure to vernolate was apparently converted to another (their #4) compound within 312 hours.

Jacobsohn and Andersen (1968) found large differences among 214 lines of wild oats in their response to diallate, triallate, and barban. They found no correlation with plant morphology but frequency distributions of response suggested that reactions are quantitatively inherited.

Fang (1969) presented the following scheme for degradation of thiocarbamate molecules in plants and animals, Figure 18-8. Fang (1969) suggested that the thiocarbamate molecule is probably hydrolyzed at the ester linkage with the formation of a mercaptan, CO_2, and an amine. The mercaptan could be converted to an alcohol and further oxidized. The sulfur of the mercaptan may be incorporated into sulfur-containing amino acids. All ^{14}C labels used in such studies give rise to $^{14}CO_2$. The $^{14}CO_2$ may then be used for photosynthesis and/or dark CO_2 fixation to yield such plant constituents as fructose, glucose, serine, threonine, alanine, aspartic acid, and glutamic acid. The metabolism of CDEC leads to lactic acid, of vernolate to

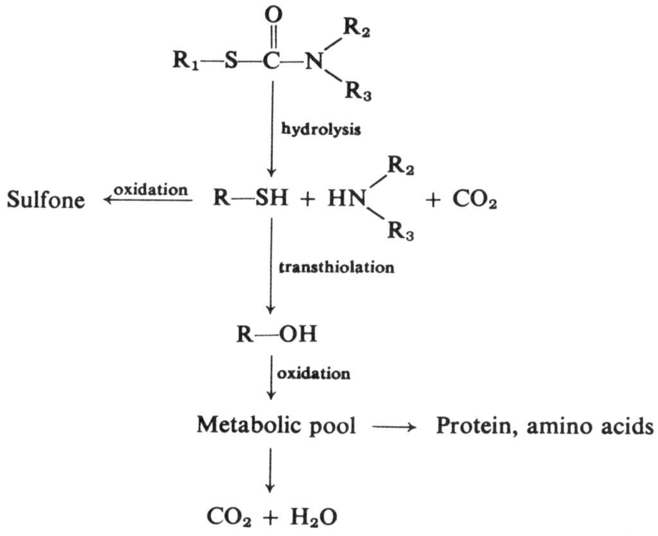

$$R_1\text{—}S\text{—}\overset{\overset{\displaystyle O}{\parallel}}{C}\text{—}N\overset{\nearrow R_2}{\searrow R_3}$$

hydrolysis

$$\text{Sulfone} \xleftarrow{\text{oxidation}} R\text{—SH} + HN\overset{\nearrow R_2}{\searrow R_3} + CO_2$$

transthiolation

$$R\text{—OH}$$

oxidation

Metabolic pool $\longrightarrow$ Protein, amino acids

$$CO_2 + H_2O$$

FIGURE 18-8. Proposed pathway for thiocarbamate herbicide degradation in higher plants (Fang, 1969) Reprinted from P. C. Kearney and D. D. Kaufman, eds., Degradation of Herbicides, p. 161, by courtesy of Marcel Dekker, Inc.

citric acid. The ready denaturing of the thiocarbamates is a disadvantage with respect to the length of herbicidal activity, an advantage as concerns residues.

Biochemical Responses

While only a limited amount of research has been carried out on metabolic pathways in thiocarbamate-treated plants, it seems that they all affect mitosis in the young shoot with relatively little effect upon root tissues. When growth is arrested in the coleoptile stage, the first leaf may force its way out through the base of the coleoptile; such a leaf may elongate to form a loop with its tip fixed within the coleoptile tip; usually such effects are fatal.

CDEC is reported to chelate copper and other metals and possibly to inhibit tyrosinase and cytochrome oxidase (Anon., 1970).

In metabolic studies on EPTC, Ashton (1963) found no significant effect upon photosynthesis in red kidney bean; inhibition of $^{14}CO_2$ fixation required high concentrations and long exposures; furthermore, the typical phytotoxic action of this herbicide is growth inhibition at a stage where photosynthesis is not yet important. EPTC increased respiration of excised embryos of maize and mung bean when the data were expressed on fresh weight basis, but when calculated on a per embryo basis the stimulation was not evident. At relatively high concentrations EPTC markedly inhibited both phosphorus uptake and oxygen consumption. Phosphorus esterification was more sensitive than oxygen utilization. Because of the high concentrations required to bring about these results, Ashton questioned their physiological significance.

Mann et al. (1965) reported that CDEC and EPTC slightly inhibited ^{14}C-1-leucine incorporation into protein in segments of barley coleoptile or hemp sesbania (*Sesbania exaltata*) hypocotyls. At 2 and 5 ppm, CDEC inhibited this reaction 25 and 29%, respectively, in barley, and 7 and 0%, respectively, in hemp sesbania. At 2 and 5 ppm, EPTC inhibited this reaction 38 and 22%, respectively, in barley and 14 and 11%, respectively, in hemp sesbania. In a similar study by Moreland et al. (1969), using soybean hypocotyl sections, CDEC at 2×10^{-4}M inhibited protein synthesis 33% and EPTC at 6×10^{-4}M inhibited protein synthesis 24%.

Moreland et al. (1969) also reported that gibberellic acid induced *de novo* α-amylase synthesis in barley half-seeds was inhibited 68% by 2×10^{-4}M CDEC and 39% by 6×10^{-4}M EPTC. They also measured RNA synthesis by ^{14}C-8-ATP or ^{14}C-6-orotic acid incorporation using soybean hypocotyl sections or maize mesocotyl sections, respectively. RNA synthesis was inhibited 22 and 13% by 2×10^{-4}M CDEC in the maize and soybean assay, respectively, and 58 and 2% by 6×10^{-4}M EPTC in the maize and soybean assay, respectively.

The effect of herbicides, including CDEC and EPTC, on ATP levels in excised soybean hypocotyls has recently been described by Gruenhagen and Moreland (1971). They found that CDEC and EPTC did not reduce tissue ATP levels significantly. However, they suggest that the failure of these herbicides to depress ATP levels could be related to the failure in penetrating the tissue effectively and the inability of the technique used to measure small but meaningful changes in ATP levels. Small changes could be of sufficient magnitude to interfere with RNA and protein synthesis. However, these herbicides could inhibit these important reactions in ways other than by interfering with the generation of energy.

Beste and Schreiber (1972), studying the antagonistic interaction between EPTC and 2,4-D, found that EPTC inhibited growth and RNA synthesis in soybean tissue. The addition of 2,4-D to EPTC was antagonistic and caused an increase in total RNA synthesis. Analysis of soybean tissue nucleic acids by MAK column chromatography showed that EPTC inhibited rRNA, D-RNA, and tightly bound RNA synthesis. The combination of 2,4-D with EPTC caused an increase in D-RNA and bound RNA synthesis compared to EPTC alone; this appears to be the basis for the antagonism. Analysis of rRNA indicated EPTC preferentially inhibited the synthesis of 18S rRNA more than 25S rRNA and that 2,4-D had no effect on the selective inhibition.

Chen, Seaman, and Ashton (1968) studied the herbicidal action of molinate, an effective inhibitor of barnyardgrass in rice, upon the RNA content of barnyardgrass coleoptiles. Although the total RNA content of the treated coleoptiles differed little from controls, when subcellular organelles were separated, the soluble RNA from molinate-treated coleoptiles was only 30% of that of controls. The molinate treatment did not appear to alter the RNA content of the nuclei, chloroplasts, mitochondria, or ribosomes. Since the reduced fraction, the soluble RNA (transfer RNA) is required in protein synthesis, molinate may be acting as an inhibitor of this important process.

Mode of Action

The thio- and dithiocarbamates have the thiocarbamate grouping in common

$$NH_2 - \overset{\overset{\displaystyle O}{\|}}{C} - SH;$$

the dithiocarbamate has a second sulfur on the carbamate carbon

$$NH_2 - \overset{\overset{\displaystyle S}{\|}}{C} - SH.$$

In addition to these common groups this class of compounds has many substituent atoms, carbons, or bondings that are responsible for their separate unique selectivities. These involve allyl bonding on the carbamate carbon (CDEC, diallate, triallate), ring structures on the carbamate N (molinate, cycloate), and a variety of alkyl chains on the thio sulfur. Another common property is their vapor pressure; they are all volatile; some are herbicidal by virtue of their volatility (diallate, triallate).

In general, these herbicides inhibit the growth of shoots of germinating seedlings to a greater degree than the roots. It is not clear whether this is related to some inherent difference between these two organs or the fact that they are more readily absorbed by the shoot than the roots. The major phytotoxic symptom induced by these compounds is the abnormal growth and emergence of the leaves from the coleoptiles of grasses. At high rates, the leaves may not emerge from the coleoptiles; at lower rates they may emerge but they may remain longitudinally rolled or emerge through the base of the coleoptile. Such a leaf may elongate to form a loop with its tip fixed within the coleoptile tip. In broadleaf weeds the phytotoxic symptoms are less specific; growth may be generally inhibited and the leaves may be cupped with necrotic tissue around the edges.

These soil-applied herbicides appear to be absorbed more readily by the emerging seedling shoots than by the roots. Although they have been shown to be readily translocated via the apoplastic system there is some evidence that they also move symplastically.

The molecular fate of these herbicides in higher plants appears to involve hydrolysis at the ester linkage with the formation of CO_2, a mercaptan, and an amine. These later two compounds are further metabolized into normal metabolites and CO_2.

The thio- and dithiocarbamate herbicides seem to alter normal plant metabolism in a variety of ways, e.g., photosynthesis, respiration, oxidative phosphorylation, protein synthesis, and nucleic acid metabolism. At the present time, it is not possible to indicate the relative importance of these various processes to the herbicidal action. However, the primary site of action may reside in nucleic acid metabolism or protein synthesis, although their influence on the other processes may be involved in the ultimate death of the plant.

REFERENCES

Anon. 1970. *Herbicide Handbook of the Weed Science Society of America.* W. F. Humphrey Press, Inc. Geneva, N.Y. 368 pp.

Appleby, A. P., W. R. Furtick, and S. C. Fang. 1965. Soil placement studies with EPTC on *Avena sativa. Weed Res.* 5:115–122.

Ashton, F. M. 1963. Effect of EPTC on photosynthesis, respiration, and oxidative phosphorylation. *Weeds* 11:295–297.

Banting, J. D. 1967. Factor affecting the activity of diallate and triallate. *Weed Res.* 7:302–315.

Banting, J. D. 1970. Effect of diallate and triallate on wild oat and wheat cells. *Weed Sci.* 18:80–84.

Beste, C. E. and M. M. Schreiber. 1972. RNA synthesis as the basis for EPTC and 2,4-D antagonism. *Weed Sci.* 20:8–11.

Bourke, J. B. and S. C. Fang. 1968. The metabolism of ^{14}C-vernolate in soybean seedlings. *Weed Sci.* 16:290–292.

Chen, T. M., D. E. Seaman, and F. M. Ashton. 1968. Herbicidal action of molinate in barnyardgrass and rice. *Weed Sci.* 16:28–31.

Dawson, J. H. 1963a. Development of barnyardgrass seedlings and their response to EPTC. *Weeds* 11:60–66.

Dawson, J. H. 1963b. Effects of EPTC on barnyardgrass seeds. *Weeds* 11:184–186.

Dawson, J. H. 1971. Response of sugarbeets and weeds to cycloate, propachlor, and pyrazon. *Weed Sci.* 19:162–165.

Fang, S. C. 1969. Thiolcarbamates, pp. 147–164. In P. C. Kearney and D. D. Kaufman, Eds. *Degradation of Herbicides.* Marcel Dekker, New York.

Gentner, W. A. 1966. The influence of EPTC on external foliage wax deposition. *Weed Sci.* 14:27–30.

Gray, R. A. and A. J. Weierich. 1969. Importance of root, shoot, and seed exposure on the herbicidal activity of EPTC. *Weed Sci.* **17**:223–229.

Gruenhagen, R. D. and D. E. Moreland. 1971. Effect of herbicides in ATP levels in excised soybean hypocotyls. *Weed Sci.* **19**:319–323.

Jacobsohn, R. and R. N. Andersen. 1968. Differential response of wild oat lines to diallate, triallate and barban. *Weed Sci.* **16**:491–494.

Kearney, P. C. and D. D. Kaufman. 1969. *Degradation of Herbicides.* Marcel Dekker, Inc., New York. 394 pp.

Mann, J. D., L. S. Jordan, and B. E. Day. 1965. A survey of herbicides for their effect upon protein synthesis. *Plant Physiol.* **40**:840–843.

Moreland, D. E., S. S. Malhotra, R. D. Gruenhagen, and E. H. Shokraii. 1969. Effects of herbicides on RNA and protein synthesis. *Weed Sci.* **17**:556–563.

Nalewaja, J. D. 1968. Uptake and translocation of diallate in wheat, barley, flax, and wild oat. *Weed Sci.* **16**:309–312.

Nalewaja, J. D., R. Behrens, and A. R. Schmid. 1964. Uptake, translocation, and fate of ^{14}C-EPTC in alfalfa. *Weeds* **12**:269–272.

Oliver, L. R., G. N. Prendeville, and M. M. Schreiber. 1968. Species differences in site of root uptake and tolerance to EPTC. *Weed Sci.* **16**:534–537.

Parker, C. 1963. Factors affecting the selectivity of 2,3-dichloroallyl diisopropylthiolcarbamate (Di-allate) against *Avena* spp. in wheat and barley. *Weed Res.* **3**:259–276.

Prendeville, G. N., L. R. Oliver, and M. M. Schreiber. 1968. Species differences in site of shoot uptake and tolerance to EPTC. *Weed Sci.* **16**:538–540.

Wilkinson, R. E. and W. S. Hardcastle. 1969. EPTC effects on sicklepod petiolar fatty acids. *Weed Sci.* **17**:335–338.

Yamaguchi, S. 1961. Absorption and distribution of ^{35}S-EPTC. *Weeds* **9**:374–380.

CHAPTER 19

Triazines

In 1952, the J. R. Geigy S.A., Basle, Switzerland, started investigations with triazine derivatives as potential herbicides. Their working hypothesis was based on the fact that in certain pharmaceuticals and vat-dyestuffs, the insertion of the triazine or urea molecule into relatively complex compounds yielded compounds with similar properties. Furthermore, the herbicidal properties of the substituted urea compounds had been demonstrated (Knüsli, 1970). The herbicidal properties of chlorazine were reported by Gast et al. (1955) and Antognini and Day (1955). Subsequently numerous triazine derivatives have been synthesized and screened for their herbicidal properties. Table 19-1 lists many of these. They are used selectively in certain crops as well as for non-selective weed control. Their greatest use has been as a selective herbicide in maize and a non-selective herbicide on industrial sites. Although most of them are applied to the soil certain ones are highly active as foliar applications. Recently several other chemical companies have started developing triazine herbicides.

Several excellent reviews have been written on various aspects of the triazine herbicides, i.e., history (Knüsli, 1970), uses (Gast, 1970), degradation (Knuesli et al., 1969), molecular structure-function (Gysin, 1960, 1971), comprehensive (Gysin and Knüsli, 1960), and mode of action (Ashton, 1965). Volume 32 of *Residue Reviews* contains 15 chapters dealing with triazine-soil interactions.

Growth and Plant Structure

The triazine herbicides have been shown to inhibit the growth of all organs of intact plants. This has generally been attributed to a deficiency of photosynthate which is necessary for growth and is caused by a blockage of photosynthesis. Atrazine has been shown to inhibit the growth of chlorella (Gramlich and Frans, 1964; Ashton et al., 1966); however, this growth

310

TABLE 19-1. Common Name, Chemical Name, and Chemical Structure of the Triazine Herbicides

Common name	Chemical name	R_1	R_2	R_3
Ametryne	2-(ethylamino)-4-(isopropylamino)-6-(methylthio)-s-triazine	—SCH$_3$	—NH·iso-C$_3$H$_7$	—NHC$_2$H$_5$
Atratone	2-(ethylamino)-4-(isopropylamino)-6-methoxy-s-triazine	—OCH$_3$	—NH·iso-C$_3$H$_7$	—NHC$_2$H$_5$
Atrazine	2-chloro-4-(ethylamino)-6-(isopropylamino)-s-triazine	—Cl	—NH·iso-C$_3$H$_7$	—NHC$_2$H$_5$
Chlorazine	2-chloro-4,6-*bis*(diethylamino)-s-triazine	—Cl	—N(C$_2$H$_5$)$_2$	—N(C$_2$H$_5$)$_2$
Desmetryne	2-(isopropylamino)-4-(methylamino)-6-(methylthio)-s-triazine	—SCH$_3$	—NH·iso-C$_3$H$_7$	—NHCH$_3$
Ipazine	2-chloro-4-(diethylamino)-6-(isopropylamino)-s-triazine	—Cl	—NH·iso-C$_3$H$_7$	—N(C$_2$H$_5$)$_2$
Prometone	2,4-*bis*(isopropylamino)-6-methoxy-s-triazine	—OCH$_3$	—NH·iso-C$_3$H$_7$	—NH·iso-C$_3$H$_7$
Prometryne	2,4-*bis*(isopropylamino)-6-(methylthio)-s-triazine	—SCH$_3$	—NH·iso-C$_3$H$_7$	—NH·iso-C$_3$H$_7$
Propazine	2-chloro-4,6-*bis*(isopropylamino)-s-triazine	—Cl	—NH·iso-C$_3$H$_7$	—NH·iso-C$_3$H$_7$
Simazine	2-chloro-4,6-*bis*(ethylamino)-s-triazine	—Cl	—NHC$_2$H$_5$	—NHC$_2$H$_5$
Simetone	2,4-*bis*(ethylamino)-6-methoxy-s-triazine	—OCH$_3$	—NHC$_2$H$_5$	—NHC$_2$H$_5$
Simetryne	2,4-*bis*(ethylamino)-6-(methylthio)-s-triazine	—SCH$_3$	—NHC$_2$H$_5$	—NHC$_2$H$_5$
Terbutryn	2-(*tert*-butylamino)-4-(ethylamino)-6-(methylthio)-s-triazine	—SCH$_3$	—NH·*tert*-C$_4$H$_9$	—NHC$_2$H$_5$
Trietazine	2-chloro-4-(diethylamino)-6-(ethylamino)-s-triazine	—Cl	—N(C$_2$H$_5$)$_2$	—NHC$_2$H$_5$

inhibition could be overcome by the addition of glucose to the culture media (Ashton et al., 1966). However, at subtoxic concentrations certain triazine herbicides have been shown to stimulate growth. Lorenzoni (1962) showed that the rate of growth and dry weight accumulation in maize increased with increasing concentrations of simazine from 0.25 to 2.0 mg/dmg. Freney (1965) reported that 1.0 ppm simazine in soil increased the dry weight and growth of maize tops by 27% when measured 29 days after planting. The addition of nitrogen was required to obtain these results in nitrogen-deficient soils.

Lorenzoni (1962) reported that low concentrations of simazine accelerated the germination of maize. *Chenopodium humile* seed germination was also increased by increasing concentrations of simazine from 1 ppm upward (Jordan and Day, 1968). Tas (1961) did not observe any effect of simazine on germination of several species in the concentration range of 10^{-6} to 10^{-2}M.

Although foliar chlorosis followed by necrosis is the usual phytotoxic symptom of the triazines, increased chlorophyll content of leaves of certain species has also been reported. The chlorophyll content of leaves of maize, millet, and oats germinated on filter paper treated with simazine or atrazine increased initially (one-leaf stage) but later decreased to the level of the controls (Mastakov and Prohorcik, 1962). However, when treated at the one-leaf stage there was a decrease in chlorophyll content of the leaves of millet and wheat; oat was unchanged. Glabiszewski et al. (1966) reported that the chlorophyll content of leaves of oat seedlings treated with atrazine, simazine, or propazine decreased after 9 days. Oats germinated in soil containing 6 kg/ha of propazine, prometryne, or prometone produced leaves with a chlorophyll content lower than the controls (Glabiszewski et al., 1967). Foy and Bisalputra (1964) reported that sublethal concentrations of prometryne caused thicker and darker green leaves in cotton. Goren and Monselise (1965) treated two soil types in which citrus was growing with 0.8 and 1.6 kg/ha of atrazine or simazine in November. The effect was measured the following summer. In the heavy soil, simazine increased the chlorophyll content of the leaves whereas atrazine decreased their chlorophyll content. In the light soil, both herbicides increased the chlorophyll content of the leaves. Freeman et al. (1966) reported that when they applied atrazine, 3.34 and 5.62 kg/ha, or simazine, 2.25 and 5.62 kg/ha, to soil in which raspberries were grown, atrazine reduced the chlorophyll content of the leaves but simazine had no effect. Agha and Anderson (1967) treated soil containing several different types of fruit trees with 4 to 12 lb/A of simazine over a three-year period. Chlorophyll in both sour cherry and apple leaves increased at the 4-pound rate, a 28% increase was reported for sour cherry. At the highest level of simazine the chlorophyll content of the leaves decreased in both of these species. Sour cherry and apple were considered resistant to simazine. However, two susceptible species, sweet cherry and *Prunus mahaleb*, showed a decrease in chlorophyll content of the leaves at all levels of simazine. Unfortunately the herbicide content of the leaves was not determined in any of these studies. However, it would appear that relatively low concentrations of the triazine herbicides in leaves are associated with elevated chlorophyll content, intermediate concentrations with no change, and high concentrations with chlorosis and necrosis. The darker green color of resistant crops treated with the triazine herbicides has frequently been observed.

Walker and Zelitch (1963) reported that atrazine prevents the opening of stomata in the light, induces closure of already open stomata in the light to the same aperture by inhibiting the opening reaction.

Ashton et al. (1963a) made a detailed study of the histological changes in bean induced by atrazine. They found (1) the vacuolation of cells of developing leaves was accelerated in plants treated with atrazine or placed in the dark as compared to control plants kept in the light, (2) the chloroplasts of developing leaves and mature primary leaves disintegrated in plants treated with atrazine and kept in the light but remain unchanged in plants placed in the dark with or without atrazine, (3) the airspace-system of the mature primary leaves was reduced in plants treated with atrazine and kept in the light but not in plants placed in the dark with or without atrazine, (4) the integrity of the ectoplast and tonoplast was modified in plants treated with atrazine and kept in the light but not in plants placed in the dark with or without atrazine, and (5) atrazine caused a cessation of cambial activity and a lesser thickness of the cell walls of sieve and tracheary elements of the stem of plants kept in the light but not of plants placed in the dark with or without atrazine (Figure 19-1). They suggested that those changes which occurred in atrazine-treated plants in the light but did not occur in atrazine-treated or control plants in the dark were not caused by atrazine *per se* or by a block in photosynthesis but rather by the combined effects of atrazine and light. They further proposed that this interaction caused the production of a secondary toxic substance (or substances), perhaps a "free radical(s)."

Foy and Bisalputra (1964) studied the effect of prometryne on leaves of cotton and reported that at first the leaves became thickened as a result of increased development of the spongy mesophyll with concomitant reduction of intercellular spaces followed by disorganization and breakdown of chloroplasts and finally collapse of the mesophyll cells.

Effects of atrazine on the fine structure of chloroplasts was investigated in bean by Ashton et al. (1963b) and in barnyardgrass (*Echinochloa crusgalli*) by Hill et al. (1968). Ashton et al. (1963b) reported that the fine structure of the chloroplasts was grossly altered by treatment with atrazine when the plants were kept in the light, but atrazine had no effect in the dark. The following progressive changes occurred: (1) they became spherical rather than discoid, (2) starch disappeared from the lamellar system, (3) the frets or parts of

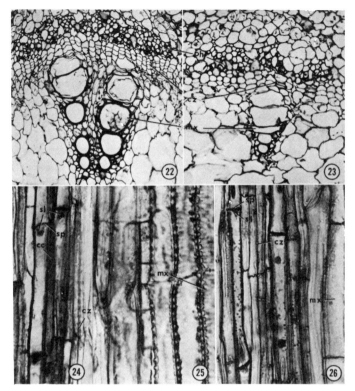

FIGURE 19-1.   Vascular tissues of stems of untreated and atrazine-treated plants. Plates 22, 23, transections of stems of
untreated and treated plants, respectively. Plates 24, 25, longi-
sections of stems of untreated plant showing phloem and xylem,
respectively. Plates 26, longisection of small portion of vascular
bundle of treated plant showing phloem and xylem: cc, companion
cell; cz, cambial zone; mx, metaxylem; ph, phloem; sl, slime
plug; sp, sieve plate. (Ashton et al., 1963a.)

frets were destroyed, leading to the disorganization of the granal arrangement, (4) the compartments of the grana swelled, and (5) the chloroplast
envelope and the swollen compartments ultimately disintegrated (Figure
19-2). They proposed that these changes were brought about by the formation
of a toxic substance or substances involving the interaction of atrazine and
light in the presence of chlorophyll. Hill et al. (1968), on studies with barnyardgrass, reported similar results except that the alteration of the chloroplast
occurred much sooner than in bean and furthermore they showed that the
ultrastructural changes preceded any macroscopic symptoms.

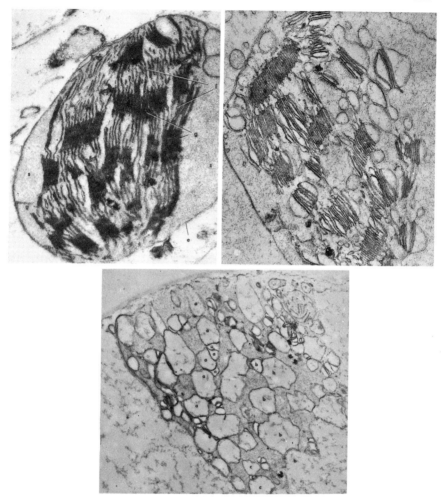

FIGURE 19-2. Effect of atrazine on the ultrastructure of bean chloroplasts. Control (upper left), moderate injury (upper right), and severe injury (lower).

Ashton et al. (1966) treated a synchronous culture of *Chlorella vulgaris* with atrazine, atrazine plus glucose, or glucose. Atrazine stopped growth and chlorophyll formation; these were counteracted by glucose. Atrazine did not cause any observable abnormalities in cell organelles including chloroplasts. Perhaps the proposed phytotoxic substance was leached from the cells into the culture medium and therefore did not accumulate in the chloroplasts in sufficient quantity to cause injury.

The cytogenetic effects of atrazine to grain sorghum was studied by Liang et al. (1967). The herbicide was applied to the soil at the rate of 2.7 kg/ha and the pollen mother cells collected at several development stages. Many microsporocytes were affected. Abnormalities consisted of multinucleate cells, bridges, and increased chromosome numbers, some of which were not multiples of the basic number. Dyads and quartets often contained micronuclei. Atrazine apparently interfered with meiotic stability. There was no effect on yield.

Absorption and Translocation

Over 50 reports have been published on the absorption and/or translocation of the triazine herbicides; only a limited number of these can be cited. The absorption of atrazine by roots of intact soybean plants from an aqueous solution was shown to consist of two phases (Vostral et al., 1970). There was a rapid initial uptake period which occurred within the first 30 minutes followed by a slow continuous uptake, Figure 19-3. Herbicide absorption rates during the first 30 minutes were 10-fold or more, greater than those observed later in the 24-hour period. The rate of absorption was also shown to increase with increasing root temperatures and herbicide concentrations. These results suggest that the rapid initial uptake may have been movement into the apoplast of the roots. It may be that a rapid initial absorption of

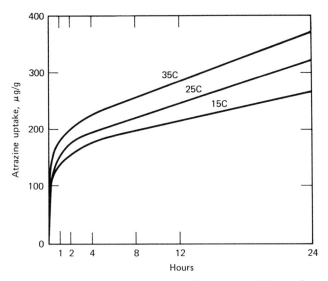

FIGURE 19-3. Atrazine uptake from 0.04-mM atrazine solution at three root temperatures for 24 hr expressed as μg per g dry wt of plants. (Vostral et al., 1970.)

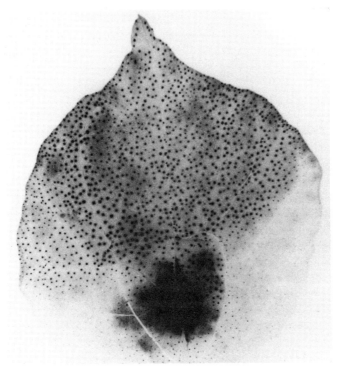

FIGURE 19-4. Autoradiogram of a cotton leaf showing pronounced accumulation of ^{14}C in the lysigenous glands of cotton 72
hours after drop treatment with 10 μl (23.4 μg, 0.05 μc) of
ipazine-2,4,6-^{14}C. Also note the wedge-shaped pattern of
radioactivity distal to the point of application resulting from
apoplastic movement during transpiration. (Foy, 1964.)

many herbicides could be demonstrated if workers used much shorter initial
sampling times than are usually used in time-course studies. Davis et al.
(1959a) reported that simazine was rapidly absorbed by roots of maize,
cotton, and cucumber and translocated to the leaves. Some radioactivity was
detected in the leaves of cucumber within $\frac{1}{2}$ hour. Autoradiographs showed
rather uniform distribution of ^{14}C in maize, accumulation at the edges of the
leaves of cucumber, and accumulation in the lysigenous glands of cotton
leaves. Foy (1964) observed the accumulation of ipazine and other triazines
in the lysigenous glands of cotton, Figure 19-4. He suggested that this
accumulation phenomenon was dependent upon the degree of lipid solubility
of the compound. Sikka and Davis (1968) reported the accumulation of
prometryne in the lysigenous glands of cotton leaves. Davis et al. (1959a) also

provided evidence that the differential patterns of distribution of simazine in leaves of maize, cotton, and cucumber were due in part to the differential rates of simazine degradation in the three species and the autoradiographs showed the distribution of the degradation products and simazine rather than simazine *per se*. Schneider (1959) stated in a discussion paper that simazine is taken up by the roots of plants and translocated upward to the leaves and that all of the triazines are taken up by roots of plants. Additional evidence that shows that the triazine herbicides are readily absorbed by roots and translocated via the apoplastic system to the shoots of many plants may be found in the following papers: (1) simazine-(Montgomery and Freed, 1961; Foy and Castelfranco, 1961; Sheets, 1961; Grover, 1962; Plaisted and Ryskiewich, 1962; Biswas, 1964; Foy, 1964; Lund-Hoie, 1969), (2) atrazine-(Montgomery and Freed, 1961; Minshall, 1962; Biswas, 1964; Foy, 1964; Davis et al., 1965; Wax and Behrens, 1965; Vostral and Buchholtz, 1967), (3) propazine-(Foy and Castelfranco, 1961; Grover, 1962; Biswas, 1964; Foy, 1964; Biswas and Hemphill, 1965), (4) prometone-(Biswas, 1964; Foy, 1964), (5) propmetryne-(Sikka and Davis, 1968), (6) trietazine-(Foy, 1964), and (7) ipazine-(Foy, 1964).

In certain grasses, such as oats, which do not readily degrade the triazines, there is progressive accumulation of the herbicide from the tip to the base of the blade following root applications. In most broadleaf species accumulation of the triazine herbicides occurs at the margins of the leaves following root applications. The development of the phytotoxic symptoms follow these same patterns. Such distribution patterns are characteristic of apoplastic translocation.

Several workers have shown that the rate of absorption and translocation of the triazine herbicides from roots to shoots is proportional to amount of water absorbed and/or the transpiration rate (Sheets, 1961; Foy and Castelfranco, 1961; Grover, 1962; Davis et al., 1965; Wax and Behrens, 1965; Vostral et al., 1970; Lund-Hoie, 1969). This evidence supports the concept of apoplastic translocation of the triazine herbicides.

Atrazine has been shown to reduce the transpiration rate (Wills et al., 1963; Smith and Buchholtz, 1962, 1964; Sikka et al., 1964; Graham and Buchholtz, 1968). The site of atrazine inhibition of transpiration appears to be in the leaf and to be associated with a high carbon dioxide concentration in the stomatal cavity. High carbon dioxide concentrations in the stomatal cavity are probably caused by an inhibition of photosynthesis by atrazine. The inhibition of transpiration results in a decrease in atrazine absorption and translocation. Shimabukuro and Linck (1967) agree that within the first few days the inhibition of absorption and translocation by atrazine is probably due to its effect on transpiration; however, thereafter they suggest that the inhibition may be due to reduced absorption by the roots brought about by a reduction in carbohydrate concentration, resulting in injury to root tissues.

Graham and Buchholtz (1968) reported that when roots of intact plants were exposed to an atrazine solution and subsequently placed in water, the roots and stem acted as a reservoir for continued atrazine movement to the leaves for at least 8 hours after removal of the plants from the atrazine solution. Some leakage of atrazine out of the roots into the water medium also occurred. This leakage provides additional evidence for the involvement of apparent free space of roots in atrazine absorption.

An application of potassium nitrate or urea to the soil of detopped tomato plants increased the rate of exudation from the stumps over three-fold and increased the concentration of atrazine in this augmented exudate from 9 to 40%. Atrazine applied to the soil was detected within 10 minutes and approached its maximum concentration by the end of 3 hours; potassium nitrate accelerated the increase in concentration. Increasing the soil temperature from 10° to 30°C increased the rate of exudation, and potassium nitrate or urea increased the concentration of atrazine. Similar results were found for atratone, propazine, and prometone and the concentration of the herbicides in the exudate increased directly with their solubility in water (Minshall, 1969).

Although the resistance of certain species of higher plants to the triazine herbicides has usually been attributed to their ability to rapidly degrade the herbicide to non-toxic forms, at least two reports have suggested that absorption and translocation may be involved in some cases. Freeman et al. (1964) reported that the resistant species white pine (*Pinus strobus*) contains only one-third as much radioactivity in its needles following the application of ^{14}C-simazine to soil as the susceptible species red pine (*Pinus resinosa*). The ratio of distribution of ^{14}C between the roots and the top was 1 to 1.4 in red pine and 1 to 8.4 in white pine. Furthermore, it was suggested that mycorrhizae may also enhance the resistance of white pine to simazine, since the absorption of ^{14}C in non-inoculated plants was more than double that of inoculated plants. Unfortunately degradation of the herbicide was not determined in these studies; however, Dhillon et al. (1968) showed that simazine was degraded to a limited extent by red pine seedlings. So it is still possible that differential rates of degradation of simazine may be involved in this selectivity. Sikka and Davis (1968) studied the differential absorption, translocation, and metabolism of prometryne by cotton, a resistant species, and soybean, a susceptible species. They attributed the resistance of cotton to poor translocation, accumulation in the lysigenous glands, and metabolism. Two days following root treatment with the herbicide 63% of the radioactivity was in the roots and 37% in the shoots, whereas with soybeans 23% was in the roots and 77% in the shoots.

Research on the absorption and translocation of the triazine herbicides following foliar application has been fairly limited as compared to studies of root application. Davis et al. (1959a) reported that almost no absorption of simazine occurred through intact leaves of maize, cotton, or cucumber; however, simazine did enter when the cuticle was broken. No basipetal translocation of atrazine occurred following its application to leaves of yellow nutsedge (*Cyperus esculentus*) (Donnalley and Rahn, 1961). The very comprehensive study by Foy (1964) of the foliar absorption and translocation of simazine, atrazine, propazine, trietazine, ipazine, and prometone by maize, soybean, tomato, cotton, oat, and sorghum showed: (1) foliar penetration and acute toxicity appeared to be directly correlated with the water solubility of the compound, (2) there was no appreciable movement of any of the herbicides downward or out of the treated leaf in any of the species, (3) a surfactant tended to enhance their activity, and (4) acropetal (apoplastic) movement of all compounds occurred readily. These data strongly reinforce our concept of the almost exclusive apoplastic translocations of the triazine herbicides.

Considerable interest developed during the mid-1960's for the use of atrazine as a postemergence spray for weeds in maize and sorghum. Since atrazine is not translocated from treated leaves, a foliar application is essentially a contact spray unless the herbicide reaches the soil, is absorbed by the roots, and translocated throughout the plant. Although atrazine has some postemergence activity, the use of additives in the formulation increases this activity markedly (Ilniki et al., 1965). Dexter et al. (1966) evaluated the effect of twenty surfactants in combination with atrazine for the control of large crabgrass (*Digitaria sanguinalis*) in sorghum. Surface tension was not the main factor governing the phytotoxicity of foliar-applied atrazine solutions. Translocation of atrazine from sorghum and large crabgrass leaves was very slight or non-existent, even when 1% Surfactant WK (dodecyl ether of polyethylene glycol) was added to the atrazine solution. Therefore, surfactants appear to increase the phytotoxicity of atrazine by increasing foliar absorption. Non-phytotoxic oils have also been used to increase the foliar activity of atrazine. Thompson and Slife (1969) suggest that the foliar application of atrazine plus non-phytotoxic oil or surfactant to giant foxtail (*Setaria faberii*) requires root absorption as well as foliar absorption for complete control. The foliarly-absorbed atrazine weakens the plant by killing the expanded leaves but not those unrolled, thereby making the plant more easily killed by root-absorbed atrazine. The foliar-applied atrazine is washed into the soil by rainfall and absorbed primarily by the coronal roots.

The absorption and translocation of the triazine herbicides by aquatic plants has also been investigated. Simazine was absorbed by both roots and shoots of the submerged waterstargrass (*Heteranthera dubia*) and translocated to the untreated portion (root or shoot) of the plant (Funderburk and Lawrence, 1963a). However, ametryne and prometryne were translocated only slightly from shoot to root and no movement from root to shoot could be detected in this species (Funderburk and Lawrence, 1963b). In the emersed alligator weed (*Alternanthera philoxeroides*) they found little movement of foliar-applied simazine, ametryne, or prometryne but these compounds were readily absorbed by the roots and translocated to the shoots. The translocation of simazine in curlyleaf pondweed (*Potamogeton crispus*) appears to be quite different; Sutton and Bingham (1968) reported that it did not move out of the roots but did show slight basipetal movement from foliar treatment. In the non-vascular aquatic plant *Chara vulgaris* very little movement of simazine occurred (Evrard, 1967). These results show that the translocation of the triazine herbicides in aquatic plants varies greatly with species and the particular compound under consideration.

Haskell and Rogers (1960) reported that the distribution of simazine in seeds following a 12- to 72-hour immersion in the herbicide solution varied with time and species. In maize it was found in the cotyledon, radical, and epicotyl after 48 hours; in soybean it was distributed throughout the seed within 12 hours; and no simazine entered the seed of jimsonweed (*Datura stromonium*). Birdsrape mustard (*Brassica rapa*) seeds were treated with an aqueous suspension of atrazine for 0.58 to 36 hours (Hocombe, 1968). Dry seeds and seeds preimbibed in water were used. After the herbicide treatment period the soaking solutions were analyzed for atrazine and the seeds grown in untreated soil. Hocombe concluded that the initial uptake of atrazine was independent of water uptake and was primarily confined to the seed coat. Reider and Buchholtz (1970) determined the uptake of atrazine by soybean seeds. They found that there was a direct proportional relationship between uptake and concentration; raising the temperature from $10°$ to $30°C$ increased the uptake. Uptake rates were similar in living and dead seeds. The concentration of atrazine within the seed was 6.7 times greater than that of the external solution. They concluded that the uptake was largely a physical process.

Molecular Fate

The molecular transformations of the triazine herbicides by chemical, physical, and biological means has recently been reviewed by Knuesli et al. (1969). We shall deal here only with those shown to occur in higher plants.

There are two degradative reactions of the triazine herbicides which have been demonstrated to occur in higher plants: (1) hydroxylation of the number two position with dechlorination, demethoxylation, or demethylthiolation depending on the parent substitution; (2) dealkylation of the alkyl side chains with their subsequent oxidation. Cleavage of the triazine ring may occur in higher plants; however, the research results are not in agreement on this point.

Roth (1957) was the first to observe that simazine was rapidly degraded; he used expressed maize sap. After 15, 39, and 100 hours; 55, 82, and 97% of the added simazine was degraded. Heating the sap to 80°C for 2 hours destroyed its ability to degrade the herbicide. It was therefore concluded that the degradative factor was thermolabile. He also reported that although sap from maize, a resistant species, rapidly degraded simazine, the sap from a susceptible species, wheat, was inactive. He used chemical isolation and spectrophotometric methods, whereas most subsequent workers used radioactive tracer techniques and chromatography. Following the Roth (1957) report several laboratories worked simultaneously on the characterization of maize-resistant factor. Montgomery and Freed (1960) suggested that one of the first intermediates in the degradation of atrazine was hydroxyatrazine. Castelfranco et al. (1961) reported that an extract of maize was able to convert simazine to the corresponding 2-hydroxy analogue. The active constituent was dialysable, soluble in 90% acetone, extractable with ether and ethyl acetate, and totally destroyed by ashing. In a crude extract the active constituent was destroyed by boiling; these results agree with those of Roth (1957). However, after purification by acetone precipitation the activity of the active constituent was not decreased by boiling. They concluded that the maize-resistant factor was not a protein (enzyme). They were not able to find this resistant factor in oats, a susceptible species. Roth and Knulsi (1961) and Hamilton and Moreland (1962) independently characterized the maize-resistant factor as 2,4-dihydroxy-3-keto-7-methoxy-1,4-benzoxazine (benzoxazine) which is originally present in the plant in the form of its 2-glucoside (I).

$$CH_3O\text{—}\underset{\displaystyle}{\bigcirc}\text{—}OC_6H_{11}O_5$$

(I)

Castelfranco and Brown (1962) showed that pyridine and hydroxylamine were capable of converting simazine to hydroxysimazine. Although the rate of chemical hydrolysis of the chlorotriazines to their corresponding hydroxy derivatives can be directly correlated to the concentration of benzoxazine there are some complications in considering it the only mechanism for carrying out this reaction. For example, grain sorghum and johnsongrass (*Sorghum halepense*) have been shown to contain the hydroxy derivatives but appear to be devoid of benzoxazine. Moreover, benzoxazine has been detected in susceptible species. Therefore, it appears that although the benzoxazine hydrolysis system is an important degradation mechanism for forming 2-hydroxy derivatives from the triazine herbicide other significant mechanisms may be operative (Negi et al., 1964; Funderburk and Davis, 1963; Hamilton, 1964a, 1964b; Palmer and Crogan, 1965).

Many studies have shown the *in vivo* hydrolysis of the chlorotriazines (simazine, atrazine, ipazine, and propazine) to their respective hydroxy forms by a variety of species (Hamilton and Moreland, 1962, 1963; Montgomery and Freed, 1964; Plaisted and Thornton, 1964; Negi et al., 1964; Hamilton, 1964a, b; Hurter, 1966; Lund-Hoie, 1969). However, Dhillon et al. (1968) were not able to detect the presence of hydroxysimazine from red pine seedlings treated with simazine and Shimabukuro et al. (1966) and Shimabukuro (1968) were not able to detect hydroxyatrazine from peas or sorghum treated with atrazine. It appears that the 2-hydroxylation of the chlorotriazine herbicides is not universal for all plants.

The benzoxazine system of maize does not appear to be active on the methylthiotriazines (Montgomery and Freed, 1964; Mueller and Payot, 1965; Mueller, unpublished); however, the corresponding hydroxy derivative has been reported to be present in a number of species (Montgomery and Freed, 1964; Mueller, 1966; Plaisted and Thornton, 1964; Whitenberg, 1965; Barba, 1967; Sikka and Davis, 1968).

Degradation of the triazine herbicides by higher plants may also involve the substitutions at carbon atoms 4 and 6. The delayed appearance of the phytotoxic symptoms following the application of chlorazine to soil followed by an oat bioassay in a time-course study was suggested to be the result of the conversion of chlorazine to trietazine and/or simazine (Sheets et al., 1962). If one ethyl group was lost from one amino group trietazine would be formed; whereas, if one ethyl group was lost from both amino substituents, simazine would be the product. Trietazine and simazine are much more phytotoxic in soil applications than chlorazine. Ipazine caused no apparent damage to red pine for 80 days; however, in the next 20 days toxicity symptoms developed very rapidly (Kozlowski, 1965). These results may be explained by the conversion of ipazine to atrazine by the removal of one ethyl group from the amino group. In both of these examples, the molecular conversions may have been brought about by microorganisms in the soil, rather than the higher plant, but they could conceivably occur in higher plants as well. Direct proof for the oxidative degradation of the side chains has been provided by the use of molecules labeled with ^{14}C in the side chains. Funderburk and Davis (1963) reported that maize, cotton, and soybeans yielded appreciable amounts of [14]CO_2 when grown in solutions containing side-chain-labeled simazine. Mueller and Payot (1965) found up to 70% of the ^{14}C from side-chain-labeled atrazine released as [14]CO_2 by maize plants. This demonstrates rapid and extensive N-dealkylation and oxidation of the substituents at carbon atoms 4 and 6.

Additional evidence for N-dealkylation of the triazine herbicides by higher plants has been the identification of the dealkylated triazine molecule. Mueller and Payot (1965) showed that pea plants were able to remove one of the isopropyl groups from prometryne leaving the rest of the molecule intact. Shimabukuro et al. (1966) found that the predominant degradation product of atrazine in peas was 2-chloro-4-amino-6-isopropylamino-*s*-triazine (ethyl group removed) and trace amounts of 2-chloro-4-ethylamino-6-amino-*s*-triazine (isopropyl group removed) were also found (Shimabukuro, 1967a). These two degradation products of atrazine were also detected in soybean, wheat, maize, and sorghum (Shimabukuro, 1967b). Sorghum dealkylated either *N*-alkyl side chain of atrazine readily, but peas, soybean, and wheat predominantly dealkylated the ethyl alkyl group. These two intermediates are phytotoxic but less toxic than atrazine. Shimabukuro (1968) reported that the metabolic products of atrazine in maize and sorghum were not the same. In maize, atrazine was metabolized via both the 2-hydroxylation and N-dealkylation pathways, while sorghum metabolized atrazine via the N-dealkylation pathway only. Hydroxyatrazine was also N-dealkylated by both maize and sorghum. The formation of 2-hydroxy-4-ethylamino-6-amino-*s*-triazine from simazine by maize has been observed by Montgomery et al. (1968). They also found that this metabolite could be formed from hydroxysimazine by maize and cotton. Hurter (1966, 1967) reported that simazine is completely N-dealkylated and identified 2-hydroxy-4,6-*bis*(amino)-*s*-triazine as degradation products in experiments with *Coix lacrima jobi* and cogongrass (*Imperata cylindrica*).

Roeth and Lavy (1971) found that sudangrass and sorghum metabolized atrazine primarily to 2-chloro-4-amino-6-(isopropylamino)-*s*-triazine and 2-chloro-4-amino-6-(ethylamino)-*s*-triazine which are only partially detoxified compounds. Maize metabolized atrazine to 2-hydroxy-4-(ethylamino)-6-(isopropylamino)-*s*-triazine which is non-phytotoxic. Schultz and Tweedy (1971) studied atrazine degradation using the ^{14}C-ethyl-labeled compound. More incorporation of the ^{14}C-ethyl side chain into free amino acids was found in the roots and shoots of a resistant maize species than in a susceptible one. More radioactivity was found in roots of susceptible plants than in resistant ones; this trend was reversed in the shoots. Thompson et al. (1971) found the order of tolerance of five monocot species (maize > fall panicum and large crabgrass > giant foxtail > oats) to be identical with their ability to metabolize atrazine. Six hours after treatment these five species metabolized 96, 44, 50, 17, and 2%, respectively; of the ^{14}C-atrazine absorbed from a 10 ppm solution and translocated to the foliage, leaving concentrations of 2.2, 34.8, 30.1, 59.8, and 66.3 mμmoles of atrazine per gram of fresh weight of shoots. Oliver and Schreiber (1971) showed that two new varieties of green foxtail, robust white and robust purple, are the most tolerant to atrazine and propazine of five foxtails studied. Shimabukuro and Swanson (1969) have reviewed atrazine degradation in various species as related to selectivity.

The cleavage of the triazine ring may also occur in higher plants; however, several reports suggest that it does not. Most of those reports which indicate any cleavage show that it proceeds at a relatively slow rate. It is also conceivable that if the ring is indeed cleaved the fragments are rapidly incorporated into insoluble natural products which would not be detected by the usual analytical techniques used.

Ring cleavage experiments have involved the use of ^{14}C-ring-labeled compounds and measurement of evolved [14]CO_2. Several species have been used including maize, sorghum, oats, peas, soybeans, and cotton. Generally less than 2.5% of the absorbed radioactivity was released as [14]CO_2 within a week from simazine, atrazine, or propazine (Davis et al., 1959b; Foy and Castelfranco, 1960; Ragab and McCollum, 1961; Montgomery and Freed, 1961; Foy, 1962; Funderburk and Davis, 1963; Montgomery and Freed, 1964; Mueller and Pagot, 1965). It is not clearly evident that all of these workers used a suitable control, testing for the carryover of the intact molecule to the [14]CO_2 fractions in the absence of the plant. Funderburk and Davis (1963) indicated that they encountered considerable difficulty because radioactivity was collected from the control. They contributed this to ^{14}C-simazine which was somewhat volatile and ultimately contaminated their $BaCO_3$. However, Montgomery and Freed (1961) did not find this a problem in their studies with simazine and atrazine. Two other experimental variables may alter the results from such studies; these are light and CO_2 concentration. The release of [14]CO_2 from ^{14}C-ring-labeled simazine by maize in the dark was about 7 times greater than when the plants were grown in the light (Montgomery and Freed, 1961), Table 19-2. However, with atrazine it was only less than twice as high in the dark as in the light. In part, the increased release of CO_2 in the dark may be due to its lack of fixation by photosynthesis. However, it must also be remembered that these compounds inhibit CO_2 fixation in the light. Foy (1962) reported that there was no measurable amount of [14]CO_2 released from sorghum treated with ring-labeled propazine when carbon dioxide free air was passed over the plants but a considerable amount was released when regular air was used. These results suggest that all of the [14]CO_2 from the metabolized triazine ring is immediately fixed by the plant in the absence of exogenous CO_2, but does not rule out the possibility that exogenous CO_2 is required for the metabolism of the triazine ring. However, Montgomery and Freed (1961) used CO_2 free air in their studies of the metabolism of ^{14}C-ring-labeled simazine and atrazine by maize and found that [14]CO_2 was released, Table 19-2.

There have been a number of reports in which [14]CO_2 release has not been detected following treatment with ^{14}C-ring-labeled triazine herbicides. Some

TABLE 19-2. Metabolism of ^{14}C-Ring-Labeled Simazine and Atrazine by Corn Plants as Measured by the Release of $^{14}CO_2$ (Montgomery and Freed, 1961)

Experiment	Duration, hrs	μg triazine metabolized	μg triazine remaining in plant	% triazine metabolized of that taken up
simazine 1	50	0.48		
2	48	0.19	2.00	8.7
3 (in dark)	68	0.91	0.53	63.2
atrazine 1	54	0.67	3.05	18.0
2	72	0.30	1.49	16.8
3 (in dark)	70	0.28	0.73	27.7

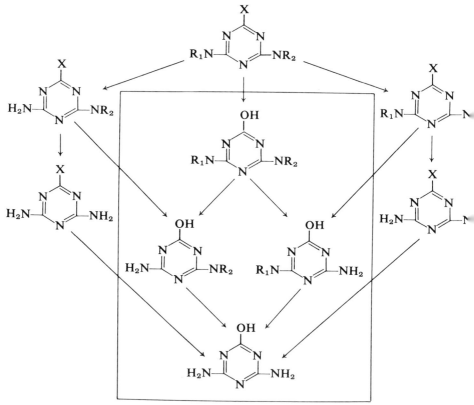

FIGURE 19-5. Proposed pathway for the degradation of triazine herbicides in higher plants. The total scheme is for those plants which have both the 2-hydroxy and N-dealkylation degradation systems. The scheme outside of the box is for those plants which have only the N-dealkylation system.

of these are: atrazine in maize (Davis et al., 1965; Shimabukuro, 1967b), atrazine in sorghum (Shimabukuro, 1967b), ipazine in cotton (Hamilton and Moreland, 1963), prometryne in cotton (Whitenberg, 1965), prometryne in carrots (Montgomery and Freed, 1964), and prometryne in maize (Montgomery and Freed, 1964).

A proposed pathway for the degradation of triazine herbicides in higher plants is presented in Figure 19-5.

Biochemical Response

The inhibition of photosynthesis by the triazine herbicides is a well-established fact. It has been demonstrated by many workers using various techniques in isolated chloroplasts, unicellular algae, and intact higher plants. Indirect evidence of a photosynthetic block was provided by Gast (1958) when he found that starch synthesis in leaves of *Coleus blumei* was blocked by simazine but the addition of sucrose counteracted this effect. Exer (1958) and Moreland et al. (1958, 1959) showed that simazine inhibited the Hill reaction in isolated chloroplasts. Additional studies by Exer and associates provided the data for Table 19-3 which shows the effect of several triazines and other herbicides on the Hill reaction and Table 19-4 which shows the effects of other triazines and possible triazine metabolites on the Hill reaction (Gysin and Knusli, 1960). Moreland and Hill (1962) also reported several other triazines inhibited in the Hill reaction. Additional research on the effect of the triazines on the Hill reaction have been provided by Exer (1961), Good (1961), Bishop (1962), Gagarina et al. (1968), and Shimabukuro and Swanson (1968).

Moreland and Hill (1962) showed that the inhibition of the Hill reaction by simazine was reversible. That is, when the chloroplasts were treated with 6×10^{-5}M or 4×10^{-6}M simazine the inhibition was 50 and 85%, respectively; however, when they were washed following the herbicide treatment, values of 96 and 94%, respectively, were obtained. They also observed that the degree of inhibition of the Hill reaction by chloroplasts treated with simazine from various species (maize, turnip greens, soybeans) was similar although their *in vivo* susceptibility to the toxic action of simazine is quite different. Therefore, the selective action of simazine on these species cannot result from differences in the chloroplasts.

Bishop (1962) suggests that although inhibition of the Hill reaction nearly always implies a poisoning of the mechanism for oxygen production, more rigorous tests are required to make this explanation certain. One such relatively easy test can be made with algae adapted to do photoreduction with hydrogen. In *Scenedesmus*, and other hydrogenase-containing algae, oxygen evolution can be practically eliminated by specific inhibitors of this step while

TABLE 19-3. Effects of 2-halo-4,6-*bis*(alkylated amino)-*s*-triazines and Some Known Herbicides in the Inhibition of the Hill Reaction (Exer, from Gysin and Knusli, 1960)

Z	X	Y	V^a
Cl	NH_2	NH_2	(10^{-4}M → 0–10% inhibition)
Cl	$NHCH_3$	$NHCH_3$	0.002–0.01
Cl	NHC_2H_5	NHC_2H_5	1
Br	NHC_2H_5	NHC_2H_5	2–2.4
Cl	$NH \cdot iso\text{-}C_3H_7$	$NH \cdot iso\text{-}C_3H_7$	1.03–3
Br	$NH \cdot iso\text{-}C_3H_7$	$NH \cdot iso\text{-}C_3H_7$	1.8
Cl	$NH \cdot n\text{-}C_3H_7$	$NH \cdot n\text{-}C_3H_7$	0.4–0.67
Br	$NH \cdot n\text{-}C_3H_7$	$NH \cdot n\text{-}C_3H_7$	1.7–2
Cl	$NH \cdot n\text{-}C_4H_9$	$NH \cdot n\text{-}C_4H_9$	(10^{-4}M → 0 inhibition)
Cl	NH_2	NHC_2H_5	0.2
Cl	$NHCH_3$	$NH \cdot iso\text{-}C_3H_7$	0.27–0.44
Cl	NHC_2H_5	$NH \cdot iso\text{-}C_3H_7$	1.8–2.1
Cl	NHC_2H_5	$NH \cdot n\text{-}C_4H_9$	3.6–3.8
Cl	NHC_2H_5	$N(C_2H_5)_2$	0.0025–0.008
Cl	$NH \cdot iso\text{-}C_3H_7$	$N(C_2H_5)_2$	0.013–0.048
Cl	$N(C_2H_5)_2$	$N(C_2H_5)_2$	0.021–0.04
Cl	NH—⟨ ⟩—Cl	NH—⟨ ⟩—Cl	< 0.001

3-(3′-chlorophenyl)-1,1-dimethylurea (monuron)	0.7–1
3-(3′,4′-dichlorophenyl)-1,1-dimethylurea (diuron)	4.1–6
2,4-dichlorophenoxyacetic acid (2,4-D)	(10^{-4}M → 0 inhibition)
isopropyl-*N*-(3-chlorophenyl)-carbamate (CIPC)	0.0067–0.0075
maleic hydrazide (MH)	(10^{-4}M → 10% inhibition)
aminotriazole (AT)	(10^{-4}M → 0 inhibition)

^{a} V = concentration of simazine giving 50% inhibition divided by the concentration of test substance giving 50% inhibition. Concentration of simazine giving 50% inhibition = 7×10^{-7}M.

TABLE 19-4.   Effects of Some Triazines and Possible Triazine Metabolites in the Inhibition of the Hill Reaction (Gysin and Knusli, 1960)

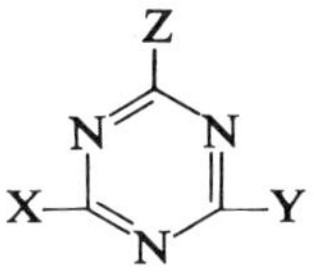

$$
\begin{array}{c}
Z \\
\diagup \diagdown \\
N \quad N \\
X \diagdown \diagup Y \\
N
\end{array}
$$

Z	X	Y	V^a
OCH_3	NHC_2H_5	NHC_2H_5	1.3–1.5
SCH_3	NHC_2H_5	NHC_2H_5	4.5–5
OCH_3	$NH \cdot iso\text{-}C_3H_7$	$NH \cdot iso\text{-}C_3H_7$	0.44–0.62
SCH_3	$NH \cdot iso\text{-}C_3H_7$	$NH \cdot iso\text{-}C_3H_7$	6.1–9.1
OCH_3	$NHCH_3$	$NH \cdot iso\text{-}C_3H_7$	0.18–0.21
SCH_3	$NHCH_3$	$NH \cdot iso\text{-}C_3H_7$	3.9–4.4
OCH_3	NHC_2H_5	$NH \cdot iso\text{-}C_3H_7$	1.3–1.5
SCH_3	NHC_2H_5	$NH \cdot iso\text{-}C_3H_7$	6.1
H	NHC_2H_5	NHC_2H_5	(10^{-4}M → 0 inhibition)
CH_3	NHC_2H_5	NHC_2H_5	0.93
C_2H_5	NHC_2H_5	NHC_2H_5	0.18–0.22
$iso\text{-}C_3H_7$	NHC_2H_5	NHC_2H_5	0.23
$n\text{-}C_3H_7$	NHC_2H_5	NHC_2H_5	0.15
$n\text{-}C_4H_9$	NHC_2H_5	NHC_2H_5	0.0043
C_2H_5	$NH \cdot iso\text{-}C_3H_7$	$NH \cdot iso\text{-}C_3H_7$	0.20–0.29
NHC_2H_5	NHC_2H_5	NHC_2H_5	0.056–0.13
$N(C_2H_5)_2$	$N(C_2H_5)_2$	$N(C_2H_5)_2$	0.019–0.025
OH	NHC_2H_5	NHC_2H_5	(10^{-4}M → 0 inhibition)
OH	NHC_2H_5	$NH \cdot iso\text{-}C_3H_7$	0.0033–0.0077

[a] V = concentration of simazine giving 50% inhibition divided by the concentration of test substance giving 50% inhibition. Concentration of simazine giving 50% inhibition = 7×10^{-7}M.

the primary photochemistry and carbon dioxide reduction may continue at a relatively high rate; and furthermore, these specific inhibitors prevent the deadaptation which normally occurs at higher light intensities. The effect of simazine on such a system is shown in Figure 19-6a,b. Equal inhibition was produced at all light intensities, Figure 19-6a. Except for what appears to be a small inhibition of photoreduction, the effects of simazine disappear after adaptation to hydrogen, Figure 19-6b. These results are consistent with the concept that the triazine herbicides inhibit the oxygen-evolving system of photosynthesis.

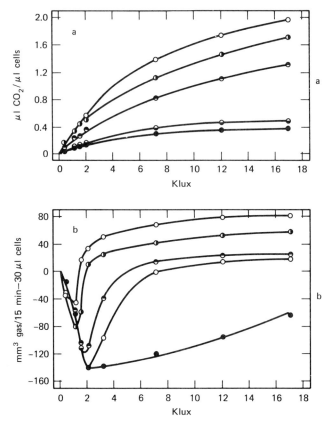

FIGURE 19-6a.   Rates of photosynthesis at various light intensities for *Scenedesmus* D_3 with and without simazine. Phosphate
buffer (*p*H 6.5); gas phase, $H_2 - 4\%$ CO_2; palladium catalyst in
center well of each vessel; temperature, $25°$; cell volume, 25
μl/vessel. Rate measurements, 10 min at each light intensity. ○,
control; ◑, 6.6×10^{-7} M simazine; ◕, 3.3×10^{-6} M simazine;
◑, 6.6×10^{-6} M simazine; ●, 6.6×10^{-5} M simazine.
(Bishop, 1962.)

FIGURE 19-6b.   Same as in Figure 19-6a except the palladium
catalyst was removed from center wells, vessels reflushed with
$H_2 - 4\%$ CO_2, and a dark incubation period of 16 hr allowed
prior to illumination. (Bishop, 1962.)

Good (1961) reported that ATP formation (photophosphorylation) was inhibited by the triazine herbicides in isolated chloroplasts when flavin mononucleotide (FMN) was used as the electron acceptor but not when N-methylphenazonium (PMS) was used as the electron acceptor. This has been confirmed by Shimabukuro and Swanson (1968). Bishop (1962) reported that simazine inhibits FMN-catalyzed photophosphorylation but not that catalyzed by vitamin K_3. This is extremely interesting since it is our only example of significant difference between the mechanism of action of the triazine and urea herbicides; diuron inhibits both FMN- and vitamin K_3-catalyzed photophosphorylation (Avron, 1960). Oxygen evolution by *Elodea* has been shown to be inhibited by simazine (Roth, 1958).

The blockage of light reactions of photosynthesis by the triazine herbicides should result in a lack of reducing power which is required for carbon dioxide fixation. The inhibition of carbon dioxide fixation by several triazine herbicides in a variety of plant species has been reported. Ashton and Zweig (1958) found that simazine inhibited carbon dioxide fixation by beans. The effect of simazine, trietazine, and simetone on CO_2 fixation by beans was investigated by Ashton et al. (1960). They found that these triazines drastically inhibited CO_2 fixation in the light. They reported that 40 hours after applying simazine to the roots of bean plants, using sand culture, CO_2 fixation was decreased by about 30% at 0.25 ppm and was almost completely blocked at 1 ppm. In a time-course study at 1 ppm, simazine inhibited CO_2 fixation about 35% at 3 hours but the trietazine and simetone were not inhibitory. However, after 6 hours the inhibition was about 95, 80, and 50% for simazine, trietazine, and simetone, respectively; at 24 hours about 98, 95, and 90%, respectively; at 48 hours CO_2 was almost completely blocked with all three herbicides. The following triazine herbicides have been reported to inhibit CO_2 fixation: simetone in beans (van Oorschot, 1964), atrazine in beans (Funderburk and Carter, 1965), atrazine in maize, cotton, and soybeans (Couch and Davis, 1966), and prometryne in cotton and soybeans (Sikka and Davis, 1969). However, in the resistant species (atrazine in maize and prometryne in cotton) the degree of inhibition was less than in the susceptible species.

Although the triazine herbicides inhibit CO_2 fixation in the light they do not inhibit non-photosynthetic CO_2 fixation (Zweig and Ashton, 1962; Couch and Davis, 1966; Sikka and Davis, 1969).

The effect of the triazine herbicides on several biochemical events other than photosynthesis have been reported. However, it is often not clear how these are related to the herbicidal properties of these compounds.

Respiration has been reported to be increased, decreased, or not to be affected by the triazine herbicides in various systems. However, most of the reports indicated an inhibition. In intact plants, atrazine (Funderburk and Davis, 1963; Olech, 1966), and ipazine (Nasyrova et al., 1968) have been reported to inhibit respiration. Sinzar (1967) observed that simazine inhibited respiration in tomato but stimulated it in maize. Davis (1968) and Olech (1967) reported that atrazine had no effect on respiration. Atrazine did not affect the rate of respiration of excised embryos of red kidney bean plants (Ashton and Uribe, 1962). In excised barley roots using various substrates, Palmer and Allen (1962) observed a general trend of stimulation of respiration. Olech (1966) concluded that the inhibition of respiration was indirect and caused by a lowered level of assimilates as a result of the inhibition of photosynthesis. With isolated mitochondria, Foy and Penner (1965) reported that atrazine inhibited respiration but Davis (1968) observed no effect. Lotlikar et al. (1968) reported that simazine uncoupled oxidative phosphorylation and Voinilo et al. (1967) obtained similar results with atrazine. Voinilo et al. (1968) suggested that the site of action was between cytochrome b and c. Thompson et al. (1969) reported that both state 3 and 4 of respiration were inhibited by prometryne but state 3 was inhibited the most. Prometryne caused a decrease in second state 3 oxidation of both malate and succinate, but the decrease was overcome by 2,4-dinitrophenol indicating a response similar to that of oligomycin (McDaniel and Frans, 1969).

Several investigators have reported that the triazine herbicides reduce the amount of glucose, fructose, and/or sucrose in a variety of species (Glabiszewski et al., 1966; Swietochowski et al., 1966; Ploszynski and Zurawski, 1967; Glabiszewski et al., 1967; Timofeeva, 1967). A reduction in sugars would be one of the first effects one would expect to observe from photosynthesis inhibitors. The metabolism of ^{14}C-sucrose by excised leaves of beans treated with atrazine was investigated by Ashton and Uribe (1962). They found that atrazine caused a marked increase in the relative radioactivity of aspartic acid and glutamic acid, while that of serine, alanine, and glyceric acid decreased. Malic acid and citric acid were unaffected by the atrazine treatment. However, when photosynthesis was blocked by placing the plants in the dark rather than treating with atrazine, there was no effect on sucrose metabolism. They concluded that atrazine not only blocks photosynthesis but also interferes with other metabolic processes. In similar studies using ^{14}C-serine they came to the same conclusion.

Nitrogen metabolism changes induced by the triazine herbicides are influenced by the amount of herbicide applied, the amount and form of nitrogen supplied to the plant, environmental growth conditions, the species of plant, and in certain cases whether the data are calculated on a percent by weight or per plant basis. This problem has been investigated rather thoroughly since some of the research has suggested that subtoxic levels of the triazine herbicides increase growth and the amount of nitrogen and/or protein in the plant. However, in spite of the several investigations which have been conducted on this problem there are still differences of opinion on whether the triazine herbicides increase the amount of nitrogen in plants. This points out the complexity of the problem and the resulting difficulty of interpreting the results.

Bartley (1957) reported that simazine increased the growth and green color of maize. Gast and Grob (1960) observed that atrazine or simazine increased the protein content of maize. DeVries (1963) demonstrated an increased nitrogen uptake by maize when treated with simazine and those plants showing nitrogen deficiency symptoms recovered when simazine was applied. Apple and peach trees treated with simazine plus amitrole-T (amitrole plus ammonium thiocyanate) grew more and had higher levels of leaf nitrogen than trees actually supplied with nitrogen fertilizer (Ries et al., 1963).

Ries and Gast (1965) treated maize plants grown in nutrient solutions with three levels of simazine, three levels of nitrogen as ammonium nitrate, and various combinations of these. In one experiment, where the growth conditions (low light and temperature) were not optimal, simazine at 0.5 and 3.0 ppm increased the percent of nitrogen in the shoots and total milligrams of nitrogen regardless of the nitrogen level in the solution, Table 19-5. The simazine treatments also caused a reduction of the typical nitrogen deficiency symptoms which occurred at the two lower nitrogen levels in the solution in the absence of simazine. However, as a second experiment, under more favorable growth conditions, the total milligrams of nitrogen was not increased by simazine, although the percent of nitrogen in the shoots was

TABLE 19-5.   Effect of Simazine on Growth and N Content of Maize (Ries and Gast, 1965)

Treatments		Total dry weight, g[b]	Shoot/root ratio[b]	N in shoot, %	Total N in plants, mg[c]
N applied, mg[a]	simazine level, ppm				
190	0.0	7.7	3.3	1.50	106
190	0.5	6.7	5.2	3.40	201
190	3.0	4.5	6.3	3.90	163
760	0.0	8.7	5.2	2.46	192
760	0.5	6.7	7.8	4.38	285
760	3.0	5.6	7.5	4.68	249
3040	0.0	4.6	6.5	3.22	188
3040	0.5	4.6	6.3	6.50	282
3040	3.0	2.0	6.1	7.52	141

[a] In addition, approximately 90 mg of N was available from the 3 seeds planted.
[b] F value for linear effect of simazine and quadratic effect of N significant at 1% level.
[c] F value for quadratic effect of simazine significant at 1% level.

TABLE 19-6. Effect of Sublethal Concentrations of Atrazine on the Dry Weight, Percent Nitrogen, and Total Nitrogen of Maize, Cotton, and Soybean 7, 14, and 21 Days After Initiation of Treatment (Eastin and Davis, 1967)

Species	Atrazine, ppm	Dry weight, mg per plant	%N Shoot	%N Root	Mg N per plant Shoot	Root[b]	Total
				7 Days			
maize	0	1.58	3.97[a]	3.61[a]	44.0[b]	17.5	61.5[b]
	4	0.80	4.81[ab]	3.97	26.2[a]	10.3	36.5[a]
	8	0.38	5.44[b]	3.86[b]	12.4[a]	5.8	18.2[a]
cotton	0	0.45	5.26[a]	4.58[b]	19.7[b]	3.2	22.9[b]
	0.1	0.51	5.44[a]	4.78[c]	32.2[b]	4.3	27.5[b]
	0.25	0.21	6.92[b]	4.28[a]	12.4[b]	1.3	13.7[a]
soybean	0	0.70	5.32[a]	4.36[a]	29.2[b]	6.5	35.7[b]
	0.025	0.51	5.50[ab]	4.57[b]	22.6[ab]	4.8	27.2[ab]
	0.05	0.33	6.20[b]	4.92[c]	16.7[a]	2.9	19.7[a]
				14 Days			
maize	0	5.21	3.65[a]	3.38[a]	124.0[b]	61.8	186.0[b]
	4	3.24	3.86[a]	3.89[b]	90.0[b]	35.8	126.0[b]
	8	1.08	4.94[b]	3.84[b]	36.0[a]	13.1	49.0[a]

cotton	0	2.01	4.92^a	4.31^a	86.0^b	11.2	97.0^b
	0.1	1.88	5.02^{ab}	4.52^{ab}	82.0^b	11.3	93.0^b
	0.25	0.42	5.56^b	4.80^b	19.0^a	2.9	22.0^a
soybean	0	2.51	4.48^a	3.58^a	101.0^b	10.7	116.0^b
	0.025	1.79	4.64^a	3.63^a	69.0^b	10.6	80.0^b
	0.05	0.63	4.96^b	4.42^b	25.0^a	4.9	30.0^a

21 Days

maize	0	10.7	4.08^a	3.71^a	319.0^b	105.0	424.0^b
	4	11.5	4.23^a	4.06^b	362.0^b	122.0	484.0^b
	8	4.4	4.70^b	4.00^b	119.0^a	74.0	193.0^a
cotton	0	5.6	4.85^b	4.56^a	285.0^b	34.0	319.0^b
	0.1	5.7	4.60^a	4.61^a	224.0^b	36.0	260.0^b
	0.25	1.4	5.00^b	4.54^a	61.0^a	10.0	72.0^a
soybean	0	11.1	4.38^a	3.58^a	409.0^c	64.0	473.0^c
	0.025	6.5	4.36^a	3.89^b	236.0^b	43.0	279.0^b
	0.05	3.2	4.41^a	4.40^c	111.0^a	29.0	140.0^a

[a] Each value is an average of two replications. Values in a column within a single species and age followed by a common letter are not significantly different as determined by Duncan's new multiple-range test at the 5% level of protection.
[b] Roots of all plants for a given treatment were composited prior to chemical analysis; thus, no statistical analysis of these data was possible.

increased at the lower nitrogen level in the solution. These results suggest that increased levels of nitrogen in plants as a result of subtoxic amounts of simazine are most likely to be demonstrable at low levels of available nitrogen and under adverse climatic conditions.

Several other investigators have reported increases in various nitrogen components in a variety of species with several triazine herbicides (Goren and Monselise, 1965; Zurawski et al., 1965; Glabiszewski et al., 1966, 1967; Swietochowski et al., 1966). While others have reported no increases or reductions (Vorob'er and Ch'A, 1960; Chesalin and Timofeeva, 1968; Singh and West, 1967); Gramlich and Davis (1967). Solecka et al. (1969) observed that simazine initially increased the protein content of apple leaf tissue but as the season advanced the effect progressively diminished and by late June was non-existent. Lin and Anderson (1967) found that although the percent of total nitrogen, protein nitrogen, and total soluble nitrogen increased in apple seedlings treated with simazine, the increase was not due to an increase in nitrogen absorption but rather to a decrease in other constituents.

Eastin and Davis (1967) conducted an extensive series of experiments on the effect of atrazine on nitrogen metabolism of seven different species. The plants were grown in the field or in a phytotron in flats of soil or culture solution. Under all three growing conditions, whenever the percent of nitrogen was increased by the atrazine treatment, there was a corresponding decrease in dry weight. The nitrogen content on a per plant basis was unchanged or decreased by atrazine. These observations were true whether the plants were resistant, intermediate, or susceptible to atrazine. Atrazine apparently brought about an increase in percent of nitrogen by a reduction in growth of the plant rather than an increase in nitrogen uptake, Table 19-6. It is assumed that in these studies the level of nitrogen available to the plants and the climatic conditions were favorable enough so that it was not possible to demonstrate that atrazine caused an increase in the nitrogen content of the plants on a per plant basis.

Additional studies by Ries and associates have given support to the point of view that subtoxic levels of the triazine herbicides increase growth and nitrogen content of certain species (Tweedy and Ries, 1967; Ries et al., 1967; Schweizer and Ries, 1969). This appears to be associated with an increase in nitrate reductase activity; it occurs in plants grown on nitrate as the nitrogen source but not ammonia; and the magnitude of the response decreases as the nitrate concentration approaches optimal levels.

The effect of the triazines on nitrogen metabolism in higher plants may be summarized as follows. Many studies have shown that these herbicides increase the percent of several nitrogen fractions in plants. Although some studies have demonstrated that there is actually an increase in the amount of the nitrogen components on a per plant basis others have not. Frequently, the increase in the percent of nitrogen fractions has been attributed to a decrease in dry weight rather than a real increase in nitrogen. Apparently, it is not easy to unequivocally demonstrate that subtoxic levels of the triazine herbicides increase the actual amount of nitrogen compounds present on a per plant basis. Certain specific conditions seem to be required.

Mode of Action

The triazine herbicides have usually been shown to inhibit the growth of intact plants and this has been attributed to the blockage of photosynthesis. However, at subtoxic rates a stimulation of growth has also been reported.

The usual phytotoxic symptom of the triazine herbicides is foliar chlorosis followed by necrosis; however, increased greening of the leaves has also been shown in some species. Apparently very low concentrations of the herbicide in the leaves cause increased greening, while intermediate levels cause no change and high concentrations cause chlorosis and necrosis.

Histological studies have shown that atrazine causes accelerated vacuolation of the cells of developing leaves, reduced airspace-system of mature leaves, modified integrity of cell membranes in leaves, cessation of cambial activity, and a lesser thickness of the cell walls of sieve and tracheary elements of the stem. Prometryne was reported to cause leaves to become thickened as a result of increased development of the spongy mesophyll with concomitant reduction of intercellular spaces. Cytological investigations have shown that the fine structure of the chloroplasts is grossly altered by atrazine.

The absorption of atrazine by roots of intact soybean plants has been shown to involve a rapid initial uptake, presumably into apparent free space, followed by a slow continuous uptake. The translocation of the triazine herbicides to the shoots following root application is very rapid, usually detectable within $\frac{1}{2}$ hour or less. Increased rates of transpiration increase the translocation. The triazine herbicides are not translocated from treated leaves. There is general agreement that this class of herbicides is almost exclusively translocated via the apoplastic system.

The rate of degradation of the triazine herbicides in higher plants varies greatly with different species. In resistant species they are degraded rapidly, while in susceptible species they are degraded slowly. This is the basis for their selective use in several crops, as well as the fact that a few weed species are not controlled. Two degradative reactions of the triazines have been demonstrated to occur in higher plants. One of these reactions involves hydroxylation of the ring in the number two position with dechlorination, demethoxylation, or demethylthiolation. The other reaction is a dealkylation of the alkyl side chain(s) in the four and/or six position. Cleavage of the triazine ring may occur in higher plants; however, the research results are not in agreement on this point.

The mechanism of action of the triazine herbicides in higher plants is a blockage of photosynthesis. More specifically their site of action is within photosystem II at the photolysis of water step. However, their herbicidal activity involves more than a mere blocking of photosynthesis. The plants do not simply die from starvation due to a lack of photosynthate. The phytotoxic symptoms of the triazine herbicides are not typical of starvation and they occur more rapidly than could be accounted for by lack of photosynthate. It appears that a reaction which is probably coupled to the photolysis of water results in the formation of a secondary phytotoxic agent which is primarily responsible for their herbicidal properties.

REFERENCES

Agha, J. T. and J. L. Anderson. 1967. Effects of simazine upon pigments, soluble solids, and some oxidative enzymes of apple and cherry. *Proc. 21st West. Weed Control Conf.* pp. 44–45.

Antognini, J. and B. E. Day. 1955. Geigy 444, a preemergence and post-emergence herbicide for cotton. *Proc. 8th Southern Weed Control Conf.* pp. 92–98.

Ashton, F. M. 1965. Physiological, biochemical, and structural modifications of plants induced by atrazine and monuron. *Proc. 18th Southern Weed Control Conf.* pp. 596–602.

Ashton, F. M., T. Bisalputra, and E. B. Risley. 1966. Effect of atrazine on *Chlorella vulgaris. Am. J. Bot.* **53**:217–219.

Ashton, F. M., E. M. Gifford, and T. Bisalputra. 1963a. Structural changes in *Phaseolus vulgaris* induced by atrazine. 1. Histological changes. *Bot. Gaz.* **124**: 329–335.

Ashton, F. M., E. M. Gifford, and T. Bisalputra. 1963b. Structural changes in *Phaseolus vulgaris* induced by atrazine. 2. Effects on fine structure of chloroplasts. *Bot. Gaz.* **124**:336–343.

Ashton, F. M. and E. G. Uribe. 1962. Effect of atrazine on ^{14}C-sucrose and ^{14}C-serine metabolism. *Weeds* **10**:295–297.

Ashton, F. M. and G. Zweig. 1958. The effect of simazin, 2-chloro-4,6-*bis*(ethylamino)-*s*-triazine, on ^{14}CO$_2$ fixation in excised leaves of red kidney bean. *Plant Physiol. Abstr.* **33**:xxvi–xxvii.

Ashton, F. M., G. Zweig, and G. W. Mason. 1960. The effect of certain triazines on ^{14}CO$_2$ fixation in red kidney beans. *Weeds* **8**:448–451.

Avron, M. 1960. Photophosphorylation by swiss-chard chloroplasts. *Biochim. Biophys. Acta* **40**:257–272.

Barba, R. C. 1967. The selectivity and activity of *s*-triazine herbicides in banana plants. Ph.D. Thesis, University of Hawaii. 120 pp.

Bartley, C. E. 1957. Simazine and related triazines as herbicides. *Agr. Chem.* **12**:34–36, 113–115.

Bishop, N. I. 1962. Inhibition of the oxygen-evolving system of photosynthesis by amino-triazines. *Biochim. et Biophys. Acta* **57**:186–189.

Biswas, P. K. 1964. Absorption, diffusion, and translocation of ^{14}C-labeled triazine herbicides by peanut leaves. *Weeds* **12**:31–33.

Biswas, P. K. and D. D. Hemphill. 1965. Role of growth regulators in the uptake and metabolism of *s*-triazine herbicide by tea leaves. *Nature* **207**:215–216.

Castelfranco, P. and M. S. Brown. 1962. Purification and properties of the simazine-resistance factor of *Zea mays*. *Weeds* **10**:131–136.

Castelfranco, P., C. L. Foy, and D. B. Deutsch. 1961. Non-enzymatic detoxification of 2-chloro-4,6-*bis*(ethylamino)-*s*-triazine (simazine) by extracts of *Zea mays*. *Weeds* **9**:580–591.

Chesalin, G. A. and A. A. Timofeeva. 1968. The effect of atrazine on some characteristics of metabolism in sensitive plants. *Agrokhimiya* **5**:108–113.

Couch, R. W. and D. E. Davis. 1966. Effect of atrazine, bromacil, and diquat on $^{14}CO_2$-fixation in corn, cotton, and soybeans. *Weeds* **14**:251–255.

Davis, D. E. 1968. Atrazine effects on mitochondrial respiration. *Proc. 21st Southern Weed Conf.* p. 346.

Davis, D. E., H. H. Funderburk, and N. G. Sansing. 1959a. The absorption and translocation of ^{14}C-labeled simazin by corn, cotton, and cucumber. *Weeds* **7**:300–309.

Davis, D. E., H. H. Funderburk, and N. G. Sansing. 1959b. Absorption, translocation, degradation, and volatilization of radioactive simazine. *Proc. 12th Southern Weed Conf.* pp. 172–173.

Davis, D. E., J. V. Gramlich, and H. H. Funderburk. 1965. Atrazine absorption and degradation by corn, cotton, and soybeans. *Weeds* **13**:252–255.

De Vries, M. L. 1963. The effect of simazine on Monterey pine and corn, as influenced by lime, bases, and aluminum. *Weeds* **11**:220–222.

Dexter, A. G., O. C. Burnside, and T. L. Lavy. 1966. Factors influencing the phytotoxicity of foliar applications of atrazine. *Weeds* **14**:222–228.

Dhillon, P. S., W. R. Byrnes, and C. Merritt. 1968. Simazine distribution and degradation in red pine seedlings. *Weed Sci.* **16**:374–376.

Donnalley, W. F. and E. M. Rahn. 1961. Translocation of amitrole, atrazine, dalapon, and EPTC in northern nutgrass. *Proc. 15th Northeast Weed Control Conf.* p. 46.

Eastin, E. F. and D. E. Davis. 1967. Effects of atrazine and hydroxyatrazine on nitrogen metabolism of selected species. *Weeds* **15**:306–309.

Evrard, T. O. 1967. Movement of ^{32}P, ^{35}S, ^{45}Ca, and four ^{14}C-labeled products in *Chara vulgaris* L., a non-vascular aquatic plant. Ph.D. Thesis, Virginia Polytechnic Inst. 83 pp.

Exer, B. 1958. Über Pflanzenwachstumsregulatoren: Der Einfluss von Simazin auf den Pflanzenstoffwechsel. *Experientia* 13:136–137.

Exer, B. 1961. Inhibition of the Hill reaction by a herbicide. *Weed Res.* 1:233–244.

Foy, C. L. 1962. Detoxification of triazine herbicides by tolerant plant species. *Proc. 14th Calif. Weed Conf.* pp. 82–86.

Foy, C. L. 1964. Volatility and tracer studies with alkylamino-*s*-triazines. *Weeds* 12:103–108.

Foy, C. L. and T. Bisalputra. 1964. Anatomical effects of prometryne on leaves of glanded and glandless cotton. *Plant Physiol. Abstr.* 39:lxviii.

Foy, C. L. and P. Castelfranco. 1960. Distribution and metabolic fate of ^{14}C-labeled 2-chloro-4,6-*bis*(ethylamino)-*s*-triazine (simazine) and four related alkylamino triazines in relation to phytotoxicity. *Plant Physiol. Abstr.* 35:xxviii.

Foy, C. L. and D. Penner. 1965. Effect of inhibitors and herbicides on tricarboxylic acid cycle substrate oxidation by isolated cucumber mitochondria. *Weeds* 13:226–231.

Freeman, J. A., A. J. Renney, and H. Driediger. 1966. Influence of atrazine and simazine on leaf chlorophylls and fruit yield of raspberries. *Can. J. Pl. Sci.* 46:454–455.

Freeman, F. W., D. P. White, and M. J. Bukovac. 1964. Uptake and differential distribution of ^{14}C-labeled simazine in red and white pine seedlings. *Forest Sci.* 10:330–334.

Freney, J. R. 1965. Increased growth and uptake of nutrients by corn plants treated with low levels of simazine. *Aust. J. Agr. Res.* 16:257–263.

Funderburk, H. H. and M. C. Carter. 1965. The effect of amitrole, atrazine, dichlobenil, and paraquat on the fixation and distribution of ^{14}CO$_2$ in beans. *Proc. 18th Southern Weed Control Conf.* p. 607.

Funderburk, H. H. and D. E. Davis. 1963. The metabolism of ^{14}C-chain- and ring-labeled simazine by corn and the effect of atrazine on plant respiratory systems. *Weeds* 11:101–104.

Funderburk, H. H. and J. M. Lawrence. 1963a. Preliminary studies on the absorption of ^{14}C-labeled herbicides in fish and aquatic plants. *Weeds* 11:217–219.

Funderburk, H. H. and J. M. Lawrence. 1963b. Absorption and translocation of radioactive herbicides in submersed and emerged aquatic weeds. *Weed Res.* 3:304–311.

Gagarina, M. I., A. S. Kupriyanova, I. A. Mel'nikova, and Yu. A. Baskakov. 1968. Herbicidal derivatives of hydroxylamine. 8. The effect of methoxyamino-*s*-triazines on the Hill reaction. *Agrokhimiya* 5:114–117.

Gast, A. 1958. Über Pflanzenwachstumsregulatoren. Beiträge zur Kenntmis der phytotoxischen Wirkung von Triazinen. *Experientia* 13:134–136.

Gast, A. 1970. Use and performance of triazine herbicides on major crops and major weeds throughout the world. *Residue Rev.* 32:11–18.

Gast, A. and M. Grob. 1960. Triazines as selective herbicides. *Pest Tech.* 3:68–73.

Gast, A., E. Knuesli, and H. Gysin. 1955. Über Pflanzenwachstumsregulatoren: Chlorazin, eine phytotoxisch wirksame Substanz. *Experientia* 11:107–108.

Glabiszewski, J., M. Ploszynski, G. Szumilak, and H. Zurawski. 1966. Influence of s-triazine herbicides on dynamics of some biochemical processes of oat in test experiments. Part I. Researches on the action of simazine, atrazine, propazine, and prometryne. *Pam. Pulaw.* 21:233–257.

Glabiszewski, J., M. Ploszynski, G. Szumilak, and H. Zurawski. 1967. Influence of s-triazine herbicides on dynamics of some biochemical processes of oat in test experiments. 2. Research on the effect of propazine, prometryne, and prometone. *Pam. Pulaw.* 28:45–61.

Good, N. E. 1961. Inhibitors of the Hill reaction. *Plant Physiol.* 36:788–803.

Goren, R. and S. P. Monselise. 1965. Some physiological effects of triazines on citrus trees. *Plant Physiol. Abstr.* 40:xv.

Graham, J. C. and K. Buchholtz. 1968. Alteration of transpiration and dry matter with atrazine. *Weed Sci.* 16:389–392.

Gramlich, J. V. and D. E. Davis. 1967. Effect of atrazine on nitrogen metabolism of resistant species. *Weeds* 15:157–160.

Gramlich, J. V. and R. E. Frans. 1964. Kinetics of chlorella inhibition by herbicides. *Weeds* 12:184–189.

Grover, R. 1962. Uptake and distribution of ^{14}C-simazine and ^{14}C-propazine in some tree seedlings. *Plant Physiol.* 37:12–13.

Gysin, H. 1960. The role of chemical research in developing selective weed control practices. *Weeds* 4:541–559.

Gysin, H. 1971. The chemical structure and biological relationship of s-triazines. *Second Internat. Cong. of Pesticide Chem.* (In press).

Gysin, H. and E. Knusli. 1960. Chemistry and herbicidal properties of triazine derivatives, pp. 289–358. In R. L. Metcalf, Ed. *Advances in Pest Control Research.* Interscience Publishers, New York.

Hamilton, R. H. 1964a. A corn mutant deficient in 2,4-dihydroxy-7-methoxy-1,4-benzoxazin-3-one with an altered tolerance of atrazine. *Weeds* 12:27–30.

Hamilton, R. H. 1964b. Tolerance of several grass species to 2-chloro-s-triazine herbicides in relation to degradation and content of benzoxazinone derivatives. *J. Agr. Food Chem.* 12:14–17.

Hamilton, R. H. and D. E. Moreland. 1962. Simazine: Degradation by corn seedlings. *Science* 135:373–374.

Hamilton, R. H. and D. E. Moreland. 1963. Fate of ipazine in cotton plants. *Weeds* 11:213–217.

Haskell, D. A. and B. J. Rogers. 1960. The entry of herbicides into seeds. *Proc. 17th N. Cent. Weed Control Conf.* p. 39.

Hill, E. R., E. C. Putala, and J. Vengris. 1968. Atrazine induced ultrastructural changes of barnyardgrass chloroplasts. *Weed Sci.* **16**:377–380.

Hocombe, S. D. 1968. The uptake of atrazine by germinating seeds of turnip (*Brassica rapa*). *Weed Res.* **8**:68–71.

Hurter, J. 1966. Abbauprodukte von Simazin in Gramineen. *Experientia* **22**:741–742.

Hurter, J. 1967. The inactivation of simazine in resistant grasses. *6th Int. Congr. Pl. Prot., Vienna* pp. 398–399.

Ilnicki, R. D., W. H. Tharrington, J. F. Ellis, and E. I. Visinski. 1965. Enhancing directed postemergence treatments in corn with surfactants. *Proc. 19th Northeast Weed Control Conf.* pp. 295–299.

Jordan, L. S. and B. E. Day. 1968. Simazine, gibberellin, and light interactions on seed germination. *WSSA Abstr.* pp. 13–14.

Knüsli, E. 1970. History of the development of triazine herbicides. *Residue Rev.* **32**:1–9.

Knuesli, E., D. Berrer, G. Dupuis, and H. Esser. 1969. *s*-Triazines, pp. 51–78. In P. C. Kearney and D. D. Kaufman, Eds. *Degradation of Herbicides.* Marcel Dekker, New York.

Kozlowski, T. T. 1965. Variable toxicity of triazine herbicides. *Science* **205**: 104–105.

Liang, G. H. L., K. C. Feltner, Y. T. S. Liang, and J. L. Morrill. 1967. Cytogenetic effects and responses of agronomic characters in grain sorghum (*Sorghum vulgare* Pens.) following atrazine application. *Crop Sci.* **7**:245–248.

Lin, M. and J. L. Anderson. 1967. Influence of simazine on nitrogen fractions of apple seedlings at different nitrogen levels. *Proc. 21st West. Weed Control* p. 45.

Lorenzoni, G. G. 1962. The stimulating effect of simazine at high dilutions. *Maydica* **7**:115–124.

Lotlikar, P. D., L. F. Remmert, and V. H. Freed. 1968. Effects of 2,4-D and other herbicides in mitochondria from cabbage. *Weed Sci.* **16**:161–164.

Lund-Hoie, K. 1969. Uptake, translocation, and metabolism of simazine in Norway spruce (*Picea abies*). *Weed Res.* **9**:142–147.

Mastakov, S. M. and R. A. Prohorcik. 1962. An investigation of triazine products as regulators of plant growth. 2. The effect of simazine and atrazine on the content of chlorophyll in the leaves of crop plants. *Dokl. Akad. Nauk Beloruss. SSR* **6**:517–520.

McDaniel, J. L. and R. E. Frans. 1969. Soybean mitochondrial response to prometryne and fluometuron. *Weed Sci.* **17**:192–196.

Minshall, W. H. 1962. The effect of root pressure mechanism on the uptake of pesticides by roots. *Proc. 9th Meet. Agric. Pest. Tech. Soc. Canada* pp. 10–11.

Minshall, W. H. 1969. Effect of nitrogenous materials on the uptake of triazine herbicides. *Weed Sci.* **17**:197–201.

Montgomery, M. L., D. L. Botsford, and V. H. Freed. 1968. The dealkylation of hydroxysimazine by corn plants. *WSSA Abstr.* p. 33.

Montgomery, M. and V. H. Freed. 1960. The metabolism of atrazine by expressed juice of corn. *Res. Prog. Rept., Western Weed Control Conf.* p. 71.

Montgomery, M. and V. H. Freed. 1961. The uptake, translocation, and metabolism of simazine and atrazine by corn plants. *Weeds* **9**:231–237.

Montgomery, M. L. and V. H. Freed. 1964. Metabolism of triazine herbicides by plants. *J. Agr. and Food Chem.* **12**:11–14.

Moreland, D. E. and K. L. Hill. 1962. Interference of herbicides with the Hill reaction of isolated chloroplasts. *Weeds* **10**:229–236.

Moreland, D. E., K. L. Hill, and J. L. Hilton. 1958. Interference with the photochemical activity of isolated chloroplasts by herbicidal materials. *WSSA Abstr.* pp. 40–41.

Moreland, D. E., W. A. Gentner, J. L. Hilton, and K. L. Hill. 1959. Studies on the mechanism of herbicidal action of 2-chloro-4,6-*bis*(ethylamino)-*s*-triazine. *Plant Physiol.* **34**:432–435.

Mueller, P. W. 1966. Biochemical aspects of triazine herbicides. Proc. 463rd Meet. Biochem. Soc., Aberystwyth, Colloq. Biochem. Herb. *Biochem. J.* **101**:1–2.

Mueller, P. W. and P. H. Payot. 1965. Fate of ^{14}C-labeled triazine herbicides in plants. *Proc. IAEA Symp. Isotopes Weed Res., Vienna* pp. 61–69.

Nasyrova, T., Kh. Mirkasimova, and S. Pazilova. 1968. The effect of herbicides on respiration intensity and nitrogen metabolism in cotton and weeds. *Uzbek. Biol. Zh.* **12**:23–26.

Negi, N. S., H. H. Funderburk, and D. E. Davis. 1964. Metabolism of atrazine by susceptible and resistant plants. *Weeds* **12**:53–57.

Olech, K. 1966. Influence of some herbicides on photosynthesis and respiration of crop plants and weeds. 1. Atrazine. *Annals Univ. Mariae Curie—Sklodowska* (E) **21**:289–308.

Olech, K. 1967. The effect of herbicides on photosynthesis and respiration of crop plants and weeds. 2. Atrazine activity in relation to external conditions. *Annals Univ. Mariae Curie—Sklodowska* (E) **22**:149–166.

Oliver, L. R. and M. M. Schreiber. 1971. Differential selectivity of herbicides on six *Setaria* Taxa. *Weed Sci.* **19**:428–431.

Palmer, R. D. and M. S. Allen. 1962. The respiration of excised barley roots in the presence of different levels of simazine and certain substrates. *Proc. 15th Southern Weed Conf.* p. 271.

Palmer, R. D. and C. O. Crogan. 1965. Tolerance of corn lines to atrazine in relation to content of benzoazinone derivative, 2-glucoside. *Weeds* **13**:219–222.

Plaisted, P. H. and D. P. Ryskiewich. 1962. The uptake and metabolism of ^{14}C-labeled simazin by young corn plants. *Plant Physiol. Abstr.* **37**:xxv.

Plaisted, P. H. and M. L. Thornton. 1964. A method for separating some triazine degradation products from plants. *Contrib. Boyce Thompson Inst.* **22**:399–403.

Ploszynski, M. and H. Zurawski. 1967. The compensating action of simazine and atrazine on some meadow species in relation to the content of free amino acids and sugars. *Pam. Pulaw.* **28**:63–77.

Ragab, M. T. H. and J. P. McCollum. 1961. Degradation of ^{14}C-labeled simazine by plants and soil microorganisms. *Weeds* **9**:72–84.

Reider, G. and K. D. Buchholtz. 1970. Uptake of herbicides by soybean seed. *Weed Sci.* **18**:101–105.

Ries, S. K., H. Chmiel, D. R. Dilley, and P. Filner. 1967. The increase in nitrate reductase activity and protein content of plants treated with simazine. *Proc. Natn. Acad. Sci. U.S.* **58**:526–532.

Ries, S. K. and A. Gast. 1965. The effect of simazine on nitrogenous components of corn. *Weeds* **13**:272–273.

Ries, S. K., R. P. Larsen, and A. L. Kenworthy. 1963. The apparent influence of simazine on nitrogen nutrition of peach and apple trees. *Weeds* **11**:270–273.

Roeth, F. W. and T. L. Lavy. 1971. Atrazine translocation and metabolism in sudangrass, sorghum and corn. *Weed Sci.* **19**:98–101.

Roth, W. 1957. Etude comparee de la reaction du Mais et du Ble a la Simazine, substance herbicide. *Compt. Rend.* **245**:942–944.

Roth, H. W. 1958. Substances regulatrices de la croissance vegetale. Etude de l'action de la Simazine sur la physiologie d'Elodea. *Experientia* **14**:137–138.

Roth, W. and E. Knuesli. 1961. Beitrag zur Kenntnis der Resistenzphanomene eingetner Pflanzen gegenuber dem phytotoxischen wirkstoff Simazin. *Experientia* **17**:312–313.

Schneider, F. O. 1959. A discussion of the mode of action, tolerance and soil type effects of the triazines. *Proc. 13th Northeast Weed Control Conf.* pp. 416–420.

Schultz, D. P. and B. G. Tweedy. 1971. Incorporation of atrazine side-chain into amino acids of corn. *Weed Sci.* **19**:133–134.

Schweizer, C. J. and S. K. Ries. 1969. Protein content of seed: increase improves growth and yield. *Science* **165**:73–76.

Sheets, T. J. 1961. Uptake and distribution of simazine by oat and cotton seedlings. *Weeds* **9**:1–13.

Sheets, T. J., A. S. Crafts, and H. R. Drever. 1962. Influence of soil properties on the phytotoxicities of the *s*-triazine herbicides. *J. Agr. and Food Chem.* **10**:458–462.

Shimabukuro, R. H. 1967a. Significance of atrazine dealkylation in root and shoot of pea plants. *J. Agr. and Food Chem.* **15**:557–562.

Shimabukuro, R. H. 1967b. Atrazine metabolism and herbicidal selectivity. *Plant Physiol.* **42**:1269–1276.

Shimabukuro, R. H. 1968. Atrazine metabolism in resistant corn and sorghum. *Plant Physiol.* **43**:1925–1930.

Shimabukuro, R. H., R. E. Kadunce, and D. S. Frear. 1966a. Degradation of atrazine by dealkylation in mature pea plants. *WSSA Abstr.* pp. 37–38.

Shimabukuro, R. H., R. E. Kadunce, and D. S. Frear. 1966b. Dealkylation of atrazine in mature pea plants. *J. Agr. and Food Chem.* **14**:392–395.

Shimabukuro, R. H. and A. J. Linck. 1967. Root absorption and translocation of atrazine in oats. *Weeds* **15**:175–178.

Shimabukuro, R. H. and H. R. Swanson. 1968. Atrazine metabolism, selectivity, and mode of action. *Abstr. Pap. 155th Natn. Meet. Am. Chem. Soc.* A24.

Shimabukuro, R. H. and H. R. Swanson. 1969. Atrazine metabolism, selectivity and mode of action. *J. Agr. and Food Chem.* **17**:199–205.

Sikka, H. C. and D. E. Davis. 1968. Absorption, translocation, and metabolism of prometryne in cotton and soybean. *Weed Sci.* **16**:474–477.

Sikka, H. C., D. E. Davis, and H. H. Funderburk, Jr. 1964. The effect of various types of herbicides on transpiration rate of soybeans (*Glycine max* v. Merr.). *Proc. 17th Southern Weed Conf.* pp. 340–350.

Sikka, H. C. and D. E. Davis. 1969. Effect of prometryne on $^{14}CO_2$ fixation in cotton and soybean. *Weed Sci.* **17**:122–123.

Singh, R. P. and S. H. West. 1967. Influence of simazine on chloroplast ribonucleic acid and protein metabolism. *Weeds* **15**:31–34.

Sinzar, B. 1967. The influence of simazine on the respiration of maize and tomato plants at various growth stages. *Abstr. 6th Int. Congr. Pl. Prot., Vienna* pp. 403–404.

Smith, D. and K. P. Buchholtz. 1962. Transpiration rate reduced in plants with atrazine. *Science* **136**:263–264.

Smith, D. and K. P. Buchholtz. 1964. Modification of plant transpiration rate with chemicals. *Plant Physiol.* **39**:572–578.

Solecka, M., H. Profic, and D. F. Millkan. 1969. Some biochemical effects in apple leaf tissue associated with the use of simazine and amitrole. *J. Am. Soc. Hort. Sci.* **93**:55–57.

Sutton, D. L. and S. W. Bingham. 1968. Translocation patterns of simazine in *Potamogeton crispus* L. *Proc. 22nd Northeast Weed Control Conf.* pp. 357–361.

Swietochowski, B., M. Ptoszynski, and H. Zurawski. 1966. Influence of simazine on changes in content of free amino acids and sugars in seedlings of *Avena sativa* and *Lepidium sativum*. *Pam. Pulaw.* **21**:211–226.

Tas, R. S. 1961. Contribution to the study of the physiological effect of herbicides. 1. The effect of some herbicides on the germination of seed. *Proc. 13th Internat. Symp. Phytofarm. Phytiatrie Gent.* pp. 1465–1477.

Thompson, L. and F. W. Slife. 1969. Foliar and root absorption of atrazine applied postemergence to giant foxtail. *Weed Sci.* **17**:251–256.

Thompson, L., Jr., J. M. Houghton, F. W. Slife, and H. J. Butler. 1971. Metabolism of atrazine by fall panicum and large crabgrass. *Weed Sci.* **19**:409–412.

Thompson, O. C., B. Truelove, and D. E. Davis. 1969. Effect of the herbicide prometryne [2,4-*bis*(isopropylamino)-6-(methylthio)-*s*-triazine] on mitochondria. *J. Agr. and Food Chem.* **17**:997–999.

Timofeeva, A. A. 1967. The effect of atrazine on the dynamics of soluble carbohydrates in mustard. *Agrokhimiya* **4**:123–124.

Tweedy, J. A. and S. K. Ries. 1967. Effect of simazine on nitrate reductase activity in corn. *Plant Physiol.* **42**:280–282.

Van Oorschot, J. L. P. 1964. Some effects of diquat and simetone on CO_2 uptake and transpiration of *Phaseolus vulgaris. Proc. 7th Brit. Weed Control Conf.* pp. 321–324.

Voinilo, V. A., V. P. Deeva, and S. M. Mashtakov. 1967. The change in the oxidative phosphorylation of pea mitochondria under the effects of chemically dissimilar herbicides. *Dokl. Akad. Nauk Belorussk. SSR* **11**:638–642.

Voinilo, V. A., V. P. Deeva, and S. M. Mashtakov. 1968. Qualitative changes in respiratory drain of plant mitochondria under the effect of chemical growth regulators. *Dokl. Akad. Nauk Belorussk. SSR* **12**:460–462.

Vorob'er, F. K. and J. P. Ch'A. 1960. Effect of simazine and 2,4-D on nitrogen metabolism of plants. *Doklady Moskov. Sel'skhorkhoz. Akad. K. A. Timirjozeva* **57**:63–69.

Vostral, H. J. and K. P. Buchholtz. 1967. Root temperature and transpiration influence on atrazine absorption and translocation. *WSSA Abstr.* pp. 39–40.

Vostral, H. J., K. P. Buchholtz, and C. A. Kust. 1970. Effect of root temperature on absorption and translocation of atrazine in soybeans. *Weed Sci.* **18**:115–117.

Walker, D. A. and I. Zelitch. 1963. Some aspects of metabolic inhibitors, temperature, and anaerobic conditions on stomatal movement. *Plant Physiol.* **38**:390–396.

Wax, L. M. and R. Behrens. 1965. Absorption and translocation of atrazine in quackgrass. *Weeds* **13**:107–109.

Whitenberg, D. C. 1965. Fate of prometryne in cotton plants. *Weeds* **13**:68–70.

Wills, G. D., D. E. Davis, and H. H. Funderburk. 1963. The effect of atrazine on transpiration in corn, cotton, and soybeans. *Weeds* **11**:253–255.

Zurawski, H., R. Baranowski, J. Bors, and M. Ploszynski. 1965. Influence of some herbicides on speed of taking up P^{32} by plants. *Symp. Use of Isotopes in Weed Res., Vienna* p. 6.

Zweig, G. and F. M. Ashton. 1962. The effect of 2-chloro-4-ethylamino-6-isoprople-amino-*s*-triazine (atrazine) on distribution of ^{14}C-compounds following $^{14}CO_2$ fixation in excised kidney bean leaves. *Expt. Bot.* **13**:5–11.

CHAPTER 20

Triazoles

During the early 1950's, the American Chemical Paint Co. (presently, Amchem Products, Inc.), supplied experimental samples of amitrole to several investigators to evaluate its potential as a herbicide and defoliant. In 1953, Shaw and Swanson (1953) reported its phytotoxicity to several species and Hall et al. (1953) showed its defoliation characteristics in cotton. Both investigators showed that it inhibited growth of higher plants. Behrens (1953) presented a summary of its biological properties and suggested that it was a potent new herbicide. Much of the information presented by Behrens appears to have been subsequently reported in detail by Hall et al. (1954). It was first introduced as a herbicide by Amchem Products, Inc. (U.S. Patent No. 2,670,282) in 1954. Although it has been used for weed control in several crops it is currently not registered for use in any crop in the United States. It is being used for weed control in non-cropped areas. Since it is used as a foliar spray and decomposes rapidly in moist, warm soils, it is frequently used in combination with a long lasting soil-active herbicide. The amitrole controls the emerged vegetation and the soil-active herbicide controls the weeds which germinate later. The addition of ammonium thiocyanate increases the foliar activity of amitrole and the combination is referred to as Amitrol-T.

Amitrole

Amitrole (3-amino-s-triazole) is a water-soluble heterocyclic nitrogen compound composed of a five-member ring having three nitrogen and two carbon atoms in the ring as well as an amino group substitution on the three position of the ring. It is insoluble in non-polar organic solvents.

A number of 1,2,4-triazole derivatives have been synthesized and evaluated as herbicides since the introduction of amitrole. Some of the types of compounds are: (1) ureidotriazoles (Kaiser and Peters, 1955; Levy and Menoret, 1959), (2) substituted phenoxyalkanecarboxyl-3-amino-1,2,4-triazoles (Norris, 1961), (3) 4-chloro-2-butynyl-(1,2,4-triazol-3-yl)-carbamates (Hopkins and Puller, 1964), and (4) α-hydroxy-β,β,β-trichloroethylamino-1,2,4-triazoles (Matolcsy and Gomaa, 1968). None of these have been shown to have significantly greater biological activity in higher plants than amitrole.

Growth and Plant Structure

The most striking symptom of amitrole phytotoxicity is the development of albino leaves and shoots. Hall et al. (1954) showed that low concentrations of amitrole stimulated growth, but at higher concentrations it inhibited growth and caused chlorosis. The chlorosis may be temporary if the dosage is sublethal, or permanent resulting in death when the dosage is high enough. Experiments in cotton seedlings showed that the sprayed cotyledons receiving concentrations of 650 and 840 ppm turned chlorotic and remained chlorotic, those receiving 8.4, 84, 210, and 420 ppm recovered their normal green color. Amitrole appeared to cause chlorophyll destruction and impaired chlorophyll development in tissues formed at the time or subsequent to the absorption of the chemical. The inhibition of chlorophyll development was proportional to the concentration and the age of the tissues at the time of treatment. The possibility that the restriction of chlorophyll formation was due to immobilization of Mg, Fe, Mn, N, P, or K by amitrole appeared unlikely; the inhibitory effect appeared to be prior to the protochlorophyll stage. Additional aspects of the effect of amitrole on chlorophyll development will be discussed later.

In addition to the albinism which appears in developing leaves and shoots, the mature leaves to which the herbicide is applied begin to undergo senescence. Depending on the species and amitrole concentration, this often involves a yellowing or browning of the leaves, followed by death. The stems of Pacific poison oak (*Rhus diversiloba*) are dead within two months following an amitrole application to the foliage and stem in mid-summer. These observations suggest that the mechanism involved in these responses are not the result of albinism of the foliage.

Hall et al. (1954) sprayed 8- to 10-leaf stage cotton plants with a series of 11 concentrations of amitrole from 9 to 10,000 ppm and made critical observations of the phytotoxic symptoms 2 months later. Table 20-1 gives the results of this experiment; amitrole progressively inhibited main stem elongation, development of axillary branches, leaf size, flowering, and fruiting. Presumably root growth was also inhibited. Amitrole was also shown to progressively inhibit the elongation of Avena coleoptile sections from 0.84 to 21 ppm and markedly inhibit their growth at 42 ppm and above. These investigators also determined the effect of amitrole on root and hypocotyl growth of cucumber and cotton seedlings germinated in the herbicide solution on filter paper and found that at concentrations up to 21 ppm there was a slight stimulation of elongation of the primary root, above 21 ppm root growth was inhibited and hypocotyl growth was inhibited even at the lowest concentration, 0.84 ppm.

Jackson (1961) investigated the effect of amitrole on the elongation of root hairs of red top (*Agrostis alba*) during periods of 2 hours and found a concentration of $10^{-3} M$ or more inhibited their growth during the last 30 minutes of treatment. Eliasson (1962) reported amitrole inhibited the root growth of trembling aspen (*Populus tremuloides*) treated with amitrole on their mature leaves in quantities of 0.1 mg per plant or more or at concentrations from $3 \times 10^{-5} M$ in the root culture medium. In both cases the response was delayed about 24 hours. Only high concentrations applied to the roots caused a considerably more rapid response. The onset of inhibition was accompanied by the appearance of necroses in the elongating part of the root, while the meristem and mature part of the root showed no visible injuries. The inhibition of root elongation occurred before foliar chlorosis. The inhibition of shoot and root growth in *Lens culinaris* by amitrole has also been shown by Gaschen (1963). Therefore, amitrole has been shown to inhibit growth of several plant organs and appears to be a general growth inhibitor, with the degree of inhibition increasing with increasing concentrations until it becomes lethal.

The germination of seeds harvested from yellow nutsedge (*Cyperus esculentus*) plants treated with 2 lb/A of amitrole was inhibited from 16 to 30% depending on time of harvest (Hill et al., 1963).

The presence of albino leaves following the treatment of plants with amitrole could be due to the destruction of chloroplast pigments or an inhibition of their synthesis. However, the fact that mature leaves which are treated with moderate rates of amitrole usually do not become albinized whereas the newly developing leaves do, suggests that pigment synthesis is involved rather than pigment destruction. This albinism could involve an inhibition of chloroplast development which in turn could interfere with pigment synthesis or an inhibition of pigment synthesis *per se* without an effect on chloroplast development. It should be emphasized that the newly developing leaves from plants treated with amitrole are albino, lacking all pigments, and not merely chlorotic, lacking chlorophyll. It may be noted that the amitrole symptom is often referred to as chlorotic in this chapter; this

TABLE 20-1.   Effects of amitrole on Growth of the Cotton Treated at the 8- to 10-Leaf Stage 2 months after Treatment (Hall et al., 1954)

Concentration amitrole, mg per liter	Average height main stem, inches	Remarks
0 (controls)	52.2	Cotton normal for spring-summer greenhouse-grown cotton
9	51.0	Plants normal in color, fruiting, branching and other growth characteristics
19	51.6	Plants normal in color and fruiting, vegetative and fruiting branches less than controls
39	48.0	Plants apparently normal except for lack of axillary branches
78	45.0	Size reduction noticeable, otherwise normal vegetatively and reproductively. Sparse branching at base of plant
156	42.0	Old leaves with necrotic areas, otherwise plants almost normal except for reduced size
312	36.4	Old leaves partially mottled. New leaves smaller and lighter colored than controls, shortened internodes in upper plant. Fruiting slightly inhibited
625	36.0	Old foliage partially mottled and dried. New leaves almost normal in color but reduced in size. Reduced flowering and fruiting only at top of plant

is an attempt to maintain the original researchers terminology. The discussion to follow will primarily deal with the effect of amitrole on chloroplast development and its effect on pigment synthesis will be taken up in the section on biochemical responses, although the two phenomena are closely associated.

Several workers have reported that plastids were completely absent or very few in number in chlorotic tissue from plants treated with amitrole. However, in most of this research the tissues were examined with the light microscope. Most investigations utilizing the electron microscope have found plastids although their structure was grossly altered. The early work of Rogers (1957a, 1957b) reported that plastids were absent from chlorotic maize tissue from maize plants treated with amitrole. However, in later research by Jacobson and Rogers (1961) plastids were found in this same

TABLE 20-1   (*continued*)

Concentration amitrole mg per liter	Average height main stem, inches	Remarks
1250	31.0	Old leaves partially mottled and dried. New leaves small in size but almost normal in color. Growing point region and axillaries chlorotic. Small, abnormal fruiting at top of plant
2500	27.6	Most of old leaves had abscised, mottled or dried. Growing point inactive, extremely short internodes, white axillaries. Leaves small and abnormal. No flowering or fruiting
5000	24.0	Terminal meristem dead, axillaries white and dead. Little growth after application. Main stem still alive at base. No intact fruits. Squares abscised
10,000	23.3	Terminal meristem dead and main stem dead almost to base. Axillaries white and dead. Leaves and squares abscised. Plant practically dead for all purposes

tissue but the plastids from the amitrole-treated plants were grossly altered in their internal structure. Pyfrom et al. (1957) observed that chlorotic leaf tissue of barley and potato plants treated with amitrole contained few plastids and these were shrunken and misshapen. Naylor (1964) failed to find plastids in *Elodea* plants treated with amitrole. Similar results were obtained by George (1968) in *Eupatorium adorstum* and *Rhoea discolor* but several weeks later chloroplasts developed. Cytological examination of chlorotic leaves from amitrole-treated plants revealed the presence of proteinous structures which might be interpreted as being deformed plastids in pine (*Pinus* spp.) (Shive and Hansen, 1958) and in wheat (Weier and Imam, 1965).

In a series of three papers Bartels and coworkers reported on the ultrastructural changes induced in plastids by amitrole (Bartels, 1965; Bartels et al., 1967; Bartels and Weier, 1969). In the first report (Bartels, 1965), wheat was grown in petri dishes containing filter paper wet with 10 ml of water with or without $10^{-4}M$ amitrole both in the light and in the dark. Seven days after treatment leaf tissue was harvested and prepared for examination with an electron microscope. Amitrole blocked the light-induced plastid development as evidenced by the absence of the normal lamellar system. If a membrane system was present in the plastids from the treated plants, it was either concentrically arranged or was highly disorganized in contrast to the normal grana from the control plants. In addition, some plastids from the light-grown treated plants contained an electron dense body, which may be equivalent to the prolamellar body. However, membrane formation in proplastids from dark-grown plants was not affected by amitrole. A subsequent ultrastructural study (Bartels and Weier, 1969) using similar techniques revealed that the plastids from light-grown wheat plants treated with amitrole lacked normal grana-fret membrane systems as well as chloroplast ribosomes. The lack of grana membranes in these chloroplasts may be caused by the absence of the chloroplast ribosomes. Ribosomes are known to be intimately involved in protein synthesis; therefore the protein portion of the chloroplast membranes could probably not be synthesized. Figure 20-1 shows the effect of amitrole on chloroplast development in the light. In contrast to these results, dark-grown plants treated with amitrole contained proplastids with ribosomes. These observations confirm the ultracentrifugation study of Bartels et al. (1967) which showed that amitrole caused the complete loss of $70S$ chloroplastic ribosomes and $18S$ Fraction I protein in light-grown plants. They suggest that amitrole may be interfering with some aspect of chloroplastic nucleic acid metabolism. This series of investigations illustrates the importance and value of a structure-function approach to studies of the mechanism of action of herbicides.

Putala (1967) reported that sub-lethal concentrations of amitrole prevented maturation of proplastids in the apical meristem of tomato without inhibiting growth or interfering with the differentiation of other cellular organelles. The severity of this effect was directly proportional to the amount of amitrole applied. Amitrole caused the proplastids to retain vesicles which would normally differentiate into granal discs, while chloroplasts which were already differentiated formed disorganized granal discs along with scattered vesicles.

Absorption and Translocation

There have been numerous investigations in the absorption and translocation of amitrole in a wide variety of species. Most of these studies utilized ^{14}C-amitrole labeled in the five position in the ring. In general, amitrole has been

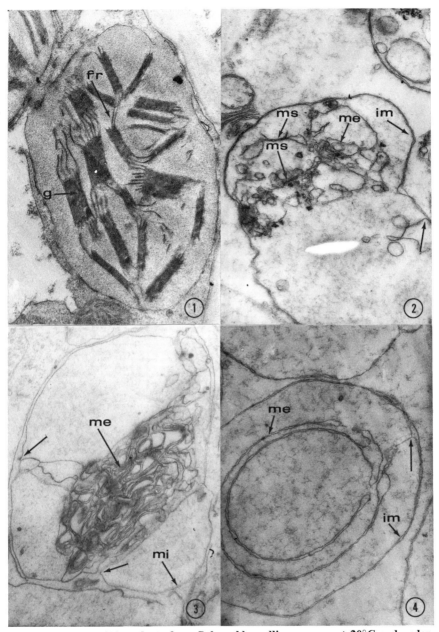

FIGURE 20-1. Chloroplasts from 7-day-old seedlings grown at 20°C and under 600 ft-c of light. KMnO₄ fixation. Plate 1—Normal chloroplast from untreated leaf center with grana (g) and frets (fr). Plate 2—Part of a treated plastid with disorganized membranes (me). Some of the membranes appear as grana-like structures (ms). Plate 3—Plastid from treated tissue having an aggregate of highly disorganized membranes (me). Plate 4—Portion of a plastid from treated plant with concentric arranged membranes (me) attached to inner plastid envelope (im) (see arrow). (Bartels and Weier, 1969.)

shown to be absorbed and translocated readily in plants via both the symplastic and apoplastic systems. Some investigators have suggested that it actually circulates in plants. Since amitrole is readily converted in plants to other metabolites without loss of the ^{14}C-label it is not clearly evident whether translocation patterns presented represent amitrole, its metabolites or both.

The pioneering research on amitrole by Hall et al. (1954) reported on absorption and translocation using patterns of chlorosis as the indexes. They concluded that amitrole was readily absorbed by roots and aerial organs of the cotton plant. Since its translocation to the shoot following a root application was not blocked by bark-girdling, this transport was via the xylem. Foliar applications resulted in predominately upward movement to the meristematic tissues, followed by transport to other plant parts. When plants were bark-girdled half-way up the main stem and amitrole applied to leaves above the girdle, amitrole symptoms were not found below the girdle. When amitrole was applied to leaves below the girdle, amitrole symptoms were not found above the girdle. It was concluded that amitrole moves either up or down in the stem in living cells of the phloem following foliar application. Hauser and Thompson (1954) reported severe foliar chlorosis in new growth following the application of amitrole to the roots, shoots, or tubers of purple nutsedge (*Cyperus rotundus*) and foliage of johnsongrass (*Sorghum halepense*), indicating absorption and translocation of amitrole. They also found that excised rhizomes from amitrole-treated johnsongrass plants which were planted developed severely chlorotic shoots, showing translocation into the rhizomes.

Although the above reports yielded considerable information on the absorption and translocation of amitrole, the availability of ^{14}C-labeled amitrole added considerable refinement. Rogers (1957c) found that radioactivity was translocated throughout each plant within 70 hours after the application of ^{14}C-amitrole to the axils of soybean or Canada thistle (*Cirsium arvense*) plants or on the base of the lower leaf of johnsongrass plants. Some radioactivity was even localized in the hairs on soybean stems and found to be translocated some 6 inches through the rhizome from a treated Canada thistle plant into an attached plant. Radioactivity was also present in associated aphids which presumably sucked it from the phloem. Bondarenko (1957) also studied the absorption and translocation of ^{14}C-amitrole in Canada thistle and observed that the radioactivity from the foliar-applied herbicide moved from the treated leaf into all other leaves of the plant within 1 hour and was concentrated mainly in the midrib. It was also found uniformly distributed in small quantities throughout the aerial and underground parts of the stem and in the roots to a depth of about 1 foot. Progressively with time, the radioactivity tended to concentrate in the young leaves and near the stem tip. Radioactivity was also present in associated aphids. During the following decade many other workers found that amitrole or its metabolites were translocated very readily in many plant species (Anderson, 1958; Massini, 1958; Racusen, 1958; Penot, 1960; Miller and Hall, 1961; Hill et al., 1963; Shimabukuro and Linck, 1964; Herrett and Bagley, 1964; Volger, 1964; Leonard and Hull, 1965; Leonard et al., 1965; Volger, 1965; Smith and Davies, 1965; Leonard et al., 1966; Forde, 1966a; Yamaguchi and Islam, 1967; Lund-Hoie and Bayer, 1968).

Ammonium thiocyanate (NH_4SCN) has been used in amitrole formulations to increase its effectiveness. It has been reported to influence the absorption and/or translocation as well as the metabolism of amitrole. Donnalley and Ries (1964) reported that NH_4SCN applied prior to or with amitrole did not influence the absorption of ^{14}C-amitrole but did increase the amount of ^{14}C translocated in quackgrass (*Agropyron repens*). NH_4SCN applied one day after amitrole had no effect. They suggested that it may reduce the binding of amitrole at the site of application. In contrast to these findings, Forde (1966b) showed that NH_4SCN retarded the translocation of amitrole in quackgrass. He suggested that the immobilizing influence of NH_4SCN on amitrole is exerted in the treated leaf rather than on long distance transport. Van der Zweep (1965) found that NH_4SCN reduced the necrotic foliar toxicity of amitrole in bean while increasing the shoot inhibition effect. He further showed that NH_4SCN increased the dispersal of label from ^{14}C-amitrole within the treated leaf with increasing concentrations of NH_4SCN. However, the addition of ^6N-benzyladenine to amitrole produced a more pronounced synergistic effect than did NH_4SCN.

Freed and Montgomery (1958) reported that a surfactant greatly increased the absorption of amitrole. A non-phytotoxic paraffinic oil has also been shown to increase the effectiveness of amitrole (Aya and Ries, 1968). High relative humidities increased the translocation of amitrole from a foliar application into the roots (Leonard et al., 1965).

Herrett and Linck (1961) have postulated that amitrole is translocated out of the treated leaf as the glucose-adduct and movement of this compound in significant quantities does not occur until 24 hours after absorption of amitrole by the leaf. Shimabukuro and Linck (1964) found that amitrole translocation from the treated primary leaves of bean over a period of 11 days was highest at a petiole temperature of $10°C$ and this is in contrast to previous reports for the translocation of sugar (Swanson and Bohning, 1951; Bohning et al., 1952). Over a period of 48 hours, however, translocation of amitrole was highest at a petiole temperature of $22°C$ and is parallel to the translocation of sugar. These results are in agreement with the hypothesis that amitrole forms a glucose-adduct before being translocated from the primary leaf to other parts of the bean plant. Herrett and Bagley (1964) suggested that the translocatable form of amitrole from light–starved leaves is a compound which they refer to as unknown III; such conditions are unfavorable for the translocation of amitrole. However, Lund-Hoie and Bayer (1968) reported that although amitrole is metabolized into three compounds in ponderosa pine (*Pinus ponderosa*) and white fir (*Abies concolor*), amitrole *per se* was the mobile, toxic compound in both species and was translocated in both symplast and apoplast. Furthermore, Carter (1969) considers the glucose-adduct of amitrole to be simply an interesting artifact and suggests that the unknown III of Herrett and Bagley (1964) may well be an artifact. It is therefore evident that considerably more research must be conducted before we can unequivocally state the translocatable form of amitrole.

Molecular Fate

The molecular fate of amitrole has recently been reviewed by Carter (1969). The formation of conjugates between amitrole and endogenous plant constituents appears to be the major metabolic alteration of amitrole in higher plants. Most of these are less toxic than amitrole and their formation can be considered to be a detoxication mechanism. However, a metabolite referred to as unknown III by Herrett and Bagley (1964) was herbicidally more active than amitrole. Ring cleavage with the subsequent evolution of $^{14}CO_2$ from ^{14}C-5-amitrole has been reported. However, Carter (1969) does not consider the evidence conclusive for rapid and extensive ring cleavage.

A discussion of the formation of amitrole conjugates in higher plants is considerably complicated by the varying terminology used by different workers. Most researchers have reported the formation of two major conjugates or metabolites. One of these which is a condensation product between amitrole and serine (3-amino-1,2,4-triazolylalanine, see Figure 20-2 for structure) is probably the compound referred to as "X" by Racunsen (1958) and Miller and Hall (1961), "ATX" by Massini (1959), "compound 1" by Carter and Naylor (1961), "Unknown II" by Herrett and Linck (1961) and Smith et al. (1968a, 1968b), "X2" by Lund-Hoie and Bayer (1968) and "3-ATAL" by Carter (1969). We have chosen to use the term 3-ATAL for this compound. According to Racunsen (1958) 3-ATAL gives positive reactions with ninhydrin reagents and azo dyes, behaves as a zwitterion during electrophoresis at different pH's, and is stable in $6N$ HCl for 5 hours at 100°C.

A second major conjugate or metabolite has been found to be formed in most studies on the molecular fate of amitrole in higher plants. However the evidence suggests that it is not the same compound in all the studies. This second compound has been referred to as "Y" by Racunsen (1958) and Miller and Hall (1961), "compound 2" by Carter and Naylor (1961),

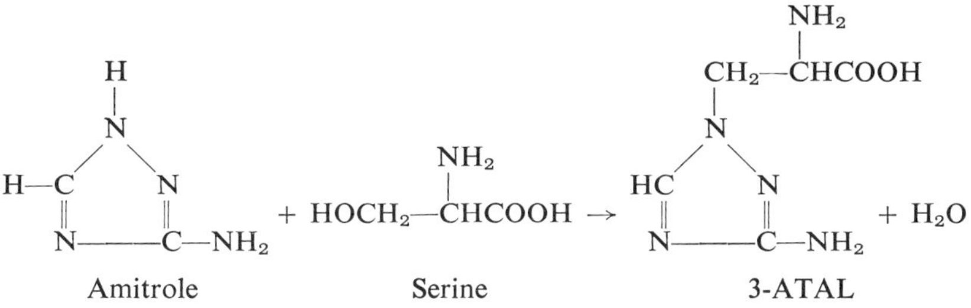

$$\underset{\text{Amitrole}}{\overset{\displaystyle H}{\underset{\displaystyle N}{\underset{\displaystyle N}{\overset{\displaystyle \|}{\underset{\displaystyle C}{\overset{\displaystyle}{\underset{\displaystyle}{}}}}}}} } + \underset{\text{Serine}}{\overset{NH_2}{HOCH_2-CHCOOH}} \rightarrow \underset{\text{3-ATAL}}{\overset{\overset{NH_2}{|}}{\overset{CH_2-CHCOOH}{\underset{N}{\cdots}}}} + H_2O$$

FIGURE 20-2.   Condensation of amitrole and serine to form 3-ATAL (3-amino-1,2,4-triazolylalanine) (Carter, 1969). Reprinted from P. C. Kearney and D. D. Kaufman, eds., *Degradation of Herbicides*, p. 192, by courtesy of Marcel Dekker, Inc.

"Unknown I" by Herrett and Linck (1961) and Smith et al. (1968a, 1968b), "X1" by Lund-Hoie and Bayer (1968) and "ATY" by Carter (1969). According to Racunsen (1958), Y, like 3-ATAL, reacts with azo dyes and is stable to hot HCl but does not react with ninhydrin agents. However, Unknown I of Herrett and Linck (1961) is not stable to hot HCl. Therefore Y and Unknown I must be different compounds.

A third metabolite was reported to be formed by Herrett and Bagley (1964) and Lund-Hoie and Bayer (1968). The latter authors concluded that Unknown III of Herrett and Bagley (1964) was not the same as their compound X3 since X3 was very heat stable and not destroyed by the column chromatography whereas Unknown III is.

Carter and Naylor (1961) observed thirteen radioactive compounds derived from ^{14}C-5-amitrole, none of which appeared to be normal metabolites arising from $^{14}CO_2$ fixation. They suggested that these were probably a variety of amitrole conjugates.

Apparently amitrole forms a number of conjugates with endogenous compounds of higher plants. Two of these are usually formed in relatively large amounts. One of the two is usually 3-ATAL, while the other one may vary in its chemical composition depending on the species and environmental conditions. Other minor metabolites are also often detected.

Sund et al. (1960) found that certain purines tend to overcome the effect of amitrole on the growth and chlorosis of higher plants. Riboflavin was effective in this respect. Furthermore, Sund and Little (1960) reported that amitrole had an inhibitory effect on riboflavin production in the mold, *Eremothecium Ashbyii*. However, Castelfranco et al. (1963) showed that amitrole decomposed in the presence of riboflavin and light. These results suggest that the role of riboflavin as a protective agent for amitrole injury may be related to the photo-inactivation of amitrole rather than an inhibition of riboflavin synthesis by amitrole.

Biochemical Responses

Amitrole has a pronounced effect on chloroplast development in young leaves and as a result the photosynthetic apparatus does not become functional. The resulting lack of photosynthate could cause many secondary alterations in metabolism.

Respiration of leaf disks from intact cotton plants treated by dipping the leaves in solutions of amitrole was determined by Hall et al. (1954). The concentrations of amitrole used were 10^{-3}, 10^{-2}, and $10^{-1}M$ and the leaves sampled at 3, 24, 48 and 72 hours after treatment. Amitrole increased the oxygen uptake from 20 to 60% over the control within 3 hours following treatment. After 48 hours, oxygen uptake was inhibited at the two higher concentrations and decreased below that of the control at the 72-hour measurement. The decrease in respiration at the higher concentrations after 48 hours coincided with rapid blade dehydration and the initiation of visual abscission. Respiration was stimulated throughout the experimental period at the lowest concentration of amitrole. Miller and Hall (1957) found that a $4.8 \times 10^{-2}M$ solution of amitrole sprayed on cotton leaves caused an initial stimulation at 4 hours but it returned to normal after 26 hours. McWhorter and Porter (1960) reported that amitrole had little effect on the respiration of detached leaves of maize for periods up to 4 days after treatment when treated by vacuum infiltration and/or applied to leaf surface. However, when treated via the roots in sand culture at $9.5 \times 10^{-4}M$, respiration of leaf sections was increased about 20% after 24 hours. Subsequent research showed that the chlorotic tissue had a respiratory quotient of 0.82 (McWhorter and Porter, 1960). Seeds attached to these plants had greater fat content and a more active lipoxidase system than did untreated plants. Untreated plants had about 50% less total carbohydrates but twice the aldolase activity of treated plants. They postulated that chlorotic plants were metabolizing fats as a major respiratory substrate while untreated plants were metabolizing carbohydrates. McWhorter (1963) also observed that chlorotic maize leaves contained increased concentrations of manganese, free ammonia, total nitrogen and generally total carbohydrates. Protein content of chlorotic tissue was usually reduced whereas free amino acids were increased. Carter and Naylor (1961) reported that amitrole had little if any effect on the metabolism of ^{14}C-U-glucose, ^{14}C-1-succinate, or dark $^{14}CO_2$ fixation in beans. Lotlikar et al. (1968) reported that concentrations even as high as $1 \times 10^{-3}M$ did not significantly alter oxygen uptake, phosphorous esterification, or the P to O ratio of isolated mitochondria. Therefore, it is doubtful that amitrol has a significant direct effect on respiration.

Hall et al. (1954) studied the effect of amitrole on carbohydrate composition of cotton plants by spraying plants in the field to thoroughly wet the foliage without excessive runoff and analyzing the plant parts 24 and 48 hours after treatment. Concentrations of the spray solutions were 1.3×10^{-2}, 3.0×10^{-2}, and $6.0 \times 10^{-2} M$. Within 48 hours, the aerial organs lost approximately half of the original reducing sugars and sucrose; a slight increase in starch was essentially balanced by an equivalent loss in the hemi-cellulose fraction. The result was a decrease in total carbohydrates. Fractionation of the treated plants showed that soluble sugars decreased in both upper and basal shoot, but reserve carbohydrates increased in the upper plant and decreased in the basal parts following treatment.

The protein content of light-grown amitrole-treated plants decreases, with a compensating increase in free amino acids (McWhorter, 1963; Bartels and Wolf, 1965). This would suggest an increase in protein hydrolysis and/or a decrease in protein synthesis. Amitrole has been shown to have no effect on the activity of proteinases or dipeptidases isolated from squash cotyledons or little if any effect on their normal increase during germination (Ashton et al., 1968; Tsay and Ashton, 1971). Bartels and Wolf (1965) reported that wheat seedlings germinated in a petri dish containing a $10^{-4} M$ solution of amitrole for 3 days inhibited the incorporation of ^{14}C-labeled glycine into protein in the shoot by about 50%. However, Mann et al. (1965) and Moreland et al. (1969) concluded that amitrole did not inhibit the incorporation of ^{14}C-leucine into protein in stem or coleoptile sections. Brown and Carter (1968) also were unable to find a direct effect of amitrole on protein synthesis. These results would seem to indicate that amitrole does not inhibit the protein synthesis reactions *per se* but rather inhibits the formation of a critical component of protein synthesis.

Sund et al. (1960) found that purines would alleviate the growth inhibitory effect of amitrole in bean and tomato. Castelfranco et al. (1963) showed that riboflavin would reduce the effects of amitrole in maize. However, Naylor (1964) reported that purines would not reverse chlorosis and growth inhibition caused by amitrole in higher plants. Bartels and Wolf (1965) suggested that amitrole may intefere with purine synthesis and thereby affect the nucleic acid content of the plants. Consequently, a lower level of RNA would occur which would decrease the plants' ability to synthesize protein, and result in the accumulation of amino acids.

The incorporation of ^{32}P into the acid-insoluble fraction (nucleic acids, phospholipids, phosphoproteins) of both roots and shoots of barley decreased within 24 hours following amitrole treatment, whereas the acid-soluble fraction (nucleotides, sugar phosphates, inorganic phosphates) increased (Wort and Loughman, 1961). On the basis of these results and other data collected in these experiments, Wort and Loughman (1961) concluded that the effect of amitrole on phosphate metabolism is not concerned with oxidative phosphorylation or glycolysis, but is confined to the processes involved in the incorporation of phosphate into the components of the acid-insoluble fraction, possibly the nucleic acids. However, Schweizer and Rogers (1964) reported that amitrole did not alter the RNA content of maize roots or coleoptiles but did decrease the acid-soluble nucleotide content of the coleoptiles. However, Bartles and Wolf (1965) investigated the effect of amitrole on the nucleic acids of wheat seedlings and found that amitrole had no effect on the DNA and acid-soluble nucleotide content, but caused a 15% decrease in the RNA content. An analysis of the RNA revealed that the quantity of purines and pyrimidines was reduced, but the base composition of the RNA was unaffected. Amitrole was found to inhibit the incorporation of labeled precursors into RNA, DNA, and acid-soluble nucleotides. Bartels et al. (1967) reported that amitrole inhibits the formation of normal chloroplasts and causes a complete loss of 70S chloroplastic ribosomes and 18S Fraction I protein in light-grown plants. In more recent studies Bartels and Hyde (1970) have shown that light-grown, amitrole-treated plants lack detectable quantities of chloroplast DNA whereas amitrole-treated, dark-grown plants contain plastid DNA. Since amitrole had no observable effect on etioplast DNA of amitrole-treated, dark-grown plants, amitrole must require light to inhibit the formation or accumulation of chloroplast DNA in the leaf. They suggest that amitrole brings about a metabolic change in the developing plastids of light-grown plants which either destroys the chloroplast DNA or blocks the synthesis of chloroplast enzymes or structural components necessary for chloroplast DNA synthesis.

In microorganisms amitrole inhibits the growth by competitive inhibition of imidazoleglycerol phosphate dehydratase (IGP dehydratase), an enzyme of histidine biosynthesis (Hilton et al., 1965; Klopotonski and Wiater, 1965). Since we have chosen to restrict our coverage to higher plants in this text, the reader is referred to the Carter (1969) review for information on this subject.

In higher plants, however, McWhorter and Hilton (1967) and Hilton (1969) suggested that histidine biosynthesis could not account for the herbicidal action of amitrole. These conclusions were based on two observations. Firstly, an exogenous supply of histidine was not able to counteract the toxic action of amitrole in higher plants, whereas histidine was able to accomplish this in microorganisms. Secondly, in young seedlings endogenous histidine remained higher in amitrole treated maize plants than in untreated seedlings until after seed reserves were exhausted. Amitrole inhibits growth during this period of high histidine content. In two recent papers, Wiater et al. (1971a, 1971b) demonstrated a histidine biosynthetic pathway in barley and oats comparable to that in microorganisms. *In vivo*, the plant IGP dehydratase activity was more strongly inhibited by amitrole than the same enzyme from microorganisms. In commenting on these two papers, Hilton (1972) states, "Although most of the early work on amitrole-histidine interactions concerned microorganisms, these papers indicate that the inhibitions are equally valid for green plants. In my opinion, the inhibition of histidine biosynthesis is a contributing factor to the herbicidal action of amitrole in established plants (those having exhausted seed reserves). But it is not by itself sufficient to explain all of the effects of amitrole. The contribution of this inhibition to herbicidal action is perhaps masked by additional sites of inhibition which are equally sensitive."

Mode of Action

Although amitrole has been shown to stimulate growth at low concentrations and inhibit growth at high concentrations, the most obvious phytotoxic symptom is the albino appearance of the foliage which develops after the herbicide application. This lack of foliar pigmentation appears to result from an interference of amitrole with development of chloroplasts. Amitrole also caused senescence and death of mature leaves and other plant parts.

Amitrole is readily absorbed by both leaves and roots and translocates throughout the plant. Translocation proceeds via both the symplastic and apoplastic systems. It is one of our most mobile herbicides in higher plants and may actually circulate in the plant. Some workers have suggested that amitrole *per se* is not the mobile form but rather that a glucose-amitrole conjugate is the mobile form.

The major fate of amitrole in higher plants appears to involve the formation of conjugates with various amino acids and sugars, especially serine and glucose. Thirteen different radioactive compounds have been isolated following the application of ^{14}C-5-amitrole to bean plants and many of these have been suggested to be various amitrole conjugates. Ring cleavage with subsequent $^{14}CO_2$ evolution from ^{14}C-5-amitrole has been reported, however there have been questions as to the importance of this.

A wide variety of biochemical alterations have been reported to be induced by amitrole including effects on carbohydrate, lipid, nitrogen, and other areas of metabolism. However, it would appear that the action of amitrole is more likely to be closely associated with the biochemistry of chloroplast development than these areas of general metabolism. This is because the evident phytotoxic symptom of amitrole is foliar albinism which is the result of a blockage in chloroplast development. Amitrole has been shown to block the formation of 70S chloroplast ribosomes, 18S Fraction I protein, and chloroplast DNA in light-grown plants. However, the blockage of chloroplast development does not explain the phytotoxic symptoms free of albinism which may occur in mature leaves or other plant parts within a few days or weeks following amitrole application at relatively high rates.

REFERENCES

Anderson, O. 1958. Studies on the absorption and translocation of amitrole (3-amino-1,2,4-triazole) by nut grass (*Cyperus rotundus* L.). *Weeds* **6**:370–385.

Ashton, F. M., D. Penner, and S. Hoffman. 1968. Effect of several herbicides on proteolytic activity of squash seedlings. *Weed Sci.* **16**:169–171.

Aya, F. O. and S. K. Ries. 1968. Influence of oils on the toxicity of amitrole to quackgrass. *Weed Sci.* **16**:288–290.

Bartels, P. G. 1965. Effect of amitrole on the ultrastructure of plastids in seedlings. *Plant Cell Physiol.* **6**:227–230.

Bartels, P. G. and A. Hyde. 1970. Buoyant density studies of chloroplast and nuclear deoxyribonucleic acid from control and 3-amino-1,2,4-triazole-treated wheat seedlings, *Triticum vulgare*. *Plant Physiol.* **46**:825–830.

Bartels, P. G., K. Matsuda, A. Siegel, and T. E. Weier. 1967. Chloroplastic ribosome formation: inhibition by 3-amino-1,2,4-triazole. *Plant Physiol.* **42**:736–741.

Bartels, P. G. and Weier, T. E. 1969. The effect of 3-amino-1,2,4-triazole on the ultrastructure of plastids of *Triticum vulgare* seedlings. *Am. J. Bot.* **56**:1–7.

Bartels, P. G. and F. T. Wolf. 1965. The effect of amitrole upon nucleic acid and protein metabolism of wheat seedlings. *Physiol. Plant.* **18**:805–812.

Behrens, R. 1953. Amino triazole. Proc. *10th North Central Weed Control Conf.* p. 61.

Bohning, R. H., C. A. Swanson, and A. J. Linck. 1952. The effect of hypocotyl temperature on translocation of carbohydrates from bean leaves. *Plant Physiol.* **27**:417–421.

Bondarenko, D. D. 1957. The absorption and translocation of ^{14}C-labeled ATA and 2,4-D in Canada thistle. *Proc. 14th North Central Weed Control Conf.* pp. 9–10.

Brown, J. C. and M. C. Carter. 1968. Influence of amitrole upon protein metabolism in bean plants. *Weed Sci.* **16**:222–226.

Carter, M. C. 1969. Amitrole, pp. 187–206. In P. C. Kearney and D. D. Kaufman, *Degradation of Herbicides*. Marcel Dekker, Inc. New York.

Carter, M. C. and A. W. Naylor. 1961. Studies on an unknown metabolic product of 3-amino-1,2,4-triazole. *Physiol. Plant.* **14**:20–27.

Castelfranco, P., A. Oppenheim, and S. Yamaguchi. 1963. Riboflavin-mediated photodecomposition of amitrole in relation to chlorosis. *Weeds* **11**:111–115.

Donnalley, W. F. and S. K. Ries. 1964. Amitrole translocation in *Agropyron repens* increased by the addition of ammonium thiocyanate. *Science* **145**:497–498.

Eliasson, L. 1962. The response of aspen roots to 3-amino-1,2,4-triazole. *Physiol. Plant.* **15**:229–238.

Forde, B. J. 1966a. Translocation in grasses. 2. Perennial ryegrass and couch-grass. *N. Z. Jour. Bot.* **4**:496–514.

Forde, B. J. 1966b. Translocation patterns of amitrole and ammonium thiocyanate in quackgrass. *Weeds* **14**:178–179.

Freed, V. H. and M. Montgomery. 1958. The effect of surfactants on foliar absorption of 3-amino-1,2,4-triazole. *Weeds* **6**:386–389.

Gaschen, M. 1963. Action of 3-amino-1,2,4-triazole on growth and elongation in *Lens culinaris*. *Ber. Schweiz Bot Ges.* **73**:227–275.

George, K. 1968. Observations on the regeneration of plastids. *Curr. Sci.* **37**:531–532.

Hall, W. C., G. B. Trochelut, and H. C. Lane. 1953. Chemical defoliation and regrowth inhibition in cotton. *Texas Agr. Expt. Sta. Bull.* **759**:1–25.

Hall, W. C., S. P. Johnson, and C. L. Leinweber. 1954. Amino triazole—A new abscission chemical and growth inhibitor. *Texas Agr. Exp. Sta. Bull.* **789**:1–15.

Hauser, F. W. and J. Thompson. 1954. Effects of 3-amino-1,2,4-triazole and derivatives on nutgrass and johnsongrass. *J. Agr. and Food Chem.* **2**:680–681.

Herrett, R. A. and W. P. Bagley. 1964. The metabolism and translocation of 3-amino-1,2,4-triazole by Canada thistle. *J. Agric. and Food Chem.* **12**:17–20.

Herrett, R. A. and A. J. Linck. 1961. The metabolism of 3-amino-1,2,4-triazole by Canada thistle and field bindweed and the possible relation to its herbicidal action. *Physiol. Plant.* **14**:767–776.

Hill, E. R., W. H. Lackman, and D. N. Maynerd. 1963. Translocation of amitrole in yellow nutsedge and its effect on seed germination. *Weeds* **11**:165–166.

Hilton, J. L. 1969. Inhibitions of growth and metabolism of 3-amino-1,2,4-triazole (amitrole). *J. Agr. and Food Chem.* **17**:182–198.

Hilton, J. L. 1972. Personal communication.

Hilton, J. L., P. C. Kearney, and B. N. Ames. 1965. The mode of action of the herbicide 3-amino-1,2,4-triazole (amitrole) inhibition of an enzyme of histidine biosynthesis. *Arch. Biochem. Biophys.* **112**:544–547.

Hopkins, T. R. and J. W. Puller, 1964. 4-halo-2-butynyl-1(,2,4-triazole-3-yl)-carbamates. U.S. Patent No. 3,132,150.

Jackson, W. T. 1961. Effect of 3-amino-1,2,4-triazole and L-histidine on rate of elongation of root hairs of *Agrostis alba*. *Weeds* **9**:437–442.

Jacobson, A. B. and B. J. Rogers. 1961. Observations concerning aminotriazole effect on maize chloroplasts using the electron microscope. *Plant Physiol. Abstr.* **36**:xl.

Kaiser, D. W. and G. A. Peters. 1955. Ureidotriazoles. U.S. Patent No. 2,723,274.

Klopotonski, T. and A. Wiater. 1965. Synergism of amitrole and phosphate on the inhibition of yeast IGP dehydrase. *Arch. Biochem. Biophys.* **112**:562–566.

Leonard, O. A., D. E. Bayer and R. K. Glenn. 1966. Translocation of herbicides and assimilates in red maple and white ash. *Bot. Gaz.* **127**:193–201.

Leonard, O. A., R. K. Glenn, and D. E. Bayer. 1965. Studies on the cut-surface method. 1. Translocation in blue oak and madrone. *Weeds* 13:346–351.

Leonard, O. A. and R. J. Hull. 1965. Translocation of ^{14}C-labelled substances and [32]PO$_4$ in mistletoe-infected and uninfected conifers and dicotyledonous trees. *Symp. Use of Isotopes in Weed Res.* Vienna, p. 17.

Levy, R. and J. Menoret. 1959. Herbicides. French Patent No. 1,193,374.

Lotlikar, P. D., L. F. Remmert, and V. H. Freed. 1968. Effects of 2,4-D and other herbicides on oxidative phosphorylation in mitochondria from cabbage. *Weed Sci.* 16:161–165.

Lund-Hoie, K. and D. E. Bayer. 1968. Absorption, translocation and metabolism of 3-amino-1,2,4-triazole in *Pinus ponderosa* and *Abies concolor*. *Physiol. Plant.* 21:196–212.

Mann, J. D., L. S. Jordan, and B. E. Day. 1965. A survey of herbicides for their effect upon protein synthesis. *Plant. Physiol.* 40:840–843.

Massini, P. 1958. Uptake and translocation of 3-amino- and 3-hydroxy-1,2,4-triazole in plants. *Acta. Botan. Neerl.* 7:524–530.

Massini, P. 1959. Synthesis of 3-amino-1,2,4-triazolyl alanine from 3-amino-1,2,4-triazole in plants. *Biochim. et Biophys. Acta* 36:548–549.

Matolcsy, G. and E. A. A. Gomaa. 1968. The herbicidal effect of some α-hydroxy-β,β,β-trichloroethylamino-1,2,4-triazoles. *Weed Res.* 8:1–7.

McWhorter, C. G. 1963. Effects of 3-amino-1,2,4-triazole on some chemical constituents of *Zea mays*. *Physiol. Plant.* 16:31–39.

McWhorter, C. G. and J. L. Hilton. 1967. Alterations in amino acid content caused by 3-amino-1,2,4-triazole. *Physiol. Plant.* 20:30–40.

McWhorter, C. G. and W. K. Porter. 1960. Some effects of amitrol on the respiratory activities of *Zea mays*. *Weeds* 8:29–38.

Miller, C. S. and W. C. Hall. 1957. Effects of amino triazole salts and derivatives on cotton defoliation, growth inhibition and respiration. *Weeds* 5:218–226.

Miller, C. S. and W. C. Hall. 1961. Absorption and metabolism of amino-triazole in cotton. *J. Agr. and Food Chem.* 9:210–212.

Moreland, D. E., G. S. Malhotra, R. D. Gruenhagen, and E. H. Shokraii. 1969. Effects of herbicides on RNA and protein synthesis. *Weed Sci.* 17:556–563.

Naylor, A. W. 1964. Complexes of 3-amino-1,2,4-triazole in plant metabolism. *J. Agr. and Food Chem.* 12:21–25.

Norris, J. 1961. Acyl aminotriazole compounds and herbicide compositions. British Patent No. 883,732.

Penot, M. 1960. Circulation de L'aminotriazole apres apport foliaire. *Acad. Des. Sci. Compt. Rend.* 250:1325–1327.

Putala, E. C. 1967. A study of the ultrastructure and histochemistry of the apoplastidic effect of 3-amino-1,2,4-triazole on the shoot apex of tomato. Ph.D. Thesis, Univ. Calif., Berkeley, 51 pp.

Pyfrom, H. T., D. Appleman, and W. G. Heim. 1957. Catalase and chlorophyll depression by 3-amino-1,2,4-triazole. *Plant Physiol.* **32**:674–676.

Racusen, D. 1958. The metabolism and translocation of 3-aminotriazole in plants. *Arch. Biochem. and Biophys.* **74**:106–113.

Rogers, B. J. 1957a. Chlorosis and growth effects as induced by the herbicide 3-amino,1,2,4-triazole. *Proc. 14th North Central Weed Control Conf.* p. 9.

Rogers, B. J. 1957b. Chlorosis in corn as induced by the herbicide 3-amino-1,2,4-triazole. *Plant Physiol. Abstr.* **32**:vi–vii.

Rogers, B. J. 1957c. Translocation and fate of amino triazole in plants. *Weeds* **5**:5–11.

Schweizer, E. E. and B. J. Rogers. 1964. Effects of amitrole on acid-soluble nucleotides and ribonucleic acid in corn. *Weeds* **12**:310–311.

Shaw, W. C. and C. R. Swanson. 1953. The relation of structural configuration to the herbicidal properties and phytotoxicity of several carbamates and other chemicals. *Weeds* **2**:43–65.

Shimabukuro, R. H. and A. J. Linck. 1964. The interaction of temperature and carbohydrate concentration with absorption and translocation of 3-amino-1,2,4-triazole. *Physiol. Plant.* **17**:100–106.

Shive, C. J. and H. L. Hansen. 1958. Some anatomical responses of conifer needles to 3-amino-1,2,4-triazole. *Hormolog* **2**:9–10.

Smith, L. W., D. E. Bayer, and C. L. Foy. 1968a. Metabolism of amitrole in excised leaves of Canada thistle ecotypes and beans. *Weed Sci.* **16**:523–527.

Smith, L. W., D. E. Bayer, and C. L. Foy. 1968b. Influence of environmental and chemical factors on amitrole metabolism in excised leaves. *Weed Sci.* **16**:527–530.

Smith, L. W. and P. J. Davies. 1965. The translocation and distribution of three labelled herbicides in *Paspalum distichum* L. *Weed Res.* **5**:343–347.

Sund, K. A., F. C. Putala, and H. N. Little. 1960. Reduction of 3-amino-1,2,4-triazole phytotoxicity in tomato plants. *J. Agric. and Food Chem.* **8**:210–212.

Sund, K. A. and H. N. Little. 1960. Effect of 3-amino-1,2,4-triazole on the synthesis of riboflavin. *Science* **132**:622.

Swanson, C. A. and R. H. Bohning. 1951. The effects of petiole temperature on the translocation of carbohydrates from bean leaves. *Plant Physiol.* **26**:557–564.

Tsay, R. and F. M. Ashton. 1971. Effect of several herbicides on dipeptidase activity of squash cotyledons. *Weed Sci.* **19**:682–684.

van der Zweep, W. 1965. Laboratory trials on the interaction between ammonium thiocyanate and N^6-benzyladenine, respectively, and amitrole. Ergebn. 6. dt. Arbeitsbesprechung über Fragen der Unkrautbiologie u.-bekampfung, Hohenheim, 1965. *Z. PflKrankh. PflPatt. PflSchutz.* (Sonderh 3), 123–127.

Volger, C. 1964. Contributions to research on the translocation of amitrole in *Pteridium aquilinum* (L.) Kahn. *16th Internat. Symp. Fytofarm. Fytiatrie, Gent* pp. 663–676.

Volger, C. 1965. The translocation of aminotriazol in bracken. Ergebn. 6. dt. Abeitsbesprechung über Fragen der Unkrautbiologie u.-bekampfung. Hohenheim, 1965. *PfKrankh. PflPath. PflSchutz* (Sonderh 3), 129–138.

Weier, T. E. and A. A. Imam. 1965. Some cytological effects of amitrole. *Am. J. Bot.* **52**:631.

Wiater, A., K. Krajewska-Grynkiewicz, and T. Klopotowski. 1971a. Histidine biosynthesis and its regulation in higher plants. *Acta Biochim. Polonica* **18**:299–307.

Wiater, A., T. Klopotowski, and G. Bagdasarian. 1971b. Synergistic inhibition of plant imidazoleglycerol phosphate dehydratase by aminotriazole and phosphate. *Acta Biochim. Polonica* **18**:309–314.

Wort, D. J. and B. C. Loughman. 1961. The effect of 3-amino-1,2,4-triazole on the uptake, retention, distribution and utilization of P-32 by young barley plants. *Can. J. Bot.* **29**:339–351.

Yamaguchi, S. and A. S. Islam. 1967. Translocation of eight ^{14}C-labeled amino acids and three herbicides in two varieties of barley. *Hilgardia* **38**:207–229.

CHAPTER 21

Ureas

$$H \quad O \quad R_2$$
$$\backslash \quad \| \quad /$$
$$N-C-N$$
$$/ \qquad \backslash$$
$$R_1 \qquad R_3$$

Several substituted urea compounds were included in the survey of almost 1100 compounds by Thompson et al. (1946) for growth-inhibitory activities. Although some of these substituted urea compounds inhibited growth their full herbicidal potential was not clearly evident because of the nature of the tests. It should be remembered that this research was conducted at the dawn of organic herbicides. However, it does point out the necessity of carefully designed experiments for the screening of compounds for herbicidal activity. The student of weed science should also be cognizant of the seventeen papers in addition to that of Thompson et al. (1946) which were published in volume 107 of *Botanical Gazette* because of their historical significance. These papers cover much of the early research basic to weed science. The research was conducted but could not be published because of the wartime security policies of World War II.

DCU was the first substituted urea to be used commercially for weed control. It was recommended as a preemergence material, toxic to grasses and selective in certain broadleaf plants. It has been superceded by other herbicides and is not currently registered for use in crops in the United States. Bucha and Todd (1951) described the herbicidal potential of monuron. Over the intervening years thousands of substituted urea compounds have been tested as herbicides and many are available for commercial use.

Most urea herbicides are relatively nonselective and are usually applied to the soil; however certain ones are active through the foliage. Foliar activity is often increased by formulation, i.e., addition of surfactants or non-phytotoxic oils. Selectivity may be obtained in certain crops by taking advantage of the water solubility and adsorptive properties of the herbicide

TABLE 21-1. Urea Type Herbicides in General Use

Common name	Chemical name	R_1	R_2	R_3
buturon	3-(p-chlorophenyl)-1-methyl-1-(1-methyl-2-propynyl)urea	Cl—⟨phenyl⟩—	CH_3—	CH_3—CH— / C≡CH
chlorbromuron	3-(4-bromo-3-chlorophenyl)-1-methoxy-1-methylurea	**Br**—⟨phenyl⟩—, Cl	CH_3—	CH_3O—
chloroxuron	3-[p-(p-chlorophenoxy)phenyl]-1,1-dimethylurea	Cl—⟨phenyl⟩—O—⟨phenyl⟩—	CH_3—	CH_3—
cycluron	3-cyclooctyl-1,1-dimethylurea	⟨cyclooctyl⟩—	CH_3—	CH_3—
DCU	1,3-bis(2,2,2-trichloro-1-hydroxyethyl)urea	Cl—C—C—OH with Cl, H, Cl	H—	Cl—C—C—OH with Cl, H, Cl
diuron[a]	3-(3,4-dichlorophenyl)-1,1-dimethylurea	Cl—⟨phenyl⟩—, Cl	CH_3—	CH_3—
fenuron	1,1-dimethyl-3-phenylurea	⟨phenyl⟩—	CH_3—	CH_3—
fenuron TCA[b]	1,1-dimethyl-3-phenylurea mono(trichloroacetate)	⟨phenyl⟩—	CH_3—	CH_3—

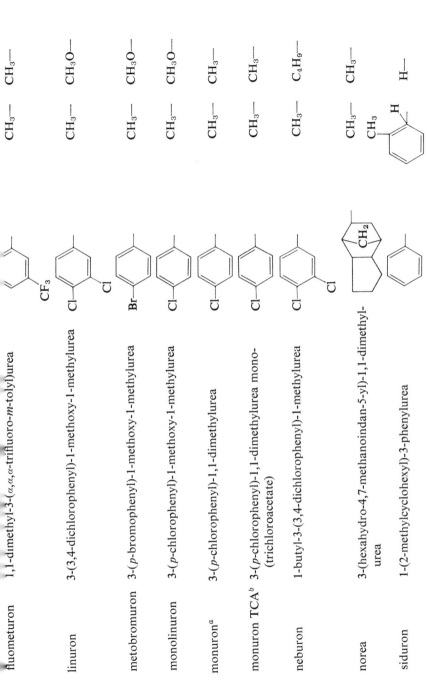

			R₂	R₃
fluometuron	1,1-dimethyl-3-(α,α,α-trifluoro-*m*-tolyl)urea		CH_3-	CH_3-
linuron	3-(3,4-dichlorophenyl)-1-methoxy-1-methylurea		CH_3-	CH_3O-
metobromuron	3-(*p*-bromophenyl)-1-methoxy-1-methylurea		CH_3-	CH_3O-
monolinuron	3-(*p*-chlorophenyl)-1-methoxy-1-methylurea		CH_3-	CH_3O-
monuron[a]	3-(*p*-chlorophenyl)-1,1-dimethylurea		CH_3-	CH_3-
monuron TCA[b]	3-(*p*-chlorophenyl)-1,1-dimethylurea mono-(trichloroacetate)		CH_3-	CH_3-
neburon	1-butyl-3-(3,4-dichlorophenyl)-1-methylurea		CH_3-	C_4H_9-
norea	3-(hexahydro-4,7-methanoindan-5-yl)-1,1-dimethyl-urea		CH_3-	CH_3-
siduron	1-(2-methylcyclohexyl)-3-phenylurea			$H-$

a Monuron and diuron were previously designated as CMU and DCMU respectively; the photosynthetic biologists continue to use the older terminology.
b Fenuron TCA and monuron TCA have the same structure as the parent urea compound but in addition contain mono (trichloroacetate) [·CCH₃COOH] "loosely" bound to positions R₂ and R₃.

369

in conjunction with soil characteristics. Shallow-rooted annual weeds may be controlled in deep-rooted perennial crops by using a compound which has low water solubility and high adsorptive properties in a soil of relatively high adsorptive capacity, i.e. annual weeds in tree crops. Conversely, deep-rooted weeds may be controlled in shallow-rooted crops by using a compound which has high water solubility and low adsorptive properties in a soil of relatively low adsorptive capacity, i.e. woody plants on range land. In this latter case, sufficient rain must follow the application to leach the herbicide below the rooting zone of the crop. The water solubilities of herbicides are given in Table A-3 in the appendix.

In addition to the above basis of selectivity, certain species of plants appear to have an inherent resistance to some of the urea herbicides. Citrus species, turkey mullein (*Eremocarpus setigerus*) and common groundsel (*Senecio vulgaris*) are examples of plants which are resistant to monuron. Linuron can be used selectively for annual weed control in carrots as a post-emergence treatment. There are several other examples of resistance to the urea herbicides as evidenced by the several registrations. The bases of these selectivities are usually restricted absorption and translocation or rapid inactivation by the crop plant.

It has also been shown that all urea herbicides are not equally phytotoxic. Sheets and Crafts (1957) reported that six times as much fenuron was required to inhibit the growth of oats 50% relative to diuron, when applied via the nutrient solution, and four times as much fenuron as monuron was required, (See Table 21.2). However, when these four herbicides were mixed into Yolo clay loam and oats planted, monuron and fenuron were about six times as toxic as DMU indicating that the capacity of the soil to adsorb these compounds alters their relative toxicity tremendously.

Substituted ureas have proven effective in the control of aquatic weeds. Walsh and Grow (1971), using representative species from six genera of

TABLE 21-2. Comparisons of the Toxicity to Oats of 4 Urea Herbicides in Nutrient Solution (Sheets and Crafts, 1957)

Chemical	Gm of chemicals $\times$ 10^{-5} required to reduce growth 50%
DMU[a]	1.8
diuron	1.9
monuron	3.0
fenuron	12.1

[a] 3-(3,4-dichlorophenyl)-1-methylurea.

marine unicellular algae, proved that four substituted urea herbicides depressed concentration of carbohydrate in all species at salinities ranging from 5 to 30 parts per thousand (ppt). *Dunaliella tertiolecta* was most resistant; its carbohydrate content decreased 9.2% at 5 ppt and 17.9% at 30 ppt. Chlorococcum was most susceptible; its carbohydrate content decreased 49.1% at 5 ppt, and 65.6% at 30 ppt salinity.

Growth and Plant Structure

In the initial paper in *Science* introducing monuron to the scientific community Bucha and Todd (1951) stated, "The initial effect generally is leaftip dieback, beginning on the older leaves. This is followed by progressive chlorosis and retardation of growth, ending in the death of the plant." This simple treatment is most descriptive and clearly showed that the urea herbicides were distinctly different from the previously described auxin-like herbicides which caused epinasty or the carbamate herbicides which caused malformations of meristems.

Minshall (1957) described the symptoms of monuron injury to bean, tomato, spinach, maize, and barley as water-soaked blotch, silver blotch, indeterminate gray blotch, wilt, petiole collapse, stem collapse, rapid yellowing, abscission, and partial chlorosis. The specific symptom(s) was influenced by plant species, dose, environmental conditions, and time after treatment. The "acute" symptom, as water-soaked blotch, was preceded by the development of light green areas. The water-soaked blotch required two to three days to develop in intact potted plants, but could be demonstrated somewhat sooner in culture solution or in excised shoots placed in the herbicide solution. Conditions favoring high transpiration rates also accelerated the development of the symptom. The internal foliar concentration required for this acute symptom was in excess of 100 μg/g fresh weight. The "chronic" symptoms of wilt, stem collapse, silver blotch, indeterminate gray blotch, rapid yellowing, and abscission required several days and developed at internal foliar concentrations of less than 50 μg/g fresh weight. Two types of yellowing or chlorosis were described, (1) rapid yellowing which resembled that of normal senescence but developed more rapidly and was either localized close to the primary veins or affected large areas of the leaf blade, (2) partial chlorosis which affected the entire leaf blade and was light green or yellowish green in color. Figures 21-1 and 21-2 from Minshall (1957) illustrate these phytotoxic symptoms.

The action spectrum of monuron toxicity to oats and beans was determined by Ashton (1965) applying monuron via the roots using sand culture. He reported that the action spectrum was similar to the absorption spectrum of chlorophyll indicating that chlorophyll was the principal absorbing pigment involved in monuron injury. No injury was observed in the dark and the

FIGURE 21-1.   (*1*) Early stage of intercostal water-soak blotch in bean. Blotches were light green in color. Photographed 53 hours after excised shoot placed on 32 ppm monuron. Plant was grown under slatted shades in greenhouse in August. (*2*) Same leaf as in (*1*), photographed 4 days later. Blotches were brown in color. (*3*) Marginal water-soak blotch in bean. (*4*) Early stage of indeterminate gray blotch in bean. Blotches were grayish in color. Photographed 6 days after intact plant was watered with 32 ppm monuron. Plant was grown under slatted shades in greenhouse in August. (*5*) Early stage of silver blotch in bean. Blotches were silvery or whitish in color. Intact potted plant watered with 1 ppm monuron for 7 days while held under slatted shades in greenhouse in August. On 8th day plant placed in full sunlight. Photograph taken on the 9th day. This blotch developed after plant was placed in sunlight. (*6*) Silver blotch. A second leaf from plant receiving identical treatment and photographed at the same time as leaf in (*5*). Foliage shaded a part of this leaf during exposure to sunlight. (Minshall, 1957.)

372

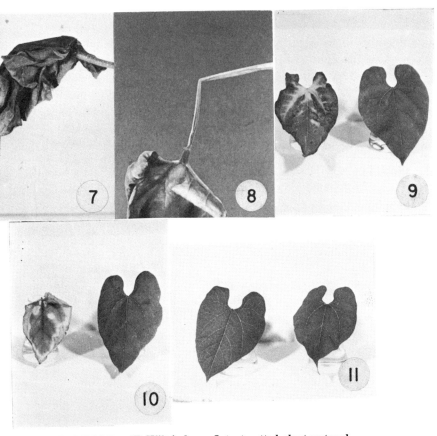

FIGURE 21-2. (7) Wilt in bean. Intact potted plant watered with 8 ppm monuron for 7 days while held under slatted shades in greenhouse in August. On 8th day plant placed in full sunlight. Photograph taken on 9th day. The wilt developed after plant placed in sunlight. (8) Petiole collapse in bean. Intact potted plant watered with 32 ppm monuron while held under slatted shades in greenhouse in August. Photographed on the 9th day. (9) and (10) Rapid yellowing in bean. Leaves to right of each figure were water controls. Leaves to left were excised from plants grown out-of-doors in September, treated with monuron, then held at temperature of 82°F and light intensity of 700 ft-c. Photographed 7 days after treatment. Treated leaf in (9) was from a 4-weeks-old plant and contained 25 μg/g monuron. Treated leaf in (1) was from a 4-weeks-old plant and contained 200 μg/g monuron. (11) Partial chlorosis in bean. Leaf to left was a water control. Leaf to right was excised from a 2-weeks-old plant grown in greenhouse under slatted shades in September, treated with 25 μg/g monuron, then held at temperature of 82°F and light intensity of 700 ft-c. Photographed 7 days after treatment. (Minshall, 1957.)

higher the light intensity the greater the injury. In red pine (*Pinus resinosa*) seedlings, treated with monuron via the roots, the needles remained green for 30 days but after 40 days the green color began to fade (Sasaki and Kozlowski, 1966). In the unicellular green algae, *Chlorella pyrenoidosa* and *Euglena gracilis*, fluometuron reduced the chlorophyll content (Sikka and Pramer, 1967).

Non-photosynthetic tissues also appear to be adversely affected by monuron and perhaps other substituted urea herbicides. Muzik et al. (1954) showed monuron could kill excised roots of *Stizolobium deeringianum* grown on a synthetic media containing 100 or 200 mg/L of the herbicide and inhibit root growth at 25 or 50 mg/L. The elongation of detached roots of pea was inhibited by monuron to about the same degree as attached roots of maize and timothy, indicating that the toxic action of monuron was exerted directly on the attached roots (Minshall, 1960). However, in the chlorophyll-containing roots of *Hydrocharis morsusranae* a much lower concentration of the herbicide was required to give equivalent inhibition. Voderberg (1961) also reported that monuron inhibited the growth of roots of several species. The study of Jordan et al. (1966) using non-photosynthetic tobacco callus tissue grown in the dark with an exogenous supply of sucrose clearly showed that monuron can inhibit the growth of organs other than roots by a mechanism not involving photosynthesis.

Grant (1964) observed that monuron produced abnormal meiotic cells in root tips and Wuu and Grant (1966) reported that monuron and linuron were highly effective in inducing chromosome aberrations in the C_1 generation of barley.

Absorption and Translocation

Bucha and Todd (1951) concluded from their original greenhouse studies with monuron that it was readily absorbed by the root system and translocated to the leaves. Field observations by McCall (1952) provided further evidence that monuron was more effective as a soil application than as a foliar application and this led to the assumption that entrance of the herbicide into the plant was primarily through the roots. However, Muzik et al. (1954) were able to show that monuron was absorbed by leaves, stems, and roots of *Stizolobium deeringianum* in sufficient quantity to cause injury. Although uptake by the roots was by far the more effective, immersion of one-half of the root system in 200 mg/L of the herbicide for 15 minutes was sufficient to cause subsequent death. They concluded that the movement of monuron was primarily toward the shoot and that the path was through the xylem via the transpiration stream. They did not have radioactively labeled monuron available and used such classical methods as girdling the phloem, cutting the xylem and reducing transpiration in conjunction with toxicological symptoms. Concurrently, workers at E. I. duPont de Nemours and Company (Haun and Peterson, 1954) using ^{14}C-labeled monuron on several species found that absorption by roots was very rapid but downward translocation from a foliar application was nil. Numerous other investigators have studied the absorption and translocation of monuron in a variety of species; some of these are: Crafts, 1959, 1962, 1967; Crafts and Yamaguchi, 1958, 1960; Fang et al., 1955; Leonard et al., 1966; Minshall, 1954; Muzik et al., 1954; Pereira et al., 1963; Smith and Sheets, 1966; Yamaguchi and Islam, 1967. The unanimous conclusion is that monuron is absorbed rapidly by roots but at much slower rates by other organs and it is almost exclusively translocated via the apoplastic system. Monuron was the first herbicide known to be translocated only by the apoplastic system and its translocation pattern has been used as a standard to indicate this type of transport by other compounds.

However, there is one problem involved in accepting the exclusive apoplastic translocation of monuron and that is the site of action is generally considered to be in the chloroplast. Since the chloroplasts are surrounded by protoplasm monuron must move through the plasmalemma and protoplasm to reach the site of action. However, since the distances from the cell wall to the chloroplasts are quite short this movement probably occurs passively by simple diffusion. It seems possible too that this movement may be at concentrations that cannot be detected by autoradiography.

In an elegant series of experiments by Donaldson (1967) it was found that monuron uptake by roots was passive (vs. active) by diffusion and that it was not influenced by low temperatures, anaerobic conditions or metabolic inhibitors. Absorption of exchangeable monuron increased rapidly for the first hour but after that time no further absorption occurred. Non-exchangeable monuron absorption exhibited a steady increase with time. Between 80 and 90% of the absorbed monuron was present in the exchangeable fraction. About 0.6 times as much monuron was present in 1 gram of roots as in 1 ml of external solution. As absorption by diffusion would not result in accumulation, this indicates that monuron penetrates approximately 60% of the root volume. It was suggested, since monuron is relatively nonpolar, that monuron can passively penetrate the membranes of the cortical cells of the root. From these it moves into the stele and thence upward in the transpiration stream. This would explain how monuron is able to bypass the Casparian strip of the endodermis. Histoautoradiographic studies revealed that diuron was almost entirely in the cell walls, particularly those of the stele.

Pickering (1965) investigated the foliar penetration pathway of monuron by histoautoradiographic techniques and concluded that it entered the leaf and became evenly distributed throughout the treated tissue. However, it was present at high concentrations in the vascular tissue but only at low concentrations in the chlorenchyma tissue. It appeared to enter the symplast with difficulty and translocation was restricted to the xylem.

Day (1955) found that monuron, as used to control weeds under citrus crops, is absorbed by the citrus trees and translocated apoplastically. However, the fruits, which transpire very little, and are developed by nutrients from the assimilate stream, were found to contain little or none of the monuron. Rose cuttings were also able to tolerate the urea herbicides which are used commercially in their production.

The absorption and translocation of diuron from leaves or roots of several species show that its pattern of movement is also primarily restricted to the apoplast (Bayer and Yamaguchi, 1965; Leonard et al., 1966; Leonard and Glenn, 1968). Nishimoto et al. (1967) studied the site of uptake of diuron by germinating seedlings by treating bands of soil above and/or below the location of the seed whereas Prendeville et al. (1967) investigated the same problem by using a double plastic pot technique. Both investigations showed that diuron was primarily absorbed by the roots and little, if any, absorbed by the emerging shoot.

Strang and Rogers (1971) found that ^{14}C-diuron accumulates to marked concentrations in the lysigenous or pigment glands of cotton leaves and in trichomes; this binding of diuron may be a major factor in the tolerance of cotton to diuron.

The absorption and translocation of linuron appears to vary considerably with different species although the pattern of distribution suggests primarily apoplastic movement. Borner (1964, 1965) reported that linuron was translocated very rapidly to the shoots following an application to the roots of *Sinapis arvensis* but transport was much slower when bean was used. Within 9 days *S. arvensis* was almost dead whereas bean was not significantly damaged. In contrast to the finding with diuron reported above which showed that it was primarily absorbed by roots rather than the emerging shoot, Knake and Wax (1968) reported that linuron reduced the fresh weight of giant foxtail (*Setaria faberii*) when the herbicide was placed in the shoot zone but no reduction occurred when the herbicide was placed in the root zone. These results suggest that linuron is absorbed more readily by the emerging shoot than by the root system. The uptake of linuron by excised roots of soybean was found to be temperature dependent, between 5° and 30°C uptake increased with temperature, suggesting active uptake (Moody and Buchholtz, 1968). However, Donaldson (1967) reported that monuron absorption by roots was not temperature dependent and suggested passive uptake. Additional studies by Moody et al. (1970) showed that about 40% of the linuron taken up by excised roots of soybean during the first hour moved out of the roots within 30 minutes when placed in water and an additional 10 to 20% was released from the roots during the next 3.5 hours. The use of non-radioactive linuron solution instead of water as the release solution did not alter the results. They concluded that the herbicide molecules released were those that had diffused out of free space rather than those released from uptake or adsorptive sites. Taylor and Warren (1968) observed that linuron moved much slower than chlorpropham, amitrole or 2,4-D through vertical 6 mm long sections of bean petiole sections. Chloramben did not move at all. However, after 12 hours the rate of linuron transport increased markedly. Pretreatment of the petiole sections with dinoseb, 10^{-5} to 10^{-3}M, or KN_3, 10^{-4} to 10^{-2}M, stimulated the movement. They concluded that these metabolic inhibitors increased membrane permeability or reduced active retention.

Chloroxuron absorption and translocation have been investigated by Aebi and Ebner (1962) and Geissbühler et al. (1963a). They reported that chloroxuron was readily absorbed by roots and translocated into the stem and leaves. However there were species differences, smallflower galinsoga (*Galinsoga paraviflora*) absorbed and translocated about twice as much chloroxuron as wild buckwheat (*Polygonum convolvulus*). The translocation rate was reduced by increased relative humidity, lack of light and lower temperatures in both species. From a leaf application, there was no translocation into the neighboring leaves but some transport occurred within the treated leaf from the site of application at the base of the leaf toward the apex. All of these findings support the concept of translocation via the apoplastic system.

Fluometuron absorption, translocation, and metabolism was studied in cotton, a resistant species, and cucumber, a susceptible species (Rogers and Funderburk, 1967a, 1967b, 1968a, 1968b). Following root uptake from a culture solution, fluometuron was found to be accumulated at the leaf margins with only a limited amount in the root of cucumber; while in cotton, uniform distribution throughout the plant was observed except for some accumulation in the lysigenous glands of the leaf. There was no basipetal translocation of fluometuron from a foliar application in either species, but cucumber absorbed more of the herbicide than cotton. Four degradation products were found in cotton and two in cucumber. They concluded that although differential absorption did not account for the differential selectivity of these two species to fluometuron, differential translocation and metabolism may be involved. In citrus, Goren (1969) found that fluometuron was not present in sufficient quantity in the leaves to affect photosynthesis, respiration, dry weight or chlorophyll content following high rates of application to the soil. Autoradiograph studies showed that the radioactivity, presumably fluometuron, accumulated in the roots and lower stem following an application to the roots. However, photosynthesis in leaf disks was inhibited 70% by 80 mg/L of fluometuron, showing that if the herbicide had been present in the leaves of the intact plant, photosynthesis would have been inhibited. He concluded that the resistance of citrus to fluometuron either involves the lack of translocation or demethylation or both.

Comparative studies on the translocation of chloroxuron, metabromuron and fluometuron in bean are most revealing, see Figure 21-3 (Voss and Geissbühler, 1966). It is clear that these three urea-type herbicides have markedly different mobilities in spite of the fact that they all translocated in the apoplastic system. Whereas the movement of chloroxuron to the shoot was restricted in this relatively short-term experiment of 24 hours; both fluometuron and metabromuron were rapidly translocated upward from the roots. However, the distribution of fluometuron and metabromuron are quite different. Metabromuron was mainly confined to the veins with restricted movement therefrom, whereas fluometuron was primarily in the interveinal areas with very little remaining in the veins.

Splittstoesser and Hopen (1968) reported that siduron is rapidly absorbed by roots and translocated via the apoplastic system to the shoot but not from the shoots to the roots.

It is apparent from the above reports that the chemical structure and species of the plant are critical factors in determining the degree of absorption and the pattern of translocation of the urea-type herbicides. Selectivity, based on differential absorption and/or translocation, has been obtained in certain species with some urea-type herbicides. These have usually been soil applications in which the herbicide translocation to the shoots is limited.

Rieder et al. (1970) reported that linuron may be taken up by soybean seeds from an aqueous solution. There was a direct proportional relationship between uptake and herbicide concentration in the soaking solution. Raising the temperature from 10° to 30°C increased the uptake of linuron. After 48 hours at 30°C, 83% of the linuron in the original solution was absorbed. Uptake rate was similar in living and dead seeds. It was concluded that the uptake by seeds was largely a physical process that required seed hydration.

Molecular Fate

The molecular fate of the substituted urea herbicides has been recently reviewed by Geissbühler (1969).

Demethylation appears to be the primary detoxification mechanism for most urea herbicides since the removal of a methyl group from a dimethylated urea reduces its phytotoxicity and the removal of the second methyl group makes it essentially non-phytotoxic. Geissbühler et al. (1963b) showed that chloroxuron was demethylated to form N'-(4-chlorophenoxy)-phenyl-N-methylurea and N'-(4-chlorophenoxy)-phenylurea by leaves and roots of

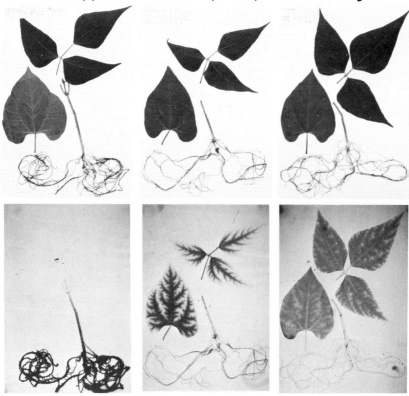

CHLOROXURON METOBROMURON FLUOMETURON
2.5ppm herbicide (equal specific activity)

FIGURE 21-3. Comparative uptake by roots and translocation to aerial parts of ^{14}C-labeled chloroxuron, metobromuron, and fluometuron in French dwarf bean seedlings, cultured in nutrient solution supplied with 2.5 ppm (and equal concentrations of radioactivity) of each herbicide. Plants kept in herbicide solution for 24 hours and then cultured in regular nutrient solution for an additional 24 hours. Top: mounted plants. Bottom: radioautographs of same plants. (Voss and Geissbühler, 1966.) Reprinted from P. C. Kearney and D. D. Kaufman, eds., *Degradation of Herbicides*, p. 86, by courtesy of Marcel Dekker, Inc.

maize, bean, and broad bean. These compounds are intermediates and not the end degradation product. They also reported that wild buckwheat, a resistant species, was able to metabolize chloroxuron more rapidly than small-flower galinsoga, a susceptible species. The major pathway of fluometuron and metabromuron degradation was considered to be stepwise demethylation by Voss and Geissbühler (1966) in cotton roots and potato leaves. Rogers and Funderburk (1968b) also found that fluometuron undergoes demethylation in cotton and cucumber. Both the roots and shoots of cotton are able to demethylate fluometuron but the shoots are more active (Rogers and Funderburk, 1968a). Rubin and Eshel (1971) tested the phytotoxicity of fluometuron and three analogues, one lacking one methyl group, one lacking both methyl groups and one a trifluoro-*m*-toluidine on cotton, foxtail, and redroot pigweed. Fluometuron was the most active of the four compounds, the demethylated derivative and the aniline analogue were the least toxic. Photosynthesis in cotton and redroot pigweed was inhibited by fluometuron, to a lesser degree by the monomethyl analogue and not at all by the demethyl and aniline analogues. Cotton was the most tolerant of the three species studied. The demethylation of monuron and/or diuron by several species has been demonstrated although the rate of degradation varied with species (Smith and Sheets, 1967; Swanson and Swanson, 1968; and Onley et al., 1968). Frear (1968) and Frear and Swanson (1969) isolated a microsomal enzyme system from several species which was able to demethylate monuron, diuron, fluometuron and N'-(4-chlorophenyl)-N-methylurea. The required cofactors were molecular oxygen and NADPH or NADH.

Demethoxylation of the urea herbicides containing a methoxy group (linuron, monolinuron, metabromuron, chlorbromuron) has not been studied as thoroughly as demethylation. Voss and Geissbühler (1966) reported that metabromuron undergoes both demethylation and demethoxylation to give 4-bromophenylurea in potatoes. Borner (1964, 1965) presented evidence that linuron was not metabolized within 14 days in bean and *Sinapsis arvensis*. However, Kuratle (1968) observed the degradation of linuron in carrot and common ragweed (*Ambrosia artemisiifolia*). Carrots, a resistant species, had 89% of the total radioactivity in a non-toxic form while common ragweed, a susceptible species, had only 13% in non-toxic form. A small but measurable amount of 3-(3,4-dichlorophenyl)-1-methoxyurea and 3,4-dichloroaniline were found as degradation products of linuron in maize, soybean and large crabgrass (*Digitaria sanguinalis*) showing that it also undergoes demethoxylation (Nashed and Ilnichi, 1970). Voss and Geissbühler (1966) suggested that the demethoxylation may actually be a two-step reaction involving a hydroxylation with the removal of the methyl group and subsequent dehydroxylation, however the hydroxylamine intermediates were not detected.

No evidence is available as to whether substitutions other than methyl or methoxy groups on the R_2 or R_3 positions (see Table 21-1) of other urea herbicides (buturon, DCU, neburon) are metabolized by higher plants. However, Splittstoesser and Hopen (1968) reported that siduron was not degraded by barley or crabgrass.

Following the dealkylation and/or dealkoxylation of the urea herbicides, the resultant substituted urea derivatives (R—NH—CO—NH$_2$) are subject to further degradation in higher plants. They undergo hydrolysis which involves a deaminization and decarboxylation yielding the corresponding aniline (R—NH$_2$). However, the extent of this reaction and how generally it occurs is open to question since the aniline derivative is not always found and when present the concentrations are relatively low (Geissbuhler, 1969), Table 21-3. These results could also be explained by further rapid transformations of the anilines, such as by conjugation or oxidation, which would prevent accumulation of the aniline. The corresponding aniline of the following compounds have been reported for certain species; monuron (Smith and Sheets 1967, Swanson and Swanson 1968); diuron (Onley et al. 1968); fluometuron (Voss and Geissbühler 1966, Rogers and Funderburk 1967b, 1968a, 1968b); chlorbromuron (Nashed et al. 1970); linuron (Nashed and Ilnichi 1970).

Onley et al. (1968) identified 3,4-dichloronitrobenzene (R—NO$_2$) from the extract of a plant treated with diuron suggesting that the aniline derivative may undergo oxidation.

Conjugation of the urea herbicides with normal cellulose constituents was first suggested by Fang et al. (1955) in their studies on the metabolism of monuron. They later reported that monuron appeared to be conjugated with a low molecular weight protein or peptide (Freed et al., 1961). In most of the subsequent studies on the metabolism of the urea herbicides the possibility of such conjugates has been ignored. However, recently additional reports have appeared supporting this point of view. Katz (1967) reported that appreciable amounts of radioactivity were not acetone extractable from plants treated with linuron but were released by alkaline digestion or treatment with the proteolytic enzyme ficin. In studies with maize, soybean, and large crabgrass, Nashed and Ilnichi (1970) found that from 15 to 25% of the absorbed linuron was non-extractable with acetone but was released by alkaline hydrolysis. Nashed et al. (1970) also presented evidence for conjugation of chlorbromuron. Voss and Geissbuhler (1966) indicated that fluometuron and metabromuron formed unidentified conjugates in cotton and wheat, respectively. However, some of these conjugates may involve degradation products of the herbicide rather than the intact herbicide molecule.

Camper and Moreland (1971) having established that diuron is bound by protein, used bovine serum albumin in studies to find the influence of temperature, pH, ionic strength and modification of the protein on sorption expressed as moles of chemical bound per mole of protein. They found that free amino groups of the serum albumin were involved in the binding. Studies with diuron derivatives suggested that the amide nitrogen and the carbonyl oxygen of the phenylamide are involved; conformation of the protein may control the extent of binding, increased chlorination of the phenyl ring correlated with increased binding.

TABLE 21-3.   Demethylation and Hydrolysis of Urea Herbicides after Application of Labeled Compounds to Different Plant Species (Geissbuhler, 1969)

^{14}C-labeled herbicide	Plant species	Mode of application	Exposure time	Recognized metabolites[a]	Ref.[b]
monuron ring-^{14}C	cotton soybean	{ nutrient solution	5 days	U, M, D ⩾ A U > M, D, (A)	83 83
	cotton plantain soybean maize	{ leaf disks incubat- ed with solution	8 hours	U < M, D ⩾ A U, M, D ⩾ A U, M ⩾ D ⩾ A U ⩾ M > D, (A)	85 85 85 85
diuron carbonyl-^{14}C	oat soybean maize cotton	{ nutrient solution	5 days	U < M > D U ⩽ M > D U > M ⩾ D U < M ⩽ D	83 83 83 83
	cotton plantain soybean maize	{ leaf disks incuba- ted with solution	8 hours	U > M > D U < M, D D > M, (D) D ⩾ M, (D)	85 85 85 85
diuron ring-^{14}C	maize	nutrient sol.	0–6 days	U ⩾ M, D ⩾ A	84
methyl-^{14}C carbonyl-^{14}C	maize	leaves	0–6 days	U, unknowns	84
fluometuron trifluoro- methyl-^{14}C	cotton cucumber	{ nutrient solution	0–4 days	U ⩽ M, D ⩾ A U < M > D, (A)	86 86
	cotton maize	{ nutrient solution	9–17 days	U ⩽ M < D ⩾ A U > M, D, (A)	44 44

[a] Relative amounts of metalobites as measured upon termination of the experiment: U, structurally unchanged herbicides =R—NH—CO—N(CH$_3$)$_2$; M, monomethylated derivative =R—NH—CO—NHCH$_3$; D, demethylated derivative =R—NH—CO—NH$_2$; A, corresponding aniline =R—NH$_2$; ( ) metabolite searched for, but not observed.
[b] See orginal paper for references.
Reprinted from P. C. Kearney and D. D. Kaufman, eds, *Degradation of Herbicides*, p. 97., by courtesy of Marcel Dekker, Inc.

Figure 21-4 presents a suggested pathway for degradation of urea herbicides in higher plants.

Biochemical Responses

The development of the current concept of the mechanism of action of the urea-type herbicides is historically very informative. It is perhaps our best

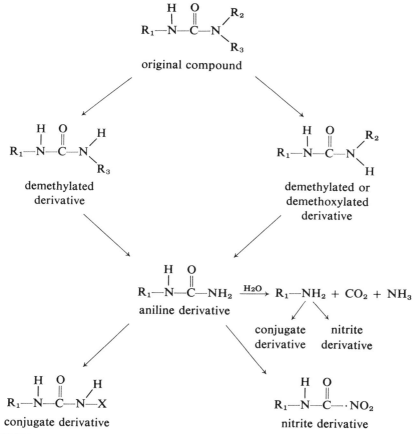

FIGURE 21-4. Proposed pathway for degradation of urea herbicides in higher plants, where usually R_1 = ring structure, R_2 = —CH_3, and R_3 = —CH_3 or —OCH_3 (See Table 21-1).

example of the step-wise approach for localization of the primary biochemical lesion or site of action of a herbicide starting with morphological and physiological clues and following them with biochemical approaches. The philosophy of this approach is discussed in Chapter 7.

Bucha and Todd (1951) observed that the initial response to monuron is generally leaf-tip dieback followed by progressive chlorosis and retardation of growth. They also suggested that the herbicide was absorbed by the roots and translocated to the leaves. Christoph and Fisk (1954) also reported that one of the early symptoms of monuron toxicity is leaf chlorosis. In addition to these reports, the studies of Muzik et al. (1954) placed the site of toxic action of monuron in the leaves.

Cooke (1955) reported that monuron brought about a large decrease in the sugar content of plants and suggested that the mechanism of action was through an interference with photosynthesis. This report along with the known fact that phenylurethane, whose structure is somewhat like that of monuron, had been shown to inhibit the Hill reaction (MacDowall, 1949), must have stimulated the investigations on the effect of monuron on the Hill reaction. Cooke (1956), Wessels and Van der Veen (1956) and Spikes (1956) reported that monuron inhibited the Hill reaction. Ashton et al. (1961) found that when a 10 ppm solution of monuron was applied to the roots of bean plants, carbon dioxide fixations by leaves in the light were inhibited 93.5% within 2 hours. Lower concentrations were less effective in inhibiting photosynthesis but longer exposure times produced greater inhibition. The distribution of radioactivity in various compounds following $^{14}CO_2$ exposure in light and dark of monuron-treated plants demonstrated that monuron almost completely blocked the pathway of photosynthetic carbon dioxide fixation but did not affect the dark carbon dioxide fixation pathway. Subsequently there have been many papers on the effect of monuron and other urea herbicides on various aspects of photosynthesis. Several papers have reviewed many of these reports (van Overbeek 1962; Black and Myers 1966; Davis 1966; Moreland 1967; Zweig 1969; Caseley 1970; Hoffmann 1971).

It is generally considered that green plant photosynthesis involves two light reactions, photosystem I and photosystem II (Figure 6-2). Most of the research places the most sensitive site of the urea herbicides action on photosynthesis in the light system II, involving oxygen evolution. The research of Bishop (1958) is particularly relevant in which these herbicides had no effect on photoreduction reactions of the hydrogen adapted alga, *Scenedesmus*. This is also supported by electron spin resonance studies (Treharne et al. 1963). However, some researches including that of Asahi and Jagendorf (1963) have shown that photosystem I is also inhibitable by monuron but that it requires a much higher concentration of the herbicide than the inhibition of photosystem II. Since photosystem II preceeds photosystem I and light system II is completely blocked at concentrations which do not affect photosystem I, it is unlikely that the inhibition of photosystem I is a significant factor in the herbicidal properties of the urea herbicides. There are many additional papers dealing with the effect of the urea herbicides on the light reactions of photosynthesis; some of these are Jagendorf and Avron (1959), Bamberger et al. (1963), Gingras et al. (1963), Homann and Gaffron (1963), Izawa and Good (1965), Izawa (1968), Makeeva-Gur'yanova and Chkanikov (1968), McMahon and Borograd (1968), Moreland and Blackmon (1968), Nishida (1968), and Zweig et al. (1968). In summary, the primary site of action of the urea herbicides appears to be located in photosystem II and in or near the oxygen evolving step, Figures 6-2 and 21-5.

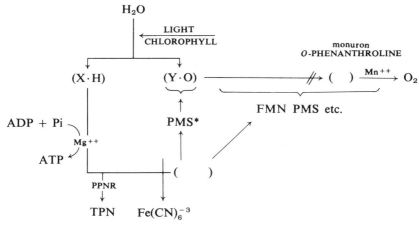

FIGURE 21-5. Scheme for electron- and oxyen-transport pathways in chloroplasts. TPN stands for triphosphopyridine nucleotide, and PPNR for the enzyme photosynthetic pyridine nucleotide reductase (14) which is required for the reduction of TPN in the Hill reaction. PMS stands for phenazine methosulfate; the asterisk on the PMS internal to the usual cycle indicates that a white-light-induced form of PMS is probably the one most responsible for resistance to monuron (Jagendorf and Avron, 1959).

Several investigators have suggested that interference with the light reactions of photosynthesis and the subsequent inhibition of carbon dioxide fixation does not adequately explain the light-dependent phytotoxic symptoms of the urea herbicides. In other words, the plants do not merely starve from lack of photosynthate. Growth studies with *Chlorella* and *Euglena* have shown that monuron is toxic in the light but not in the dark, even when the growth medium contains an energy source (Hoffmann, unpublished). Jagendorf and Avron (1959) reported that monuron affected the Hill reactions, and most types of photosynthetic phosphorylation, and placed the site of action between $Y \cdot O$ and oxygen evolution, Figure 21-5. Furthermore, they suggested that monuron should cause some pile-up of $Y \cdot OH$ and perhaps this might contribute to damaging photooxidations. Sweetser and Todd (1961) placed the site of action of monuron at this same location and suggested that the herbicidal action of monuron is due to the build-up of a phytotoxic substance in the oxygen-liberating pathway of photosynthesis. Their evidence was obtained by measuring the effect of monuron on the growth rate of *Chlorella* at various carbon dioxide concentrations and in the presence or absence of light, Figure 21-6. Although the exact nature of this phytotoxic substance is unknown, it appears that it is the agent responsible for leaf injury in the light at relatively low concentrations of the urea herbicides, rather than the herbicide *per se*. At higher concentrations which inhibit the

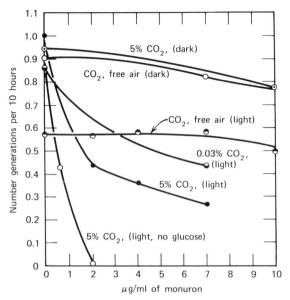

FIGURE 21-6. Effect of CO_2 concentration on the inhibitory action of monuron on the growth of *Chlorella pyrenoidosa* (Sweetser and Todd, 1961.)

growth of non-photosynthetic tissue an entirely different mechanism may be functioning.

St. John (1971) using chlorella has found that diuron inhibited the photochemical but not the oxidative production of ATP; diuron equally inhibited growth and ATP synthesis; although levels of ATP and growth were suppressed, the cells still maintained ATP levels in balance with growth.

Mode of Action

The phytotoxic symptoms of the urea herbicides are primarily observable in the leaves; however, in some instances changes in non-photosynthetic tissues are induced. The foliar symptoms are of two types, the acute or the chronic. The acute symptoms are associated with relatively high concentrations of the herbicide in the leaf and appear within a few days. Initially light green areas appear, these areas then take on a water-soaked appearance, and finally become necrotic. The chronic symptoms occur at relatively low foliar concentrations and require several days to develop. These include wilting of the leaves, the appearance silver and/or indeterminate gray blotches and rapid yellowing. In grasses, these foliar symptoms commence at the tip of the leaf and proceed basipetally. A second type of chlorosis may also occur; this has been referred to as partial chlorosis in which the leaf merely becomes light or yellowish green in color. Presumably this partial chlorosis occurs at very low foliar concentrations of herbicide. The urea-type herbicides have also been shown to inhibit the growth of non-photosynthetic tissues; however this has usually required high concentrations.

Most of the urea herbicides are readily absorbed by the roots and rapidly translocated to the upper plant parts via the apoplastic system. Applications to the leaves are also translocated apoplastically with little if any being exported from the treated leaf. However, the actual amount absorbed and translocated from roots to shoot varies greatly with various compounds. Furthermore, differences in absorption and translocation between species have been sufficient with specific urea herbicides to allow their selective use in certain crops.

The urea herbicides are subject to degradation by higher plants, however a given herbicide may be degraded at different rates by various species. This is the basis of certain selective uses of some of the urea herbicides. Demethylation is the primary detoxification mechanism for the urea herbicides. A microsomal enzyme system has been isolated which is able to carry out the demethylation of the urea herbicides; molecular oxygen and NADPH or NADH are required cofactors for this reaction. Demethoxylation also occurs in those compounds containing a methoxy group. Following the dealkylation and/or dealkoxylation the molecule may be subject to hydrolysis, which would involve a deaminization and decarboxylation yielding the corresponding aniline. This aniline may be subject to oxidation to yield the corresponding nitrite or to conjugation with normal cellular constituents.

Inhibition of the Hill reaction of photosynthesis is generally acknowledged to indicate the primary site of action of the urea herbicides. This involves the site of oxygen evolution in photosystem II. This prevents the formation of ATP and NADPH which are required for carbon dioxide fixation. However, many workers do not believe that this explains the light-dependent phytotoxic symptoms of the urea herbicides. In other words the plants do not merely starve from lack of photosynthate. Rather it has been postulated that a secondary phytotoxic substance is formed in the oxygen-liberating pathway of photosynthesis. The nature of this secondary phytotoxic substance has not been determined.

REFERENCES

Aebi, H. and L. Ebner. 1962. The fate of the herbicide N-4(p-chlorophenoxy)-phenyl-N',N'-dimethylurea in soils and plants. *6th Brit. Weed Contr. Conf.* pp. 479–488.

Asahi, T. and A. T. Jagendorf. 1963. A spinach enzyme functioning to reverse the inhibition of cyclic electron flow by *p*-chlorophenyl-1,1-dimethylurea at high concentrations. *Arch. Biochem. Biophys.* **100**:531–541.

Ashton, F. M. 1965. Relationship between light and toxicity symptoms caused by atrazine and monuron. *Weeds* **13**:164–168.

Ashton, F. M., E. G. Uribe, and G. Zweig. 1961. Effect of monuron on $^{14}CO_2$ fixation by red kidney beans. *Weeds* **9**:575–579.

Bamberger, E. S., C. C. Black, C. A. Fewson, and M. Gibbs. 1963. Inhibition studies on carbon dioxide fixation, ATP formation, TPN reduction by spinach chloroplasts. *Plant Physiol.* **38**:483–487.

Bayer, D. E. and S. Yamaguchi. 1965. Absorption of ^{14}C-diuron. *Weeds* **13**:232–235.

Bishop, N. I. 1958. The influence of the herbicide, DCMU, on the oxygen evolving system of photosynthesis. *Biochim. et Biophys. Acta* **27**:205–206.

Black, C. C. Jr. and L. Meyers. 1966. Some biochemical aspects of the mechanisms of herbicidal activity. *Weeds* **14**:331–338.

Börner, H. 1964. Unterschungen mit C^{14}-mariertem Afalon (3,4-dichlorphenyl-metroxymethylharnstoff). *Z. Pflanzenkrankh. Sonderheft* **2**:41–42.

Börner, H. 1965. Ruchstandsbestimmungen von Afalon [*N*-(3,4-dichlorphenyl)-*N'*-methoxy-*N'*-methylharnstoff] und Aresin [*N*-(4-chlorphenyl)-*N'*-methoxy-*N'*-methylharnstoff] in Erntegut von Mohren, Buschbohnen. *Z. Pflanzenkrankh.* **72**:449–457.

Bucha, H. C. and C. W. Todd. 1951. 3-(*p*-chlorophenyl)-1,1-dimethylurea—a new herbicide. *Science* **114**:493–494.

Camper, N. D. and D. E. Moreland. 1971. Sorption of substituted phenyl-amides onto Bovine serum albumin. *Weed Sci.* **19**:269–273.

Caseley, J. 1970. Herbicide activity involving light. *Pestic. Sci.* **1**:28–32.

Christoph, R. J. and E. L. Fisk. 1954. Responses of plants to the herbicide 3-(*p*-chlorophenyl)-1,1-dimethylurea (CMU). *Bot. Gaz.* **116**:1–14.

Cooke, A. R. 1955. Effect of CMU on the biochemical composition of several legumes. *North Central Weed Control Conf. Res. Rept.* pp. 181–182.

Cooke, A. R. 1956. A possible mechanism of action of the urea-type herbicides. *Weeds* **4**:397–398.

Crafts, A. S. 1959. Further studies on comparative mobility of labeled herbicides. *Plant Physiol.* **34**:613–620.

Crafts, A. S. 1962. Use of labeled compounds in herbicide research. *Intern. J. Appl. Radiation Isotopes* **13**:407–415.

Crafts, A. S. 1967. Bidirectional movement of labeled tracers in soybean seedlings. *Hilgardia* **37**:625–638.

Crafts, A. S. and S. Yamaguchi. 1958. Comparative tests on the uptake and distribution of labeled herbicides by *Zebrina pendula* and *Tradescantia fluminensis*. *Hilgardia* **27**:421–454.

Crafts, A. S. and S. Yamaguchi. 1960. Absorption of herbicides by roots. *Am. J. Bot.* **47**:248–255.

Davis, E. A. 1966. The role of starvation in fenuron injury to shrub live oak. *Weeds* **14**:10–17.

Day, B. E. 1955. Urea herbicides for citrus weed control. *Calif. Citrog.* **40**:398, 408, 410–412, 414.

Donaldson, T. W. 1967. Absorption of the herbicides 2,4-D and monuron by barley roots. Ph.D. Thesis, Univ. Calif., Davis, 124 pp.

Fang, S. C., V. H. Freed, N. H. Johnson, and D. R. Coffee. 1955. Absorption, translocation, and metabolism of radioactive 3-(*p*-chlorophenyl)-1,1-dimethylurea (CMU) by bean plants. *J. Agr. and Food Chem.* **3**:400–402.

Frear, D. S. 1968. Microsomal *N*-demethylation, by a cotton leaf oxidase system of 3-(*p*-chlorophenyl)-1,1-dimethylurea (monuron). *Science* **162**:674–675.

Frear, D. S. and H. R. Swanson. 1969. Metabolism of urea herbicides in plants: isolation and characterization of a microsomal *N*-demethylase from cotton. *Abstr. Pap. 157th Nat. Meet. Am. Chem. Soc.*, AGFD 15.

Freed, V. H., M. Montgomery, and M. Kief. 1961. The metabolism of certain herbicides by plants—a factor in their biological activity. *N. East Weed Cont. Conf. Proc.* **15**:6–16.

Geissbühler, H. 1964. The behavior of urea herbicides in plants. *16de Internat. Symp. Fytofarm. Fytiatrie, Gent.* pp. 704–718.

Geissbühler, H. 1969. The substituted ureas. pp. 79–111. In P. C. Kearney and D. D. Kaufman, *Degradation of Herbicides*. Marcel Dekker, Inc., New York.

Geissbühler, H., C. Haselbash, H. Aebi, and L. Ebner. 1963a. The fate of *N*′-(4-chlorophenoxy)-phenyl-*N*,*N*-dimethylurea (C-1983) in soils and plants. 2. Uptake and distribution within plants. *Weed Res.* **3**:181–194.

Geissbühler, H., C. Haselbach, H. Aebi, and L. Ebner. 1963b. The fate of *N*′-(4-chlorophenoxy)-phenyl-*N*,*N*-dimethylurea (C-1983) in soils. 3. Breakdown in soils and plants. *Weed Res.* **3**:277–297.

Gingras, G., C. Lemasson and D. C. Fork. 1963. A study of the mode of action of 3-(4-chlorophenyl)-1,1-dimethylurea on photosynthesis. *Biochim. Biophys. Acta* **69**:438–440.

Goren, R. 1969. The effect of fluometuron on the behavior of citrus leaves. *Weed Res.* **9**:121–135.

Grant, W. F. 1964. Cytogenetic effects of pesticides. *Rech. Agron.* **65**:10,28.

Haun, J. R. and J. H. Peterson. 1954. Translocation of 3-(*p*-chlorophenyl)-1,1-dimethylurea in plants. *Weeds* **3**:177–187.

Hoffmann, C. E. 1971. The mode of action of bromacil and related uracils. *Second Internat. Conf. of Pest. Chem.* (in press)

Homann, P. and H. Gaffron. 1963. Flavin sensitized photoreactions: effects of 3-(*p*-chlorophenyl)-1,1-dimethylurea. *Science* **141**:905–906.

Izawa, S. 1968. Effect of Hill reaction inhibition on photosystem I, pp. 140–147. In K. Shibata, A. Takamya, A. T. Jagendorf and R. K. Fuller, Ed., *Comparative*

390 Ureas

Biochemistry and Biophysics of Photosynthesis. Univ. of Tokyo Press, Tokyo, and Univ. Park Press, State College, Penn.

Izawa, S. and N. E. Good. 1965. The number of sites sensitive to 3-(3,4-dichlorophenyl)-1,1-dimethylurea, 3-(4-chlorophenyl)-1,1-dimethylurea and 2-chloro-4-(2-propylamino)-6-ethylamino-*s*-triazine in isolated chloroplasts. *Biochim. Biophys. Acta* **102**:20–38.

Jagendorf, A. T. and M. Avron. 1959. Inhibitors of photosynthetic phosphorylation. *Arch. Biochem. and Biophys.* **80**:246–257.

Jordan, L. S., T. Murashige, J. D. Mann, and B. E. Day. 1966. Effect of photosynthesis-inhibiting herbicides on non-photosynthetic tobacco callus tissue. *Weeds* **14**:134–136.

Katz, S. E. 1967. Determination of linuron and its known and/or suspected metabolites in crop materials. *J. Ass. Off. Agr. Chem.* **50**:911–917.

Knake, E. L. and L. M. Wax. 1968. The importance of the shoot of giant foxtail for uptake of preemergence herbicides. *Weed Sci.* **16**:393–395.

Kuratle, H. 1968. The mode of action and basis for selectivity of linuron herbicide. Ph.D. Thesis, Univ. Delaware, 105 pp.

Leonard, O. A. and R. K. Glenn. 1968. Translocation of herbicides in detached bean leaves. *Weed Sci.* **16**:352–356.

Leonard, O. A., L. A. Lider and R. K. Glenn. 1966. Absorption and translocation of herbicides by Thompson Seedless (Sultanina) grape, *Vitis vinifera*. *Weed Res.* **6**:37–49.

MacDowall, F. D. H. 1949. The effects of some inhibitors of photosynthesis upon the photochemical reduction of a dye by isolated chloroplasts. *Plant Physiol.* **24**:462–480.

Makeeva-Gur'yanova, L. T. and D. I. Chkanikov. 1968. More data on the effect of urea-derivative herbicides on photosynthetic processes. *Agrokhimiya* **5**:93–98.

McCall, G. L. 1952. "CMU", new herbicide. *Agr. Chem.* **7**:40–41, 127–129.

McMahon, D. and L. Borograd. 1968. Inhibition of the formation of photosynthetic enzymes by inhibitors of photosynthesis. *Plant Physiol.* **43**:188–192.

Minshall, W. H. 1954. Translocation path and place of action of 3-(4-chlorophenyl)-1,1-dimethylurea in bean and tomato. *Canad. J. Bot.* **32**:795–798.

Minshall, W. H. 1957. Primary place of action and symptoms induced in plants by 3-(4-chlorophenyl)-1,1-dimethylurea. *Canad. J. Plant Sci.* **37**:157–166.

Minshall, W. H. 1960. Effect of 3-(4-chlorophenyl)-1,1-dimethylurea on dry matter production, transpiration, and root extension. *Canad. J. Bot.* **38**:201–216.

Moody, K. and K. P. Buchholtz. 1968. Effect of temperature and concentration on the uptake of herbicides of excised soybean roots. *WSSA Abstr.* p. 18.

Moody, K., C. A. Kust, and K. P. Buchholtz. 1970. Release of herbicides by soybean roots in culture solutions. *Weed Sci.* **18**:214–218.

Moreland, D. E. 1967. Mechanism of action of herbicides. *Ann. Rev. Plant Physiol.* **18**:365–386.

Moreland, D. E. and W. J. Blackmon. 1968. Comparative action of herbicides on electron transport and phosphorylation in mitochondria and chloroplasts. *Abstr. Pap. 156th Meet. Am. Chem. Soc.* AGFD-74.

Muzik, T. J., H. J. Cruzado and A. J. Loustalot. 1954. Studies on the absorption, translocation, and action of CMU. *Bot. Gaz.* **116**:65–73.

Nashed, R. B. and R. D. Ilnicki. 1970. Absorption, distribution, and metabolism of linuron in corn, soybean and crabgrass. *Weed Sci.* **18**:25–28.

Nashed, R. B., S. E. Katz, and R. D. Ilnicki. 1970. The metabolism of ^{14}C-chlorbromuron in corn and cucumber. *Weed Sci.* **18**:122–125.

Neptune, M. D. and H. H. Funderburk. 1968. The fate of fluometuron in corn and wheat. *Proc. 21st Southern Weed Conf.* p. 339.

Nishida, K. 1968. Effects of inhibitors on the volume change of isolated chloroplasts on illumination, pp. 97–105. In K. Shibata, A. Takamya, A. T. Jagendorf, and R. C. Fuller, Ed., *Comparative Biochemistry and Biophysics of Photosynthesis.* Univ. of Tokyo Press, Tokyo; and Univ. Park Press, State College, Penn.

Nishimoto, R. K., A. P. Appleby and W. R. Furtick. 1967. Site of uptake of preemergence herbicides. WSSA *Abstr.* pp. 46–47.

Onley, J. H., G. Yip, and M. H. Aldridge. 1968. A metabolic study of 3-(3,4-dichlorophenyl)-1,1-dimethylurea (diuron) applied to corn seedlings. *J. Agr. and Food Chem.* **16**:426–433.

Overbeek, J. van. 1962. Physiological responses of plants to herbicides. *Weeds* **10**:170–174.

Pereira, J. F., A. S. Crafts, and S. Yamaguchi. 1963. Translocation in coffee plants. *Turrialba* **13**:64–79.

Pickering, E. R. 1965. Foliar penetration pathways of 2,4-D, monuron and dalapon as revealed by historadioautography. Ph.D. Dissertation, Univ. Calif. Davis, 186 pp.

Prendeville, G. N., Y. Eshel, M. M. Schreiber, and G. F. Warren. 1967. Site of uptake of soil-applied herbicides. *Weed Res.* **7**:316–322.

Rieder, G., K. P. Buchholtz and C. A. Kust. 1970. Uptake of herbicides by soybean seeds. *Weed Sci.* **18**:101–105.

Rogers, R. L. and H. H. Funderburk. 1967a. Absorption, translocation, and metabolism of Cotoran by cotton and cucumber. *WSSA Abstr.* pp. 43–44.

Rogers, R. L. and H. H. Funderburk. 1967b. Extraction and chromatography of Cotoran and several plant metabolites. *Proc. 20th Southern Weed Conf.* p. 388.

Rogers, R. L. and H. H. Funderburk. 1968a. Fluometuron degradation in cotton. *Proc. 21st Southern Weed Conf.* p. 341.

Rogers, R. L. and H. H. Funderburk. 1968b. Physiological aspects of fluometron in cotton and cucumber. *J. Agr. and Food Chem.* **16**:434–440.

Rubin, B. and Y. Eshel. 1971. Phytotoxicity of fluometuron and its derivatives to cotton and weeds. *Weed Sci.* **19**:592–594.

Sasaki, S. and T. T. Kozlowski. 1961. Influence of herbicides on respiration of young *Pinus* seedlings. Nature **210**:439–440.

Sheets, T. J. and A. S. Crafts. 1957. The phytotoxicity of four phenylurea herbicides in soil. *Weeds* **5**:93–101.

Sikka, H. C. and D. Pramer. 1967. Some physiological effects of Cotoran on unicellular algae. *WSSA Abstr.* p. 54.

Smith, J. W. and T. J. Sheets. 1966. Uptake, distribution and metabolism of diuron and monuron by soybean and cotton. *WSSA Abstr.* p. 39.

Smith, J. W. and T. J. Sheets. 1967. Uptake, distribution and metabolism of monuron and diuron by several plants. *J. Agr. and Food Chem.* **15**:577–581.

St. John, J. B. 1971. Comparative effects of diuron and chlorpropham on ATP levels in chlorella. *Weed Sci.* **19**:274–276.

Strang, R. H. and R. L. Rogers. 1971. A microradiographic study of ^{14}C-diuron absorption by cotton. *Weed Sci.* **19**:355–362.

Spikes, J. B. 1956. Effects of substituted ureas on the photochemical activity of isolated chloroplasts. *Plant Physiol. Abstr.* **31**:xxxii.

Splittstoesser, W. E. and H. J. Hopen. 1968. Metabolism of siduron by barley and crabgrass. *Weed Sci.* **16**:305–308.

Swanson, C. R., G. G. Still and H. R. Swanson. 1967. Monuron degradation and recovery of monuron-inhibited photosynthetic activity. *WSSA Abstr.*, pp. 60–61.

Swanson, C. R. and H. R. Swanson. 1968. Metabolic fate of monuron and diuron in isolated leaf discs. *Weed Sci.* **16**:137–143.

Sweetser, P. B. and C. W. Todd. 1961. The effect of monuron on oxygen liberation in photosynthesis. *Biochim. et Biophys. Acta* **51**:504–508.

Taylor, T. D. and G. F. Warren. 1968. Herbicide transport as influenced by certain metabolic inhibitors. *WSSA Abstr.* pp. 11–12.

Thompson, H. E., C. P. Swanson and A. G. Norman. 1946. New growth-regulating compounds. I. Summary of growth-inhibitory activities of some organic compounds. *Bot. Gaz.* **107**:476–507.

Treharne, R. W., T. E. Brown, and L. P. Vernon. 1963. Separation of two light-induced electron-spin-resonance signals in several algal species. *Biochim. et Biophys. Acta* **75**:324–332.

Voderberg, K. 1961. The dependence of herbicide action on root growth different plants, an external factor. *Nachrbl. dtsch. PflSchDnst* **15**:68–70.

Voss, G. and Geissbühler, H. 1966. The uptake, translocation and metabolism of fluometuron and metobromuron in plants. *Proc. 8th Brit. Weed Control Conf.* pp. 266–268.

Walsch, G. E. and T. E. Grow. 1971. Depression of carbohydrate in marine algae by urea herbicides. *Weed Sci.* **19**:568–570.

Wessels, J. S. C. and R. Van der Veen. 1956. The action of some derivatives of phenylurethan and of 3-phenyl-1,1-dimethylurea on the Hill reaction. *Biochim. et Biophys. Acta* **19**:548–549.

Wuu, K. D. and W. F. Grant. 1966. Morphological and somatic chromosomal aberrations induced by pesticides in barley (*Hordeum vulgare*). *Can. J. Genet. Cytol.* **8**:481–501.

Yamaguchi, S. and A. S. Islam. 1967. Translocation of eight ^{14}C-labelled amino acids and three herbicides in two varieties of barley. *Hilgardia* **38**:207–229.

Zweig, G. 1969. Mode-of-action of photosynthetic inhibitor herbicides. *Residue Rev.* **25**:69–79.

Zweig, G., J. E. Hitt and R. McMahon. 1968. Effect of certain quinones, diquat, and diuron on *Chlorella pyrenoidosa* Chick (Emerson strain). *Weed Sci.* **16**:69–73.

CHAPTER 22

Unclassified Herbicides

The foregoing chapters have included information on all herbicides that could be grouped into units that provide sufficient information to make up chapters. There are still many compounds that do not submit to such classification, and these are covered in this chapter. Some of the compounds in this chapter have been used for many years; some are relatively new. To be included in the complex testing programs required to meet the demands of present-day registration they must have high phytotoxicity, they must be producible with relative ease, and they must be susceptible to patent protection. In spite of rising costs of synthesis and production, industry is continuing and intensifying programs of searching for new and better herbicides. As evidenced during the past decade, great rewards lie ahead for those companies who have the confidence and financial backing required to carry through the syntheses, screening, and development necessary to bring a herbicide to market.

Acrolein

$$CH_2=CH-CHO$$
acrolein

Acrolein is the common name of this compound; it is also a completely acceptable chemical name. Acrylaldehyde and 2-propenal are also acceptable chemical names. Acrolein is a pungent, colorless liquid, extremely irritating to the eyes; it has been used as a tear gas. It is highly volatile, and between 3.0 and 31.0% by weight in air, is explosive; the vapor will flash below 0°F. For these reasons acrolein is difficult to handle and is used only by licensed operators who are adequately equipped to apply it. It is used principally to control aquatic weeds.

Acrolein is a general plant-cell toxicant of high reactivity related to its unsaturation. It destroys the integrity of plant cell membranes and reacts with various enzyme systems. Plants treated with acrolein become flaccid,

394

disintegrate within a few hours, and float downstream; large masses of vegetation are not released, as with chaining, which may clog irrigation structures. In water at 80°F or above, cessation of oxygen evolution may be seen within minutes. In cooler water the effects become evident more slowly. Temperature is important and at 60°F the dosage must be double that at 80°F.

Acrolein is toxic to most common aquatic weeds found in lakes and streams. Algae, including *Chara* spp., *Cladophora* spp., *Hydrodictyon* spp., and *Spirogyra* species, have been killed and all forms seem to be susceptible. Of the common higher plants, waterstarwort (*Callitriche verna*), coontail (*Ceratophyllum* spp.), water weed (*Anacharis canadensis*), waterstargrass (*Heteranthera dubia*), naiad (*Najas* spp.), horned pondweed (*Zannichellia palustris*), and many pondweeds (*Potomogeton* spp.) have also been controlled.

Emersed forms such as waterlettuce (*Pistia stratiates*), waterhyacinth (*Eichornia crassipes*), California waterprimrose (*Jussiaea californica*), water pennywort (*Hydrocotyle umbellata*), and watermilfoil (*Myriophyllum* spp.) are more difficult to control with acrolein than the species listed in the preceding paragraph; higher dosages are suggested where these occur.

The amount of acrolein required for a given situation is determined by the amount of water flowing in the stream, water temperature, velocity of flow, density of weed growth, and species of weeds involved. Because the material is pumped into the water over a given time, in flowing water a blanket of treated water moves along the channel. The length of the blanket depends on velocity of flow and time of application; density of the blanket depends on the rate of application. The amount of toxicant bathing a given mass of weeds is then a product of (time) × (concentration), and within rather wide limits this product should approach a constant value. Acrolein is absorbed by plants and hence the blanket of treated water becomes less and less concentrated as it moves along the channel. The chemical should be added periodically as the water moves so that the treated blanket maintains a certain minimum concentration. In contrast to Benoclor and solvent naphthas which are required at concentrations of 300 to 700 ppm, acrolein is effective at 50 ppm or less. It may be effective for distances up to 15 miles, whereas blankets of the above herbicides require reinforcing each mile or two. This is of tremendous advantage in the very extensive irrigation ditch systems of the western United States. Because of its greater efficiency, acrolein may be used in large main canals, whereas the chlorinated and methylated benzenes are effective only in laterals of limited capacity.

In ponds and lakes the herbicide may be applied by boat using a boom extending over the side or behind to introduce the chemical into the water. In irrigation reservoirs the water should be drawn to its lowest limit and the chemical pumped into the incoming stream as the reservoir is refilled. Introduction of the chemical should continue throughout the refilling process.

Acrolein may be introduced into small ponds and along swimming beaches by addition at a number of points not to exceed 30 feet apart. The water should remain static for several days after the treatment.

Most fish are killed by acrolein applied at its usual concentration; however, in static water they may swim away from the treatment area. For this reason, if the fish are to be retained, treatment should be made slowly and in spot treatments, or by treating narrow strips so that fish may escape to untreated water. If the whole area of water is to be treated, application may be made to one portion allowing fish to escape and then after 48 hours, the free area treated, allowing the fish to return to the earlier treated area.

Acrolein is volatile and reactive. It apparently does not present a residue problem in crops. Data obtained on 19 crops indicated that degradation eliminated all residue in a marketable crop by harvest time (Shell Chem. Corp., 1959).

Bensulide

$$\langle\!\!\!\bigcirc\!\!\!\rangle\!\!-\!SO_2\!-\!NH\!-\!CH_2\!-\!CH_2\!-\!S\!-\!\overset{\overset{S}{\uparrow}}{P}\!-\!\left(O\!\overset{CH_3}{\underset{CH_3}{C}H}\right)_{\!2}$$

O,O-diisopropyl phosphorodithioate S-ester with N-(2-mercaptoethyl) benzenesulfonamide.

Bensulide in pure form is a liquid, slightly soluble in water, miscible with methylisobutyl ketone. It is formulated as an emulsifiable liquid and it is applied at a rate of 10 to 20 lb/A in 120 to 150 gal of water. Bensulide is one of a number of preemergence herbicides used to kill weeds in turf, principally crabgrass (*Digitaria* spp.). It is customarily applied in the fall before the germination of winter and spring growing weeds. It is rather tightly bound on colloids in the top 0 to 2 inches of soil and persists long enough to control crabgrass and other summer weeds. Crabgrass, annual bluegrass (*Poa annua*), redroot pigweed (*Amaranthus retroflexus*), barnyardgrass (*Echinochloa crus-galli*), common lambsquarters (*Chenopodium album*), shepherdspurse (*Capsella bursa-pastoris*), goosegrass (*Eleusine indica*), and deadnettle (*Lamium* spp.) in grass and dichondra lawns are the weeds commonly brought under control. Bensulide is also used as a preplant soil-incorporation treatment in cotton and several vegetable crops.

Cutter et al. (1968) studied the effects of bensulide on the growth, morphology, and anatomy of oat roots. Oat seedlings were grown in a dilute Hoagland's solution for a period of 7 days and the roots were measured at daily intervals. Figure 22-1 shows the growth of roots in culture solution alone, and in four bensulide concentrations; obviously root growth was

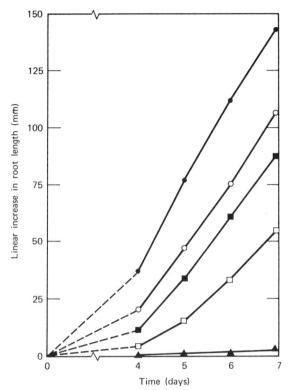

FIGURE 22-1.   The effect of various concentrations of
bensulide on increase in length of oat roots during a
7-day period: ●, control; ○, 5 × 10^{-7} M; ■, 10^{-6} M;
□, 5 × 10^{-6} M; ▲, 10^{-5} M.   (Cutter et al., 1968.)

inhibited. Roots were sometimes curved and root hairs were often present
close to the root tip; epidermal cells were elongated radially; short pitted
tracheary elements differentiated close to the root tips; mitosis was not
completely inhibited. Ashton et al. (1968) found that bensulide reduced
proteolytic activity in squash seedlings to 35% of normal at 5 × 10^{-5}M
concentration.

Bingham (1967) found that bensulide prevented rooting from stolon nodes
of bermudagrass (*Cynodon dactylon*). Inhibition was limited to the regions on
the stolon which were in direct contact with treated soil. Root initiation was
inhibited and the roots that grew did not elongate beyond ½ cm; they did not
become attached to the soil so that they could function. The root tips became
enlarged and microscopic study proved that cell division had ceased. Growth

of roots established in soils of different bensulide concentrations was greatly reduced at 20 ppm or more. Under field conditions, bensulide caused a delay in rooting of stolons, but growth rates were normal after 12 weeks.

Mazur et al. (1969) used a bioassay involving measurement of shoot and root growth of seedlings of 22 plant species growing on filter paper in petri dishes containing 5 ml of suspensions of bensulide. The five species most sensitive to bensulide were barnyardgrass, browntop millet, crabgrass, oats, and sudangrass (*Sorghum sudanense*). These five species were then tested in soil-herbicide mixtures containing 1, 10, 100, and 1000 ppm. Oats and browntop millet proved to be as sensitive to bensulide in the soil cultures as in the petri dish tests; other species had lowered sensitivity. Root lengths of oats and browntop millet were reduced by bensulide at 1 ppm; more at 10, 100, and particularly at 1000 ppm.

DCPA

dimethyl tetrachlorophthalate

DCPA is a white, crystalline, odorless compound, very slightly soluble in water, more soluble in organic solvents. It is formulated as a 50% wettable powder and applied as a suspension in water. DCPA is a highly active compound, selective against annual grasses and some broadleaf weeds in turf and in a number of vegetable crops. It is used preemergence to weeds.

DCPA is absorbed from the soil, but not by foliage; it is not translocated in plants. Its phytotoxicity is expressed on germinating seeds. It is not considered, to be metabolized in plants. Utter (1960) has suggested that it is absorbed by the coleoptiles of grass seedlings.

Bingham (1968) found DCPA to inhibit rooting from nodes of bermudagrass. Histological examination of affected root tips proved that cell division had ceased while cell enlargement continued for some time. Affected cells became excessively large and irregular in shape. In maize radicle, cell size was unchanged but treated tissue contained six times as many dinucleate cells as untreated tissue; numerous nuclei in treated-tissue cells were at metaphase.

In onion, DCPA reduced the rate of root growth but did not completely stop growth. Nuclei in treated cells were large, granular, and oval in shape.

Nishimoto and Warren (1971a) found DCPA to inhibit root growth more than shoot growth in maize and sorghum; maize shoots were more tolerant than sorghum shoots when separately exposed. Cucumber roots tolerated DCPA when treated to concentrations of 1, 2, and 4 times $10^{-5}M$; hypocotyl length was reduced only when the shoot tissue was exposed, ^{14}C-DCPA was readily absorbed by the hypocotyl of cucumber and moved into the foliage. The meristematic region of the sorghum shoot before emergence was more sensitive than other regions of the shoot to DCPA. Injury from other application sites depended upon uptake from the treated area and movement to the meristematic region (Nishimoto and Warren, 1971b).

Bayer et al. (1965) studied the host-parasite relations of alfalfa and dodder as affected by application of DCPA. Treatments in March, under California conditions, lasted through the season when applied preemergence to dodder germination. December treatment failed to provide season-long control and the May treatment was not satisfactory. Residues of DCPA in alfalfa treated for dodder control were below the 1 ppm level except in the first cutting after treatment.

In greenhouse studies, Bayer et al. (1965) found that alfalfa sprayed with DCPA was not parasitized by dodder, but young stems growing after the DCPA spray would support dodder. Dodder exposed to treated foliage enlarged to twice or three times normal, became greenish yellow and the haustoria would not penetrate the alfalfa stems. Dodder treated at the 5 lb/A rate showed signs of recovery in 5 weeks time. Treated at 10 lb/A, the tendrils were short and stubby; at 20 lb/A, the tendrils and haustoria were brown and dying. Bayer et al. (1965) could find no evidence for systemic toxicity to dodder through alfalfa.

De Hertogh et al. (1962) used DCPA among some eight herbicides to find a satisfactory preemergence material for use in flue-cured tobacco. It was tested on three different localities in North Carolina. It had a slight depressing effect on reducing sugars, and no adverse effect in smoke and taste tests.

Mazur et al. (1969) used seedlings of 22 plant species for bioassay of residues of three herbicides, including DCPA in turf soils; the latter was used as a suspension of a 75% wettable powder. The five species giving the greatest response to DCPA were browntop millet, buckwheat, foxtail (*Setaria* spp.), foxtail millet (*Setaria italica*), and sorghum. Crabgrass, the species toward which all three herbicides were directed in turf, was among the top five species controlled by bensulide, but it was not among the top five species controlled by DCPA. The same order of sensitivity to DCPA was shown when these herbicides were tested in soil cultures; reductions at 1000 ppm were somewhat less in the soils.

DCPA was among some 11 herbicides used by Smith and Callahan (1968) in studies on phytotoxicity of chemicals to emerald zoysia sod plugs, used to establish a zoysia turf. All of the herbicides provided excellent preemergence control of crabgrass and goosegrass throughout the growing season following treatments. DCPA at 12 lb/A gave 14.8% injury to zoysia 3 months after treatment; 23.1% after 11 months. Zoysia had 83.0% ground cover 17 months after the DCPA treatment.

Fullerton et al. (1970) tested the effects of DCPA on winter injury of newly-established bermudagrass. Using the Tifgreen variety on an Arkansas soil, they found that DCPA applied to newly-sprigged bermudagrass before June 15 did not influence winter injury. Treatment in July, August, or September to plots established in June caused about 15% winter injury. Injury in this range may be less detrimental than competition from weeds on non-treated plots. Treatment in August resulted in 60% winter injury to the bermudagrass but nitrates, amino acids, and proteins were higher than in controls. The authors suggest that the heavy winter injury may indicate a less winter-hardy turfgrass condition as a result of DCPA treatment; the grass plants entered winter in a more vegetative condition.

Endothall

7-oxabicyclo[2.2.1]heptane-2,3-dicarboxylic acid

Endothall is a white, crystalline, odorless solid. It is an acid and forms water-soluble salts with sodium, potassium, and amine bases. It is used as a general-purpose herbicide and as an aquatic weed herbicide. It is also useful in sugar beets and certain legume seed crops.

Endothall may be applied pre- or postemergence; it requires soil moisture to be effective in the soil. It has been used as a total postplanting surface spray, a prelisting total surface spray, a postplanting band spray over the seed row, and as a band treatment by soil incorporation, prior to or during planting. It is active against both grass and broadleaf weeds but some of the more drought-resistant weeds of arid regions require heavy dosage. Endothall kills by contact; foliage, stem, and root tissues are affected. It is readily absorbed by roots and translocated to a limited extent to the foliar plant parts via the xylem; it is not phloem-mobile. Since it responds to activation with ammonium sulfate it must penetrate the cuticle as the undissociated parent molecule. It has no formative effects and its toxication consists of rapid penetration, desiccation, and browning of the foliage.

Endothall causes callose formation in sieve tubes (Dunning, 1958; Maestri, 1967). It prevents vein loading (Leonard and Glenn, 1968); possibly this may explain its failure to move in the phloem. Thomas and Seaman (1968) using ^{14}C-labeled endothall recorded symplastic movement in the aquatic plant, American pondweed (*Potamogeton nodosus*). There was no movement of this tracer from roots to tops of these plants, possibly because there is no movement of the transpiration stream. They found that endothall not only injures roots but is strongly absorbed; it may not be able to penetrate into the xylem.

Endothall breaks down quite rapidly in natural bodies of water; little is known of its metabolism in plants. Hiltibran (1962), using the flax seed root growth bioassay, found that endothall applied at from 0.3 to 10.0 ppm in the field was not detectable after about $2\frac{1}{2}$ days, the maximum period of detection was 4 days. Dissipation in a series of ponds took place in 48 hours where 1.0 ppm was applied; 72 to 96 hours where 5.0 and 10.0 ppm were used. In some lake waters, tested in aquaria, dosages of 5.0 and 10.0 ppm reached the 0.1 ppm level in from 9 to 61 days.

In further studies Hiltibran (1963) found the sodium salt of endothall to be quite selective on aquatic weeds. Sago pondweed (*Potamogeton pectinatus*), leafy pondweed (*P. foliosus*), and curlyleaf pondweed (*P. crispus*) were susceptible to both liquid and granular preparations. American pondweed required a liquid formulation; waterwillow (*Justicia americana*) and *Jussiaea repens*, two emergent species, were defoliated but not killed. Common cattail (*Typha latifolia*) and narrowleaf cattail (*T. angustifolia*) were susceptible; cabomba (*Cabomba caroliniana*) and *Chara* species were not killed.

Penner (1968) studying the herbicidal influence on amylase activity in barley and squash seedlings, found that endothall, among four potent herbicides, prevented development of amylase activity in the distal halves (embryo lacking) of barley. These four herbicides did not inhibit gibberellic acid-induced amylase synthesis in de-embryonated halved barley seeds. Herbicidal action on the degradation of reserve materials during germination may be the cause of differences in seedling susceptibility to herbicidal injury formed in seedlings dependent on stored carbohydrates.

The proteolytic activity of the cotyledons of squash increases during germination. The development of this proteolytic activity has been shown to be inhibited about 60% 3 days after planting the seeds into vermiculite treated with endothall (Ashton et al., 1968). Penner and Ashton (1968) used dichlobenil, endothall, and bromoxynil to determine their influence on kinin control of proteolytic activity. None of these herbicides affected the activity *per se* of isolated proteolytic enzymes. Endothall at the concentrations used (10^{-3} to 10^{-5}M) reduced the development of the proteolytic activity in the cotyledons of intact embryos but had little effect on excised cotyledons. The addition of benzyladenine did not reverse the endothall-caused inhibition. The effect of endothall with and without benzyladenine on the development of proteolytic activity in excised cotyledons and cotyledons from intact plants was very similar to the effect of actinomycin D. Since actinomycin D selectively inhibits the DNA-directed RNA (mRNA) synthesis, these results suggest that endothall may interfere with mRNA synthesis. Delayed introduction into the culture solution reduced the ability of this herbicide to inhibit proteolytic activity development.

One advantage of endothall over many aquatic weed killers is its low toxicity to fish. A median tolerance limit of 9 species of warm-water fish ranged from 95 to 150 ppm (Walker, 1963). Smallmouth bass, green sunfish, and mosquito fish, present in the reservoirs that Yeo (1970) treated, were not distressed by his treatments which did not exceed 4.0 ppm.

Walker (1963) used the disodium salt of endothall and a compound described as the di-N,N'-dimethyl-o-cocoamine salt of endothall in studies on the control of some 19 species of aquatic weeds and the effects on 15 fish-food organisms. As noted above, the disodium endothall is relatively non-toxic to fish. The median tolerance limits for the cocoamine salt of endothall ranged from about 0.06 to 0.3 ppm for five species of fish. This formulation applied at concentrations of 0.3 to 1.0 ppm, toxic to fish, was useful as a tool to control aquatic weeds and to renovate stunted fish populations.

Liquid formulations of endothall were best in controlling algal mats, floating and emergent weeds. Granular formulations proved most effective on submersed rooted plants. Residues of the cocoamine salt were of short duration. Disappearance depended upon time and concentration. Detectable residues disappeared within 8 days following applications of 0.3 ppm and within 2 weeks for 0.6 ppm; 1.0 to 3.0 ppm took up to 25 days to disappear. Fish flesh showed no traces of this compound at sublethal concentrations. Intraperitoneal injection of endothall into fish resulted in disturbed osmoregulation. Total numbers of fish-food organisms were reduced following endothall treatments at 5.0 to 10.0 ppm; application of the cocoamine salt at 1.0 ppm resulted in increases in numbers and weight of 15 bottom-dwelling fish-food organisms.

Fenac

(2,3,6-trichlorophenyl) acetic acid

Fenac in its pure state is a white, odorless, crystalline powder, slightly soluble in water but relatively soluble in organic solvents. It is available as a water-soluble sodium salt, emulsifiable ester, granular, and wettable powder formulation.

Fenac has growth regulating properties causing epinasty, bud inhibition, and bud necrosis. However, since perennial weeds may be controlled by application to the soil, and since translocation is principally apoplastic, fenac is most effective applied through the soil. It is used preemergence to control seedling annual and perennial weeds. Its principal uses are in sugar cane, as an aquatic weed killer, and for perennial weed control in non-cropped areas. It may be used postemergence against perennial broadleaf weeds on non-cropped lands at 10 to 20 lb/A. It has proved to be especially useful against puncturevine (*Tribulus terrestris*) in the western states. In granular form, it is very effective for controlling certain aquatic weeds in static water, i.e., lakes and reservoirs. The sodium salt formulation is supplied as a liquid concentrate for application in water at 20 to 100 gpa or as a granular form on attaclay as carrier.

Fenac is adsorbed on soils and resists leaching. Its breakdown by micro-organisms is slow; it persists from 1 to 2 years when used at rates of 10 lb/A or above.

Fenac has been found useful for controlling quackgrass (*Agropyron repens*) when applied in late spring and plowed or disked under within a week. Field bindweed (*Convolvulus arvensis*) has also been effectively controlled by fenac. At rates of 10 to 40 lb/A, fenac has been effective in the control of some of our most persistent perennial weeds. Soil type, rainfall, time and rate of application, and weed species are all involved in the effects of fenac application.

Robinson (1961) tested some 15 herbicides in a search for a satisfactory control for witchweed (*Striga asiatica*). Earlier tests had proven that preplant soil-incorporated materials were more consistent in the killing of witchweed than pre- or postemergence treatments alone. In 1959 fenac was included in the studies and it proved to be the most promising of all of the materials used. Shaw et al. (1962), in a comprehensive report on witchweed control, stated that maize receiving fenac as a preplanting soil-incorporated treatment, and using 2,4-D as a postemergence spray as needed, has given effective full-season control of witchweed. Tobacco and soybeans cannot be grown in rotation with maize for 2 years after use of fenac because of residues in the soil.

Funderburk and Lawrence (1963) studied the uptake and distribution of ^{14}C-labeled fenac in fish and aquatic plants. In waterstargrass, fenac was absorbed by both roots and shoots and translocated to the untreated portion (shoot or root) of plants that had their roots isolated from their shoots by a seal of silicone stopcock grease. The experiments ran for 1 week and although acropetal movement from roots was obvious in their autographs, symplastic transport from shoots to roots was weak; it could have resulted from cell to cell movement by protoplasmic streaming within the 7-day experiment period.

Fluorodifen

p-nitrophenyl α,α,α-trifluoro-2-nitro-*p*-tolyl ether

Fluorodifen is a yellow, odorless, crystalline compound with a water solubility of less than 2.0 ppm. It is readily soluble in a number of organic solvents. It is available as a technical 3.0 lb/gal emulsifiable concentrate and as a 15% granular material. It is a selective contact herbicide applied preemergence at time of planting or later at any time before emergence of weeds and crop. It is used on non-food crop soybeans in the U.S.A. It has been tested and found useful both pre- and postemergence on rice and wheat; preemergence on cotton, peanuts, maize, dwarf beans, peas, and sorghum. It controls a wide range of broadleaf and grass weeds including barnyardgrass, cattails, goosegrass, and foxtails; it is effective in suppressing nutsedge (Ebner et al., 1968).

Fluorodifen was applied to cotyledons of soybean seedlings producing a visible burn; applied to young or mature leaves it caused a local cleaving of the veins. Application to the stem results in local browning but does not affect subsequent development. As maize shoots emerge through the treated surface soil, a scorched effect producing a broken line of spots across each of the first three or four leaves may occur; the plants recover and yield is not reduced. Culture-solution-grown soybean plants will absorb and translocate fluorodifen from the roots to the foliage. Leaf veins become translucent and changes in morphology occur. Roots in the culture medium are reduced in growth and so is the whole plant in time (Ebner et al., 1968).

Walter et al. (1970) investigated the movement of fluorodifen in plants. They applied fluorodifen to both roots and leaves of culture solution plants of soybean, grain sorghum, peanut, and tall morningglory (*Ipomoea purpurea*). The herbicide was absorbed by the treated tissues, but only limited translocation into other plant parts was detected. From root application apoplastic translocation of fluorodifen was found to be greatest in tall morningglory and grain sorghum. Ebner et al. (1968) also reported that fluorodifen was not translocated symplastically to any appreciable extent in plants.

Walter et al. (1968) determined the fluorodifen residues in preemergence treated soybeans that had received dosages of 2 to 6 lb/A. The herbicide was partitioned into the hexane portion of a 1-to-1 mixture of acetone and hexane by adding distilled water. The hexane portion was chromatographed on a silica gel column; interfering compounds were eluted with hexane. The herbicide was eluted with acetone and quantified by electron-capture gas chromatography. Sensitivity was 0.05 ppm and recovery better than 85%. No residues were found in soybeans treated as noted above.

Eastin (1969) studied the movement and fate of fluorodifen in peanut seedlings. Using autoradiography and liquid scintillation counting they found that this herbicide is rapidly absorbed from nutrient solution but only 6.5% of the activity absorbed is translocated to the shoot after 144 hours; acropetal movement was confined to the stem and petioles. Metabolism of fluorodifen proceeded rapidly, the major pathway involving hydrolysis of the ether linkage and reduction of the 4-nitrophenol to form 4-aminophenol. Other metabolites found amounted to less than 1.0% of the radioactivity applied. Data indicated that the major detoxication takes place in the roots of peanut seedlings with the metabolites being translocated to the shoot. Radioactivity remaining in the fibrous residue was possibly conjugated with insoluble plant constituents.

Eastin (1971a) used ^{14}C-fluorodifen labeled in the 1 position of the p-nitrophenyl ring (fluorodifen-1-^{14}C) and on the trifluoromethyl carbon (fluorodifen-$^{14}CF_3$) in studies on fate in peanuts. He found rapid absorption but limited acropetal movement after root treatment for 48 hours followed by 96 hours in nutrient solution. The major products of fluorodifen-1-^{14}C degradation were p-nitrophenol and an unknown number 1, possibly a conjugate of p-nitrophenol. Some p-nitrophenyl-α,α,α-trifluoro-2-amino-p-tolyl ether and several minor unknowns were detected. From breakdown of fluorodifen-$^{14}CF_3$, Eastin (1971a) found a second unknown which he thought might be a conjugate of 2-amino-4-trifluoromethylphenol. Figure 22-2 shows Eastin's scheme for fluorodifen degradation in peanut seedlings. Following these steps he proposes that the respective phenols are conjugated with natural plant substances to form water-soluble conjugates.

In cucumber seedlings, Eastin (1971b) found that the degradation pathway of fluorodifen-1-^{14}C was similar to that in peanut but slower. The major degradation products were p-nitrophenol and an unknown (1). More rapid absorption and translocation coupled with slower degradation, Eastin proposes, may contribute to the susceptibility of cucumber to fluorodifen.

In studies on fluorodifen metabolism in peanut roots, Eastin (1971c) found a continued uptake of fluorodifen and a steady breakdown throughout his experimental period from 0 to 72 hours; as percent of total, fluorodifen went from 73.5% at 2 hours to 12.0% at 72 hours. Meanwhile, 38.1% was in unknown 1 and 44.3% in p-nitrophenol.

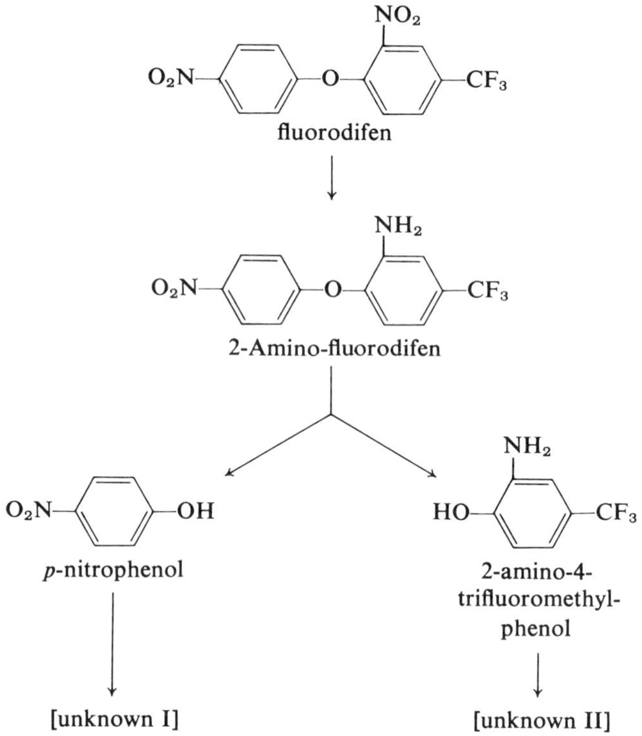

FIGURE 22-2.   Proposed major pathway for fluorodifen degradation in peanut seedlings (Eastin, 1971a).

Fluorodifen breaks down quite rapidly in soil. Ebner et al. (1968) cited an example where an application at 8 kg/ha resulted in 10.8 ppm in the surface 5 cm, 10 days after application; after 60 days the concentration had lowered to 1.9 ppm.

MH

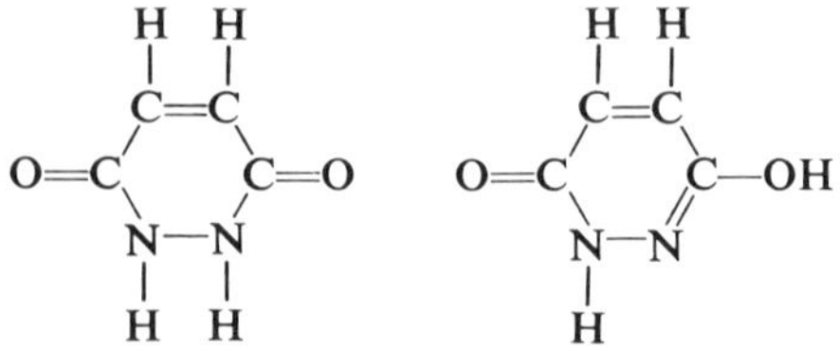

1,2-dihydro-3,6-pyridazinedione   6-hydroxy-3-(2H)-pyridazinone

MH, also known as maleic hydrazide, was introduced in 1949 (Schoene and Hoffman, 1949) as a growth inhibitor on tomato plants and White (1950) reported a delaying action on blossoming. Currier and Crafts (1950) first announced the selective inhibition and eventual killing of grass by MH. As a non-selective grass herbicide MH gave promise of being the long-sought toxicant to complement 2,4-D for general weed control. Further study, however, proved these two compounds to be antagonistic.

MH has been used successfully to inhibit growth of turf plants for the purpose of reducing the labor of maintaining highways. Foote and Himmelman (1967, 1971) reported tests carried out in Minnesota for this purpose. They used dosages of 8, 10, and 12 lb/A and treated narrow strips on fence lines. Applied early in the season when the grasses were 2 to 4 inches tall, the 12 lb rate gave significant reduction in plant height, reducing the variation in height among the turf grasses. It did not reduce the number of leaves on the plants but did greatly reduce the number of influorescences, which improved the appearance. There was some discoloration of some turf species, especially redtop grass and white clover. Foote and Himmelman (1967) concluded that their treatments were effective in reducing or eliminating mowing or hand clipping in hard-to-mow fence lines and guard rails. MH had the advantages of not eliminating all vegetation, of not moving laterally, and of not being conducive of erosion. Timing proved to be important and inclusion of 2,4-D was advantageous where 2,4-D susceptible weeds were present.

Buchholtz (1953) showed that MH treatment of quackgrass in the spring followed within a few days by plowing was effective in controlling this perennial grass for a whole season. Where annual grasses reduced yield of maize, a preemergence application of a urea or triazine completed the control and tillage became unnecessary.

Youngner and Goodin (1961) used MH successfully to control flowering in kikuyugrass (*Pennisetum clandestinum*) by applying solutions to foliage at weekly intervals. On the small cultures used, untreated controls produced 176 flowers; plants treated with 500, 1000, and 2000 ppm MH produced 15, 7, and 1 flowers, respectively. These treatments had no serious injurious effect upon vegetative growth.

In addition to inhibition of growth, MH was found to induce dormancy. It was used on nursery stock, cane berries, and a great array of ornamentals to control storage life and to inhibit growth once dormancy was broken. It was also used to cause male sterility, to prevent suckering in flue-cured tobacco plants, and to prevent bolting of root crops.

MH proved to have many interesting effects on plants. In addition to retardation of growth, it was found that MH at concentrations of 0.2 to 0.4% caused thickening of leaves (Crafts, 1961). Girolami (1951) found that MH caused callusing and obliteration of sieve tubes, an effect that resulted in injury. Detailed studies proved that MH interfered with cell division; growth of embryos is inhibited and maturation may be precocious. Leaves, mature at time of application, may survive for weeks; young leaves cease growth and metabolism slows. Cytological studies showed cessation of mitosis of broadbean (*Vicia faba*) roots with breakage of chromosomes. In maize, mitosis slowed, chromosomes were broken, and chromosomes and fragments were lost; nucleolar volume increased and persisting nucleoli were found throughout the entire mitotic cycle. Continued studies confirmed many of these observations on the morphological and physiological effects of MH on plants (Crafts, 1961).

MH did not follow the response of penetration to pH found with dinitro compounds and 2,4-D. Tested through a range from pH 2 to 10, the greatest uptake appeared to occur at pH 7. Water-soluble formulations of MH are more effective than ester forms of low water solubility. High humidity promotes absorption. MH apparently enters the plant along an aqueous pathway. Oil carriers for MH were less effective than water. Aromatic oil, sodium chlorate, TCA, and dalapon as supplements to MH aided in the control of perennial grasses. MH-induced dormancy was most prolonged in old, well-established cultures of bermudagrass; in some it lasted for over a year. Simulated tillage increased effectiveness of MH; possibly microbiological activity in the soil contributes to this response (Crafts, 1961).

MH is readily absorbed by plant foliage, moved to the vascular bundles, and translocated symplastically to sinks in roots, shoot tips, young leaves, and buds. Absorbed by roots, it moves slowly into vascular channels and translocates to foliage via the xylem. MH becomes fixed within the plant and is not metabolized; however, it is rapidly broken down by soil microorganisms (WSSA, Herbicide Handbook, 1970).

Isenberg et al. (1951) found, using the triphenyl tetrazolium test, that in onion tissues, both light-grown and etiolated, treatment with maleic hydrazide at 1000 and 2000 ppm inhibited one or more of the dehydrogenases. In the light-grown plants the amounts of triphenylformazan formed indicated a lowering in dehydrogenases in the tissues. In etiolated plant tissues even greater reductions were found where succinate, malate, and pyruvate were the substrates. Thus MH seems to affect respiration through the inhibition of one or more dehydrogenases; the rate of these responses is governed by the rate of absorption into the plant.

MH has also been shown to induce the exudation of droplets of sticky material which, upon analysis, proved to be fructosans. Sucrose accumulated and leaves gradually became gorged with starch. MH treatment of maize resulted in increases in peroxidase, phosphatase, and polyphenolase. Inhibited seedlings had increased catalase and cytochrome oxidase (Crafts, 1961).

The mode of action of MH has not been thoroughly elaborated. Crafts

(1961) suggested that, in the case of maintained dormancy of perennial grasses, depletion of plant foods might prove lethal. On the other hand, evidence for disruption of vascular tissues might result in disturbed metabolism and death. Crafts and Yamaguchi (1960) showed that MH enters the roots of barley plants quite slowly; movement to foliage as shown by autoradiographs of ^{14}C-labeled MH took 2 days; treatment with DNP at 10^{-4}M speeded up the process somewhat.

Nooden (1967) reported, before the annual meeting of the American Society of Plant Physiologists on the mechanism of action of MH. Citing Callaghan and Grun (1961), to the effect that MH is bound to nuclear material in roots, he reported that MH at 1mM does not inhibit cell enlargement but blocks mitosis. MH does not seem to act as a base analogue or sulfhydryl inhibitor. ^{14}C-MH is absorbed by maize or pea roots and bound to some acid- and (80%) ethanol-insoluble material. Except for traces of two derivatives, the 80% ethanol-soluble ^{14}C in roots is MH. The 80% ethanol-insoluble ^{14}C is not released by mild acid hydrolysis but most of it is solubilized by 2N HCl at 90° for 2 hours. Large amounts of ^{14}C can also be released by 0.5N KOH. Treatment with RNase, DNase, and cellulase did not solubilize the ^{14}C, although proteolytic enzymes released some. Paper chromatography showed that the ^{14}C released is in MH and one additional compound. The binding of ^{14}C-MH to the ethanol-insoluble material was inhibited by DNP but not by puromycin, fluorouracil, or fluorodeoxyuridine. MH (1mM) inhibits DNA synthesis in whole intact maize roots in about 16 hours and RNA synthesis a few hours later. These findings would seem to rationalize the very slow passage of MH through roots to a binding of the molecule by some energy-requiring process to ethanol-insoluble substance, possibly a protein. This is far from a satisfactory explanation of its ready uptake and rapid translocation by foliar organs, nor does it provide a clue to its dormancy-inducing properties. Biswas et al. (1966), finding that MH increased the protein hydrolysate in tea without increasing growth, suggested that the growth inhibition might result from failure to synthesize growth hormones responsible for cell enlargement or that it might have a direct effect on the cell wall, reducing plasticity while allowing protein synthesis to continue.

In a later paper, Nooden (1969) reported additional information on the mode of action of MH. In his experiments, MH inhibited maize root elongation through an effect on cell division apparently without affecting cell enlargement. A decrease in the rate of elongation became apparent only after a lag period of over 14 hours. MH failed to inhibit growth of roots of maize seeds given large doses of γ-irradiation that stop cell division so that growth takes place by cell enlargement alone. Nor did MH counteract the cell enlargement induced by IAA in maize coleoptile sections. Tests on a number of compounds known to alleviate MH inhibition in other tissues proved that none was able to prevent the inhibition of maize root elongation by MH. Although MH has long been suspected of inhibiting plant growth mainly through its effect on cell division rather than cell enlargement, past evidence has not been clear. Nooden points out that, at least in the case of maize roots, MH inhibits cell division but not enlargement. Since cell enlargement is known to require active metabolism, including ATP and protein synthesis, it follows that MH apparently does not significantly interfere with these.

Mylone

$$H_2-C \overset{S}{\diagup} C=S$$
$$H_3C-N \diagdown \underset{C}{\diagup} N-CH_3$$
$$\underset{H_2}{|}$$

3,5-dimethyltetrahydro-1,3,5,2H-thiadiazine-2-thione

Mylone is a white crystalline solid with a melting point of 99.5°C, a flash point of 280°F, and a very low water solubility. Mylone, introduced as Crag Brand 974, has proved effective as a weed killer (Anon., 1956) against annual weeds in seed beds and potting soils when applied preemergence to the weeds, incorporated in the soil, and irrigated to effect a water seal. Used in this way, at 300 lb/A it controlled Florida pusley, crabgrass, and bermudagrass. Mylone applied preplanting to chrysanthemum nurseries proved effective in controlling all weeds.

To sterilize potting soil, Foret (1959) recommends treating 6-inch layers by spraying over the soil, adding more soil, and spraying successively until the pile is complete. Then the pile is covered with a polythene tarp and left for 3 weeks. The soil should be moist enough for planting when treated. When completed, the treatment effectively eliminates weed seeds, nematodes, and many fungi and other soil-borne pests. The following weeds have been reported controlled: crabgrass, henbit (*Lamium amplexicaule*), pigweed (*Amaranthus* spp.), foxtail, common purslane (*Portulaca oleracea*), common lambsquarters, common chickweed (*Stellaria media*), carpetweed (*Mollugo verticillata*), pepperweed (*Lepidium* spp.), mustard (*Brassica* spp.), coffeeweed (*Daubentonia texana*), field bindweed, bermudagrass, and johnsongrass (*Sorghum halepense*). Nutsedge has not been completely eliminated in all cases.

Tests have shown that upon contact with moist soil Mylone breaks down rapidly, releasing gaseous and water-soluble active ingredients into the soil. These come in contact with the weed seeds, nematodes, soil fungi, and insects and kill them. Tests at the Pineapple Research Station (Anderson and

Okimoto, 1953) have reported that fungi may be killed by the fumes of Mylone confined to the culture flask.

The breakdown products of Mylone (methyl isothiocyanate, formaldehyde, hydrogen sulfide, and monomethylamine) interacting at the site of release in the soil with the various organisms are responsible for the potent action of the chemical (Carbide and Carbon Tech. Info. Lith., 1957). The effect of soil pH seems to be negligible but soil temperature is important. Warm soil temperatures are favorable to active release of the toxic products and their rapid dissipation from the soil. A 3-week period should always be allowed for breakdown and dissipation of the chemical. Planting sooner may result in injury to crop seedlings.

Nitrofen

2,4-dichlorophenyl p-nitrophenyl ether

Nitrofen, long known as Tok, is a free-flowing solid of dark brown color and almost insoluble in water. It is sold in the form of an emulsifiable liquid for application mixed with water at 50 to 75 gpa. It is used to control many annual weeds by preemergence application in cole crops and asparagus; postemergence in carrots, celery, transplanted cole crops, and rice.

Matsunaka (1970) gave a tabular list of suggested or recommended herbicides for use in transplanted rice in the Philippines and Japan. In the Philippines, recommendations from the International Rice Research Institute include nitrofen granular plus the diisopropyl ester of 2,4-D at 2.0 + 0.5 kg/ha and nitrofen granular followed by MCPA at 2.0 + 0.8 kg/ha; both to be applied 3 days after planting but the MCPA to be applied 22 days after planting, postemergence, on broadleaf weeds.

Nitrofen causes a burning of mature foliage and malformation of developing leaves when applied to the shoot tip of a plant. Nitrofen is a contact toxicant and it kills seedlings before emergence. As a postemergence herbicide on cole crops and rice, it is selective by differential wetting (Pereira, 1970).

Pereira's (1970) literature review indicated differences in the response of cultivars of cabbage to postemergence applications of nitrofen. Pereira chose Hybelle, a tolerant variety, and Rio Verde, a susceptible one, for his work. No differences were found in germination of Hybelle or Rio Verde seeds in the dark or in the light. Both cultivars translocated nitrofen at the same rate as measured by autoradiography with ^{14}C-nitrofen. There was no evidence for systemic distribution by either phloem or xylem transport. Nitrofen induced stomatal closure and decreased transpiration in both cultivars, resulting in increased leaf temperatures.

To test penetration of nitrofen into cabbage leaves, Pereira rubbed the leaves with glass wool to remove the cuticle; this increased phytotoxicity. Rio Verde plants with glaucous bloom were more tolerant than similar plants with bluish bloom (less wax). Plants grown in shade (less cuticle) were more susceptible than plants grown in full sunlight, regardless of the cultivar. Hybelle plants had more wax per unit leaf surface than the Rio Verde plants and wax deposition increased as the leaves matured. Both cultivars were equally tolerant to nitrofen after the sixth week of age. The Rio Verde leaves absorbed ^{14}C-nitrofen twice as fast as the leaves of Hybelle plants. Pereira concluded that the mechanism of selectivity to nitrofen is a matter of herbicide wetting and penetration.

In attendant studies, Pereira (1970) found that cabbage plants increased in susceptibility to nitrofen as they were kept in the dark prior to or after nitrofen treatment. Both phytotoxicity and its enhancement in the dark were alleviated temporarily by treatment with exogenous sugar: sucrose, glucose, or fructose. Lowering the water potential of cabbage leaves increased susceptibility to nitrofen but this effect could be reversed by exogenous sucrose.

Nitrofen inhibited non-cyclic photophosphorylation and electron transport in isolated spinach chloroplasts. Respiration, as reflected by increased oxygen uptake, was increased by incubation of cabbage leaf sections in nitrofen. This herbicide promoted membrane permeability of red beet root sections and this effect was increased by DMSO (dimethylsulfoxide). This result, in turn, was completely overcome by treatment with sucrose or carbowax at 0.2M concentration.

From Pereira's studies, it seems evident that selectivity of nitrofen between young cabbage plants of different cultivars is a matter of differential wetting and penetration. The severe contact action observed evidently results from loss of membrane integrity of the leaf cells that are killed. The responses of rice and watergrass from preemergence treatments certainly cannot be of the same nature because the weed seeds fail to germinate normally. Here it would seem that Pereira's observed effect upon respiration might be pertinent; possibly the chemical enters the seeds of watergrass and so disturbs the respiratory mechanism that energy for growth fails and the embryos die without emerging.

Hawton and Stobbe (1971a) studied the selectivity of nitrofen between wild turnip (*Brassica campestris*), redroot pigweed, and green foxtail (*Setaria viridis*) using a replicated dosage-response series. On an ED_{50} basis, setting wild turnip equal to 1.0, green foxtail had a value of 5.8 and redroot pigweed 63.3. Considering spray retention, differential penetration, and relative phytotoxicity, it was estimated that green foxtail was 9 times more susceptible to nitrofen than wild turnip, and redroot pigweed 99 times more susceptible.

Hawton and Stobbe (1971b) continued their studies using ^{14}C-nitrofen. There was only limited translocation in rape, redroot pigweed, and green foxtail. Plants under high light intensity produced several labeled compounds of different molecular size and Rf values. At least two of the compounds were lipid-nitrofen conjugates or nitrofen polymers; other compounds found were smaller molecules, presumably products of cleavage at the ether linkage.

Picloram

4-amino-3,5,6-trichloropicolinic acid

Picloram, in the pure state, is a white powder with a chlorine-like odor. Picloram is on the market as the potassium salt in a 2 lb/gal formulation. It is also available as a mixture with 2,4-D; both the picloram and the 2,4-D occur as the triisopropanolamine salts in a water-soluble formulation. There is also a product in the form of beads that contains picloram and boron compounds; this formulation is applied dry as a general soil sterilant for use on non-cropped areas.

Picloram is an extremely mobile compound, readily absorbed by both foliage and roots and translocated in both the phloem and the xylem. At low levels picloram affects leaf shape. The tips of new leaves may develop into narrow extensions of the midrib and thickening of the mesophyll may occur. In some species, distinct puckering of young leaves may develop. With increased dosage, cupping and stunting of the leaves are observed and terminal growth ceases. Tissue proliferation along the stem may take place, first at the stem tip, then at the nodes, and finally throughout the length of the stem. Meanwhile epinasty, bending, and splitting of the stem occur and the roots deteriorate; the plant soon dies. Plant species vary in their responses to picloram; certain mustard and spurge (*Euphorbia*) species are quite resistant. On many susceptible plants picloram may be 100 times as potent as 2,4-D; its ready uptake by leaves and roots and increased translocation make it much more effective against woody species in particular.

Meyer (1970) reported that picloram inhibited elongation of hypocotyls and roots of honey mesquite (*Prosopis juliflora*) when grown in solutions at 1 or 2 ppm. This concentration continued to repress growth and by 5 days abnormal growth was taking place in stem cortex. Spraying of 4-month-old seedlings caused curling of the stem tips, death of the growing points, cracking of stems, and production of lateral roots; phellem cells enlarged radially, inner cortex and phloem parenchyma proliferated, and xylem vessels matured without enlarging. Starch disappeared within 8 days.

Scifres and McCarty (1968) found that picloram caused proliferative growth followed by destruction of phloem parenchyma, sieve elements, and companion cells. Destruction of the vascular tissue accompanied proliferation of mesophyll; these studies were made on leaf tissue of Western ironweed (*Vernonia baldwini*). Martin et al. (1970) describe bulb-like enlargements on hypocotyls of fourwing saltbush (*Atriplex canescens*) occurring beneath the cotyledonary node and at the soil line from picloram in the soil.

Bachelard and Ayling (1971) found that picloram:2,4-D mixture and the potassium salt of picloram disrupted chloroplast structure in leaf discs and the integrity of all membranes in stem tips of *Eucalyptus viminalis*.

Lee (1970), reporting from Oregon, found that seed production in several grass species was reduced by picloram at dosages of 1.0 lb/A or above, especially when applied in the spring; seed germination and seedling characteristics were not adversely affected. Chang and Foy (1971a) found that picloram had both promotive and inhibitory effects on growth depending on stage of growth, dosage, etc. Like other auxin-type growth regulators, picloram enhances stem elongation, inhibits root growth, induces cell wall loosening, and causes stem curvature and other formative effects. Chang and Foy (1971a) suggest that the promotive effects of picloram may result from increased synthesis of RNA and DNA. Picloram decreased root growth severely, resulting in decreased water uptake and evapotranspiration, results that Chang and Foy (1971a) thought might have resulted from aberrant development of vascular tissues. Picloram inhibited differentiation of leaf tissues and shoot apices, a result they attributed to its ready symplastic transport.

Swanson and Baur (1969) studied the absorption of picloram by potato tuber tissue. Absorption increased with concentration (5×10^{-4}, 1×10^{-3}, and 5×10^{-3}M) and time (3, 6, 12, and 24 hr). Maximum uptake occurred at pH 4.0; very little was absorbed at pH 7.0 and 8.0. Both uptake and leakage were temperature dependent. More than 90% of picloram absorbed from 5×10^{-3}M concentration was lost to fresh buffer solution within 12 hours.

Baur et al. (1971), using detached oak leaves in absorption studies, found that uptake of picloram in the presence of equimolar concentrations of 2,4,5-T exceeded that found for picloram alone by about 150%. This may explain the advantage for mixtures of picloram and 2,4,5-T found in control of certain woody plants (Robison, 1967).

Ivens (1971) found maximum effectiveness of picloram:2,4-D (Tordon 101) to coincide with the end of shoot extension when new leaves were fully expanded but still succulent; the test species were *Acacia drepanolobium* and *Tarchonanthus camphoratus*, the locality Kenya.

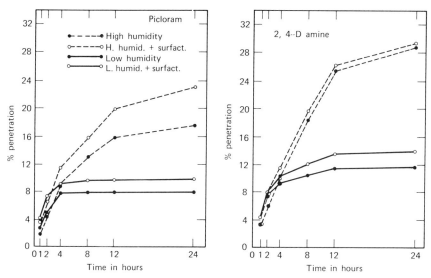

FIGURE 22-3. Penetration of picloram and the dimethylamine of 2,4-D with and without added surfactant (1% Atlex 210) as a percentage of the dose applied (200 μg) from the abaxial surface of aspen poplar leaves under low and high humidity. (Sharma and Vanden Born, 1970.)

Picloram penetrates leaves about like 2,4-D; its great differences from 2,4-D lie in its very free mobility, its ready penetration of roots, and its very high phytotoxicity. Sharma and Vanden Born (1970) studied foliar penetration of picloram and 2,4-D in quaking aspen (*Populus tremuloides*) and balsam poplar (*P. balsamifera*). As has been found with 2,4-D, they observed that temperature and humidity are the principal physical factors that determine the rate and amount of foliar penetration. Figure 22-3 shows the effects of both humidity and surfactant and Table 22-1 provides data on the effects of temperature and humidity. These are the sorts of results one may expect in tests of this kind. They emphasize the fact that temperature, humidity, and the addition of a surfactant can all affect the amount of a given dose that actually enters a leaf.

Agbakoba and Goodin (1969) studied the effect of plant maturity of field bindweed on absorption and translocation of ^{14}C-labeled 2,4-D and picloram. Using 5-week-old and 7-week-old seedlings and 16-week-old vegetatively-propagated plants, they made single-leaf treatments and assayed for ^{14}C after a 48-hour treatment period. Picloram absorption was similar in the two seedling stages; the 16-week-old plants absorbed more than did the seedlings. More picloram than 2,4-D was absorbed at all stages of growth. Translocation in seedlings exceeded that in the 16-week-old plants; there was no difference

TABLE 22-1. Penetration of picloram and 2,4-D amine with added surfactant (1%), as a percentage of the dose applied (200 μg in 40 μl), 12 hr after application to the adaxial and abaxial surfaces of aspen poplar leaves at three temperatures (Sharma and Vanden Born, 1790)

Relative humidity	Temperature (C)	Picloram		2,4-D amine		
		Adaxial	Abaxial	Adaxial	Abaxial	Means
Low	10	3.8	6.4	5.7	7.6	5.9
	25.5	9.5	9.9	11.7	13.9	11.2
	40.5	15.2	15.8	21.2	21.4	18.4
High	10	6.6	8.0	7.3	10.7	8.1
	25.5	15.0	20.6	25.6	26.5	21.9
	40.5	35.0	35.4	44.8	45.6	40.2
Means		14.2	16.0	19.4	20.9	
		L.S.D. 5% = 4.1				

in the amounts of 2,4-D and picloram translocated by the adult plants but more 2,4-D than picloram moved in the seedlings.

Sharma (1971) has conducted research on the absorption and translocation of picloram in Canada thistle (*Cirsium arvense*), soybean, and barley. Table 22-2 presents some of his results. Subsequent studies with these three species showed that Canada thistle and barley each translocated about 35% of the

TABLE 22-2. Translocation of ^{14}C-picloram following Application to a Single Leaf. Data are Presented as % of Total Activity Recovered (Sharma, 1971)

Days after treatment	Fraction	Canada thistle	Soybean	Barley
1	leaf washings	31.8	20.2	48.5
	treated leaf	42.2	31.7	29.5
	plant parts	26.0	41.8	22.0
4	leaf washings	18.1	13.1	40.3
	treated leaf	45.1	20.2	27.3
	plant parts	36.8	66.7	32.4
10	leaf washings	15.2	12.4	36.6
	treated leaf	49.5	15.8	27.2
	plant parts	35.3	71.9	36.2
20	leaf washings	14.3	11.4	33.2
	treated leaf	51.3	15.4	29.9
	plant parts	34.4	73.2	36.9

applied dose out of the treated leaf; soybean translocated about twice as much. Small amounts of the picloram were exuded from roots to the surrounding soil (Sharma et al., 1971).

Picloram is extremely stable in plants. Sharma (1971) found less than 1% of the total dose applied to single leaves of the plants to be degraded to $^{14}CO_2$ over a period of 20 days.

Picloram has a number of biochemical effects on plants. Foy and Penner (1965) had established the fact that picloram inhibited oxygen uptake by cucumber mitochondria. Using succinate as substrate they found a stepwise inhibition within the concentration range of 10^{-3} to $10^{-5}M$; the α-ketoglutarate was less sensitive. Since both mitochondria swelling and ATPase activity are associated with oxidative phosphorylation and respiration in mitochondria, Chang and Foy (1971b) studied the effects of picloram on these two responses. Picloram at $10^{-5}M$ caused swelling of barley and safflower mitochondria when these had been previously contracted with ATP. They further found that picloram at $10^{-3}M$ increased ATPase activity by 28% above the control in safflower mitochondria. More rapid swelling of contracted mitochondria induced by ATP in picloram solution seems to be consistent with a stimulated ATPase activity since mitochondria contracted by ATP became further swollen by hydrolyzing ATP, resulting in active accumulation of ions.

Malhotra and Hanson (1970), using the picloram-sensitive plants, soybean and cucumber, and the resistant species, barley, wheat, and maize, studied changes in nucleic acid metabolism. The total RNA and DNA content of tissue correlated inversely with picloram resistance; the resistant plants were low in nucleic acids whereas sensitive plants were high. The increase in nucleic acids in the sensitive species 24 hours after picloram treatment seemed associated with lower levels of RNase and DNase. Malhotra and Hanson (1970) found that the endogenous levels of RNase and DNase in the resistant species was much higher than that of sensitive species. The specific activities of bound RNase was inversely correlated with herbicide sensitivity; resistant species were higher than sensitive species in bound nucleases. There was no such correlation for the free enzyme. After picloram treatment, the bound nucleases decreased in the resistant species whereas, in general, the opposite was true for sensitive ones. Picloram seemed not to have significant effect on bound nucleases in maize and cucumber seedlings. The free nucleases showed a very small increase resulting from picloram. The authors surmised that picloram may have a similar effect in promoting nucleic acid synthesis in both resistant and sensitive species. The presence of higher levels of native bound nucleases in resistant species may prevent accumulation of nucleic acids and the nucleic acids may be degraded as soon as they are synthesized. Even though picloram reduces the levels of bound nucleases, the net concentration in the resistant species is still greater than that in the sensitive species. They conclude, however, that since the role of nucleases in cell metabolism is still unknown, the correlation between resistance and high levels of bound nucleases may be entirely fortuitous.

The above discussion indicates the difficulty in determining the exact mode of action of any growth-regulating herbicide. As explained in Chapter 6, to find direct interactions of herbicides with specific enzymes in plants is very difficult. Again it seems possible that picloram, through effects on nucleic acid synthesis and metabolism, may regulate protein synthesis in cells and affect enzymes and enzyme systems in several or many ways. To relate the phytotoxicity of picloram to any specific enzyme seems practically impossible.

Picloram is very stable in soils. Keys and Friesen (1968) found that picloram persisted in Canadian prairie soils. Using dosages ranging from $\frac{1}{2}$ to 48 oz/A, 20 to 30% remained after 1 year; the 8 oz/A dosage declined to about 10% after 24 months and to 6% or less after 35 months. Activity under their conditions was concentrated in the 0- to 6-inch level and movement into subsoil in percolating water was greatest in soils of low organic matter. Their sampling only extended to 9 inches in depth.

Pyrazon

5-amino-4-chloro-2-phenyl-3(2H)-pyridazinone

Pyrazon is a light-brown odorless powder of low water solubility. It is used as a suspension in water at 30 to 40 gpa broadcast or in bands. It is a general preemergence or early postemergence herbicide used principally in sugar beets. Under conditions where irrigation is required, pyrazon may be applied preemergence in bands and sprinkled or as a shallow-preplant soil-incorporated treatment where furrow irrigation is practiced. Dosage rates range from 2 to 4 lb/A.

Pyrazon slows growth and produces stunted, abnormal plants at sublethal dosage; at normal dosage, most common annual weeds associated with sugar beet production are killed. Developed, growing weeds wilt soon after postemergence treatment with pyrazon; both fresh and dry weight are reduced within a few days; moisture loss from treated plants increases while uptake by roots is reduced. Thus the susceptible weeds are rapidly desiccated and die (Frank and Switzer, 1969a).

Rodebush and Anderson (1970) treated bean plants during germination with pyrazon at 100 ppm. After 5 days, chlorosis started to appear at the leaf margins, progressed inward, and intensified until the leaves were dead. Microscopic study showed that chloroplast form and arrangement were altered; chloroplasts became round rather than discoid, were free of starch, and aggregated into clumps. Fine structure studies (Anderson and Schaelling, 1970) revealed that grana formation in chloroplasts was inhibited, thylakoids became swollen and perforated, ultimately disintegrating; lipid globules increased in size and numbers. The writers propose that these changes result from inhibition of thylakoid structural protein formation.

Although pyrazon is absorbed by leaves, it is not well translocated symplastically. It is readily absorbed by roots and distributed throughout the plant via the xylem. Fischer (1962) recommends treatment at the cotyledon stage of the weeds; optimum conditions for use involve calm weather, diffuse light (overcast), temperatures around 70°F, and relative humidity between 40 and 70%. Frank (1967) found the two- to four-leaf stage of beet to be the most tolerant to postemergence treatment. Lhoste et al. (1963) described the "open cotyledon" stage of sugar beet plants as one during which the seedlings are injured by postemergence treatment.

Dawson (1971) reported that injury to beets, induced by pyrazon, was greater in 1966 than in 1967. In April 1966 the average daily maximum temperature was 67°F; in 1967 it was 58°F. Total rainfall in April 1966 was 0.1 inches; in April 1967 it was 0.9 inches. Apparently the warm temperature and low rainfall at the time of planting and emergence in 1966 was responsible for the greater injury. Dawson concluded that the relationship between rates of uptake and detoxication governing the tolerance of sugar beets to pyrazon is temperature-dependent. Koren and Ashton (1969) found a similar relation.

Koren and Ashton (1971) studied the effect of temperature and soil moisture on the toxicity of pyrazon to sugar beet seedlings. High temperatures during or after germination increased the reaction of sugar beet seedlings to pyrazon; variation in soil moisture had little effect. Sugar beet seeds absorbed three times more pyrazon at 35°C than at 18.3°C. During imbibition more than 90% of the pyrazon absorbed by sugar beet fruits was concentrated in the pericarp; this pyrazon did not move into the seedling during or after germination. Apparently the pericarp served as a barrier to movement of pyrazon into the sugar beet seedlings; no such barrier existed in seeds of lambsquarters, a susceptible weed. Apparently the pericarp of the sugar beet fruit constitutes a physical mechanism that provides, at least partially, protection to the sugar beet seed during a period when the biochemical inactivation of the chemical by foliage is not yet operative.

Fischer (1962) proposes, as a result of his studies, that pyrazon breaks down to 5-amino-4-chloro-3(2H)-pyridazone; the phenyl group is split off, rendering the compound inactive as a herbicide. Frank and Switzer (1969b) found that translocation from roots to foliage was uniform in mature and developing leaves of lambsquarters and sugar beets. Pyrazon accumulated into the leaves of lambsquarters, but it was soon metabolized in sugar beet leaves. Roots and foliage of sugar beets both decomposed pyrazon, whereas such breakdown occurred only in roots of lambsquarters.

Stephenson and Ries (1967) studied the movement and metabolism of pyrazon in red beet, a tolerant species; foxtail millet, a species of intermediate reaction; and tomato, a susceptible plant. Using pyrazon labeled with tritium in the phenyl ring, they found the greatest absorption by roots and export to foliage took place in tomato; the least in red beet. A radioactive metabolite of ^{3}H-pyrazon was found in the red beet foliage; none occurred in the shoots of the other two species. Since this metabolite contained the labeled phenyl ring it must have differed from the compound which Fischer (1962) proposed as a metabolite of pyrazon.

In a subsequent report Ries et al. (1968) found what they considered to be an N-glucosyl metabolite of pyrazon in red beet. By gas–liquid chromatography and infrared spectroscopy they proved that pyrazon was present in the metabolite. When the carbohydrate portion of the metabolite was studied by infrared spectroscopy and gas chromatography it turned out to be glucose. The authors concluded that the metabolite which occurs in red beet shoots, but is absent from the roots, is N-2-chloro-4-phenyl-3(2H)-pyridazinone glucosamine. They postulated that the tolerance of beets results from the metabolic conversion of pyrazon to this metabolite.

Stephenson and Ries (1968) studied the metabolism of pyrazon in sugar beets and soil; pyrazon was ^{3}H-labeled in the phenyl ring or ^{14}C-labeled in the 4 and 5 positions. The total recoverable radioactivity decreased with time; ^{3}H-compounds more than ^{14}C-compounds. One metabolite was found in the soil; it proved to be 5-amino-4-chloro-3(2H)-pyridazinone. In the plants, three derivatives were identified, N-(2-chloro-4-phenyl-3(2H)-pyridazinone)-glucosamine, the compound noted above as present in the soil, and a complex of the latter with plant constituents. The dephenylated pyridazinone appeared in the soil within one week; it was detectable in the shoot 2 weeks later; probably it came from the soil.

Selectivity of pyrazon appeared to be associated with the rate of metabolic breakdown in the foliage of the weed; accumulation occurred in the weed whereas metabolism kept pace with uptake in the leaves of sugar beet. Whereas the rates of breakdown in the roots was similar in sugar beets and lambsquarters, in the former breakdown was eight times greater in the sugar beet leaves (Frank and Switzer, 1969b).

Stephenson et al. (1971) found that conversion of pyrazon to N-2-chloro-4-phenyl-3(2H)-pyridazinone, which they termed N-glucosyl pyrazon, takes place to greater extent in light than in the dark. Pyrazon conjugation with glucose is dependent upon the carbohydrate status of the leaf tissue; reduced formation in the dark could be restored by infiltrating the leaf tissue with sucrose.

Eshel (1969) determined the influence of pyrazon on photosynthesis. This herbicide strongly inhibited O_2 evolution from leaf disks of some ten weed and crop species; in most of these a concentration of $5 \times 10^{-5}M$ reduced photosynthesis approximately 50%. In sugar beet, a species resistant to the action of pyrazon, it took a concentration 50% greater to bring photosynthesis to the 50% level. After application to the foliage of tomato and pigweed, phytotoxicity of pyrazon occurred in the light but not in the dark and feeding of glucose through the culture medium partially reversed this effect of light; Eshel (1969) concluded that pyrazon, as an inhibitor of photosynthesis, resembles the triazines and substituted ureas. Thus Eshel proposes that inhibition of photosynthesis is the mode of action of pyrazon.

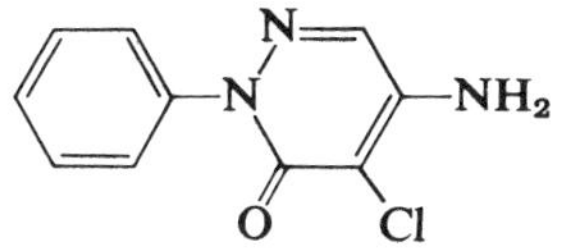

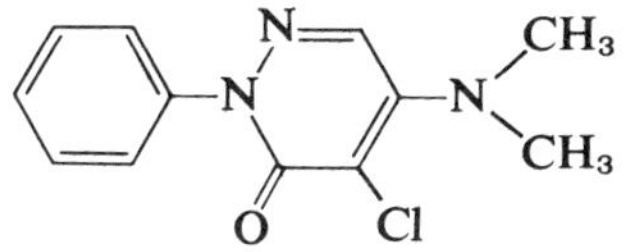

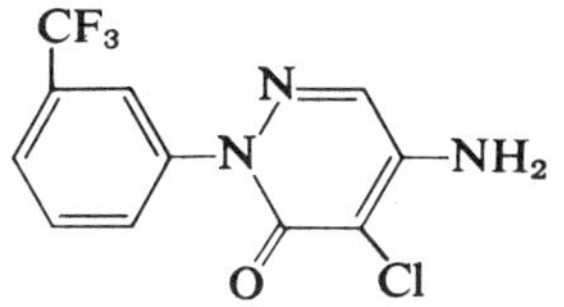

FIGURE 22-4. Structure of four phytotoxic substituted pyridazinones used in Hill reaction studies (Hilton et al., 1969).

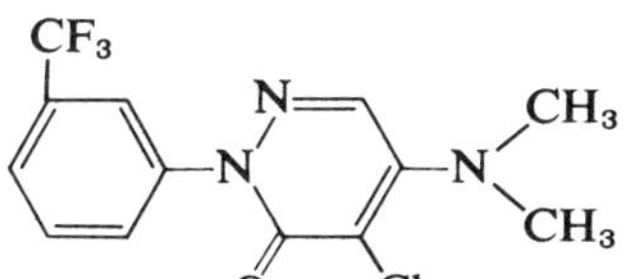

Hilton et al. (1969) studied the modes of action of the four following compounds (Figure 22-4) and compared them with atrazine, dichlormate, and amitrole. The four substituted pyridazinone compounds inhibited the Hill reaction and photosynthesis in barley seedlings. These inhibitions seemed to account for the phytotoxicity of pyrazon; they also account for the action of triazine, phenylurea, and substituted uracil herbicides. Pyrazon is selective in beets because of its rapid biological detoxication in the beet plant by glucosylation of the 5-amino group; by contrast, the non-toxic dephenylated compound is produced in the soil and may be absorbed and translocated into the foliage. As shown in Table 23-3, the pyridazinone chemicals are weaker inhibitors than atrazine. Compound 6706, containing a dimethylamino in the 5 position plus a trifluoro group in a metaposition on the phenyl ring, is a new herbicide which retains the action of pyrazon but is also more resistant to metabolic detoxication in plants and produced chlorosis in the same way that amitrole and dichlormate do. Both substitutions are necessary to give the combined phytotoxic action which proves to be 100 to 1000 times greater than that of pyrazon (Table 22-3).

TABLE 22-3. **Herbicide Inhibition of Ferricyanide Reduction by the Hill Reaction in Isolated Chloroplasts (Hilton et al., 1969)**

Herbicide[a]	Molar concn for 50% inhibition
Atrazine	4.6×10^{-7}
6706	4.4×10^{-6}
Pyrazon	7.0×10^{-6}
9785	1.4×10^{-5}
9774	1.4×10^{-5}
Dichlormate	4.0×10^{-4}
Amitrole	No inhibition at 3.0×10^{-3}

[a] The I_{50} value for 3-(3,4-dichlorophenyl)-1,1-dimethylurea (diuron) in these tests was 1.4×10^{-7} M. It may be used to relate results in this table to those in earlier literature.

Figure 22-5, from Hilton et al. (1969), shows the relative effectiveness of some compounds cited in Table 22-3 in inhibition of photosynthesis. When barley seedlings were treated with pyrazon and compounds 9785, 9774, a combination of 9785 plus 9774, and compound 6706, and the treatment solution around the roots replaced by fresh nutrient solution after 6 or 7 hours, all of the plants recovered except those treated with 6706; these continued to decline and did not recover. When Hilton et al. (1969) examined the pigments in the leaves of barley treated with pyrazon, amitrole, dichlormate, and compound 6706, they found that there was a qualitative similarity in the activity of 6706, amitrole, and dichlormate; the chlorophyll and carotinoid bands were missing from the sample treated with 6707. They also found that 6706 treatment leads to anthocyanine accumulation in wheat

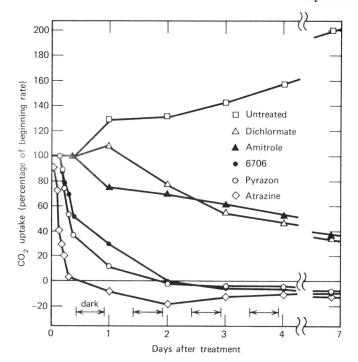

FIGURE 22-5. Inhibition of photosynthesis in barley seedlings after exposure of roots to herbicides. A nutrient solution containing 2×10^{-5} M atrazine, pyrazon, 6706, or dichlormate or 10^{-3} M amitrole, was given at zero time to seedlings 10 days old. (Hilton et al., 1969.)

seedlings; plastids in plants treated with 6706 were shrunken and misshapen. Table 22-4 shows the chlorophyll content of 3-day-old shoots of mustard and barley seedlings; it illustrates well the great activity of compound 6706 in preventing chlorophyll production in these plants.

Eshel and Sompolinsky (1970) compared the selectivity in sugar beet of pyrazon and benzthiazuron, a substituted urea [*N*-(2-benzthiazolyl)-*N*-methyl urea]. The latter had a five-fold greater inhibitory effect on photosynthesis than pyrazon; a ten-fold greater inhibitory effect on growth of sugar beet growing in treated culture solution. Treated preemergence in soil, sugar beet was more tolerant to benzthiazuron than to pyrazon. In soil, pyrazon leached more freely than benzthiazuron; this may explain the above difference in selectivity.

Smith and Meggitt (1970) found evidence for microbial breakdown of pyrazon in soil. However, in 10 weeks less than 10% of this herbicide was dephenylated in a sandy loam soil at 21°C.

TABLE 22-4. Chlorophyll Content of Shoots of 3-day-old Mustard and Barley Seedlings after Treatment with Substituted Pyridazinones (Hilton et al., 1969)

Treatment		Chlorophyll content as percentage of control[a]	
Chemical	Concn	Mustard	Barley
		%	%
Pyrazon	10^{-4}M	87	76
9785	10^{-4}M	71	77
9774	10^{-4}M	8	54
9785 + 9774	10^{-4}M each	10	37
6706	10^{-4}M	5	2
Pyrazon	10^{-5}M	91	83
9785	10^{-5}M	97	86
9774	10^{-5}M	95	96
9785 + 9774	10^{-5}M each	80	80
6706	10^{-5}M	4	1

[a] Control values (100%) were 0.408 and 0.170 mg of chlorophyll per g fresh weight for mustard and barley, respectively.

Pyriclor

2,3,5-trichloro-4-pyridinol

Pyriclor was announced in 1964 as a potential grass killer. It occurs as a white solid of high stability. It has low water solubility; however, its potassium salt is readily water soluble. It has been made available in liquid and granular forms. Pyriclor is especially useful for annual and perennial grass control and it will kill seedlings of many broadleaf weeds. Early spring treatments from before emergence to the boot stage of grasses have proved effective.

Pyriclor is readily absorbed and translocated by foliage and by roots; hence it is systemic. At high rates, pyriclor solution provides rapid contact action. At lower rates, chlorosis may be observed on most plant species; at sublethal concentrations this is reversible and hence the plants will recover.

In agricultural use, dosages of 2 to 6 oz/A preemergence have given effective control on annual grasses and many broadleaf weeds; rates up to 2 lb/A have given more persistent control. Pyriclor has been used on sugar cane, banana, rubber, and similar plantations in the tropics. Flax and sorghum are additional crops upon which pyriclor has proved useful.

Buchholtz (1968) found pyriclor effective in the control of quackgrass if applied 1 to 8 months before plowing and planting a crop; 1 to 4 lb/A were used.

Moomaw and Kim (1968) carried out glasshouse and field experiments with pyriclor in rice at the International Rice Research Institute in the Philippines. It inhibited weed seed germination and weed growth under static and percolating water status, but allowed rice to germinate and grow normally. Slight to severe chlorosis followed pyriclor treatment on rice within 5 days. Symptoms were more pronounced when the coleoptiles were 0.5 to 1.0 cm above the water surface than when the shoots were 3 to 5 cm high; symptoms disappeared after 1 week.

The induction of chlorosis by pyriclor has interested biologists who have sought to relate the gross appearance of yellowing with the ultrastructure of chloroplasts. Geronimo and Herr (1970) treated tobacco by adding pyriclor to the culture solution around the roots. Leaf samples were taken through a series of times starting at 2 hours and ending at 144 hours after treatment; these were prepared and viewed under an electron microscope. Such treatment of tobacco plants led to progressive disruption of chloroplast ultrastructure. Initial changes include swelling to a spherical form, swelling of the fret system, and loss of starch. Later all starch disappeared. The fret membrane system swelled and the membranes of the granal discs became disrupted; the chloroplast envelope finally ruptured.

The shape and size of mitochondria were not altered but these organelles seemed to increase in numbers as chloroplasts were altered in structure. When visible chlorosis appeared changes in chloroplast ultrastructure were under way. Staining responses of the leaf tissue led the writers to postulate that pyriclor or its metabolites inhibit oxidative phosphorylation.

Killion and Frans (1969) investigated the effect of pyriclor on oxygen uptake and oxidative phosphorylation using mitochondria isolated from hypocotyls of etiolated soybean seedlings. Pyriclor inhibited both coupled and uncoupled reactions and the authors suggest that pyriclor either acts on a non-phosphorylating intermediate close to the electron carrier chain, or with a component of the electron carrier chain, or both. DNP did not reverse the inhibition caused by pyriclor.

Since pyriclor causes chlorosis that seems related to disruption of chloroplast ultrastructure, it seems natural that photosynthesis would be inhibited. Meikle (1970) measured the effect of pyriclor on photosynthesis using isolated chloroplasts of spinach. CO_2 fixation, O_2 evolution, non-cyclic photophosphorylation, and cyclic photophosphorylation were also determined. These processes bring about the formation of assimilatory power, the generation of ATP, and $NADHP_2$. CO_2 assimilation and cyclic phosphorylation catalyzed by FMN were both inhibited at high levels of pyriclor. O_2 evolution and non-cyclic photophosphorylation were strongly inhibited. Meikle (1970) presented evidence for the conclusion that the primary site of action is the O_2-evolving system and/or pigment system 2 of the photosynthetic apparatus.

Gruenhagen and Moreland (1971) found that pyriclor reduces ATP levels and inhibits RNA and protein syntheses. Figure 6-3 shows the ATP contents of excised soybean hypocotyls incubated with 0.2mM ioxynil and 0.6mM chlorpropham, propanil, and pyriclor; Figure 22-6 shows the course of ^{14}C-orotic acid incorporation into RNA by excised soybean hypocotyls incubated with 0.2mM ioxynil and 0.6mM chlorpropham, propanil, and pyriclor. While not so effective as ioxynil, pyriclor has an appreciable ability to lower ATP levels and RNA and protein synthesis.

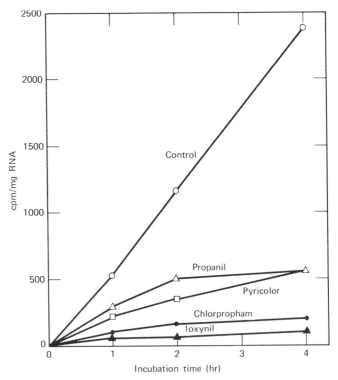

FIGURE 22-6.   Incorporation of ^{11}C-orotic acid into RNA of excised soybean hypocotyls incubated with 0.2 mM ioxynil and 0.6 mM chlorpropham, propanil, and pyriclor. (Gruenhagen and Moreland, 1971.)

Uracils

Some five substituted uracil compounds have proven to be effective herbicides; three are in production and are widely used (Figure 22-7). The uracil nucleus is the same in all of these compounds, the substitutions differ; terbacil has a chlorine in the 5 position; isocil, bromacil, and 733 have a bromine in the 5 position; lenacil has a cyclohexyl ring in the 3 position and a trimethylene ring in the 5 and 6 positions.

These compounds are all soil-borne toxicants, readily absorbed by plant roots and translocated apoplastically to the leaves. They are all toxic at very low concentrations in leaves. They all block the Hill reaction and interfere with a step in the photosynthetic pathway close to oxygen evolution. Thus

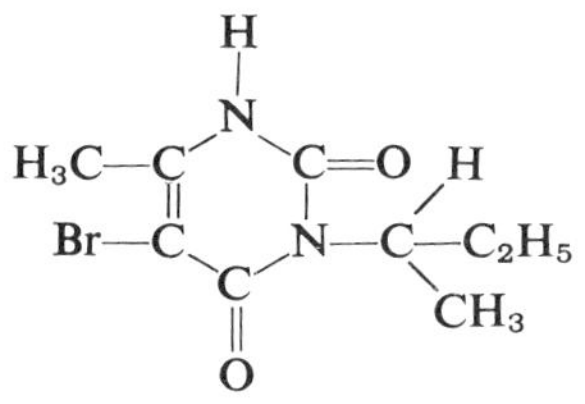

bromacil
5-bromo-3-*sec*-butyl-6-methyluracil

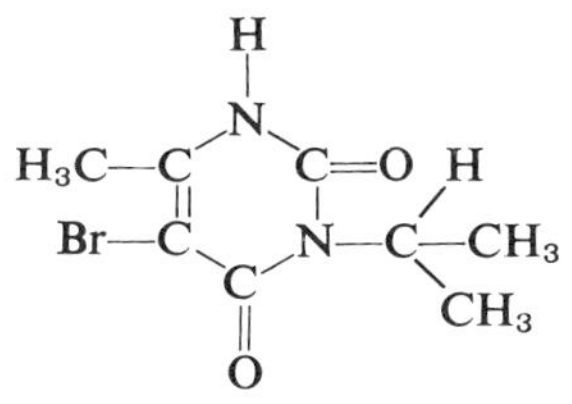

isocil
5-bromo-3-isopropyl-6-methyluracil

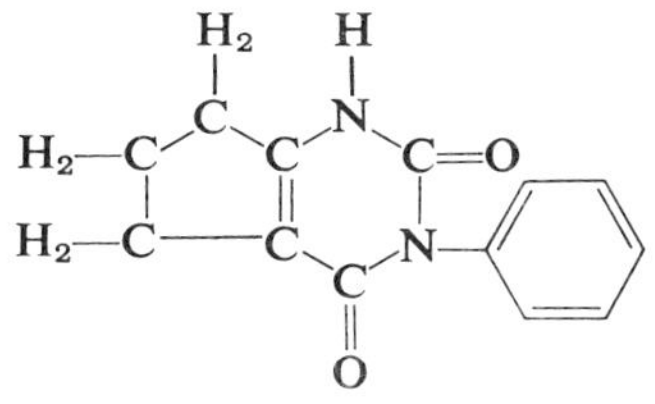

lenacil
3-cyclohexyl-6,7-dihydro-1H-
cyclopentapyrimidine-2,4(3H,5H)-dione

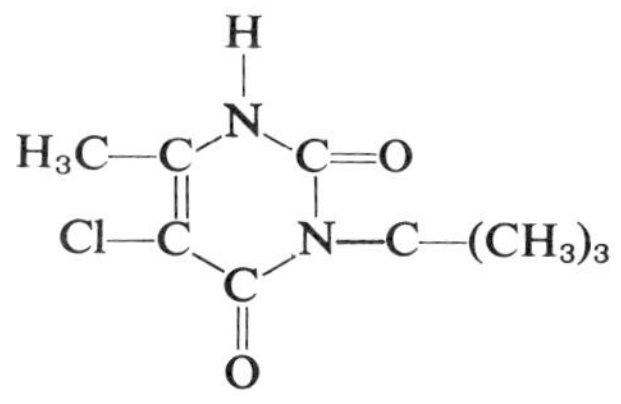

terbacil
3-*tert*-butyl-5-chloro-6-methyluracil

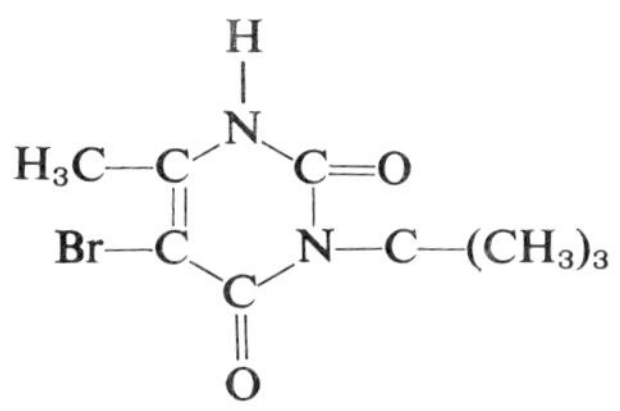

DP-733
3-*tert*-butyl-5-bromo-6-methyluracil

FIGURE 22-7.   Common name, chemical name, and chemical structure of the uracil herbicides.

they resemble the substituted urea herbicides in their site and mode of action (Hilton et al., 1964).

Bromacil and isocil are total or general herbicides used to control all weeds in non-agricultural areas. At carefully controlled low dosage rates they may be used selectively in citrus crops and pineapple. Terbacil is selective, controlling general weed growth in citrus, pineapple, sugar cane, mint, apple, peach, and asparagus crops. Lenacil is selective and is used in sugar beet crops in Europe. DuPont 733 is selective and has proved to be useful in citrus, sugar cane, and asparagus.

Bromacil is a white, crystalline, odorless solid in its original state and very slightly soluble in water. It is available as an emulsifiable concentrate or as granules. It is usually applied as a spray or as granules to the soil surface or early postemergence of weeds. It is readily absorbed by roots and translocated via the xylem; it inhibits the Hill reaction and stops photosynthesis. It is less strongly adsorbed by soil than the substituted ureas and hence will leach somewhat more.

Bromacil is most commonly used as a general herbicide against annual and perennial grasses and broadleaf weeds and some woody plants. It is used selectively in orange, grapefruit, and lemon orchards and preemergence against seedling weeds in pineapple. Bromacil has proved very useful for complete control of weeds on highway verges, along railroad rights-of-way, on airfields, and similar situations.

Terbacil is a white, crystalline, odorless solid, and soluble in water to about 710 ppm. It is available as a wettable powder. It is applied in suspension in water and requires constant agitation in the spray tank.

Terbacil is moved apoplastically in the xylem; it is adsorbed even less firmly than bromacil and hence will leach into the deeper soil horizons when flood irrigated or frequently moistened by rain (Skroch et al., 1971).

Terbacil is used to kill general weed growth in apples, peaches, citrus, mint, and sugar cane. On mint and sugar cane, preemergence applications may be made by aircraft. Annual weeds are controlled by 0.8 to 3.2 lb/A; against bermudagrass, quackgrass, nutsedge, horsenettle (*Solanum carolinense*), and red sorrel (*Rumex acetosella*) 3.2 to 8.0 lb/A are required; some tropical grasses are only partially controlled at the 8 lb/A dosage rate.

Lenacil resembles bromacil and terbacil in physical properties.

The uracils are much less firmly adsorbed by soil than many herbicides; thus they may be used to control certain perennial herbaceous and woody plants. They are broken down by microbiological activity in soils; loss by photodecomposition or volatilization is negligible. Uracils persist usually for one season when applied at sterilant rates; the half-life of ^{14}C-2-labeled bromacil and terbacil was around 5 to 6 months when applied at 4 lb/A in a silt loam soil. Persistence in soil depends upon soil type, rainfall, temperature, and other soil factors and must be determined experimentally.

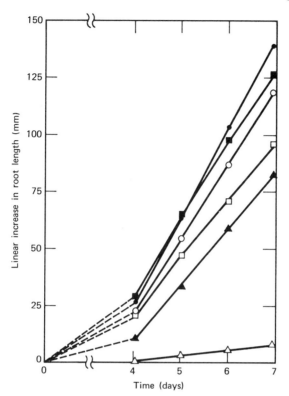

FIGURE 22-8. The effect of various concentrations of bromacil on increase in length of oat roots during a 7-day period: ◒, control; ○, 10^{-6} M; ■, 10^{-5} M; □, 10^{-4} M; ▲, 5×10^{-4} M; △, 10^{-3} M. (Ashton et al., 1969.)

Bromacil was used by Ashton et al. (1969) for studies on the effects of the uracils on growth, morphology, anatomy, and cytology of the oat plant. They grew plants in culture solution in glass tubes using herbicide solutions at 10^{-3}, 5×10^{-4}, 10^{-4}, 10^{-5}, and 10^{-6}M. Figure 22-8 shows the effects of these treatments on linear increase in root length with time over a period of 7 days; at 10^{-3}M bromacil was almost completely inhibitory. This inhibition of oat root was largely restricted to the terminal 0.5 mm segment just behind the meristem.

Severely inhibited oat root tips were swollen, with necrotic cells in the root tip meristem, epidermis, and procambium; parts of the epidermis and central metaxylem cells had a dark staining reaction, indicating disturbed metabolism. Vacuolation was precocious, and in places in the meristem cell wall formation between daughter cells was incomplete, resulting in multiple nuclei; in other places cell walls were unusually thick. Electron microscope studies showed a great number of cell-wall anomalies, incomplete cell walls, fragmented and bulbous or branching cell walls; plasmodesmata were sparse or lacking.

In leaves, development of the chloroplast grana and fret system was inhibited; normal increase in the number of loculi per granum was prevented. Bromacil also prevented the normal increase in width of grana but length of grana was greater than normal. Loculi of the grana and fret vesicles progressively swelled and there was modification of the chloroplast envelope; the authors concluded that these effects appeared to be associated with loss of integrity of membranes.

Ray et al. (1971) found that when purple nutsedge (*Cyperus rotundus*) plants having daughter plants connected by rhizomes were treated so that the daughter plants were protected from contamination, terbacil moved from mother to daughter plants via the connecting rhizomes. Since the treated plants were mowed to a height of 5 cm there was some foliage in the form of leaf bases on the sprayed plants. Under these conditions, terbacil absorbed by the foliage and roots of the mother plant could move via the xylem to the daughter plant; this is very probably the route by which the herbicide moved. This may well explain the excellent control of nutsedge by terbacil reported by Ray and Wilcox (1969).

The uracils are potent inhibitors of photosynthesis, a factor that undoubtedly is related to their herbicidal activity (Hilton et al., 1969; Hoffman, 1971). They have no effect on bacteria, fungi, and other non-photosynthetic organisms except at concentrations one to two orders of magnitude above those that affect photosynthesis; dark growth of tobacco callus tissue having an organic energy source, likewise, was unaffected at the dosages that inhibit photosynthesis.

Comparing the uracils with the urea herbicides, Hoffman (1971) proposes that a block in the pathway between the chloroplast and O_2 evolution (the OH side) could cause the accumulation of a phytotoxic product, possibly a reactive free radical; addition of hydrogen or other reducing agents allows the photosynthesis to proceed, but without O_2 evolution. When chlorella is grown in the presence of CO_2 and treated with monuron in the light with a source of organic energy, the growth is much less than in the dark. If CO_2 is removed from the system, the inhibitory effect of monuron is reversed. CO_2 had no effect in the dark, in its absence both photosynthetic pathways are blocked, and there is no accumulation of a free-radical component.

Bromacil, isocil, and terbacil have been shown to be effective Hill reaction inhibitors (Hoffman, 1971). Combinations of uracils with ureas show additive effects. And a monuron-resistant strain of Euglena showed similar resistance; it was not affected by 40 ppm of bromacil and only partially inhibited at 100 ppm. The wild strain was markedly inhibited by 2 ppm and completely stopped by 10 ppm; neither strain was affected in the dark. Hoffman (1971) points out that this antiphotosynthesis effect alone is not sufficient to account for the total phytotoxic action of the uracils.

Hewitt and Notten (1966) found that bromacil and isocil inhibit the induction of nitrate reductase by nitrate and molybdenum in leaf tissues. These herbicides were more inhibitory than several antimetabolites previously used. Inhibition was severe at a concentration of 1 μg/ml in the infiltrating solutions.

Hogue and Warren (1966) found that bromacil prolonged the healthy condition of tomato leaf disks maintained in light (6 days vs. 3 days for controls). Senescense was also retarded in detached leaflets in light by spraying or allowing uptake through the cut petiole; leaflets remaining on the plants were killed by the same concentration of bromacil.

In metabolic studies on the fate of ^{14}C-2-labeled bromacil and terbacil in soils and orange trees, Gardiner et al. (1969) found that 25 to 32% of the total applied ^{14}C activity was lost from soil as $^{14}CO_2$ in 6- to 9-week exposures. When orange seedlings were maintained for 4 weeks on a nutrient solution containing 10 ppm ^{14}C-bromacil, bromacil was the main residue in the plants. 5-bromo-3-s-butyl-6-hydroxymethyluracil and a few related minor metabolites were identified in soil and plants. No 5-bromouracil or 5-chlorouracil was detected.

Regardless of the universality of the antiphotosynthesis effect of the urea, triazine, and uracil herbicides, this effect alone is not sufficient to guarantee a useful and effective herbicide. Adequate stability, proper solubility, availability in the soil to plants, and a potential for transport to and distribution throughout leafy tissues, without being metabolized, are additional factors. Bromacil, terbacil, and lenacil seem to have these essential features.

REFERENCES

Agbakoba, C. S. O. and J. R. Goodin. 1969. Effect of stage of growth of field bindweed on absorption and translocation of ^{14}C-labeled 2,4-D and picloram. *Weed Sci.* **17**:436–438.

Anderson, A. J. and M. Okimoto. 1953. Laboratory studies of effectiveness of 3,5-dimethyltetrahydro-1,3,5,2-thiadiazine-2-thione against certain plant parasitic fungi and nematodes. *Phytopath.* **43**:465.

Anderson, J. L. and J. P. Schaelling. 1970. Effects of pyrazon on bean chloroplast ultrastructure. *Weed Sci.* **18**:455–459.

Anon. 1956. Crag Brand 974. For the control of soil fungi, nematodes and weeds. *Carbide and Carbon Chemical Co. Tech. Info. Litho.* F40041. February.

Ashton, F. M., E. G. Cutter, and D. Huffstutter. 1969. Growth and structural modifications of oats induced by bromacil. *Weed Res.* 9:198–204.

Ashton, F. M., D. Penner, and S. Hoffman. 1968. Effect of several herbicides on proteolytic activity of squash seedlings. *Weed Sci.* 16:169–171.

Bachelard, E. P. and R. D. Ayling. 1971. The effects of picloram and 2,4-D on plant cell membranes. *Weed Res.* 11:31–36.

Baur, J. R., R. W. Bovey, R. D. Baker, and I. Riley. 1971. Absorption and penetration of picloram and 2,4,5-T into detached live oak leaves. *Weed Sci.* 19:138–141.

Bayer, D. E., E. C. Hoffman, and C. L. Foy. 1965. DCPA in host-parasite relations of alfalfa and dodder. *Weeds* 13:92–95.

Bingham, S. W. 1967. Influence of herbicides on root development of bermudagrass. *Weeds* 15:363–365.

Bingham, S. W. 1968. Effect of DCPA on anatomy and cytology of roots. *Weed Sci.* 16:449–452.

Biswas, P. K., O. Hall, and D. B. Mayberry. 1966. Effects of maleic hydrazide on the synthesis of protein and metabolism of amino acids in tea. *Weeds* 14:329–331.

Buchholtz, K. P. 1953. Control of quackgrass and annual weeds in corn with herbicides. *Proc. 10th North Central Weed Control Conf.* p. 2.

Buchholtz, K. P. 1968. Control of quackgrass with pyrichlor. *Weed Sci.* 16:439–441.

Callaghan, J. J. and P. Grun. 1961. Incorporation of ^{14}C-labeled maleic hydrazide into the root-tip cells of *Allium cernum, Vicia faba* and *Tradescantia paludosa. J. Biophys. Bioch. Cytol.* 10:567–575.

Chang, I. and C. L. Foy. 1971a. Effect of picloram on germination and seedling development in four species. *Weed Sci.* 19:58–64.

Chang, I. and C. L. Foy. 1971b. Effects of picloram on mitochondrial swelling and ATP use. *Weed Sci.* 19:54–58.

Crafts, A. S. 1961. *The Chemistry and Mode of Action of Herbicides.* Interscience Publishers, New York, N.Y. 269 pp.

Crafts, A. S. and S. Yamaguchi. 1960. Absorption of herbicides by roots. *Amer. J. Bot.* 47:248–255.

Currier, H. B. and A. S. Crafts. 1950. Maleic hydrazide, a selective herbicide. *Science* 111:152–153.

Cutter, E. G., F. M. Ashton, and D. Huffstutter. 1968. The effects of bensulide on the growth, morphology, and anatomy of oat roots. *Weed Res.* 8:346–352.

Dawson, J. H. 1971. Response of sugarbeets and weeds to cycloate, propachlor, and pyrazon. *Weed Sci.* 19:162–165.

Dunning, J. J. 1958. Factors influencing the formation of callose in plant cells. Doctoral dissertation, Univ. of Calif., Davis. 137 pp.

Eastin, E. F. 1969. Movement and fate of *p*-nitrophenyl-α,α,α-trifluoro-2-nitro-*p*-tolyl ether-1'-^{14}C in peanut seedlings. *Plant Physiol.* 44:1397–1401.

Eastin, E. F. 1971a. Fate of fluorodifen in resistant peanut seedlings. *Weed Sci.* **19**:261–265.

Eastin, E. F. 1971b. Movement and fate of fluorodifen-1-^{14}C in cucumber seedlings. *Weed Res.* **11**:63–68.

Eastin, E. F. 1971c. Degradation of fluorodifen-1′-^{14}C by peanut seedling roots. *Weed Res.* **11**:120–123.

Ebner, L., D. H. Green, and P. Pande. 1968. C6989: A new selective herbicide. *9th Brit. Weed Control Conf. Proc.* pp. 1026–1032.

Eshel, Y. 1969. Effect of pyrazon on photosynthesis of various plant species. *Weed Res.* **9**:167–172.

Eshel, Y. and D. Sompolinsky. 1970. Selectivity of pyrazon and benzthiazuron in sugar beet. *Weed Res.* **10**:196–203.

Fischer, A. 1962. 1-Phenyl-4-amino-5-chlor-pyridozon-6 (PCA). *Weed Res.* **2**:177–184.

Foote, L. E. and B. F. Himmelman. 1967. Vegetation control along fence lines with maleic hydrazide. *Weeds* **15**:38–41.

Foote, L. E. and B. F. Himmelman. 1971. MH as a roadside grass retardant. *Weed Sci.* **19**:86–90.

Foret, J. A. 1959. *Station-to-Station Res. News*, **5**. No. 4. April. Union Carbide Corp.

Foy, C. L. and D. Penner. 1965. Effect of inhibitors and herbicides on tricarboxylic acid cycle substrate oxidation by isolated cucumber mitochindria. *Weeds* **13**:226–231.

Frank, R. 1967. Pyrazon, a selective herbicide for sugar beets. *Weeds* **15**:197–201.

Frank, R. and C. M. Switzer. 1969a. Effects of pyrazon on growth, photosynthesis, and respiration. *Weed Sci.* **17**:344–348.

Frank, R. and C. M. Switzer. 1969b. Absorption and translocation of pyrazon by plants. *Weed Sci.* **17**:365–370.

Fullerton, T. M., C. L. Murdock, A. E. Spooner, and R. E. Frans. 1970. Effects of DCPA on winter injury of recently-established bermudagrass. *Weed Sci.* **18**:711–714.

Funderburk, H. H. Jr. and J. M. Lawrence. 1963. Preliminary studies on the absorption of ^{14}C-labeled herbicides in fish and aquatic plants. *Weeds* **11**:217–219.

Gardiner, J. A., R. C. Rhodes, J. B. Adams, and E. J. Soboczenski. 1969. Synthesis and studies with 2-^{14}C-labeled bromacil and terbacil. *J. Agr. Food Chem.* **17**:980–986.

Geronimo, J. and J. W. Herr. 1970. Ultrastructural changes of tobacco chloroplasts induced by pyrichlor. *Weed Sci.* **18**:48–53.

Girolami, G. 1951. Private communication.

Gruenhagen, R. D. and D. E. Moreland. 1971. Effects of herbicides on ATP levels in excised soybean hypocotyls. *Weed Sci.* **19**:319–323.

Hawton, D. and E. H. Stobbe. 1971a. Selectivity of nitrofen among rape, redroot pigweed, and green foxtail. *Weed Sci.* **19**:42–44.

Hawton, D. and E. H. Stobbe. 1971b. The fate of nitrofen in rape, redroot pigweed, and green foxtail. *Weed Sci.* **19**:555–558.

Hertogh, A. A. de, J. W. Hooks, and G. C. Klingman. 1962. Herbicides on flue-cured tobacco. *Weeds* **10**:115–118.

Hewitt, E. J. and B. A. Notten. 1966. Effect of substituted uracil derivatives on induction of nitrate reductase in plants. *Biochem. J.* **101**:39–40.

Hiltibran, R. C. 1962. Duration of toxicity of endothal in water. *Weeds* **10**:17–19.

Hiltibran, R. C. 1963. Effect of endothal on aquatic plants. *Weeds* **11**:256–257.

Hilton, J. L., T. J. Monaco, D. E. Moreland, and W. A. Gentner. 1964. Mode of action of substituted uracil herbicides. *Weeds* **12**:129–131.

Hilton, J. L., A. L. Scharen, J. B. St. John, D. E. Moreland, and K. H. Norris. 1969. Modes of action of pyridazinone herbicides. *Weed Sci.* **17**:541–547.

Hoffman, C. E. 1971. The mode of action of bromacil and related uracils. *Second Internatl. Conf. of Pesticide Chem. Proc.* In press.

Hogue, E. J. and G. F. Warren. 1966. Studies on bromacil action. *WSA Abstr.* p. 42.

Isenberg, F. M. R., M. L. Odland, H. W. Popp, and C. O. Jensen. 1951. The effect of maleic hydrazide on certain dehydrogenases in tissues of onion plants. *Science* **103**:58–60.

Ivens, G. W. 1971. Seasonal differences in kill of two Kenya bush species after foliar herbicide treatment. *Weed Res.* **11**:150–158.

Juska, F. 1961. Preemergence herbicides for crabgrass control and their effects on germination of turfgrass species. *Weeds* **9**:137–144.

Keys, C. H. and H. A. Friesen. 1968. Persistence of picloram activity in soil. *Weed Sci.* **16**:341–343.

Killion, D. D. and R. E. Frans. 1969. Effect of pyriclor on mitochondrial oxidation. *Weed Sci.* **17**:468–470.

Koren, E. and F. M. Ashton. 1969. The effects of temperature on toxicity and uptake of pyrazon in sugar beets. *WSSA Abstr.* No. 32.

Koren, E. and F. M. Ashton. 1971. Effect of environmental factors on pyrazon action in sugar beets. *Weed Sci.* **19**:587–592.

Lee, W. O. 1970. Effect of picloram on production and quality of seed in several grasses. *Weed Sci.* **18**:171–173.

Leonard, O. A. and R. K. Glenn. 1968. Translocation of herbicides in detached bean leaves. *Weed Sci.* **16**:352–356.

Lhoste, J., H. D'Ille, A. Casanova, and L. A. Durgeat. 1963. Essais en plein champ de desherbage selectif des betteraves par le chloroamino-phenyl pyridazone (PCA). *Weed Res.* **3**:52–65.

Maestri, M. 1967. Structural and functional effects of endothall on plants. Doctoral dissertation. Univ. of Calif., Davis. 100 pp.

Malhotra, S. S. and J. B. Hanson. 1970. Picloram sensitivity and nucleic acids in plants. *Weed Sci.* **18**:1–4.

Martin, S. C., S. J. Shellborn, and H. M. Hull. Emergence of four-wing saltbush after spraying shrubs with picloram. *Weed Sci.* **18**:389–392.

Matsunaka, S. 1970. Weed Control in rice. Technical Papers of the *FAO International Conf. on Weed Control.* pp. 7–23.

Mazur, A. R., J. A. Jagschitz, and C. R. Skogley. 1969. Bioassay for bensulide, DCPA, and siduron in turfgrass. *Weed Sci.* **17**:31–34.

Meyer, R. E. 1970. Picloram and 2,4,5-T influence on honey mesquite morphology. *Weed Sci.* **18**:525–531.

Moomaw, J. C. and D. S. Kim. 1968. Selectivity of 2,3,5-trichloro-4-pyridino as a herbicide for direct-seeded, flooded rice. *Weed Res.* **8**:163–169.

Nishimoto, R. K. and G. F. Warren. 1971a. Site of uptake, movement, and activity of DCPA. *Weed Sci.* **19**:152–155.

Nishimoto, R. K. and G. F. Warren. 1971b. Shoot zone uptake and translocation of soil-applied herbicides. *Weed Sci.* **19**:156–161.

Nooden, L. D. 1967. Studies on the mechanism of action of maleic hydrazide. *Plant Physiol. Abstr.* **42**:S–50.

Nooden, L. D. 1969. The mode of action of maleic hydrazide. *Physiol. Plant.* **22**:260–270.

Penner, D. 1968. Herbicidal influence on amylase in barley and squash seedlings. *Weed Sci.* **16**:519–522.

Penner, D. and F. M. Ashton. 1968. Influence of dichlobenil, endothall, and bromoxynil on kinin control of proteolytic activity. *Weed Sci.* **16**:323–326.

Pereira, J. F. 1970. Some plant responses and the mechanism of selectivity of cabbage plants to nitrophen. Ph.D. dissertation. Univ. of Illinois, Urbana.

Ray, B. and M. Wilcox. 1969. Chemical fallow control of nutsedge. *Weed Res.* **9**:86–94.

Ray, B. R., M. Wilcox, W. B. Wheeler, and N. P. Thompson. 1971. Translocation of terbacil in purple nutsedge. *Weed Sci.* **19**:306–307.

Ries, S. K., M. J. Zabik, G. R. S. Stephenson, and T. M. Chen. 1968. N-glucosyl metabolite of pyrazon in red beets. *Weed Sci.* **16**:40–41.

Robinson, E. L. 1961. Soil-incorporated pre-planting herbicides for witchweed control. *Weeds* **9**:411–415.

Robison, E. D. 1967. Response of mesquite to 2,4,5-T, picloram, and 2,4,5-T/picloram combinations. *Proc. 20th Southern Weed Conf.* p. 199.

Rodebush, J. E. and J. L. Anderson. 1970. Morphological and anatomica effects of pyrazon on bean. *Weed Sci.* **18**:443–446.

Schoene, D. L. and O. L. Hoffman. 1949. Maleic hydrazide, a unique growth regulant. *Science* **109**:588–590.

Scifres, C. J. and M. K. McCarty. 1968. Reaction of Western ironweed leaf tissue to picloram. *Weed Sci.* **16**:347–349.

Sharma, M. P. 1971. Translocation of ^{14}C-picloram. Private communication.

Sharma, M. P. and W. H. Vanden Born. 1970. Foliar penetration of picloram and 2,4-D in aspen and balsam poplar. *Weed Sci.* 18:57–63.

Sharma, M. P., F. Y. Chang, and W. H. Vanden Born. 1971. Penetration and translocation of picloram in Canada Thistle. *Weed Sci.* 19:349–355.

Shaw, W. C., D. R. Shepherd, E. L. Robinson, and P. F. Sand. 1962. Advances in witchweed control. *Weeds* 10:182–192.

Shell Chemical Corp. 1959. Summary of basic data for Aqualin Herbicide. Shell Chemical Corp. 460 Park Avenue, New York 22, New York.

Skroch, W. A., T. J. Sheets, and J. W. Smith. 1971. Herbicide effectiveness, soil residues, and phytotoxicity to peach trees. *Weed Sci.* 19:257–260.

Smith, D. T. and W. F. Meggitt. 1970. Persistence and degradation of pyrazon in soil. *Weed Sci.* 18:260–264.

Smith, G. S. and L. M. Callahan. 1968. Herbicidal phytotoxicity to emerald Zoysia during establishment. *Weed Sci.* 16:312–315.

Stephenson, G. R. and S. K. Ries. 1967. The movement and metabolism of pyrazon in tolerant and susceptible species. *Weed Res.* 7:51–60.

Stephenson, G. R. and S. K. Ries. 1968. Metabolism of pyrazon in sugar beets and soil. *Weed Sci.* 17:327–331.

Stephenson, G. R., D. R. Dilley, and S. K. Ries. 1971. Influence of light and sucrose on *N*-glucosyl pyrazon formation in red beet. *Weed Sci.* 19:406–409.

Swanson, C. R. and J. R. Baur. 1969. Absorption and penetration of picloram in potato tuber discs. *Weed Sci.* 17:311–314.

Thomas, T. M. and D. E. Seaman. 1968. Translocation studies with ^{14}C-endothal in *Potamogeton nodosus* Poir. *Weed Res.* 8:321–326.

Utter, G. 1960. Private communication.

Walker, C. R. 1963. Endothal derivatives as aquatic herbicides in fishery habitats. *Weeds* 11:226–232.

Walter, J. P., E. F. Eastin, and M. G. Merkle. 1970. The persistence and movement of fluorodifen in soils and plants. *Weed Res.* 10:165–171.

Walter, J. P., M. L. Kerchersid, and M. G. Merkle. 1968. The chromatographic determination of 4-trifluoromethyl-2,4-dinitrophenyl ether residues in soybeans. *J. Agr. and Food Chem.* 16:143–144.

White, D. G. 1950. Blossoming of fruits delayed by maleic hydrazide. *Science* 111:303.

Widmer, R. E. and R. J. Stadtherr. 1961. Post-planting weed control in garden chrysanthemums. *Weeds* 9:204–208.

Wiese, A. F. and H. E. Rea. 1961. Control of field bindweed and other perennial weeds with benzoic and phenyl-acetic acids. *Weeds* 9:423–428.

Yeo, R. R. 1970. Dissipation of endothall and effects on aquatic weeds and fish. *Weed Sci.* 18:282–284.

Youngner, V. B. and J. R. Goodin. 1961. Control of *Pennisetum clandestinum*, kikuyugrass. *Weeds* 9:238–242.

Appendix

TABLE A-1. Alphabetical Listing of Herbicides by Common Name and Corresponding Trade Name, Chemical Name, and Manufacturer

Common name	Trade name	Chemical name	Manufacturer
A			
Acrolein	Aqualin	acrolein	Shell
Alachlor	Lasso	2-chloro-2',6'-diethyl-N-(methoxymethyl)acetanilide	Monsanto
Allyl alcohol	Allyl Alcohol	propen-1-ol-3	Shell
AMA	Methar	ammonium methyl arsonate	Vineland, Cleary
Ametryne	Evik, Gesapax	2-(ethylamino)-4-(isopropylamino)-6-(methylthio)-s-triazine	Geigy
Amitrole	Amizol, Weedazol	3-amino-s-triazole	Amchem, Cyanamide
Amitrole-T	Cytrol	3-amino-s-triazole plus ammonium thiocyanate	Amchem
AMS	Amate	ammonium sulfamate	duPont, Chipman
An 56477	Torpedo	N,N-bis(2-chloroethyl)-2,6-dinitro-p-toluidine	Ansul
Arsenate	Chip-Cal	tricalcium arsenate	Chipman
Arsenite, sodium	Weed killer	sodium arsenite	Many companies
Asulam	Asulox	methyl sulfanilylcarbamate	May & Baker, Rhodi
Atratone	Gesatamin	2-methoxy-4-(ethylamino)-6-(isopropylamino)-s-triazine	Geigy
Atrazine	AAtrex, Gesaprim	2-chloro-4-(ethylamino)-6-(isopropylamino)-s-triazine	Geigy
Aziprotryn	Mesoranil	2-azido-4-isopropylammo-6-methylmercapto-1,3,5-triazine	CIBA
	Brasoran		
B			
Bandane	Bandane	4,5,6,7,8,8-hexachloro-3a,4,7,7a-tetrahydro-4,7-methanoindene isomers	Velsicol

Barban	Caroyne	4-chloro-2-butynyl *m*-chlorocarbanilate	Gulf
BAS1700H	Benzazin	2-phenyl-3,1-benzoxazinone-(4)	BASF
BAS3490H	Decazolin	1-(α,α-dimethyl-β-acetoxypropionyl)-3-isopropyl-2,4-dioxodecahydroquinazoline	BASF
BAS3500H	Oxapyrazon	dimethylamineethanol salt of *N*-[1-phenyl-5-bromo-pyridazone-6-yl(4)]-oxamic acid	BASF
BAS3510H	Basagran Bentazon	3-isopropyl-2,1,3-benzothiadiazinone-(4)-2,2-dioxide	BASF
Bay 94337	Sencor	4-amino-6-*tert*-butyl-3-methylthio-4,5-dihydro-1,2,4-triazine-5-one	Bayer
Benazolin	RD-7693	4-chloro-2-oxo-benzothiazoline-3yl-acetic acid	Boots
Benefin	Balan	*N*-butyl-*N*-ethyl-α,α,α-trifluoro-2,6-dinitro-*p*-toluidine	Elanco
Bensulide	Betasan Prefar	*O,O*-diisopropyl phosphorodithioate S-ester with *N*-(2-mercaptoethyl)benzenesulfonamide	Stauffer
Benthiocarb	Bolero	*S*(4-chlorobenzyl)-*N,N*-diethylthiolcarbamate	IMC
Bentranil	Bentranil	2-phenyl-3,1-benzox-azinone-4	BASF
Benzadox	Topcide	(benzamidooxy)acetic acid	Gulf
Benzomarc	Benzomarc	*N*-benzoyl-*N*-(3,4-dichlorophenyl)-*N'*,*N'*-dimethylurea	Pechiney Progil
Benzthiazuron	Gatnon	*N*-(2-benzthiazolyl)-*N'*-methylurea	Bayer
Boron salts	Borax, Borascu	$Na_2B_4O_4 \cdot 10\ H_2O$	U.S. Borax
Bromacil	Hyvar X	5-bromo-3-*sec*-butyl-6-methyluracil	duPont
Bromofenoxim	Faneron	3,5-dibromo-4-hydroxybenzaldoxime-0-(2',4'-dinitrophenyl)-ether	CIBA
Brompyrazone	BAS2430H	5-amino-4-bromo-2-phenyl-3(2*H*)-pyridazinone	BASF

Common name	Trade name	Chemical name	Manufacturer
Bromoxynil	Brominil, Buctril	3,5-dibromo-4-hydroxybenzonitrile	Amchem, Chipman
Butachlor	Machete	N-(butoxymethyl)-2-chloro-2',6'-diethylacetanilide	Monsanto
Buturon	Etapur	3-(p-chlorophenyl)-1-methyl-1-(1-methyl-2-propynyl)urea	BASF
Butylate	Sutan	S-ethyl diisobutylthiocarbamate	Stauffer
C			
Cacodylic acid	Phytar	hydroxydimethylarsine oxide	Ansul
Calcium cyanamide	Cyanamide	Calcium cyanamide	Cyanamide
Carbetamide	Leguram	D-N-ethyllactamide carbanilate(ester)	Rhodia
CDAA	Randox	N,N-diallyl-2-chloroacetamide	Monsanto
CDEA	—	2-chloro-N,N-diethylacetamide	Monsanto
CDEC	Vegadex	2-chloroallyl diethyldithiocarbamate	Monsanto
Chloramben	Amiben, Vegiben	3-amino-2,5-dichlorobenzoic acid	Amchem
Chlorate	Several	sodium chlorate	Pennsalt, Chipman, Stauffer
Chlorazine	—	2-chloro-4,6-bis(diethylamino)-s-triazine	Geigy
Chlorazon	Chlorazon	1-phenyl-4-(α-hydroxy-β,β',β'-trichloroethyl)-amino-5-chloropyridazone-6	BASF
Chlorbromuron	Maloran	3-(4-bromo-3-chlorophenyl)-1-methoxy-1-methylurea	CIBA
Chlorbufam	BIPC, Chlorinate	butyn-1-yl-3-N-3-chlorophenylcarbamate	BASF
Chlorfluorenol	Maintain	2-chloro-9-hydroxyfluorenecarboxylic acid-9-methyl ester	U.S. Borax
Chloropicrin	Chloropicrin	chloropicrin	Dow, Monsanto

Chloroxuron	Tenoran	3-[p-(p-chlorophenoxy)phenyl]-1,1-dimethylurea	CIBA
Chlorpropham	Chloro IPC, etc.	Isopropyl m-chlorocarbanilate	Pittsburg Plate Glass
Chlorthiamid	Prefix	2,4-dichlorothiobenzamide	Shell
Chlortoluron	Dicuron	N'-(3-chloro-4-methylphenyl)N,N-dimethylurea	CIBA
CMA	Calar	calcium methane arsonate	Vineland
Copper sulfate	Bluestone	copper sulfate	Mountain Copper
			Phelps Dodge
Cycloate	Ro-Neet	S-ethyl N-ethylthiocyclohexanecarbamate	Stauffer
Cycluron	OMU	3-cyclooctyl-1,1-dimethylurea	BASF
Cypromid	Clobber	3,4'-dichlorocyclopropane-carboxanatide	Gulf
Cyprazine	Outfox	2-chloro-4-(cyclopropylamino)-6-(isopropylamino)-s-triazine	Gulf
D			
Dalapon	Dowpon	2,2-dichloropropionic acid	Dow
Dazomet	Mylone, DMTT	tetrahydro-3,5-dimethyl-2H-1,3,5-thiadiazine-2-thione	Union Carbide
DBA	DBA	sodium 2,2-dichlorobutyrate	Dow
DCPA	Dacthal	dimethyl tetrachloroterephthalate	Diamond Shamrock
DCU	Dichloral urea	1,3-bis(2,2,2-trichloro-1-hydroxyethyl)urea	Hercules
Delachlor	CP52223	2-chloro-N-(isobutoxymethyl)-2',6'-acetoxylidide	Monsanto
Desmetryne	Semeron	2-(isopropylamino)-4-(methylamino)-6-(methylthio)-s-triazine	Geigy
Diallate	Avadex	S-(2,3-dichloroallyl)diisopropylthiocarbamate	Monsanto
Dicamba	Banvel-D	3,6-dichloro-o-anisic acid	Velsicol
Dichlobenil	Casoron	2,6-dichlorobenzonitrile	Thompson, Hayward

441

Common name	Trade name	Chemical name	Manufacturer
Dichlone	Phygon	2,3-dichloro-1,4-naphthoquinone	Niagara
Dichlormate	Rowmate Sirmate	3,4-dichlorobenzylmethylcarbamate	Union Carbide
Dichlorprop	Several	2-(2,4-dichlorophenoxy)propionic acid	Boots, Amchem, Hercules
Dicryl	Dicryl	3′,4′-dichloro-2-methyl acrylanilide	Niagara
Difenoxuron	CIB3470	N-[4-(4′-methoxyphenoxy)-phenyl]-N′,N′-dimethylurea	CIBA
Dinosam	Several	2-(1-methylbutyl)-4,6-dinitrophenol	Dow, Niagara
Dinoseb	Several	2-sec-butyl-4,6-dinitrophenol	Dow, Niagara
Dinoseb-acetate	Aretite	2-sec-butyl-4,6-dinitrophenylacetate	Hoechst
Dinoterb-acetate	MC1108	2-tert-butyl-4,6-dinitrophenylacetate	Murphy Chem
Diphenamid	Dymid, Enide	N,N-dimethyl-2,2-diphenylacetamide	Elanco, Upjohn
Diphenatrile	Diphenatrile	diphenylacetonitrile	Elanco
Diquat	Reglone Diquat	6,7-dihydrodipyrido[1,2-a:2′,1′-c]pyrazinedium ion	ICI, Chipman, S.O. Cal, Niagara
Diuron	Karmex	3-(3,4-dichlorophenyl)-1,1-dimethylurea	duPont
DMPA	Zytron	O-(2,4-dichlorophenyl)-O′-methyl N-isopropyl phosphoramidothioate	Dow
DNOC	Elgetol	4,6-dinitro-o-cresol	Dow, Truffaut
DSMA	Ansar 184	disodium methanearsonate	Ansul and others
E			
Endothall	Aquathol, etc.	7-oxabicyclo[2.2.1]heptane-2,3-dicarboxylic acid	Pennsalt
EPTC	Eptam	S-ethyl dipropylthiocarbamate	Stauffer

Erbon	2-(2,4,5-trichlorophenoxy)ethyl 2,2-dichloropropionate	Dow
Euparen	N-(dichlorofluoromethylthiol)-N',N'-dimethyl-N-phenylsulfamide	Chemagro
EXD	O,O-diethyl dithiobis(thioformate)	Roberts
F		
Fenac	(2,3,6-trichlorophenyl)acetic acid	Amchem
Fenuron	1,1-dimethyl-3-phenylurea	duPont
Fenuron-TCA	1,1-dimethyl-3-phenylurea mono(trichloroacetate)	Allied Chem. Corp.
Fluometuron	1,1-dimethyl-3-(α,α,α-trifluoro-m-tolyl)urea	CIBA
Fluorodifen	p-nitrophenyl α,α,α-trifluoro-2-nitro-p-tolyl ether	CIBA
Flurenol	9-hydroxyfluorene-(9)-carbonic acid	Merck
G		
Glytac	ethyleneglycol-bis-(trichloroacetate)	Hooker
H		
HCA	1,1,1,3,3,3-hexachloro-2-propanone	General Chem.
Hexaflurate	potassium hexafluoroarsenate	Pennsalt Chem.
I		
Ioxynil	4-hydroxy-3,5-diiodobenzonitrile	Amchem
Ipazine	2-chloro-4-(diethylamino)-6-(isopropylamino-s-triazine	Geigy
Isocil	5-bromo-3-isopropyl-6-methyluracil	duPont
Isonoruron	N-[1 or 2-(3a,4,5,6,7,7a-hexahydro-4,7-methanoindanyl)]-N',N'-dimethylurea	BASF
Isopropalin	2,6-dinitro-N,N-dipropylcumidine	Elanco
K		
Karbutilate	tert-butylcarbamic acid with 3-(m-hydroxyphenyl)1,1-dimethylurea	Niagara

Baron
Euparen
Herbisan-5
Fenac, etc.
Dybar
Urab
Cotoran
Preforan
Flurenol
Tritak
HCA
Nopalmate
Certrol
G30031
Hyvar
BAS2103H
Paarlan
Tandex

Common name	Trade name	Chemical name	Manufacturer
KOCN	—	potassium cyanate	Cyanamide
L			
Lenacil	Venzar	3-cyclohexyl-6-7-dihydroxy-1*H*-cyclopentapyrimidine-2,4(3*H*,5*H*)-dione	duPont
Linuron	Afalon Lorox	3-(3,4-dichlorophenyl)-1-methoxy-1-methylurea	duPont
M			
MAA	MAA	methanearsonic acid	Ansul
MAMA	Ansar 157	monoammonium methanearsonate	Ansul
MCPA	Several	[(4-chloro-*o*-tolyl)oxy]acetic acid	Amchem, Chipman, Dow
MCPB	Several	4-[(4-chloro-*o*-tolyl)oxy]butyric acid	Amchem, Chipman
MCPES	—	2-[(4-chloro-*o*-tolyl)oxy]ethyl sodium sulfate	Union Carbide
Mecoprop	Several	2-[(4-chloro-*o*-tolyl)oxy]propionic acid	Chipman Morton Salt
Medinoterb-acetate	MC1488	2-*tert*-butyl-5 methyl-4,6-dinitrophenylacetate	Murphy Chem
Methabenz-thiazuron	Tribunil	*N*-(2-benzothiazoyl)*N*-methyl-*N*′-methylurea	Bayer
Methachlor	CP50144	2″-chloro-2,6-diethyl-*N*-(methoxymethyl)acetanilide	Monsanto
Metham	Vapam	sodium methyldithiocarbamate	Stauffer, duPont
Methazole	Probe	2-(3,4-dichlorophenyl)-4 methyl-1,2,4-oxadiazolidine-3,5-dione	Velsicol
Methoprotryne	Gesaran	2-methylmercapto-4-(3-methoxypropylamino)-6-isopropylamino-1,3,5-triazine	Geigy

Methyl bromide	—	methylbromide	Dow
Metobromuron	Patoran	3-(p-bromophenyl)-1-methoxy-1-methylurea	CIBA
Metoxuron	Dosanex	N-(3-chloro-4-methoxyphenyl)-N',N'-dimethylurea	Sandoz, Schweiz
MH	MH-30	1,2-dihydro-3,6-pyridazinedione	Uniroyal
Molinate	Ordram	S-ethyl hexahydro-1H-azepine-1-carbothioate	Stauffer
Monalide	Potablan	α,α-dimethylvalerianic-acid-p-chloranilide	Shell
Mon-0573	Mon-0573	N-(phosphonomethyl)glycine	Monsanto
Monolinuron	Aresin	3-(p-chlorophenyl)-1-methoxy-1-methylurea	duPont
Monuron	Telvar	3-(p-chlorophenyl)-1,1-dimethylurea	duPont
Monuron TCA	Urox	3-(p-chlorophenyl)-1,1-dimethylurea mono(tri-chloroacetate)	Allied Chem. Corp.
Morphamquat	PP 745	1,1'-bis-(3,5-dimethylmorpholine-carboxymethyl)-4,4'-dipyridilium dichloride	ICI
MSMA	Ansar 170	monosodium methanearsonate	Ansul
N			
Naptalam	Alanap	N-1-naphthylphthalamic acid	Uniroyal
NC3363	Chlorflurazol	2-trifluoromethyl-4,5-dichlorobenzimidazole	Fison's
Neburon	Kloben	1-butyl-3-(3,4-dichlorophenyl)-1-methylurea	duPont
NIA4562	Karsil	2-methylvaleric-3,4-dichloroanilide	Niagara
Nitralin	Planavin	4-(methylsulfonyl)2,6-dinitro-N,N-dipropylaniline	Shell
Nitrofen	Tok	2,4-dichlorophenyl p-nitrophenyl ether	Rohm & Haas
NOA	NOA	α-naphthoxymethyl acetate	Shering
Norea	Herban	3-(hexahydro-4,7-methanoindan-5-yl)-1,1-dimethylurea	Hercules
O			
OCS 21944	Glenbar	O,S-dimethyl-tetrachlorothioterephthalate	Velsicol
Oryzalin	Ryzalin	3,5-dinitro-N^4,N^4-dipropylsulfanilamide	Elanco

TABLE A-1 *(continued)*

Common name	Trade name	Chemical name	Manufacturer
P			
Paraquat	Gramoxone	1,1'-dimethyl-4,4'-bipyridinium ion	ICI, S.O. Cal
PBA	Benzak	chlorinated benzoic acid	Amchem, duPont
PCP	Penta	pentachlorophenol	Dow, Monsanto, duPont
Pebulate	Tillam	S-propyl butylethylthiocarbamate	Stauffer
Phenmedipham	Betanal	methyl m-hydroxycarbanilate m-methyl carbanilate	Shering, Nor-Am
Picloram	Tordon	4-amino-3,5,6-trichloropicolinic acid	Dow
PMA	PMA	(acetato)phenyl mercury	Linck, Cleary, Scott
Prometone	Pramitol	2,4-bis(isopropylamino)-6-methoxy-s-triazine	Geigy
Prometryne	Caparol	2,4-bis(isopropylamino)-6-(methylthio)-s-triazine	Geigy
Pronamide	Kerb	N-(1,1-dimethylpropynyl) 3,5-dichlorobenzamide	Rohm and Haas
Propachlor	Ramrod	2-chloro-N-isopropylacetanilide	Monsanto
Propanil	Rogue, Stam F-34	3',4'-dichloropropionanilide	Rohm & Haas, Monsanto
Propazine	Milogard	2-chloro-4,6-bis(isopropylamino)-s-triazine	Geigy
Propham	Chem-hoe, etc.	isopropyl carbanilate	Pittsburg Plate Glass
Proximpham	Proximpham	O-(N-phenylcarbamoyl)-propanone oxime	List, DDR
Prynachlor	Butisan	2-chloro-N-(1-methyl-2-propynyl)acetanilide	BASF
Pyrazon	Pyramin	5-amino-4-chloro-2-phenyl-3(2H)-pyridazinone	BASF
Pyriclor	Daxtron	2,3,5-trichloro-4-pyridinol	Dow
R			
RD7693	Benazolin	4-chloro-2-oxo-benzothiazoline-3-yl-acetic acid	Boots

RP17623	2-*tert*-butyl-4-(2,4-dichloro-5-isporpopyloxyphenyl)-Δ^2-1,3,4-oxadiazolin-5-one	Rhone-Poulenc
R-7465	2-(α-naphthoxy)*N*,*N*-diethylpropionamide	Stauffer
S		
SD15418	2-(4 chloro-6-ethylamino)-*s*-triazine-2-ylamino-2-propionitrile	Shell
Sesone	2-(2,4-dichlorophenoxy)ethyl sodium sulfate	Amchem
Siduron	1-(2-methylcyclohexyl)-3-phenylurea	duPont
Silvex	2-(2,4,5-trichlorophenoxy)propionic acid	Dow, Amchem
Simazine	2-chloro-4,6-*bis*(ethylamino)-*s*-triazine	Geigy
Simetone	2,4-*bis*(ethylamino)-6-methoxy-*s*-triazine	Geigy
Simetryne	2,4-*bis*(ethylamino)-6-(methylthio)-*s*-triazine	Geigy
Solan	3'-chloro-2-methyl-*p*-valerotoluidide	Niagara
GY-BON		
Pentanochlor		
Solan		
Sumital	2-*sec*-butylamino-4-ethylamino-6-methoxy-*s*-triazine	Geigy
Swep	Methyl 3,4-dichlorocarbanilate	Niagara
T		
TCA	trichloroacetic acid	Dow
TCBC	trichlorobenzyl chloride	Monsanto
Randox T		
(TCBC + CDAA)		
Terbacil	3-*tert*-butyl-5-chloro-6-methyluracil	duPont
Terbutol	2,6-di-*tert*-butyl-*p*-tolyl methylcarbamate	Hercules
Terbutryn	2-(*tert*-butylamino)-4-(ethyl-amino)-6-(methylthio)-*s*-triazine	Geigy
TH-1568A	2-amino-3-chloro-1,4-naphthoquinone	Takeda Chem. Indus.
ACNQ		
Triallate	*S*-(2,3,3-trichloroallyl)diisopropylthiocarbamate	Monsanto
Avadex BW		
Far-go		

TABLE A-1 (*continued*)

Common name	Trade name	Chemical name	Manufacturer
Tricamba	Banvel-T	3,5,6-trichloro-*o*-anisic acid	Velsicol
Trietazine	Gesafloc	2-chloro-4-(diethylamino)-6-(ethylamino)-*s*-triazine	Geigy
Trifluralin	Treflan	α,α,α-trifluoro-2,6-dinitro-*N,N*-dipropyl-*p*-toluidine	Elanco
Trimeturon	Trimeturon	1-(*p*-chlorophenyl)2,3,3-trimethylpseudourea	CIBA
Tritac	Tritac	2,3,6-trichlorobenzyloxypropanol	Hooker
2,3,6 TBA	Trysben Benzac	2,3,6-trichlorobenzoic acid	Amchem, duPont
2,4-D	Several	(2,4-dichlorophenoxy)acetic acid	Many companies
2,4-DB	Butoxone	4-(2,4-dichlorophenoxy)butyric acid	Amchem, Chipman
2,4-DEB	Sesin	2-(2,4-dichlorophenoxy)ethyl benzoate	Union Carbide
2,4-DEP	Falone	*tris*[2-(2,4-dichlorophenoxy)ethyl]phosphite	Uniroyal
2,4-DES	Sesone	sodium 2-(2,4-dichlorophenoxy)ethyl sulfate	Union Carbide
2,4,5-T		(2,4,5-trichlorophenoxy)acetic acid	Amchem, Monsanto, Dow
2,4,5-TES	Natrin	sodium 2-(2,4,5-trichlorophenoxy)ethyl sulfate	Union Carbide
U			
USB 3584	Cobex	N^3,N^3-diethyl-2,4-dinitro-6-trifluoromethyl-*m*-phenylenediamine	U.S. Borax
V			
Vernolate	Vernam	*S*-propyl dipropylthiocarbamate	Stauffer
Vorlex	Vorlex	methyl isothiocyanate	Morton Salt

TABLE A-2. Alphabetical Listing of Herbicides by Trade Name and Corresponding Common Name

Trade names	Common names
AAtrex	Atrazine
ACNQ	TH-1568A
Alanap	Naptalam
Randox	CDAA
Allyl Alcohol	Allyl Alcohol
Ammate	AMS
Amiben	Chloramben
Ansar 157	MAMA
Ansar 170	MSMA
Ansar 184	DSMA
Aqualin	Acrolein
Aquathol	Endothall
Aresin	Monolinuron
Aretite	Dinoseb-acetate
Asulox	Asulam
Avadex	Diallate
Avadex BW	Triallate
Azak	Terbutol
Balan	Benefin
Bandane	Bandane
Banvel-D	Dicamba
Banvel-T	Tricamba
Baron	Erbon
Basagran	BAS3510H
Bentranil	Bentranil
Benzak	PBA
Benzazin	BAS1700H
Benzolin	RD 7693
Benzomarc	Benzomarc
Betanal	Phenmedipham
Betasan	Bensulide
BiPC	Chlorbufam
Bladex	SD 15418
Bluestone	Copper sulfate
Bolero	Benthiocarb
Borascu, Borax	Boron salts
Brasoran	Aziprotryne
Brominil	Bromoxynil

449

Trade names	Common names
Buctril	Bromoxynil
Butisan	Prynachlor
Butoxone	2,4-DB
Calar	CMA
Cantrol	Mecoprop
Can-trol	MCPB
Caparol	Prometryne
Carbyne	Barban
Casoron	Dichlobenil
CDEA	CDEA
Certrol	Ioxynil
Chem-hoe	Propham
Chip-Cal	Arsenate
Chlorate	Chlorate
Chlorazon	Chlorazon
Chlorflurazol	NC 3363
Chlorinate	Chlorbufam
Chloro IPC	Chlorpropham
Chloropicrin	Chloropicrin
Clobber	Cypromid
Cobex	USB 3584
Cotoran	Fluometuron
CP 50144	Methachlor
CP 52223	Delachlor
Crag 974	Dazomet
Cyanamide	Calcium cyanamide
Cytrol	Amitrol-T
Dacthal	DCPA
Daxtron	Pyrichlor
DBA	DBA
Decazolin	BAS 3490H
Devrinol	R-7465
Dichlone	Dichlone
Dichloral urea	DCU
Dicuron	Chlortoluron
Difenoxuron	CIB 3470
Diphenatrile	Diphenatrile
DMTT	Dazomet
DNAP	Dinosam

TABLE A-2 (*continued*)

Trade names	Common names
DNBP	Dinoseb
Dowpon	Dalapon
Dybar	Fenuron
Dymid	Diphemamid
Eptam	EPTC
Etapur	Buturon
Euparen	Euparen
EVIK	Ametryne
Falone	2,4-DEP
Faneron	Bromofenoxim
Far-go	Triallate
Fenac	Fenac
Flurenol	Flurenol
Gatnon	Benzthiazuron
Gesafloc	Trietazine
Gesatamin	Atratone
Gesaran	Methoprotryn
Glenbar	OCS 21944
Gramoxone	Paraquat
GY-BON	Simetryne
HCA	HCA
Herban	Norea
Herbisan-5	EXD
Hyvar	Isocil
Hyvar X	Bromacil
Igran	Terbutryn
IPC	Propham
Isonoruron	Brompyrazone
Karmex	Diuron
Karsil	NIA 4562
Kerb	Pronamide
Kloben	Neburon
Kuron	Silvex
Lasso	Alachlor
Leguram	Carbetamide
Lorox	Linuron

TABLE A-2 (*continued*)

Trade names	Common names
MAA	MAA
Machete	Butachlor
Maintain	Chlorfluorenol
Maloran	Chlorbromuron
MC1108	Dinoterb-acetate
MC 1488	Medinoterb-acetate
MCPES	MCPES
MCPP	Mecoprop
Mesoranil	Aziprotryne
Methar	AMA
Methoxon	MCPA
Methylbromide	Methylbromide
Milogard	Propazine
Mon-0573	Mon-0573
Mylone	Dazomet
Natrin	2,4,5-TES
NOA	NOA
Nopalmate	Hexaflurate
OMU	Cycluron
Outfox	Cyprazine
Oxapyrazon	BAS 3500 H
Paarlan	Isopropalin
Patoran	Metobromuron
Penta	PCP
Pentanochlor	Solan
Phytar	Cacodylic acid
Planavin	Nitralin
PMA	PMA
Potassium cyanate	KOCN
PP 745	Morphamquat
Pramitol	Prometone
Prefar	Bensulide
Prefix	Chlorthiamid
Preforan	Fluorodifen
Preemerge	Dinoseb
Princep	Simazine
Probe	Methazole
Proximpham	Proximpham
Pyramin	Pyrazon

TABLE A-2 (*continued*)

Trade names	Common names
Ramrod	Propachlor
Randox	CDAA
Randox-T	TCBC + CDAA
Reglone	Diquat
Rogue	Propanil
Ro-neet	Cycloate
Ronstar	RP 17623
Rowmate	Dichlormate
Ryzalin	Oryzalin
Semeron	Desmetryne
Sencor	BAY 94337
Sesin	2,4-DEB
Sesone	2,4-DES
Sinbar	Terbacil
Sinox	DNOC
Sirmate	Dichlormate
Stam-F 34	Propanil
Sumital	GS 14254
Sutan	Butylate
Swep	Swep
Tandex	Karbutilate
TCA	TCA
Telvar	Monuron
Tenoran	Chloroxuron
Thitrol	MCPB
Tillam	Pebulate
Tok-E 25	Nitrofen
Topcide	Benzadox
Tordon	Picloram
Treflan	Trifluralin
Tribunil	Methabenzthiazuron
Trimeturon	Trimeturon
Tritac	Tritac
Tritak	Glytac
Torpedo	AN 56477
Tropotox	MCPB
Trysben	2,3,6-TBA
Tunic	VC 438
Tupersan	Siduron
2,4-D	2,4-D

TABLE A-2 (*continued*)

Trade names	Common names
2,4-DP	Dichlorprop
2,4,6-T	2,4,5-T
2,4,5-TP	Silvex
Urab	Fenuron-TCA
Urox	Monuron-TCA
Vapam	Metham
Vegedex	CDEC
Vegiben	Chloramben
Venzar	Lenacil
Vernam	Vernolate
Vorlex	Vorlex
Weedazol	Amitrole
Weedone 2,4,5-TP	Silvex
Zytron	DMPA

TABLE A-3. Physical, Chemical, and Toxicological Properties of Herbicides.[a]

Common name	M.W.	Physical state	Color	Sp. Gr.	M.P. °C	B.P. °C (mmHg)	Decomp. temp. °C	Vapor pressure mmHg	Solubility ppm				LD_{50}[b] mg/kg
									Water	Ethanol	Methanol	Acetone	
acrolein	56.1	L	C	0.841 (20/4)	−86.95	52.69	—	325.7 (30°)	25×10^4 (20°)	M	M	M	46
alachlor	269.8	S	Cr	—	39.5–41.5	D	105	0.02 (100°)	148 (25°)	So	—	So	1,800
allyl alcohol	58.1	L	C	0.855 (20/4)	—	96.9	—	32.4 (30°)	M	So	So	So	64
ametryne	227.3	S	W	—	84–85	—	—	8.4×10^{-7} (20°)	185 (20°)	So	So	So	1,110
amitrole	84.1	S	W	—	159	—	—	—	28×10^4 (25°)	26×10^4 (75°)	—	I	26,600
AMS	114.1	S	—	—	131–132	—	160	N	68.4×10^4 (25°)	—	—	—	—
atratone	211.2	S	W	—	94–96	—	—	2.9×10^{-6} (20°)	16.5×10^2 (25°)	So	So	So	—
atrazine	215.7	S	W	—	173–175	—	—	3.0×10^{-7} (20°)	33 (27°)	—	18×10^3 (27°)	—	3,080
barban	258.1	S	C	—	75	—	75	—	11 (25°)	—	—	—	1,050
benefin	335.3	S	Yo	—	65–66.5	121–122 (0.5)	—	4×10^{-7} (25°)	<1 (25°)	23.5×10^2 (25°)	—	65×10^4 (25°)	800
bensulide	397.5	L-S	—	1.224 (25/25)	34.4	—	—	—	25 (20°)	M	—	—	—
bromacil	261.1	S	W	—	158–159	—	—	—	815 (25°)	13.4×10^4 (25°)	—	16.7×10^4 (25°)	5,200
bromoxynil	276.9	S	B	—	190	—	—	—	<200 (20°)	—	—	So	250
butylate	217.4	L	A	0.940 (25/25)	—	71 (10)	—	13×10^{-3} (25°)	45 (22°)	M	—	M	4,659
cacodylic acid	138.0	S	C	—	200	—	—	—	66.7×10^4 (25°)	20.6×10^4 (20°)	—	—	830
CDAA	173.6	L	A	1.09 (25/15.6)	—	74 (0.3)	—	9.4×10^{-3} (20°)	—	50×10^4 (−18°)	—	—	750
CDEC	223.8	L	A	1.16 (25/15.5)	—	128 (1)	—	2.2×10^{-3} (200°)	92 (25°)	So	—	So	850

TABLE A-3 (*continued*)

Common name	M.W.	Physical state	Color	Sp. Gr.	B.P. M.P. °C	B.P. °C (mmHg)	Decomp. temp. °C	Vapor pressure mmHg	Solubility ppm				$LD_{50}{}^b$ mg/kg
									Water	Ethanol	Methanol	Acetone	
chloramben	206.0	S	W	—	201	—	—	—	700 (25°)	17.3×10^4 (25°)	—	23.3×10^4 (25°)	3,500
chloroxuron	290.7	S	W	—	151–152	—	—	—	2.7	—	—	So-m	3,700
chlorpropham	213.7	S	W	—	38–40	—	150	1×10^{-5} (25°)	88	So	—	M	5,000–7,500
cycloate	215.4	L	C	1.0156 (30/4)	—	145–146 (10)	—	6.2×10^{-3} (25°)	85 (22°)	—	—	—	2,000–4,100
cyprozine	227.7	S	W	—	167–168	—	—	—	6.9 (25°)	5.2×10^4 (25°)	8.7×10^4 (25°)	14.2×10^4 (25°)	1,200
2,4-D	221.0	S	W	—	135–138	160 (0.4)	—	0.4 (160°)	600 (20°)	60.1×10^4 (31°)	—	45×10^4 (33°)	300–1,000
dalapon	143.0	L	C	1.389 (22.8)	—	185–190	—	—	So	So	—	—	—
dalapon, Na salt	165.0	S	W	—	—	—	166.5	—	50.2×10^4 (25°)	—	17.9×10^4 (25°)	140 (25°)	7,570–9,330
2,4-DB	249.1	S	W	—	120–121	—	—	—	N	So	So	—	—
DCPA	332.0	S	W	—	156	—	360–370	<0.01 (40°)	0.5 (25°)	—	—	10×10^4 (25°)	>3,000
2,4-DEP	649.1	L	LB	1.433 (25/4)	—	>200 (0.1)	250–300	—	I	—	—	—	850
desmetryne	283.3	—	—	—	84–86	—	—	1.0×10^{-6} (20°)	580	So	So	So	—
diallate	270.2	L	—	—	25–30	150 (9)	—	—	14 (25°)	M	—	M	395
dicamba	221.0	S	W	—	114–116	—	—	3.75×10^{-3} (100°)	45×10^2 (25°)	92.2×10^4 (25°)	—	—	2,900
dichlobenil	172.0	S	W	—	145–146	270 (760)	—	5.5×10^{-4} (20°)	18 (20°)	5×10^4 (8°)	—	5×10^4 (8°)	>3,160
dichlormate	234.1	S	W	—	52	—	150	N (25°)	170 (25°)	—	—	So	1,870–2,140
dichlorprop	235.1	S	W	—	116–117.5	—	—	—	710 (28°)	—	—	59.5×10^4 (28°)	800
dicryl	230.1	S	W	—	121–126	—	—	—	I	—	—	20×10^4 (20°)	3,160

dinoseb	240.2	S	DB	—	32	—	1 (151.1°)	52 (25°)	48×10^4 (25°)	—	—	5–60
diphenamid	239.3	S	W	—	132–135.5	210	—	261	So	—	18.9×10^4 (27°)	686–776
diquat	184.0[c]	S	Y	Y	—	D	—	N	So	—	—	400–440
diuron	233.1	S	W	—	158–159	180–190 (D)	0.31×10^{-5} (50°)	42 (25°)	—	—	53×10^3 (27°)	3,400
DSMA	139.9	S	W	—	132–139	—	—	25.6×10^4 (20°)	—	—	—	1,800
endothall	186.2	S	W	—	144	90	—	10×10^4 (20°)	—	28×10^4 (20°)	7×10^4 (20°)	38–51
EPTC	189.3	L	Y	0.96 (25/25)	—	235	34×10^{-3} (25°)	370 (20°)	M	—	M	1,652
erbon	366.5	S	W	—	49–50	161–164 (0.5)	low	I	So	So	So	1,120
fenac	239.5	S	W	—	157–160	—	—	So-sl	So	So	So	1,780
fenuron	164.2	S	W	—	133–134	—	1.6×10^{-4} (60°)	38.5×10^2 (25°)	—	—	—	6,400
fenuron TCA	327.6	S	W	—	65–68	—	—	So-m (25°)	—	—	So-m	4,000–5,700
fluometuron	232.2	S	W	—	163–164.5	—	—	90 (25°)	So	—	So	8,900
HCA	264.8	L	C	1.741 (20)	–2	204	—	So-sl	So	So	So	1,290
hexaflurate	228.0	S	W	—	440	—	N	12.6×10^4 (20°)	—	—	—	1,200
ioxynil	370.9	S	B	—	212–213	—	—	130 (25°)	—	3.3×10^4 (25°)	$>10 \times 10^4$ (25°)	—
isopropalin	309.4	L	O	—	—	—	—	<0.5 (25°)	So	So	So	>5,000
linuron	249.1	S	W	—	93–94	—	—	75 (25°)	15×10^4 (25°)	—	50×10^4 (25°)	1,500
MAA	139.9	S	W	—	132–139	—	—	25.6×10^4	—	—	—	1,800
MCPA	200.6	S	LB	—	118–119	—	—	N	15.3×10^5	—	—	700
MCPB	249.1	S	W	—	100–101	—	—	N	So	So	So	650
mecoprop	214.6	S	B	—	94–95	—	—	600	—	—	—	—
methoprotryne	255.0	S	W	—	68–70	—	—	320 (20°)	So	So	So	—
metobromuron	259.1	S	W	—	95.5–96	—	—	330 (20°)	So	So	So	2,000

TABLE A-3 (*continued*)

Common name	M.W.	Physical state	Color	Sp. Gr.	M.P. °C	B.P. °C (mmHg)	Decomp. temp. °C	Vapor pressure mmHg	Solubility ppm				LD_{50}^{b} mg/kg
									Water	Ethanol	Methanol	Acetone	
MH	112.1	S	W	—	292	—	~300	N	60×10^2 (25°)	10×10^2 (25°)	—	10×10^2 (25°)	6,950
molinate	187.3	L	—	1.0643 (20/20)	—	202 (10)	—	5.6×10^{-3} (25°)	800 (20°)	M	M	M	720
monuron	198.6	S	W	—	174–175	—	185–200	5×10^{-7} (25°)	230 (25°)	—	—	52×10^3 (27°)	3,600
monuron TCA	362.0	S	—	—	78–81	—	90	—	1	—	—	—	2,300–3,700
naptalam	291.3	S	P	—	185	—	—	—	200 (25°)	—	—	50×10^2 (25°)	>8,200
nitralin	345.15	S	Y-O	—	151–152	—	225	$<1.5 \times 10^{-6}$ (25°)	0.6 (25°)	1.1 (20°)	—	37 (25°)	>2,000
nitrofen	284.1	S	DB	—	—	—	—	—	N	—	So	So	2,630
norea	222.3	S	W	—	171–172	—	—	—	150 (25°)	So	—	So	1,476–6,830
oryzalin	346.4	S	Y-O	—	137–138	—	—	—	85 (25°)	So	50×10^2 (25°)	So	>10,000
paraquat	186.2^{c}	S	W	—	D	D	—	—	So	N	18×10^5 (25°)	So-sl	150
PCP	266.3	S	DB	—	191	D	309	0.12 (100°)	30 (50°)	So	—	—	27–80
pebulate	203.4	L	Y	0.9555 (20/20)	—	142 (21)	—	35×10^{-3} (25°)	60	M	—	M	921–1,120
phenmedipham	300.3	S	C	—	143–144	—	—	low	<10 (20°)	—	$\sim 5 \times 10^4$ (20°)	$\sim 20 \times 10^4$ (20°)	>8,000
picloram	241.5	S	W	—	D	D	~215	6.16×10^{-7} (35°)	430 (25°)	10.5×10^3 (25°)	—	19.8×10^3 (25°)	8,200
prometone	225.3	S	W	—	91–92	—	—	2.3×10^{-6} (20°)	750 (20°)	—	$>50 \times 10^4$ (20°)	$>50 \times 10^4$ (25°)	2,980
prometryne	241.4	S	W	—	118–120	—	—	1.0×10^{-6} (20°)	48 (20°)	So	So	So	3,750
propachlor	211.7	S	B	—	67–76	110 (0.03)	170	0.03 (110°)	700 (20°)	29×10^4 (20°)	—	30.9×10^4 (20°)	710
propanil	218.0	S	LB-GB	—	85–89	—	—	—	500	So-sl	—	—	1,384–2,270
propazine	229.7	S	C	—	212–214	—	—	2.9×10^{-8} (20°)	8.6 (20°)	So-sl	So-sl	So-sl	>5,000
propham	179.2	S	B	—	87–88	Su	>150	0.074 (40°)	250 (20°)	So	So	So	5,000
pyrazon	221.6	S	LB	—	207	—	207	—	300 (20°)	—	3.54×10^4 (20°)	—	3,000

Common name	Mol. wt.	Physical state	Color	Sp. gr.	Melting point (°C)	Boiling point °C (mm)	Vapor pressure (mm Hg)	Solubility				Acute oral LD50[b] (mg/kg)
sesone	309.1	S	W	—	245	—	—	26.5×10^{4} (25°)	—	—	6.4×10^{3} (25°)	1,230–1,400
siduron	232.3	S	W	—	133–138	—	$<8 \times 10^{-4}$ (100°)	18 (25°)	16×10^{4} (25°)	—	—	>5,000
silvex	269.5	S	W	—	178.2	—	low	140 (25°)	—	$\sim 12 \times 10^{4}$ (25°)	$\sim 16 \times 10^{4}$ (25°)	650
simazine	201.7	S	W	—	225–227	—	6.1×10^{-9} (20°)	5 (20°)	—	400 (20°)	—	>5,000
simetryne	213.0	S	W	—	81–82.5	—	7.1×10^{-7} (20°)	450 (20°)	So	So	So	—
sodium arsenate	129.9	S	W	—	—	—	—	So	So-sl	So-sl	—	10–50
sodium chlorate	106.4	S	W	—	248	D	—	79×10^{4} (0°)	So	So	—	5,000
solan	239.7	S	W-Cr	—	82–86	—	—	I	—	—	46×10^{4} (20°)	>10,000
swep	220.1	S	W	—	112–114	—	—	I	—	—	—	522
2,4,5-T	255.5	S	W	—	154–155	>200	low	238 (30°)	590 (50°)	—	—	300
2,3,6-TBA	225.5	S	W	—	125–126	—	N	8.4×10^{3} (20°)	63.7×10^{4} (20°)	71.7×10^{4} (20°)	60.7×10^{4} (20°)	750–1,000
TCA	163.4	S	W	—	59	197.5	5 (77°)	13×10^{6} (20°)	—	21.4×10^{6} (25°)	85×10^{5} (25°)	5,000
TCBC	229.9	L	—	—	—	90; 93–98 (1)	—	2 (25°)	—	—	—	2,800–3,450
terbacil	216.7	S	W	—	175–177	—	4.8×10^{-7} (29.5°)	710 (25°)	—	—	—	>5,000
terbutol	277.4	S	W	—	200–201	200–201	—	6–7 (25°)	—	So	So	<7,500
terbutryn	241.0	S	W	—	104	—	9.6×10^{-7} (20°)	58 (20°)	—	—	—	>15,000
triallate	304.7	L	—	—	29–30	148–149 (9)	—	4 (25°)	So	—	So	2,400–2,980
tricamba	255.5	S	W	—	137–139	—	—	So-sl	So	So	—	1,675–2,165
trietazine	229.5	S	W	—	102–104	—	—	20	So	—	So	970
trifluralin	335.3	S	O	—	48.5–49.0	96–97 (0.18)	1.99×10^{-4} (29.5°)	<1 (27°)	74×10^{3} (27°)	—	39.5×10^{4} (27°)	3,700
vernolate	203.4	L	—	0.954 (20/20)	—	150 (30)	10.4×10^{-3} (25°)	90 (20°)	—	—	—	1,780

[a] Most of the data given in this table were obtained from the Herbicide Handbook of the Weed Science Society of America.

[b] Acute oral toxicity, most of the values refer to rats; however a few values apply to mice.

[c] Cation

Abbreviations: A = amber, B = buff, C = colorless, Cr = cream, D = decomposes, DB = dark brown, G = gas, GB = gray black, I = insoluble, L = liquid, LB = light brown, M = miscible, N = negligible, O = orange, P = purple, S = solid, So = soluble, So-m = soluble-moderate, So-sl = soluble-slight, Su = sublimes, W = white, Y = yellow.

Author Index

Subject Index

476